HOW TO LOVE STATISTICS

WHEN YOU HAVEN'T DONE THE MATH

THE UNDERLYING MATHEMATICAL AND OTHER IDEAS
THAT MAKE (MOST OF) IT WORKABLE, WONDERFUL,
AND WORTH STUDYING!

JOHN SMARTT

Green Bracken Publishing
Sydney, Australia
www.greenbracken.com
First edition: March 2026

A catalogue record for this book is available from the National Library of Australia

ISBN 978-1-7641422-0-5 (print)
ISBN 978-1-7641422-1-2 (ebook)
Includes bibliographical references

Subjects:
1. Statistics—Popular works
2. Mathematical statistics—Study and teaching
3. Mathematics—Philosophy
4. Statistics—History
5. Education—Numeracy
I. Title.
Dewey Number: 519

BISAC Subject Codes
MAT029000 — Mathematics / Probability & Statistics / General
MAT015000 — Mathematics / History & Philosophy

Thema Subject Codes
PBT — Probability and Statistics
YPMF — Educational: Mathematics and Numeracy; Statistics and Mathematics
PBX — History of Mathematics

Overlapping Lives of Some Contributors to Statistics

"One of the principal uses to which this Doctrine of Chances may be applied, is the discovering of some Truths, which cannot fail of pleasing the Mind by their Generality and Simplicity; the admirable Connexion of its Consequences will increase the Pleasure of the Discovery, and the seeming Paradoxes where with it abounds, will afford very great matter of Surprize and Entertainment to the Inquisitive."[1]

Abraham De Moivre (1667-1754)

"The only simplicity for which I would give a straw is that which is on the other side of the complex — not that which never has divined it."[2]

Oliver Wendell Holmes Jr (1841 – 1935)

"Pure mathematics is, in its way, the poetry of logical ideas."

Albert Einstein (1879-1955)

"The unity of the different applications has usually been overlooked, the more naturally because the development of the underlying mathematical theory has been much neglected."

Ronald Fisher (a founder of modern statistics) (1890-1962)

"There has been an unfortunate tendency to teach, and to include in textbooks even for students well-equipped with the necessary preliminary mathematics, mere practical directions without adequate derivation. ...Such an omission from the training of future professional statisticians and teachers of statistics must be deplored. The principle well recognized in mathematics that the proof of a theorem has a value additional to that of the result is nowhere more important than in statistics. The underlying assumptions and the limitations of applicability can scarcely be well understood in the absence of full proofs. Furthermore, when circumstances call for a variation of the problem, it is desirable that the techniques used in the proof be available for use in trying to solve the new problem."[3]

Harold Hotelling (1895 – 1973)

"You are here to learn the subtle science and exact art..."[4]

**Severus Snape (J.K. Rowling,
Harry Potter and the Philosopher's Stone, 1997)**

"Math is like going to the gym for your brain. It sharpens your mind."

Danica McKellar (1975...)

Key

Curves on the Cartesian plane

> Read this section if you don't know what equations would draw a circle, a parabola, an eclipse or a hyperbola.

So far we've been looking at straight lines, which hasn't involved raising x to a power (or, of you prefer to think of it this way, we have x raised to the power of 1). As mentioned earlier, once we start raising x to a power other than 1, the lines start being curved.

The circle

The equation $x^2 + y^2 = 1$ draws a circle, with a radius of 1 (Figure 96). If we want to show this with y as the dependent variable, the way we have with our other equations so far, we can do some simple algebra to get there.

This is our starting equation.	$x^2 + y^2 = 1$
Subtract x^2 from both sides.	$y^2 = 1 - x^2$
Take the square root of both sides, noting that there will be positive and negative values for the square root.	$y = \pm\sqrt{1 - x^2}$

A circle with a radius of 1 is called **unit circle**, and we'll meet it again in the next chapter.

> To cement your understanding
>
> Select at least one of the red points on the circle in Figure 96. Square both coordinates and add the result, to reassure yourself that the result will be 1. (There will probably be a rounding error, so your calculator may not show it as exactly 1, but it will be very close.)

will, when we get to look at differential calculus. Don't panic about that; we have already covered the key concepts. Now we just have to put them together. We will cover a few more concepts in geometry before we get there, though.

Figure 100

> Key points from this section:
>
> - The equation $x^2 + y^2 = 1$ draws a circle, with a radius of 1.
> - This is called a unit circle.
> - The equation $y = x^2$ draws a **parabola**.
> - The equation $\frac{x^2}{a^2} + \frac{y^2}{b^2} = 1$ draws an **ellipse**
> - The equation $y = \frac{1}{x}$ draws a **hyperbola**.

Where do I start?

PREFACE 7
DETAILED CONTENTS 11
PART 1: THE MATHEMATICS YOU MIGHT HAVE MISSED 17

INTRODUCTION TO PART 1 18
PRE-ALGEBRA: CONSTRUCTING MATHEMATICAL SENTENCES 22
ALGEBRA: TELLING STORIES WITH MATHEMATICAL LANGUAGE 82
GEOMETRY: MATHEMATICAL LANGUAGE IN PICTURES 167
CALCULUS: THE MATHEMATICS OF PREDICTABLE CHANGE 227
CONCLUSION TO PART 1: FERMAT'S PROOF OF HIS LAST THEOREM? ... 324
PART 2: STATISTICS & THE MATHEMATICS OF UNCERTAINTY . 330

INTRODUCTION TO PART 2 331
FIRST STREAM: GAMBLING BEYOND INTUITION 334
SECOND STREAM: AN OBSCURE PAPER BY AN OBSCURE CLERGYMAN ... 409
THIRD STREAM: ERRORS IN MEASUREMENT 490
FOURTH STREAM: THE PHYSICS OF MECHANICAL CONTRAPTIONS ... 517
CONFLUENCE: PEARSON AND FRIENDS 561
DEPTH: FISHER'S REFINEMENTS 657
NON-PARAMETRIC TESTS 710
PART 3: SIMPLICITY ON THE OTHER SIDE OF COMPLEXITY 724

PUTTING IT TOGETHER 725
THE BEGINNING 729
APPENDICES 732
ACKNOWLEDGEMENTS 775
NOTES 776
INDEX 783

Preface

Like a romantic novel, my love affair with statistics began with loathing. I had looked forward to uncovering its mysteries—perhaps too much.

When I returned to university in my early 40s, my course included a semester of statistics. Coming from the corporate world, where I'd often analysed data, I was intrigued by the chance to learn more. That might explain why I approached the subject like an annoying four-year-old, endlessly asking "why?", while being unsatisfied with the answers.

Much of the subject felt counterintuitive. The formula for standard deviation, for instance, didn't help me grasp data spread—and still doesn't. I've since learned its real strength lies elsewhere (and it's brilliant), but at the time that wasn't clear. I learned that the normal distribution (bell-shaped/Gaussian) curve described everything from **A**ptitude scores to **Z**oning approval times, yet I couldn't see why so many different things followed the same pattern—or where the formula came from. No one else seemed to know either.

It seemed statistics might be the only tertiary subject where *not* trying too hard to understand made passing easier. Reluctantly, I did what others I approached admitted to: figured out what I needed to pass, did just enough, and then promptly forgot everything after the examination.

Having a background in adult education, I recognised at the time that there could be two different, practical approaches to teaching statistics; and the course I undertook did neither, falling between them and, I felt, failing at both. One approach would be to teach statistics and to bypass the mathematics altogether, and just explain the concepts and how to apply them. The other would be to dive in and understand the mathematics in depth, so that learners could really understand what was going on. (The textbook for my course presented the formulae without explanation of where they came from.)

Since that time there have been many statistics text books written and, as a general rule, they lean towards the first approach: giving a mathematics-lite explanation of statistics, and teaching how to use the software that does the hard work.

As someone who read journal articles for professional interest, a mathematics-lite approach to statistics wasn't working for me. I wanted to know whether statistics were being used appropriately, and was unable to make those judgements without a deeper understanding. I needed a text book that took the alternative approach: one that clearly explained the mathematical basis of what I was looking at. I needed a way to understand the ideas that lay behind what was reading, and I couldn't find anything suitable.

To understand it the way I wanted to I had to understand the history: what people were thinking when they developed these ideas. I thought this would be simple. It wasn't. For instance, I wrongly believed the normal distribution curve came from Gauss (1777–1855), but it was first worked out by de Moivre (1667–1754). Yet that formula includes standard deviation—developed by Pearson (1857–1936), born after Gauss's death. Secondary sources I consulted about the history of statistics assumed that readers already understood statistics. I didn't—I was trying to use the history to help me understand the statistics themselves.

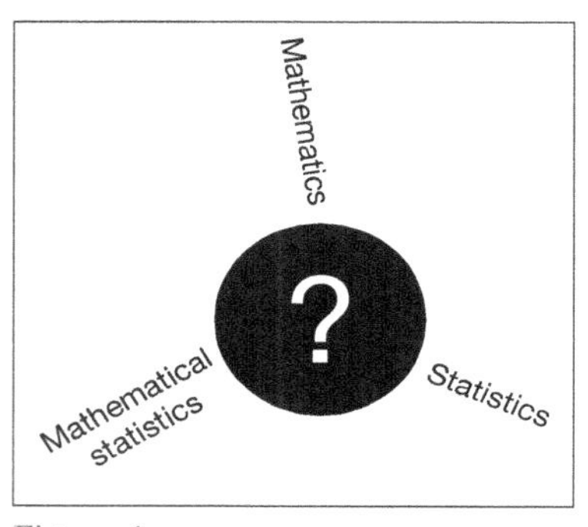

Figure 1

There was nowhere to look this up. I realised that if I was going to be able to read the book I wanted and needed, I would have to research and write it myself. I discovered that "statistics", "mathematics", and "mathematical statistics" have become separate disciplines with their own texts, and with a gap in between (Figure 1). Even experienced researchers and statisticians often couldn't access the information I sought. That means researchers may not understand the statistics they use, and even statisticians may not know why their methods work. It is hardly surprising, therefore, that half of all peer-reviewed papers may contain errors in their use of statistics (which is true across a range of disciplines)[5][6][7][8].

This book took over five years to research and write. Two things motivated me to persist over that time:

1. The harder the search became, the more vital it felt. If I couldn't find the information, others must struggle too. Statistical misunderstanding

risks the integrity of scientific research, which affects everyone. I found some widely-used techniques that are demonstrably unsound and illogical, and a formula with two accepted versions—each producing different results for the same data.

2. At the same time, I fell in love with statistics: its practicality, elegance, and the brilliance of the ideas behind it. Despite the few flawed approaches I uncovered, the journey left me in awe of most of what it does and of those who developed it. Statistics is underused, not overused. The world would benefit if more people understood it more deeply.

A note on historical (in)accuracy

This book approaches statistics by looking at what people were originally thinking when they developed statistical ideas. Unless they spelled out their thought processes in detail (which they rarely did), this has involved a lot of guesswork. Working out who developed which ideas can be difficult. The mathematical notation that was used even one hundred years ago has changed. The style of argument has also changed. The ancients did a lot with geometry and weren't familiar with algebra, and so even up until the 1700s a geometrical style of argument was sometimes used that is unfamiliar to modern readers. Mathematicians have often only briefly mentioned ideas in their work, and it can be challenging to understand the extent to which they embraced a modern understanding of these concepts. Some of what we know of people's work is only available to us as correspondence, and it is clear there were other conversations around the development of ideas that we do not have access to. Different people frequently worked on the same problem, at the same time, in different places. Patriotic sentiment has sometimes come into play (particularly in times of international hostility), and there have been concerted efforts at times to prove that certain approaches were developed by people from a particular country. It was considered acceptable, at least in the late 1600s, to pay for the privilege of having a mathematical approach named in your honour. Some mathematicians (particularly Leonhard Euler, Carl Jacobi, and Karl Pearson) were so prolific that some of their work has still never been published. The work of other mathematicians may have been published but is not available online or in English. Much of the secondary reference material that talks about who came up with what has been written by mathematicians, not historians. All of this means that secondary sources describing the origin of mathematical ideas can be quite inconsistent. Examining primary sources as well as secondary ones was sometimes illuminating and sometimes increased the confusion.

In other words, historical references and anecdotes in this book should be treated with caution. They are here to illuminate the way ideas were first formulated, to add interest, and to make it easier to grasp concepts. There is—and probably always will be—room for debate about the history.

Detailed Contents

Key 5
Where do I start? 6

PREFACE 7

DETAILED CONTENTS 11

**PART 1: THE MATHEMATICS
YOU MIGHT HAVE MISSED 17**

INTRODUCTION TO PART 1 18

PRE-ALGEBRA: CONSTRUCTING
MATHEMATICAL SENTENCES 22

*Adding and subtracting
positive and negative
numbers* 23
  Commutative property of
  addition 23
  The number line 24
  Another way to think about
  negative numbers 25
  Absolute value 26

*Equals, equations,
inequalities and brackets* .. 28
  Conventions in equations when
  there are two operators together
  30

Properties of multiplication 35
  Terms, expressions, polynomials
  36
  Brackets, and the distributive
  property of multiplication 37
  Multiplying negative numbers .. 39
  Multiples and factors 40
  Prime numbers 40

Fractions and percentages 42
  Adding and subtracting fractions
  45
  Multiplying fractions 47
  Fractions larger than 1 48
  Reciprocals 49
  Fractions and division 50

Powers 53
  Adding exponents 54
  Multiplying exponents 54
  Raising to the powers of one and
  of zero 55
  Raising to a negative power 56
  Subtracting indices 57

Roots part 1 59
  Raising to the power of a fraction
  60
  Summary of operations with
  exponents 61

Decimal fractions 63

*Ratios and irrational
numbers* 67

Categories of numbers 72

Set notation 74

*Pure mathematics,
"Pascal's" triangle, and
putting it all together* 77
  Pure mathematics 77
  "Pascal's" triangle 78
  Positive and negative numbers
  revision 79
  Prime numbers and multiples
  revision 79
  Raising numbers to powers
  revision 80

**ALGEBRA: TELLING STORIES WITH
MATHEMATICAL LANGUAGE 82**

*Algebra and algebraic
notation* 83

*Recap on priority of
mathematical operations* .. 89
  Expanding algebraic terms 90
  Factorizing 91
  Some techniques for factorizing
  91

Quadratic equations 93
  Coefficients of polynomial
  expressions 94

Quadratic formula 95

Polynomial multiplication and convolutions **100**
  Multiplying polynomials......... 100
  Introduction to convolutions.. 101

Roots part 2 .... **103**
  Operations with roots............ 103
  Square roots of negative numbers 105
  Complex numbers 106

Simultaneous equations .. **108**

Functions and variables .. **111**
  Function notation 111
  Dependent and independent variables 113
  Inverse functions 113
  Functions with more than one independent variable 114

Sequences and limits **116**

Logarithms **125**
  Logarithm identities 128

Series **131**
  Linearity 133
  Series of natural numbers 134

Geometric Series **138**

The product operator and factorials........................ **145**
  The product operator 145
  Alternative 1 for expressing the same idea 147
  Alternative 2: factorials......... 149

Combinations, permutations, and n choose r **155**

Binomial expansion **160**

GEOMETRY: MATHEMATICAL LANGUAGE IN PICTURES **167**

Basic Geometric Concepts **169**
  Points, lines, planes and spaces 169
  Angles 171
  Parallel Lines 172
  Straight-sided, two-dimensional figures.................................. 172
  Circles 174

Areas of straight-sided figures **177**

Pythagoras' theorem....... **179**

Angles, degrees and radians **182**
  Angles of a triangle................ 182
  Angles of straight-sided figures 183

Radians 183

Area of a circle and a segment......................... **187**
  Area of a segment 188

Trigonometry **190**
  The slope of a straight line 194
  Inverse trigonometric functions 195
  Reciprocal trigonometric ratios 196
  Describing the squares of trigonometric functions 197

The Cartesian plane **199**

Sloping and curved lines.. **203**
  Adding a constant to the equation for a straight line 206
  The idea of delta 207
  The circle 209
  The ellipse............................ 210
  The hyperbola 211

Trigonometry on the Cartesian plane............... **213**
  The unit-circle definition of trigonometric ratios............... 213
  Graphing the sine and cosine functions 214

Trigonometric identities .. **217**
  Trigonometric ratios of negative angles................................... 217
  Pythagoras' theorem and the unit circle.................................... 218
  Trigonometry and the alternative angle 218
  Sine and cosine of the sum of two angles................................... 219
  Other identities linking sine and cosine.................................... 219
  The law of cosines 220

Polar coordinates **223**

The complex plane........... **225**

CALCULUS: THE MATHEMATICS OF PREDICTABLE CHANGE **227**

Differential calculus **228**
  Leibniz vs prime notation....... 233

Some basic rules of differential calculus **235**
  The derivative of a polynomial expression, and the addition rule 235
  The addition rule 236
  The power rule 237
  The derivative of a reciprocal 237
  The derivative of a square root 238

The derivative of a variable plus a constant all squared 239
The minimum value of a quadratic equation 240
The closest point of a curve to a line 241
Generalised form 242

Differential calculus with trigonometry part 1 244
Derivative of the sine function 244
Derivative of the cosine function 246

More rules for differentiating combined functions 248
The product rule 248
The chain rule 249
The quotient rule 251
Simplifying polynomial fractions 252

Differential calculus with trigonometry part 2 253
Derivative of the tangent function 253
Derivative of the arc tangent function 253

l'Hôpital's rule 255
Taylor series 259
Exponential equations and Euler's constant 266
Euler's constant 268
Exponential equations with e . 270

Integral calculus 274
Advanced integration 279
The reverse power rule 279
Multiplying by a constant 279
The addition rule 280
Linearity of integral calculus . 280
Integration by parts 280
Integration by substitution 282

Multivariable calculus 285
Partial derivatives 285
Double integration 286

Changing to polar coordinates 289
Jacobian 289

Natural logarithms 292
The derivative of a natural logarithm 293

Complex numbers in trigonometry and exponentials 295
Differential equations 298
Separable differential equations 298

Some important integrals: the gamma and beta functions, and the Gaussian integral 301
Dummy variables and Eulerian integrals 301
The Gamma Function 302
Incomplete gamma functions . 304
The Gaussian Integral and gamma of one half 305
The Beta Function 310

Convolution integrals 312
Laplace and Fourier transforms 316
The Laplace transform 316
The Fourier transform 321

CONCLUSION TO PART 1: FERMAT'S PROOF OF HIS LAST THEOREM? 324
Why it doesn't work 327
Two morals of the story 327

PART 2: STATISTICS & THE MATHEMATICS OF UNCERTAINTY 330

INTRODUCTION TO PART 2 331
The predictability of randomness 331
Origins 332

FIRST STREAM: GAMBLING BEYOND INTUITION 334
Probability basics 337
Describing probability 337
Tree charts 338

The multiplication and addition rules 344
The multiplication rule 344
The addition rule 345
Beyond tree charts 347

Binomial probabilities with uneven chances 350
Probabilities with more than two options 354
Permutations with repetition . 354
Multinomial probabilities 356

Probability distributions and their terminology 360
A deeper look at variables 366
Unpacking "randomness" 367

Unpacking "independence" ... 368
Unpacking "numerical" variables 369
Describing probability distributions and their features 372
Some discrete distributions 376
Binomial distribution (again) . 377
Bernoulli distribution 380
Geometric distribution 380
Negative binomial distribution 384
Hypergeometric distribution . 386
Categorical and Multinomial distributions 390
Practical examples comparing these distributions 394
Relationships between probability distributions .. 397
Marginal and joint distributions 398
Dependent and independent probabilities 402
More complex combined distributions 404
Ratios of joint and marginal distributions 406

SECOND STREAM: AN OBSCURE PAPER BY AN OBSCURE CLERGYMAN 409
Probability beyond gambling 410
Pierre-Simon Laplace and statistics 413
Conditional probability 414
Bayes Theorem 416
The law of total probability 425
Bayesian Statistics 429
Example of a Bayesian analysis 431
Bayesian vs Frequentist statistics 435
More distributions 443
Beta distribution 443
The Poisson distribution 450
The exponential distribution .. 457
The gamma distribution 459
Distributions in practice .. 465
Recap of distributions so far .. 468
Conjugate priors in Bayesian statistics 470

Transforming random variables 472
Monotone functions 473
The Jacobian for a transformed PDF 474
Summing random variables 481

THIRD STREAM: ERRORS IN MEASUREMENT 490
Averages 492
The least squares method 497
An error probability distribution curve and the quest for a breakthrough 505
Binomial approach 506
Differential approach 508
Empirical confirmation 514

FOURTH STREAM: THE PHYSICS OF MECHANICAL CONTRAPTIONS 517
Expected value 520
"Mean" of a probability distribution 520
Expected value of continuous variables 525
Linearity and the sum of expected values 527
Product of expected values of independent variables 529
LOTUS 530
Moment generating functions 532
Proof independent random variables can be added by convolution 539
Proof that the gamma distribution is the sum of exponentials 539
Moments of probability distributions 543
Using a moment generating function to find the first moment 543
The second moment: variance 544
Alternative form of the variance 546
Table of first and second moments 548
Working with the variance 549
The third moment: skewness .. 551
The fourth moment: kurtosis .. 553
Higher moments, and matching

data to probability distributions 554
The first four moments of the gamma distribution, using the MGF 554
The Markov and Chebyshev Inequalities 556

CONFLUENCE: PEARSON AND FRIENDS 561
Standard deviation and the z score 565
Standard deviation 565
The Z score 567
Correlation and covariance 571
Correlation coefficient 571
The cautionary tale of r^2: the "coefficient of determination" 578
Other cautions with correlation 581
Generalising correlation, variance and covariance 582
Completing the formula for the normal distribution.... 588
The z score and the normal distribution 591
The area under the normal distribution curve.................. 593
Statistical significance and P values 598
The normal, the binomial, and beyond...................... 600
Moment generating function of the normal distribution 600
Moment generating function of the standardised binomial..... 602
Extending the normal distribution 604
Chi square (first extension of the normal)...................... 606
The χ2 (chi-square) distribution 610
Example of using a chi square distribution 615
MGF and moments of the χ2 (chi-square) distribution 617
Taking samples (second extension of the normal) .. 619
Mean of a sampling distribution 622
Central limit theorem 623
Common misconception........ 625
Standard deviation ("standard error") of the sample mean.... 626

The law of large numbers 630
The story of sampling so far... 632
Errors, power and confidence (third extension of the normal) 635
Types of errors 635
Type II errors and power analysis 639
Errors, sampling size and effect size.. 643
Type I errors and confidence intervals 646
Bayesian critique 649
"Student" and the t distribution (fourth extension of the normal)................... 652
On the probable error of a mean, and the variance of a sample . 654

DEPTH: FISHER'S REFINEMENTS 657
Degrees of freedom 659
An unorthodox and elegant solution 659
Expanding the concept........... 661
Fisher and the t distribution 664
Probability density function of the t distribution......................... 664
Alternative forms of the t distribution PDF 675
Where has e disappeared to in the t distribution? 675
Moment generating function and moments 677
Applying the t distribution in practice........................... 680
How many tails? 681
Single-sample t test 682
Paired t test.......................... 684
Two-sample (unpaired) t test . 684
t tests for proportions 685
Using t tests with regression and correlation 685
Correlation and t tests........... 688
ANOVA and the F distribution 690
One way ANOVA 691
The F distribution 694
Extending the idea of the ANOVA 701
Distributions at a glance.. 704
Fisher's exact test........... 707

NON-PARAMETRIC TESTS 710

Some of the non-parametric methods 712
Mean, variance and standard deviation of ranks 713
Spearman's rank correlation coefficient 715
The sign test 715
Wilcoxon signed rank test 717
Mann-Whitney U test/Wilcoxon rank-sum test 717
Cohen's kappa coefficient (another cautionary tale) . 719

PART 3: SIMPLICITY ON THE OTHER SIDE OF COMPLEXITY 724

PUTTING IT TOGETHER............ 725

THE BEGINNING 729

APPENDICES.......................... 732
Appendix 1: Using a quadratic equation to find the value of phi 733
Appendix 2: Finding the sum of squares of natural numbers .. 736
Appendix 3: Proof of Pascal's identity.................................. 738
Appendix 4: The golden ratio in a pentagram within a pentagon 740
Appendix 5: Stricter definitions of conic sections........................ 742
Appendix 6: Some more trigonometric identities 743
Appendix 7: More about the complex plane 745
Appendix 8: Polynomial long division.................................. 750
Appendix 9: Partial fraction decomposition 753
Appendix 10: The Gaussian integral without polar coordinates 758
Appendix 11: Derivation of the beta function 761
Appendix 12: Using Laplace transforms 763
Appendix 13: Proof of the Central Limit Theorem 766
Appendix 14: Proof that part of the formula for a t statistic has a normal distribution 770
Appendix 15: Proof that part of the denominator of a t statistic has a chi square distribution . 773

ACKNOWLEDGEMENTS............ 775

NOTES 776

INDEX 783

Part 1:
The
Mathematics
You Might
Have Missed

Introduction to Part 1

Whether you are a student wanting to excel, an educator trying to explain, a researcher hoping to understand, a user of research wanting to be confident or anyone with access to potentially useful data, this book will help.

Statistics is useful for everything from **A**ccounting to **Z**oology because it helps us understand more about uncertainty. Wherever data analysis can help and you would like more information than you have, statistics can probably do more than you think. The mathematics behind it is advanced, though, which makes it challenging to teach but too important to ignore.

One increasingly-popular approach is to try to teach the concepts while bypassing the mathematics[9]. This book is different: it embraces the mathematics behind the subject.

One reason for this is that learning the mathematics behind statistics can be good for the brain and bring joy to the soul. Mathematics is a symbolic language about counting (how many) and measuring (how far, how much). The main challenge with it is that the only way to understand this language is from within the language itself, and the knowledge needs to be gained sequentially. If you missed one bit the first-time around, and no one realised at the time, whatever the teacher wrote on the board after that probably looked like a strange language.

At the same time, once you learn it in sequence, it is easier than other languages because there is virtually nothing you need to memorise, it is logical, it is precise, and there are very few exceptions.

Once you have understood something in mathematics, you have it. If you forget it, you can just look it up again. Language is necessary for formulating ideas and for communicating them. If you do not have any great difficulty with mental arithmetic, then your mind works the right way to grasp higher mathematics; the rest is just understanding the language, the conventions that are used, and the arguments that people have made by using the language. This is the language that enables the thinking behind almost every aspect of our lives that separates us from a primitive existence. Understanding the language opens a new way of understanding the world as it is and designing the world as it could be.

If you feel that you do not have a mathematical brain (whatever that is), there is no need to worry. Your brain has an extraordinary capacity to change itself. In our over-specialised world, you may think that your brain is like a computer with limited memory capacity. It isn't. The more connections it makes, the more options it has for further connection.

When you were born, your body had over 85 billion nerve cells, most of which are in your brain and spine. That is more than ten times the population of the Earth. The power of your brain, though, is more about the connections between the nerve cells rather than their number. Each cell can make thousands of different connections[10]. You could probably never fill it up.

You can keep making connections throughout your life, and coming to terms with mathematics and statistics will build more[11]. As different parts of your brain are activated, the amount of blood those parts receive is altered through the operation of complex metabolic, neural, chemical, and other controls[12]. Meanwhile your brain also works holistically. Electrical activity bounces through every part of it in complicated patterns. The better each part works, the better the whole can work[13].

Dyscalculia or **acalculia** is a recognised and not uncommon neurological condition (affecting probably between 3% and 6% of the population) in which people struggle with mathematics. If you have dyscalculia, then you may struggle with this book—for now. Signs may include early trouble learning to count, struggling with mental arithmetic that most people do easily, struggling more than others to recognise by sight how many objects there are in a small collection, trouble arranging things by size, commonly mistaking one number for another and/or trouble measuring (for example, for a recipe or renovation).

The best way to think of dyscalculia seems to be that different people's brains develop different parts at different times. Einstein reportedly had delayed early speech. People with dyscalculia seem to have brains that develop mathematical capacity more slowly than others. That does not mean they cannot catch up later with the right training.

While more research is needed, it seems clear at this stage that dyscalculia is reversible in most people and that appropriate brain training can be a good place to start. The easiest way to begin is probably to acquire an app that provides training in basic mathematical skills. Once dyscalculia is resolved, you should have the ability to understand the material in this book. There is probably no need to let it defeat you.

This means the more mathematics you learn, the better you will become at learning it—and at learning other subjects. There is a demonstrated link, for example, between musical ability and mathematics. Researchers needed to use statistics to be able to say that confidently, and by using statistics, they were able to demonstrate that it is highly unlikely that these connections are secondary to underlying ability. Researchers controlled for so many **variables** (a statistical concept) that they were able to be confident it was the learning, rather than genetic ability, that made the difference[14]. This connection is so strong that if you have always wanted to learn a musical instrument and have never done so, a good time would be while you are working through this book.

Like assembling a jigsaw puzzle or learning a musical instrument, the more mathematics you do, the more the beauty emerges. At first it may seem like disjointed topics. But as you understanding deepens you will start to see patterns emerge. Mathematics is a language that relies on reason, yet it can deliver surprises that border on the awesome. And awe, as we are learning, is no small thing—it helps us to feel better. (It may even help treat clinical depression.[15] [16])

There are four additional—and even more important—reasons for approaching statistics through its mathematical underpinnings.

1. **Adults learn differently** from young children. Young children learn to count through mimicry and repetitive activity rather than a cognitive understanding of the nature of numbers, but the older we get, the more important it is for us to have a conceptual understanding if we are to truly engage with what we learn. We generally link new information to our existing understanding, then practice after that to cement what we know.

2. Because every research project is, by nature, unique, using statistical tools with insight—rather than imitation—reduces errors and increases the chance of **genuinely meaningful results**. When researchers choose their methods with understanding, they avoid the widespread mistakes found even in peer-reviewed literature. (One study found nearly half of the medical research papers it examined contained statistical errors[17]—another found something similar in psychology, even when looking at just one type of error.[18]) Informed use of software can be a powerful asset, but mindless use can be dangerous.

3. When people truly understand the statistical methods they're using, they can **prevent flawed techniques** from gaining ground. This kind of

clarity doesn't just avoid mistakes—it can stop entire fields from drifting into deeply embedded but faulty practices. You'll see examples of how and why that matters as we proceed.

4. When you understand statistics deeply, you're **free to apply it originally—even to entirely new problems**—without needing to rely on what others have done before.

Having read this book, you still might need a book covering more of the practicalities, particularly in relation to using the software that you have available. But if you want to start to understand the subject in depth then the mathematical background is the place to start.

Pre-algebra: Constructing Mathematical Sentences

Starting something new—whether it's a fitness program or making friends—can be challenging. The foundational stages of mathematics are no exception. They are, in fact, some of the hardest to master. You probably encountered this material early in your schooling, before moving on to algebra, which can be both an advantage and a disadvantage.

The advantage is that you already have some familiarity with these ideas. If algebra came easily to you, the concepts here may feel intuitive—perhaps even skippable (aside from the fun facts). The disadvantage is that if algebra was a struggle, the difficulty may have stemmed from gaps in your understanding of this earlier material. In that case, you might be surprised—and maybe even a little frustrated—to realize that some foundational topics need revisiting.

If that sounds discouraging, try thinking of this material like getting up to exercise on a cold morning. The first steps might be uncomfortable, but once you've started, you'll be glad you did. And who knows? Once you warm up, you may even find some surprisingly intriguing ideas hidden in these early concepts.

Adding and subtracting positive and negative numbers

> **Read this if** you struggled with algebra, you don't know what "commutative" means, negative numbers confuse you, you don't remember much about the number line, or you can't remember what "absolute value" means.

We can't understand the mathematics of uncertainty that lies behind statistics without understanding numbers. While basic numerical concepts may seem familiar from early childhood, many people hold views too simplistic to support the mathematics ahead.

When you first learned the word "mother", it was simple. Later, you encountered broader uses, like "necessity is the mother of invention"—a phrase too abstract for a toddler. Mathematics is similar: as understanding deepens, so does awareness of its nuances. Sometimes, we must let go of simpler ideas to grasp richer meanings.

Early on, we learn numbers occur in an ordered sequence—we count to ten. Then we learn numbers describe quantity. Next, we discover operations: we have five strawberries, add two to make seven, then take away four, leaving three.

Commutative property of addition

It doesn't matter if we start with 5 strawberries (or anything else), then add 2 then take 4 away, or if we start with 5 strawberries, take 4 away, then add 2. We'll still have three strawberries. The order of our actions doesn't matter:

$$5 + 2 - 4 = 3$$

$$5 - 4 + 2 = 3$$

This is called the **commutative property of addition and subtraction**. (You can think of this as the numbers commuting from one spot to another.) We get the same result, whatever the order of the numbers, provided we keep the + or − **sign**, or **operator**, with whatever number follows it. In the example above we kept the minus sign with the four, and that made it work.

If we tried to move the numbers around but left the signs—if we moved the 4 but left the minus sign where it was—we would get a completely different answer:

$$5 + 2 - 4 = 3$$

$$5 + 4 - 2 = 7$$

Suppose we want to change the order of the numbers again. This time we will make it

$$-4 + 2 + 5 = 3$$

This does give us the same answer but if we are thinking of objects, we can't start with −4, because we don't have any objects to take away.

At this point we encounter something we will see repeatedly in mathematics: an idea that seems completely obvious once you get it, but that isn't at all obvious until you do. We know that it isn't obvious until you see it, because people went for thousands of years without understanding it, when it could have helped them enormously. The great mathematicians of the ancient civilisations around the Mediterranean (Egyptians, Greeks, Romans and others) never discovered it. The idea may have developed in China, but it wasn't fully articulated until about 630AD in what is now India.

The number line

To understand the concept, we need to extend the idea of what a number represents.

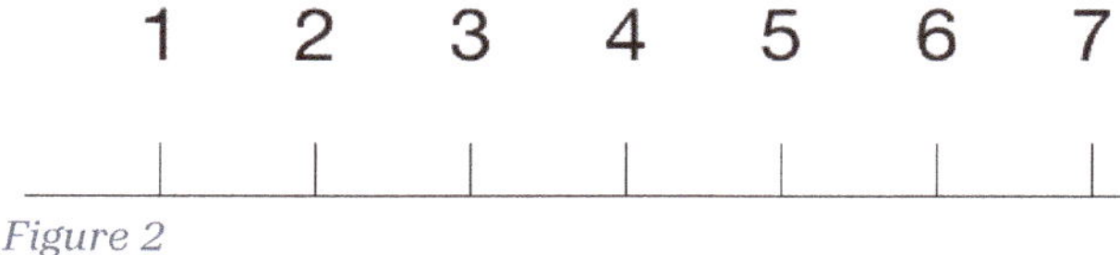

Figure 2

In ancient times if people wanted to build they would use a rope, with equally-distant knots in it (Figure 2).

They could work out a distance by counting the number of knots. It wasn't a big leap to go from there, to thinking of numbers as positions on a line, and from positions on a line to go to adding or subtracting as directions to move forward or backward along the line. The ancient civilizations around the Mediterranean never went further than this. It was probably the brilliant mathematician Brahmagupta (598-670), from what is now Rajasthan in India, who first used negative numbers in the sense they are used today. He or his colleagues invented zero, and started putting negative numbers on a line (Figure 3).

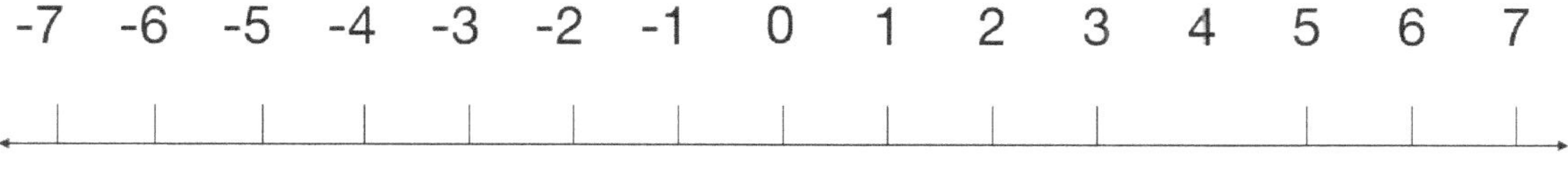

Figure 3

In terms of a number line, positive numbers always extend out to the right. We always start at 0. (In higher levels of mathematics this point is sometimes called the **origin**.) In this situation the negative or minus sign, – , is an instruction to turn around and head in the opposite direction.

So, we can interpret – 4 + 2 + 5 as being instructions to: start at 0, facing right. Now turn around and head in the opposite direction and move four spaces, which takes us to the –4 position.

Now move 2 spaces to the right, taking us to the –2 position. Now move 5 more spaces to the right, taking us to the 3 position. Which means that we get the same answer as we got by adding strawberries (3), even though we used a different way of thinking about numbers to get there—instead of thinking of numbers as counting objects, we thought about them as directions and distances on a line. Once you understand that numbers can have these different nuances, there is no need to understand what nuances are involved in any specific situation; they always just work. They are still numbers; these are just different models for thinking about them.

Another way to think about negative numbers

Numbers aren't just for counting strawberries; they're used to measure and track everything, including finances. Around 400–200 BC, the Chinese used black rods to represent assets and red rods for debts. This was probably the first use of negative numbers. Brahmagupta later talked about "fortunes" and "debts," teaching the concept through verse to aid memorisation.

Adding two bank balances (positive numbers) increases your total. But if you have money and add a debt, your total worth decreases—**adding a**

negative is the same as subtracting. If your debt exceeds your assets, you're in debt: the balance becomes negative. If someone pays off your debt (subtracting a negative), your wealth increases—**subtracting a negative is like adding a positive.**

Mathematical consistency means you don't need to lock into a single model. Whether thinking in terms of positions on a number line, financial balances, or strawberry counts, the rules stay the same. What matters is developing the mental flexibility to see numbers in different ways. Just as you can play different roles in life and still be the same person, –4 can mean both a position and a movement. It's both at once.

Absolute value

The magnitude, or size, of a number without regard to whether it is positive or negative is called the **absolute value** and marked as follows:

$$|6| = |-6|$$

$$= 6$$

The best way to think of this is as the distance from zero, regardless of the direction. The two lines on either side of the number are telling you that, in this instance, the negative sign doesn't matter; the value will always be positive.

To cement your understanding:

On a piece of paper, draw up your own number line from – 5 through 0 to 15, and use it to calculate the following;

What is 6 – 9 equal to?

What is 6 – – 9 equal to? (Remember: subtracting a negative is the same as adding a positive)

What is |6 – 9| equal to?

(Confirm that you can see why the answers were – 3, 15 and 3, respectively.)

Key points:

- Addition is commutative; the order of actions doesn't matter.

- Adding a negative number is the same as subtracting a number.

- Subtracting a negative number is the same as adding a number.

- On a number line, right is the positive direction and left is the negative one.

- Absolute value means we can ignore the sign. Just pay attention to the magnitude of the number.

Equals, equations, inequalities and brackets

Read this if: you struggled with algebra, you think the equals sign means "…has an answer of…", you don't know what an "equation" means, you don't know what an "inequality" means, or you are unsure of what any of these symbols mean: <, >, ≤, ≥, ≠, ≈

Many people think they understand the equals sign, but their view is often too narrow. This limited understanding starts in early arithmetic and works well—until algebra introduces letters instead of numbers. At that point, many students feel lost and give up.

To truly grasp statistics—not just how it works, but why—you need to understand algebra, which begins with a deeper view of the **equals sign**.

An **equation** is any line of mathematics with an equals sign. You've long known that $1 + 2 = 3$, first learning to think of it as "one plus two *has an answer of* three"—but mathematics asks you to think more broadly.

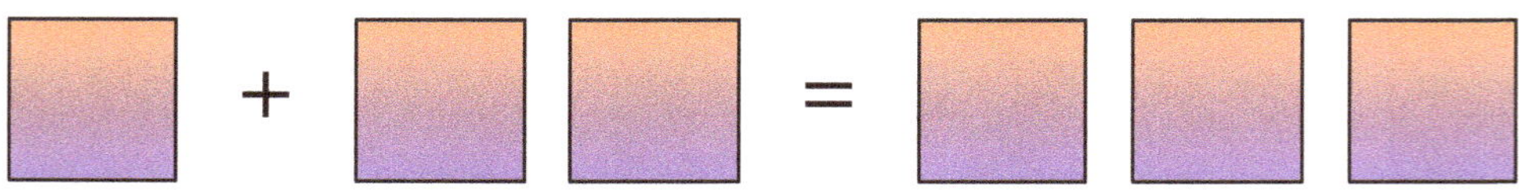

Figure 4

A better way to think of it is indicating that what is on the left is exactly the same as what is on the right: "$1 + 2 = 3$" is best understood, mathematically, as "one plus two *is exactly the same thing as* three" (Figure 4).

If $1 + 2 = 3$, then $3 = 1 + 2$; they are exactly the same thing. When we come to algebra, this distinction becomes crucial, because algebraic equations are solved by doing the same thing to each side of the equation, which works because they are the same thing even if they look different. You can think it as a set of scales. If you do the same thing to each side, it will stay balanced.

So:

$$1 + 2 = 3$$

Let's do the same thing to both sides of the equation: add 10.

$$1 + 2 + 10 = 3 + 10$$

Now each side of the equation adds to 13, but the equation still balances: it is still true, because what is on one side is the same as what is on the other side.

As long as you do the same thing to each side of the equation, the equation will still be true because what is on one side is exactly the same as what is on the other; just expressed differently.

Other variations on this sign (which you don't have to memorise, but it's good to understand) are:

< less than, for example $3 < 700$ > greater than, for example $700 > 3$	These signs are easy to understand when you think of them as the equals sign made narrower towards whichever side is smaller, and wider towards whichever side is larger.
$\leq$ less than or equal to $\geq$ greater than or equal to	These are just combining the less than or greater than signs with the equals sign. At first it may not be obvious how signs like this could be used, but you will see as we travel along they are important.
$\neq$ not equal $(1 + 1 \neq 1,200,000)$	This is just an equals sign with a slash through it.
$\approx$ approximately equal $(200,001 \approx 200,000)$	Think of this as being hand-wavey with the equals sign: not exactly equal, but close enough.

We'll discover a few more variations on the equals sign as we progress.

The combination of two expressions with a greater-than or a less-than sign between them is an **inequality**, not an equation.

$$2 + 12 + 7 > 2 + 12$$

We've seen that what you do to one side of an equation you must do to the other, however it doesn't necessarily work the same way with inequalities.

As we advance with mathematics you will have to check in each case that what happens to both sides of an inequality will keep it as a true inequality.

To cement your understanding:

Use the appropriate signs to associate these pairs of expressions:

4+3   7 (equal to)

4+3   8 (less-than, less-than-or-equal-to or not-equal-to)

4+3   6 (greater-than, greater-than-or-equal-to or not-equal-to)

6,999   7,000 (approximately equal to, less-than, less-than-or-equal-to or not-equal-to)

4+3   an elephant (not equal to)

Thinking ahead, can you think of a situation where doing something to both sides of an inequality may mean the sign has to be reversed? Don't worry if you can't; we'll get to it.

Conventions in equations when there are two operators together

Now let's work with some equations, both to get used to the idea, and to illustrate two of the issues involved with adding and subtracting positive and negative numbers. It often works out that we have two signs or operators together (meaning + or –), with only brackets between them. It might be hard to see the point of this now, but when we come to algebra, when some numbers will be replaced by letters, then understanding the way this works will become important.

The conventions are that a positive and a positive make a positive, a negative and a positive make a negative, a positive and a negative make a negative and two negatives make a positive.

The English language, in this case, works exactly the same way.

Two positives	I did do it.	Positive	It happened.
A negative and a positive	I didn't do it.	Negative	It didn't happen.
A positive and a negative	I did not do it.	Negative	It didn't happen.
A negative and a negative	I didn't not do it.	Positive	It happened.

Remember that we can think of subtracting a negative as being the same as taking away a debt: it has a positive effect on finances.

Now let's look at why it has to work this way in mathematics; it isn't just a convention that someone made up.

We could have chosen a literally unlimited range of numbers to illustrate what follows: these numbers are just chosen to illustrate a point.

Our first equation is

$$7 + 3 = 10$$

Our second is

$$3 = 24 - 21$$

If you are struggling with what follows, come back to the idea of fortunes and debts. Imagine that two people are getting married. One has \$7 in the bank. The other has a net worth of \$3, made up of \$24 in the bank, and a credit-card debt of \$21. Altogether their net worth is \$10. If it is more comfortable for you, you can think of this as thousands of dollars or millions of dollars; or of any other currency you prefer.

Replace the 3 in the first equation with $24 - 21$ from the second equation. You can do this, because they both equal the same thing. The brackets mean "do what is inside first".

$$7 + (24 - 21) = 10$$

Take 24 outside of the brackets. That won't change anything, because the order we use for addition doesn't matter.

$$7 + 24 + (-21) = 10$$

Subtract 7 from both sides. The 7 on the left disappears, because 7 minus 7 leaves 0, and -7 appears on the right. The equation will still be true. Both sides will now add up to 3, but it will still be a valid equation, because both sides are still equal. Remember, we aren't finding an "answer"; we are looking at what happens mathematically.

$$24 + (-21) = 10 - 7$$

At this stage, if you are getting confused about taking 7 from one side, this is an important lesson that will be useful as we progress: trust the mathematics. You don't need to keep track of what everything relates to as you go along. The mathematics will always work out, unless you have made a mistake.

Write the $10 - 7$ on the right-hand side as 3, because it is the same thing.

$$24 + (-21) = 3$$

Now subtract 24 from both sides.

$$+(-21) = 3 - 24$$

$$\boxed{+(-21) = -21}$$

You can see that adding negative 21 is the same thing as subtracting 21. It works out to be the same thing, which is what you would probably have expected, and is consistent with the idea of thinking of negative numbers as debts. In other words, adding a debt of \$21 is the same as taking \$21 away from the net worth.

This shows why what we noted above works: a positive and a negative immediately before a number make a negative.

To cement your understanding:

Calculate $7 + (24 - 21)$ in two different ways. First, do what is inside the brackets. Then ignore the brackets. Confirm for yourself that, in this instance, the brackets don't matter; you will still arrive at 10 (When we look at multiplication we will see that brackets are sometimes very important.)

We normally **omit positive signs** when we don't explicitly need to say that we are adding something. Just as $+(-21) = -21$, so also $+7 = 7$.

Now let's look at what happens when we have **two negative signs** together. To understand this, let's pick two more equations:

$$7 - 2 = 5$$

and

$$2 = 10 - 8$$

Now let's combine these equations.

Replace the 2 in the first equation with the $10 - 8$ from the second equation.

$$7 - (10 - 8) = 5$$

To make things a bit clearer, just for this part, we'll ignore the convention we described earlier, and show a + sign with positive numbers, instead of ignoring it.

$$+7 - (+10 - 8) = +5$$

We can get rid of the brackets (because the order we add and subtract in doesn't matter), but when we do, we have to show that the minus sign

$$+7 - +10 - -8 = +5$$

outside the brackets applies to both of the numbers inside them.

We saw earlier that a positive and a negative sign before a number is the same as just a negative one, so we can show $- + 10$ as just -10.

$$+7 - 10 - -8 = 5$$

Subtract 7 from both sides.

$$-10 - -8 = 5 - 7$$

Show $5 - 7$ on the right as -2

$$-10 - -8 = -2$$

Add 10 to both sides. The 10 on the left disappears because $-10 + 10 = 0.$ (Note that $-2 + 10 = 8$.)

$$- - 8 = -2 + 10$$

That means that minus minus eight is the same as plus eight.

$$- - 8 = +8$$

Again, in terms of debts and fortunes: taking away a debt of \$8 is the same as adding \$8 to the fortune.

This only works if our second convention is true: **two negatives** immediately prior to a number **make a positive**.

We said earlier that a negative sign means turn around and move in the opposite direction. So, if you like, you can think of two negatives as instructions to turn around in the opposite direction, then turn around in the opposite direction again, so that you are now facing to the right again.

To cement your understanding:

Calculate $7 - (10 - 8)$ in two different ways. First, do what is inside the brackets first. Then ignore the brackets. Confirm for yourself that the second time you do it you get the same result (5) if, and only if, two negatives make a positive.

Key points:

- The equals sign doesn't only mean "has an answer of", it means "both sides mean exactly the same thing".

- What you do to one side of an equation you have to do to the other, to ensure it is still true.

- There are other variations on the equals sign: greater than, less than, greater than or equal to, less than or equal to, not equal and approximately equal to.

- Two positives make a positive, a positive and a negative make a negative, a negative and a positive make a negative, and two negatives make a positive.

- Brackets mean "do what is inside the brackets first".

- You can combine two equations by substituting the value described by one into the other.

Properties of multiplication

Read this section if: you struggled with algebra, you don't understand *why* three lots of four should be the same as four lots of three, you don't know what • symbolises, you don't understand the difference between an "equation", an "expression", a "polynomial", an "inequality" and a "term", you are unsure about *why* four lots of (three plus eight) is the same as four lots of three plus four lots of eight, trying to multiply negatives and double negatives leaves you confused, you've forgotten the difference between a multiple and a factor, you can't remember what a prime number is or you don't understand much about the priority of mathematical operations.

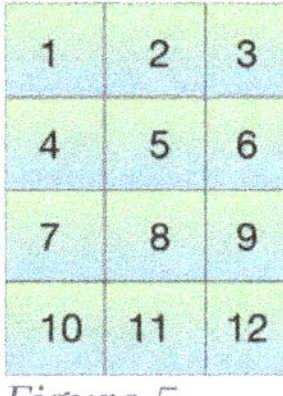

Figure 5

We've looked at addition, and multiplication is another form of it, with a twist. The difference is that, instead of adding single numbers, pieces of fruit, or anything else, we are adding the same number, a number of times.

Figure 5 shows a picture of 12 squares. We can think of this, as Figure 6 shows, as "3 lots of" 4, which is the same as $4 + 4 + 4$. We can also think of it as 4 lots of 3 (Figure 7), or $3 + 3 + 3 + 3$. So

$$3 \times 4 = 4 \times 3$$

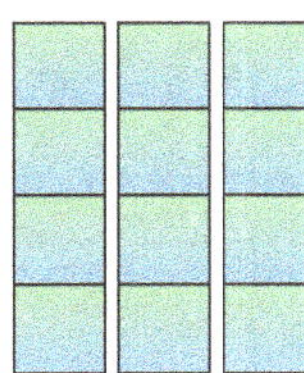

Figure 6

(same thing, which is in its own small way a bit of awe-inducing mathematics). This is called the **commutative property** of multiplication; the order doesn't matter. This still applies when more than two numbers are multiplied. We call the result when two numbers are multiplied together the **product.**

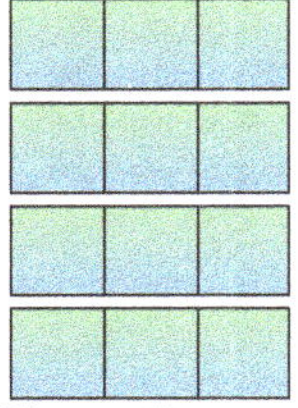

Figure 7

Another way to think about multiplication is "for each ... there are..." In this instance, for each of the three vertical columns there are four squares, or for each of the four horizontal rows there are three squares. You will find, when we come to look at probability, that it makes much more intuitive sense when you think of multiplication with the idea of "for each ... there are..." In the study of probability, that often translates into something like "for each of these four alternatives, there are three other alternatives..."

Here's something about the symbolic language of multiplication though. Early in your school career you learned to use the "x" sign for multiplication (as we did above), and you will still see it on a calculator. That is about the only place it is used. That terminology virtually disappears from the mathematical language once you advance further into it. You probably won't find a separate key for multiplication on your keyboard, and if you use a spreadsheet for multiplication you need to use the * key. When it comes to learning algebra, after years of seeing the x sign as an indicator for multiplication, suddenly it starts to be used for "the unknown quantity". (We tend to use italics for the letters in algebra, but that distinction isn't always clear.)

To avoid this confusion, from here on we will use the • symbol for multiplication.

Terms, expressions, polynomials

When we speak and write, we need to use appropriate punctuation and ordering, and to keep our letters in words, our words in sentences, and our sentences in paragraphs. It wouldn't make any sense if the last sentence read: "a a a a a a a a a a a c c c d d d d d d d d e e e e e e e e e e e e e e e e e e g g h h i i i i i i i i i i k k l n n n n n n n n n n n n n n n n o o o o o o o o o o p p p p p p p r r r r r r r r r r r r s s s s s s s s s s t t t t t t t t t t u u u u u u u w w w w w" (which is the same group of letters, just arranged differently). In mathematical language conventions of punctuation, grouping, and ordering also matter.

We saw earlier that whatever is in brackets is calculated first. When performing a calculation with multiple parts to it, multiplication and division are worked out before addition and subtraction.

For example:

$$2 + 3 \bullet 4 - 6 = 2 + (3 \bullet 4) - 6$$

That is the same as saying

$$2 + 3 \bullet 4 - 6 = 2 + 12 - 6$$

That, in turn, is the same as saying

$$2 + 3 \bullet 4 - 6 = 8$$

The three and four are multiplied together before the addition occurs, whether or not they are placed within brackets.

When writing one equation after another like this, if only one side is changing it gets clunky to keep writing "that is the same thing as saying", and to keep writing the left-hand-side out each time. So as a short cut, the left-hand side can be omitted for a series of equations that follow from one another. Subsequent lines start with an equals sign. We say this as "...which is equal to..." The usual way to write those three equations, and the way we will do it from now on in this book, is:

$$2 + 3 \bullet 4 - 6 \;=\; 2 + (3 \bullet 4) - 6$$
$$=\; 2 + 12 - 6$$
$$=\; 8$$

Looking again at these equations, if we were to just do the calculations in the order we came to them, we would usually get a completely different answer; in this case the left-hand side would equal 18.

The numbers and groups of numbers *between* addition and subtraction signs $(2, 3 \bullet 4, -6, 2, 12, -6)$ are referred to as **terms**, and a number of terms *added* or *subtracted* together is an **expression** $(2 + 3 \bullet 4 - 6)$ (Figure 8). An expression with 2 terms is called a **binomial**, and an expression with more than 2 terms is called a **polynomial**.

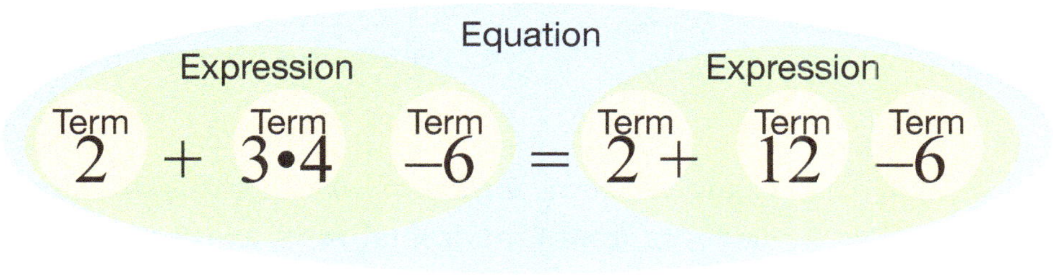

Figure 8

Brackets, and the distributive property of multiplication

(If you are struggling with this, see the How to Love Statistics YouTube video on *The distributive property of multiplication.*)

If we *don't* want to perform a multiplication before an addition or a subtraction within an expression, of course we use brackets. Once again, whatever is inside brackets is calculated before whatever is outside of them.

$$4 \bullet (3 + 5) \;=\; 4 \bullet 8$$

$$= 32$$

The 4 that is multiplying the whole contents of the brackets here can also be **distributed**, to multiply each of the terms within the brackets, without changing the value of the expression. This is called the **distributive property** of multiplication. So:

$$4 \bullet (3 + 5) \;=\; 4 \bullet 3 + 4 \bullet 5$$
$$= 12 + 20$$
$$= 32$$

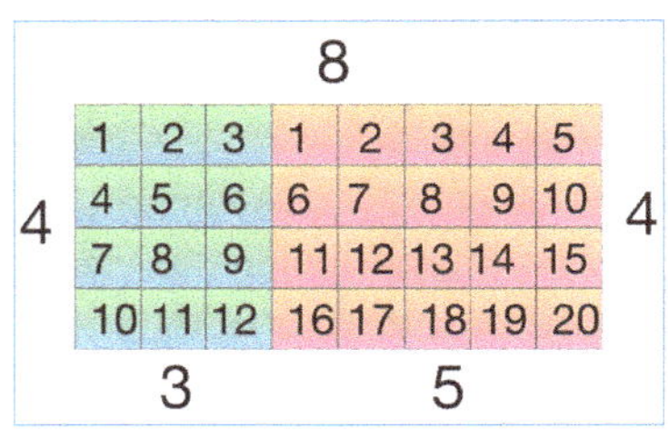

Figure 9

You can see why this is fromFigure 9.

4 lots of 8 is 32, and it is also the same as 4 lots of 3 plus 4 lots of 5.

In the equation

$$(4 + 2) \bullet (3 + 5) \;= 6 \bullet 8$$
$$= 48$$

we can multiply each of the numbers in the expression on the left-hand side of the equation together. So:

$$(4 + 2) \bullet (3 + 5) \;=\; 4 \bullet 3 + 4 \bullet 5 + 2 \bullet 3 + 2 \bullet 5$$
$$= 12 + 20 + 6 + 10$$
$$= 48$$

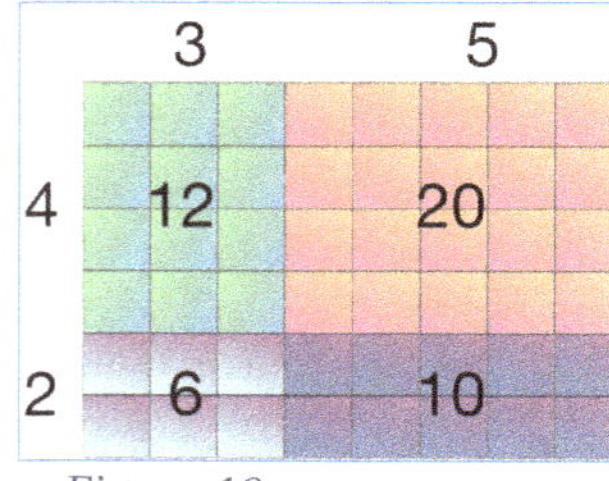

Figure 10

Figure 10 shows why this works: there are 48 squares, made up from 6 horizontal rows and 8 vertical columns. It is also made up from 3 lots of 4 squares plus 4 lot of 5 squares plus 2 lots of 3 squares plus two lots of 5 squares.

That may not seem like a particularly important point at this stage in your mathematical understanding, because it would be easier to just add the terms in brackets, then multiply. When we come to algebra, however, it will become crucial once we start putting things other than numbers into equations. Without understanding this point, most of what happens from algebra onwards won't make sense (which may be one of the reasons that some people struggle with algebra). The point is that the distributive property of multiplication isn't simply a mathematical convention; it is a recognition of the way that multiplying numbers works. The approach still applies where there are more than two terms within sets of brackets, and also where there are more than two sets of brackets in the expression.

Multiplying negative numbers

Just as with subtraction, we can think of the negative sign in multiplication as meaning "turn around and head in the other direction". The conditions are the same as with addition.

A positive multiplied by a positive makes a positive	$3 \bullet 6 = 18$
A negative multiplied by a positive makes a negative	$-3 \bullet 6 = -18$
A positive multiplied by a negative makes a negative	$6 \bullet -3 = -18$
Two negatives multiplied together make a positive	$-3 \bullet -6 = 18$

As we did with addition, let's look at some example equations, partly to see why this convention makes sense, and partly to do some more equations.

We know that $6 \bullet 3 = 18$, and we also know that $8 - 2 = 6$

Replace the 6 in the first equation with $8 - 2$ from the second equation. You can do this, because they both equal the same thing.	$(8 - 2) \bullet 3 = 18$
Use the distributive property of multiplication to distribute the 3 between the 8 and the $- 2$.	$8 \bullet 3 - 2 \bullet 3 = 18$
Note that $8 \bullet 3 = 24$	$24 - 2 \bullet 3 = 18$
Subtract 24 from both sides.	$-2 \bullet 3 = 18 - 24$
This only works if multiplying a positive by a negative gives a negative.	$\boxed{-2 \bullet 3 = -6}$

Now let's look at multiplying two negatives. We'll pick as our two example equations:

$$-2 \bullet 3 = -6$$

$$7 - 4 = 3$$

Replace the 3 in the first equation with 7 – 4 from the second equation.

$$-2 \bullet (7 - 4) = -6$$

Use the distributive property to distribute the –2 to the 7 and the –4 .

$$-2 \bullet (7 - 4) = -6$$

The brackets are here as a reminder to do what is in the brackets before anything else.

$$(-2 \bullet 7) + (-2) \bullet (-4) = -6$$

Minus 2 multiplied by 7 is minus 14

$$-14 + (-2) \bullet (-4) = -6$$

Add 14 to both sides. (Once again, we aren't trying to find an answer; this is demonstrating the way that numbers work.)

$$(-2) \bullet (-4) = -6 + 14$$

This only works if a negative multiplied by a negative gives a positive.

$$(-2) \bullet (-4) = 8$$

Multiples and factors

When you multiply two whole numbers (an **integer** is a whole number), the result is called a **multiple** of the two numbers. So 12 is a multiple of 4, while 13 and 14 are not. The two numbers that are multiplied together are called the **factors**. So, 3 and 4 are factors of 12, but 7 is not.

Prime numbers

If a number has no factors apart from 1 and itself, it is called a **prime number**. The number 1 is not considered to be a prime number. It is considered to be a category of its own, because of the unique property that you can continue to multiply any number by 1, as many times as you like, and it won't change the result.

To cement your understanding:

• If you never learned, or have forgotten, any of your "times tables", write out the ones you aren't comfortable with, working them out in your head. As an option you can memorise them, but it's actually the writing out that will help to develop your brain's sense that multiplying is the same as adding groups of numbers. If you did learn your tables up to twelve times twelve, you might like to do this exercise by calculating them (in your head) up to fifteen times fifteen.

• Calculate (3 + 5) • (7 + 4) in two different ways. First, add 3 and 5, and then add 7 and 4, then multiply the result. Second, multiply each of the numbers in each of the sets of brackets by one another, to confirm that you reach the same amount. Check that doing this both ways gives you the answer of 88.

• Of the numbers 2, 6 and 3, which are the factors and which is the multiple?

• Of the numbers 6, 7, 8 which is prime? (Confirm that you understand that it is 7).

Key points:

- Multiplication is the same as adding the same number, a specified number of times.

- The result of multiplication is called the product.

- It is commutative.

- It can be thought of as "for each… there are…"

- This book uses the • symbol to represent multiplication.

- Term: A single mathematical entity, such as a number or product of numbers and variables (e.g. $3 \bullet 3 \bullet 6$).

- Expression: A combination of terms connected by addition or subtraction, without an equality sign (e.g. $3 \bullet 3 \bullet 6 + 7 - 4$).

- Equation: A statement that two expressions are the same, connected by an equals sign (e.g. $3 \bullet 3 \bullet 6 + 7 - 4 = 57$).

- Expressions with two terms are called binomials, while expressions with more than two terms are called polynomials.

- Multiplication is performed before addition and subtraction, unless brackets are used to show that what is inside them is calculated first.

- Multiplication is distributive.

- As with addition, a positive and a positive make a positive. A positive and a negative make a negative. Two negatives make a positive.

- An integer is a whole number.

- The product of two integers is called their multiple.

- The integers that can be multiplied together to form a product are called its factors.

- A prime number is one with no factors except 1 and itself.

Fractions and percentages

Read this section if: you struggled with fractions, you can't remember what a "numerator" and "denominator" are, you've forgotten what a "common denominator" is, you don't understand *why* it is easier to multiply fractions than to add them, you didn't realise that a fraction multiplied by its reciprocal will always equal 1, you are unsure about how to work with fractions larger than 1, or you are confused about how fractions relate to division.

(If you are struggling with this, see the How to Love Statistics YouTube video on *Why Fractions are Hard*)

Back to thinking about people in ancient times using a knotted rope to measure distance, it wasn't a big leap, when shifting from 'how many knots?' (hand-spans, cubits, or whatever measure they were thinking of) to 'how much distance?'.

They realised that numbers didn't have to be an exact number of knots. If you are measuring the height of a wall, it doesn't necessarily come out as a whole number of knots. They could move along the rope between the whole numbers, and they would still have a number, or position on the line, so they needed another way to think about numbers on the line that could account for the spaces in between the marked numbers (Figure 11).

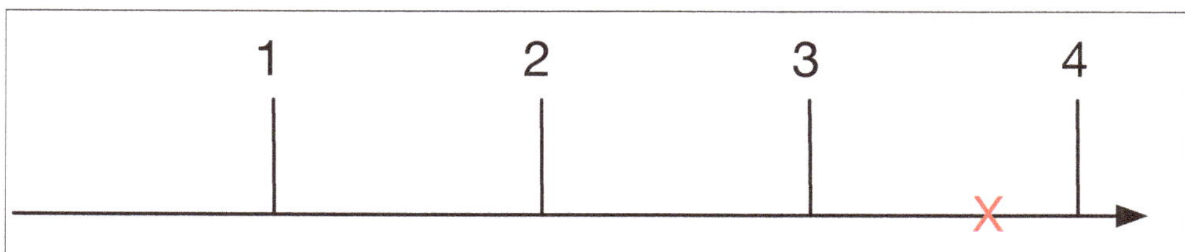

Figure 11

Likewise, if they were thinking about "how much will you pay for the wheat I have brought to the market?", whole numbers of volumes, (baskets, bushels, or whatever) wouldn't necessarily account for what they needed.

The challenge was to find a way to represent *any* position on the line between the numbers. Until you know how it is done, that isn't a simple thing to think through. The Egyptians were probably the first people to work out how to do this, in about 4000 BC, by simply writing two numbers to make another number": a fraction.

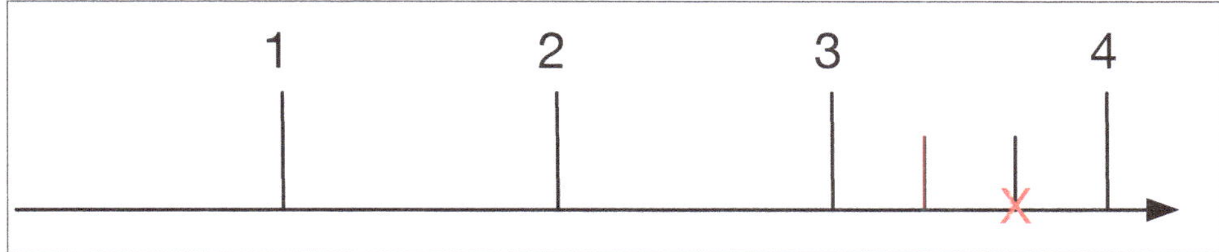

Figure 12

Supposing they wanted to indicate a distance up to the red cross in Figure 12. They achieved this by devising a way to break up the space between numbers into as many equal parts as they cared to nominate, and then to count the number of those parts that would get them to where they wanted to be. If you were to nominate three divisions between the numbers 3 and 4 on the number line, then count along two of them, that would get you precisely where you wanted to be. Of course, you might not get there precisely by dividing the segment of the line into three parts. You might need 10, 17, or any other number. So the Egyptians needed a code to tell them how many parts they were nominating to break something into, and then to number off to the point they wanted to get to. They decided to put the number of parts they were nominating on the bottom. We now call this the **denominator**. They put the number of parts they were counting on the top. We now call this the **numerator**. For added clarity, we also put a small line between the numerator and denominator (which the Egyptians didn't). So we write this fraction as two-thirds:

$$\frac{2}{3}$$

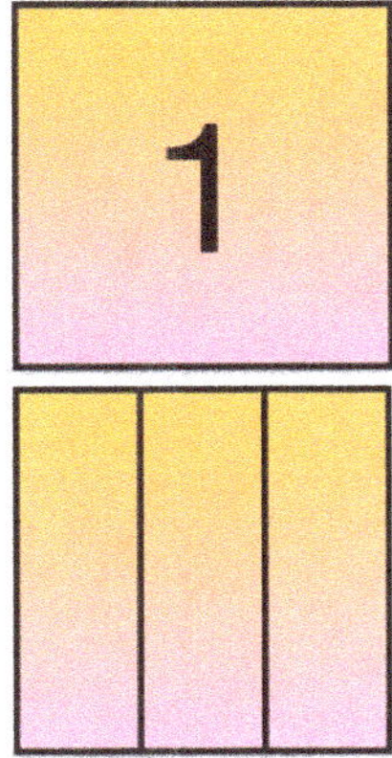

The denominator is 3: the segment is broken into three parts. The numerator is 2: we count two of them. Altogether, we call the position on the line marked in Figure 12 three and two-thirds:

$$3\frac{2}{3}$$

Let's expand on this idea by thinking about a square (Figure 13). In the first instance, we'll use it to represent the number 1. Later we'll use it to represent something else.

Figure 13

The idea of fractions is based on the idea of fracturing, fragmenting, dividing, or breaking something into smaller parts. Because we are being mathematical, we think of fractions as identical parts. So, as with the segment on a number line we just looked at, we can divide it into three (the denominator), and take 1, 2, or 3 of them (Figure 14). Three-thirds is the same as 1 whole square.

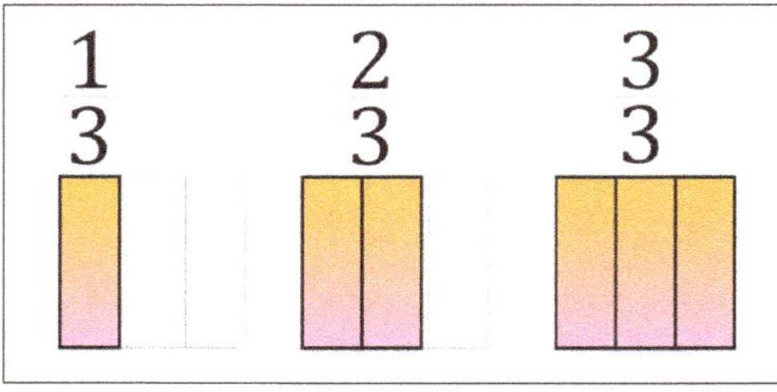

Figure 14

Now, if we divide our square into 6 equal parts (Figure 15), you can see two things: first, as the denominator gets larger (from 3 to 6), the size of each part gets smaller. We need to fit more of them into the same space.

The second thing we can see if we divide our square into 6 equal parts is:

$$\frac{2}{6} = \frac{1}{3}$$

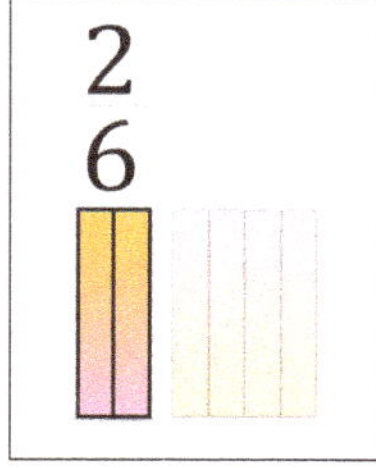

Figure 15

If you multiply or divide the numerator and the denominator by the same amount, the value of the fraction doesn't change. So if we divide both the numerator and denominator of two-sixths by two, we get one-third, which has exactly the same value

Wherever possible, we want to express fractions in the simplest form:

$$\frac{2}{6} = \frac{1}{3}$$

We do this by finding any number that is a factor of both the numerator and denominator and dividing the top and the bottom by it.

Now:

$$\frac{3}{6} = \frac{1}{2}$$

If we had divided our square into 100 pieces, then taken 50 of them, that would still be half the square.

$$\frac{50}{100} = \frac{1}{2}$$

Another way to show that a fraction has 100 as the denominator is %. The percent sign is just a form of shorthand, for "over 100". So

$$\frac{3}{4} = \frac{75}{100} \quad (multiply\ the\ numerator\ and\ denominator\ by\ 25)$$

$$= 75\%$$

We can also think of fractions as a **ratio**, or a **proportion**, as in "three quarters/seventy five percent of the crowd wore red". We could also say this as "three out of four people in the crowd wore red". We can write this as 3:4, or 75:100. It's still another way of writing a fraction.

Fractions probably convey more nuances of meaning than any other part of mathematics, which can make them some of the most difficult things to learn. More "advanced" mathematics contains symbols that appear more intimidating, and algebra allows the development of longer reasoned arguments. But in terms of understanding different implications conveyed by one simple symbol, it probably doesn't get any harder than with fractions. Fortunately, mathematics is a much more forgiving language than most spoken ones, which means that it's not usually necessary to remember all of these nuances in order to use fractions. It doesn't matter which meaning you use—it will still take you to the right place. However, unless you understand that it can carry all these meanings, some operations with fractions might not make sense.

To cement your understanding:

- In the fraction $\frac{7}{9}$ which is the denominator and which is the numerator?

- Which number is larger, $\frac{7}{9}$ or $\frac{8}{9}$?

- Which is larger, $\frac{2}{4}$ or $\frac{2}{3}$? (If you aren't sure of the answer, or can't see why it is so, use Figure 14 to work it out.)

- $\frac{6}{12} = \frac{2}{4} = \frac{1}{2}$. Can you see why?

Adding and subtracting fractions

When we add and subtract fractions, we need to make sure we are adding and subtracting the same thing. The question, 'If I have five oranges and take two apples away from them, then add three plums, how many bananas do I have?', is clearly absurd. What they have in common is that they are pieces of fruit, and asking, 'If I have five pieces of fruit

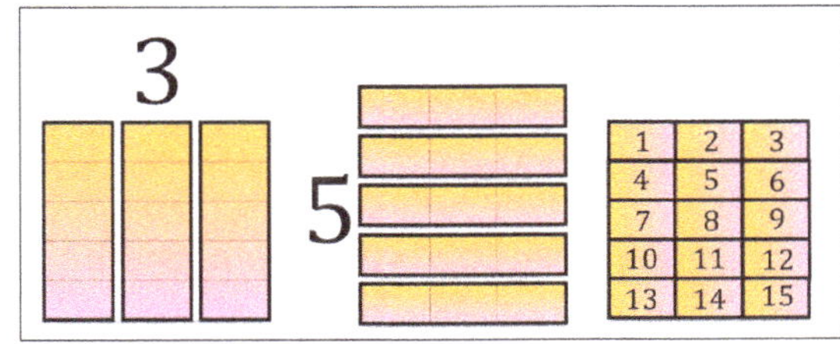

Figure 16

and take two away from them, then add three, how many pieces of fruit do I have?', makes sense. Of course, we don't need to be talking about objects (like fruit); we can just talk about whole numbers. But when we are adding and subtracting fractions, we need to work out what they have in common. We can get a picture of

$$\frac{1}{3} + \frac{2}{5}$$

by dividing our square into 3 vertical columns, and also into 5 horizontal rows (Figure 16). This has the effect of dividing our square into 15ths (3 • 5 = 15). You can see, by multiplying the numerator and the denominator by the same amount, or just by counting squares in Figure 16, that:

$$\frac{1}{3} = \frac{5}{15} \ (multiply\ the\ numerator\ and\ denominator\ by\ 5)$$

And also that

$$\frac{2}{5} = \frac{6}{15} \ (multiply\ the\ numerator\ and\ denominator\ by\ 3)$$

We can therefore say that

$$\frac{1}{3} + \frac{2}{5} = \frac{5}{15} + \frac{6}{15}$$

$$= \frac{5 + 6}{15}$$

$$= \frac{11}{15}$$

Fifteen is a common denominator; it is a number that is a multiple of both 3 and 5. There are other numbers we could use as common denominators for 3 and 5; we could have used 30, 45, or 60, for example. But if you multiply the denominators in an expression together, you will always get a common denominator.

So, the procedure for adding fractions is to multiply the denominators to find a common denominator, then multiply each numerator by whatever the denominator of that fraction was multiplied by. Then, you can add the numerators.

Subtraction works exactly the same way, remembering that subtracting is the same as adding a negative number.

$$\frac{1}{3} - \frac{2}{5} = \frac{5}{15} - \frac{6}{15}$$

$$= \frac{5-6}{15}$$

$$= -\frac{1}{15}$$

Figure 17

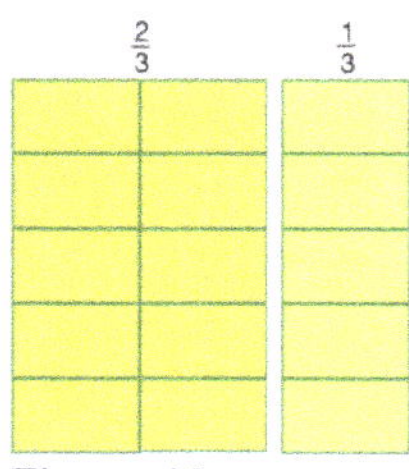

Figure 18

Figure 19

Multiplying fractions

Suppose we want to multiply two-thirds by four-fifths. We can multiply the numerators together and the denominators together, to get eight-fifteenths.

$$\frac{2}{3} \cdot \frac{4}{5} = \frac{8}{15}$$

It's actually easier than adding and subtracting fractions (which is one of the reasons some people find fractions challenging to understand.) Now let's think about *why* it works so easily. Start as we did earlier when we were adding and subtracting fractions: divide the square into 3 vertical columns and 5 horizontal rows, making 15 squares altogether (Figure 17). Now, we need four fifths *of* two-thirds. So, in Figure 18, take two of the thirds (two vertical columns), and note that they are already conveniently divided into fifths. We need four fifths *of these* (Figure 19). So altogether we've selected 8 out of the 15 squares (coloured in orange).

Let's think about what happens when we reverse the fractions:

$$\frac{4}{5} \cdot \frac{2}{3} = \frac{8}{15}$$

Figure 20

This time we are trying to find two-thirds of four-fifths. Divide the square as we did previously, but this time take the top four rows: that will be four fifths (Figure 20). These four rows are already conveniently divided into thirds, so now we take two of columns. That gives us two-thirds of four-fifths, and we have still ended up with eight-fifteenths (Figure 21).

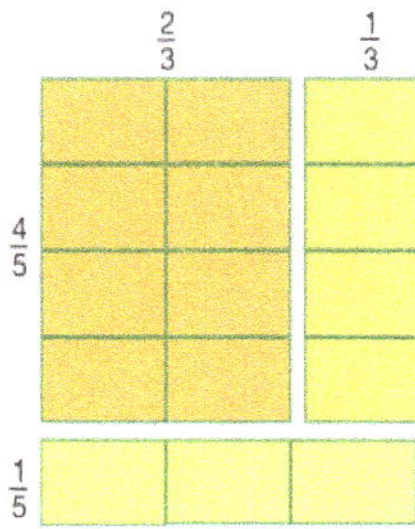

Figure 21

Either way, the large square, which represents the whole, is made up of 3 multiplied by 5. That large square gives us the denominator, and we have actually used the same process to this point that we used to find a common denominator when adding. The fraction of this that we are interested in is made up of 2 by 4 of these pieces.

To cement your understanding:

• Calculate $\frac{1}{6} \bullet \frac{3}{4}$ by multiplying the numerator and denominator.

• Draw up a square on a piece of paper. Use vertical lines to divide it into quarters (4 columns), and horizontal lines to divide it into sixths (6 rows). How many small rectangles have you created?

• Shade out 1 horizontal row in one colour or pattern. That will represent $\frac{1}{6}$ of the whole square. Now shade out 3 vertical columns. That represents $\frac{3}{4}$ of the whole square. Now work out what $\frac{3}{4}$ of $\frac{1}{6}$ is, then work out what $\frac{1}{6}$ of $\frac{3}{4}$ is. Confirm that these two expressions give you the same number of small rectangles. Note that either way, you select 3 out of the 24 small rectangles.

Fractions larger than 1

Now, instead of thinking of our square as representing 1, as we did above, we'll think of it as representing 10 (Figure 22). Not much else has to change; there is no rule that fractions must be smaller than 1 or, in fact, that they have to be fractions of 1. A fraction can be a fraction of anything, if the pieces are of equal size. It can be a fraction of an orange or of a national deficit. The number 10 is an integer (whole number), but it can also be written as a fraction:

Figure 22

$$10 = \frac{10}{1}$$

So to find two-thirds of 10, just multiply the numerator and denominator as you would with any other fraction:

$$10 \bullet \frac{2}{3} = \frac{10}{1} \bullet \frac{2}{3}$$

$$= \frac{20}{3}$$

We call a top-heavy fraction like twenty over three an **improper fraction**. (It's a strange name because it is still a perfectly legitimate fraction.)

We can think of twenty over three as 20 lots of one-third, one-third of 20, or 20 divided by 3. They all mean the same thing; fractions are very forgiving in that way. The only problem is that it can be hard to form a mental picture of twenty over three, so we usually express fractions larger than 1 as a combination of an integer and a fraction. To do that, find the largest multiple of the denominator that is less than or equal to the numerator. In this case, the largest multiple of 3 that is less than or equal to 20 is 18. We can write this as

$$\frac{20}{3} = \frac{18 + 2}{3}$$

$$= \frac{18}{3} + \frac{2}{3}$$

$$= 6\frac{2}{3}$$

It is easier to imagine six and two-thirds than to imagine twenty over three, although these two numbers are the same.

To cement your understanding:

- What is a common denominator of $\frac{5}{6}$ and $\frac{3}{7}$?

- What is $\frac{5}{6} + \frac{3}{7}$? (If you didn't get $\frac{53}{42}$ or $1\frac{11}{42}$, where did you go wrong?)

- Write $\frac{2}{5}$ as a percentage. Then add 110% to your answer. Express this as a single fraction, over 100, then as the same fraction with the smallest possible numbers in the numerator and denominator, then express this as an integer and a fraction. (If you didn't end up with $1\frac{1}{2}$ you probably need to read the material above again. It's important to get this, or the rest of mathematics will be confusing, as will statistics.)

Reciprocals

Each fraction has its own reciprocal, which means that the numerator and denominator are swapped. The fraction one-half has a reciprocal of two. Two-thirds has a reciprocal of three-halves (or one and a half). When you multiply a number by its reciprocal, the resulting answer is always 1 because it makes the numerator and the denominator equal the same number.

$$\frac{2}{3} \bullet \frac{3}{2} = \frac{2 \bullet 3}{3 \bullet 2}$$

$$= \frac{6}{6}$$

$$= 1$$

And so, for example, when you multiply two by a half you get one, whether you think of it as two lots of one half or as one half of two.

> **To cement your understanding:**
>
> • Multiply $\frac{5}{7}$ by its reciprocal. If you didn't get 1, can you see where you went wrong?
>
> • What is the reciprocal of $\frac{3}{13}$? Write this as an integer and a fraction. (If you didn't get $4\frac{1}{3}$ can you see where you went wrong?)

Fractions and division

So far, we've spent time looking at addition, subtraction and multiplication of fractions, but we haven't spent much time looking at division. That may seem strange because you probably spent a lot of time on it at school (if you can remember back that far). However, once you understand fractions, the concept of division is largely redundant.

Even though the idea of showing one number on top of the other to represent a fraction goes back at least as far as the ancient Egyptians, and the idea of putting a line between the numerator and the denominator goes back at least as far as India over 1,000 years ago, the sign you were taught to use for division ("÷") wasn't invented until 1659 by Johann Heinrich Rahn, who was writing a Swedish textbook about mathematics. It is based on a combination of both the sign for a fraction ("–") and the sign for a ratio (":"). He seems to have invented it as a way of being able to teach children about the concept of division before he thought they were ready for the idea of fractions. You won't see the sign "÷" on your computer keyboard, although you might see it on your calculator (or you might see "/" instead). That is probably the only place you will see it; you won't find it used at all when we get into higher mathematics. Yet another way to think about fractions is as a division: we divide the numerator by the denominator.

Can you remember the idea of a "remainder" that you used in school? When we looked at converting twenty over three to six and two-thirds, the equation we used was:

$$\frac{20}{3} = \frac{18 + 2}{3}$$

$$= \frac{18}{3} + \frac{2}{3}$$

$$= 6\frac{2}{3}$$

In this case, the way you first learned division, you would have said that 3 goes into twenty 6 times (making 18) with a remainder of 2. But expressing twenty over three as six and two-thirds is much more helpful than expressing it as "6 remainder 2".

You will probably never have to actually divide two fractions (unless you are sitting a school examination), but if you ever do, here is the trick to use:

Start by writing the division as a fraction, with $\frac{2}{3}$ as the numerator and $\frac{4}{5}$ as the denominator. This is the way you will always see it in this book from this point onwards.

$$\frac{2}{3} \div \frac{4}{5} = \frac{\frac{2}{3}}{\frac{4}{5}}$$

Multiply the $\frac{4}{5}$ in the denominator by its reciprocal $\left(\frac{5}{4}\right)$, and multiply the numerator by the same amount so that the value of the expression doesn't change.

$$= \frac{\frac{2}{3} \cdot \frac{5}{4}}{\frac{4}{5} \cdot \frac{5}{4}}$$

The denominator is reduced to 1, and if a fraction has 1 in the denominator it can be ignored, because dividing by 1 doesn't change anything.

$$= \frac{\frac{2}{3} \cdot \frac{5}{4}}{1}$$

What we have therefore done is, instead of dividing $\frac{2}{3}$ by $\frac{4}{5}$, we have multiplied it by the reciprocal of $\frac{4}{5}$.

$$= \frac{2}{3} \cdot \frac{5}{4}$$

$$= \frac{10}{12}$$

$$= \frac{5}{6}$$

In other words, to divide a number (expressed as a fraction or otherwise) by a fraction, multiply by the reciprocal.

As we progress, you will find that, on the one hand, tricks like this will come in handy. On the other hand, you have probably never had to divide one fraction by another in your life (without expressing it as a fraction), and you probably never will have to again. But now you know how!

To cement your understanding:

• Calculate $\frac{3}{4} \div \frac{5}{6}$. (If you didn't get $\frac{18}{20} = \frac{9}{10}$, can you see where you went wrong?)

• Now do the reverse: multiply $\frac{5}{6} \cdot \frac{9}{10}$, to confirm that the result is $\frac{3}{4}$.

Key points:

- The bottom of a fraction nominates the number of equal parts something is broken into, and is called the denominator.

- The top part counts the number of these parts we are interested in and is called the numerator.

- If you multiply or divide the numerator and the denominator by the same amount, the value of the fraction doesn't change. If you add or subtract the same thing to the numerator and denominator, you create a different fraction.

- The percentage sign is a shorthand way of writing a fraction with 100 as the denominator.

- Fractions, ratios and proportions are different ways of expressing the same idea.

- To add or subtract fractions, find a common denominator, then add or subtract the new numerators.

- To multiply fractions, multiply the numerators and denominators.

- Fractions can be larger than 1, and integers can be expressed as fractions.

- Fractions with numerators that are larger than their denominators are called improper fractions.

- Each fraction has a reciprocal, which means the numerator and denominator are reversed. When you multiply a fraction by its reciprocal, the result is always the number one.

- Fractions are also the same thing as dividing the numerator by the denominator. To divide by a fraction, multiply by its reciprocal.

Powers

Read this section if: you aren't sure what an exponent, an index and a logarithm are; you don't understand what it means to add exponents; you don't know what it means to multiply exponents; you don't understand what raising to the power of 1 or raising to the power of 0 means (or if you do know what it means, you don't understand why it works that way); you don't know what it means to raise to a negative power or you are confused about the priority of mathematical operations.

(If you are struggling with this, see the How to Love Statistics YouTube video on *Exponential Notation*.)

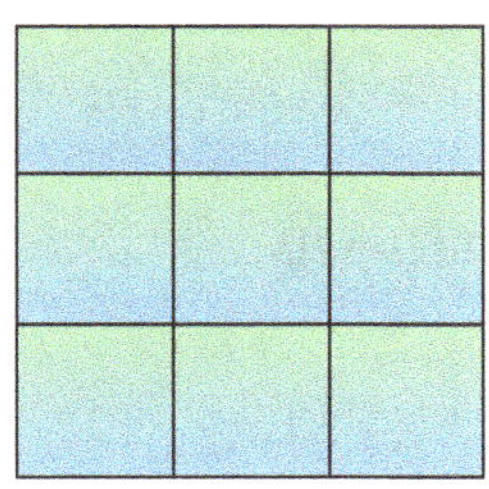

Figure 23

Just as multiplication is a shorthand way of talking about repeated addition, so raising to a power is a shorthand way of writing repeated multiplication. Just as $3 + 3 + 3 + 3$ can be shortened to $3 \bullet 4$, so $3 \bullet 3 \bullet 3 \bullet 3$ can be shortened to 3^4.

When you multiply $3 \bullet 3$ and draw up the results, it can form a square (Figure 23). This is called 3 squared, or 3 to the power of 2, and written as 3^2.

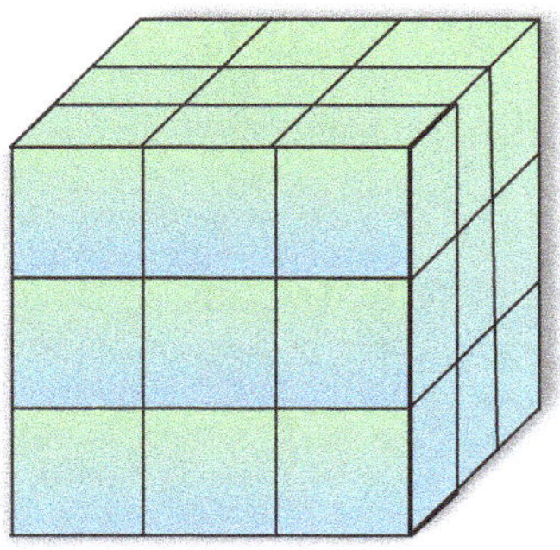

Figure 24

When you multiply $3 \bullet 3 \bullet 3$, it is raised to the power of 3 or the 3rd power, written as 3^3. This can be drawn up as a cube, so we sometimes call it three cubed (Figure 24). We can raise any number to any power we like.

We call the number being raised to a power the base. We call the number in superscript (a smaller font, raised off the line) that indicates how many times it is multiplied by itself the **exponent**. The exponent is also known as the **index** or the **logarithm**. (We'll see why when we come to look at logarithms on page 125.) We call both numbers together a base raised to a power, as in "three to the power of five" or "three to the 5th power" (Figure 25).

> **To cement your understanding:**
>
> With 9^4 which of these two numbers (9 or 4) is the index, and which is the base? Which is the exponent?

Adding exponents

When you multiply a base raised to a power by that same base also raised to a power, you can just add the exponents:

$$3^2 \cdot 3^4 = 9 \cdot 81$$
$$= 729$$

At the same time:

$$3^{2+4} = 3^6$$
$$= 729$$

You can see why this is when you remember that $3^2 \cdot 3^4$ is shorthand for:

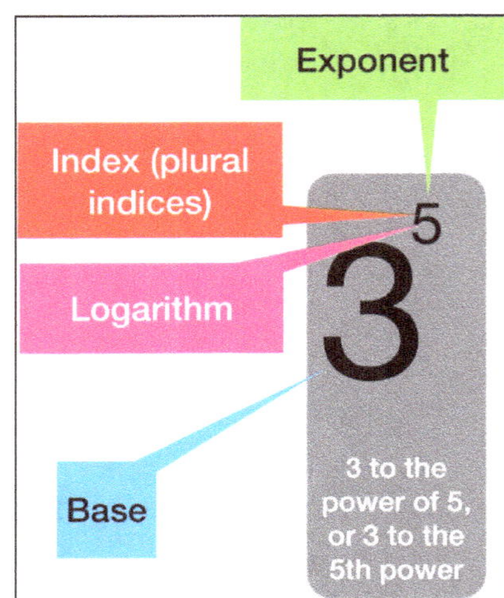

Figure 25

$$(3 \cdot 3) \cdot (3 \cdot 3 \cdot 3 \cdot 3) = 3^6$$

The brackets here don't make any difference; we can multiply the terms in any order. So we have

$$3 \cdot 3 \cdot 3 \cdot 3 \cdot 3 \cdot 3 = 3^6$$

In other words, squaring 3; then multiplying the result by 3 multiplied by itself 4 times, is the same as multiplying 3 by itself 6 times.

> **To cement your understanding** (you can use a calculator or a spreadsheet):
>
> • Calculate 2^7 in two different ways: first multiplying 2 by itself 7 times, then by treating it as $2^{3+4} = 2^3 \cdot 2^4 = 8 \cdot 16$
>
> • Write out $2^3 \cdot 2^4$ in full (using only the number 2 and the $\cdot$ symbol). Note that you have written 2 seven times, and that this is the same as adding the exponents $3 + 4$.

Multiplying exponents

If we raise something to a power, then raise the whole thing to a power, that is the same as multiplying the exponents.

Write out $(3^2)^3$ as 3^2 multiplied by itself 3 times.	$(3^2)^3 = (3^2) \bullet (3^2) \bullet (3^2)$
Write this out in full, showing each instance of 3^2 as $3 \bullet 3$	$= (3 \bullet 3) \bullet (3 \bullet 3) \bullet (3 \bullet 3)$
The brackets don't matter, because order doesn't matter in multiplication and because we are multiplying the same thing, anyway.	$= 3 \bullet 3 \bullet 3 \bullet 3 \bullet 3 \bullet 3$
We're multiplying 3 by itself 6 times, so show this with exponential notation.	$= 3^6$
6, of course, is the same as 2 multiplied by 3	$= 3^{2 \bullet 3}$

It doesn't matter what order the brackets go in:

$$(3^2)^3 = (3^3)^2$$

$$= 729$$

To cement your understanding:

• Write out $(2^4)^2$ in full (using only the number 2 and the $\bullet$ symbol). You should have written 2 out 8 times. You can use brackets if you like, but in this instance they will make no difference.

• Confirm that $(2^4)^2 = (2^2)^4$ by calculating 2 to the 4th power then squaring that; and then by squaring 2 and raising all of that to the 4th power. Check that either way, you arrived at 256.

Raising to the powers of one and of zero

There are conventions for raising numbers to the power of 1 and to the power of 0.

First, let's look at an example of raising to the power of one.

$$3^3 \bullet 3^1 = 3^{3+1}$$

$$= 3^4$$

$$= 3 \bullet 3 \bullet 3 \bullet 3$$

$$= 81$$

This only works if $3^1 = 3$, so that is the convention: **anything raised to the power of 1 equals itself.**

Now let's do something similar with raising to the power of 0.

$$3^3 \bullet 3^0 = 3^{3+0}$$

$$= 3^3$$

$$= 27$$

This only works if $3^0 = 1$

The convention, therefore, is that anything raised to the power of 0 is equal to 1. (Note: some people get confused, because when we multiply by 0 the answer is always 0. We aren't multiplying by 0 here; we are raising to the power of 0.)

These conventions turn out to be useful in a range of situations. The further we progress with mathematics, the more obvious it will be that it needs to work that way.

The idea of using a superscript to indicate raising to a power was introduced by René Descartes (1596-1690), whom we will meet later when we come to look at the Cartesian plane. It's a simple enough *notation*; the important new development was the underlying *notion* that it was important to think about raising numbers to powers. That was a key insight that allowed for virtually everything that came after it in mathematics. The person who worked out most of what follows in this section—how to work with powers—was John Wallis (1616–1703), an Anglican priest and academic from Cambridge, and then from Oxford, who worked on the Parliamentary side (with Oliver Cromwell) in the English civil war as a successful codebreaker. He was one of the key figures in the history of cryptography and is also the person who gave us the infinity symbol, ∞.

To cement your understanding:

What is 104^1 equal to? What is 104^0 equal to?

Raising to a negative power

Now let's think about what it means to raise to a negative power. Again, we'll look at what happens when we add exponents:

$$5^3 \bullet 5^{-3} = 5^{3-3}$$

$$= 5^0$$

$$= 1$$

Given that $5^3 = 125$, we could write this out again as:

$$125 \bullet 5^{-3} = 1$$

That can only work if 5^{-3} is the reciprocal of 125. That is:

$$5^{-3} = \frac{1}{125}$$

$$= \frac{1}{5^3}$$

So that is the convention: raising a number to a negative power is the same as raising the reciprocal of the base to the power.

> **To cement your understanding:**
>
> What is 6^{-2} equal to? What do you get when you multiply that by 6^2? (The answers, of course, are $\frac{1}{36}$ and 1.)

Subtracting indices

(Note: "indices" is the plural of "index" and means the same as "exponents".) We've seen what happens when we add indices and what happens when we have a negative index. Putting those together, we can say that:

$$2^{4-2} = 2^4 \cdot \frac{1}{2^2}$$

$$= \frac{2^4}{2^2}$$

$$= 2^2$$

$$= 4$$

> **To cement your understanding:**
>
> Calculate the value of 2^{4-2} in two ways: 1) by subtracting 2 from 4. 2) by calculating 2^4 and dividing it by 2^2. Confirm that whichever way you do this, the result is 4.

Key points:

- Raising to a power is a shorthand way of writing repeated multiplication.

- We call the number being raised to a power the base, and the number in superscript that indicates how many times it is multiplied by itself the exponent. The exponent is also known as the index or the logarithm.

- When you multiply a base raised to a power by that base also raised to a power, you can add the exponents.

- If we raise something to a power, then raise the whole thing to a power, that is the same as multiplying the exponents.

- Anything raised to the power of 1 equals itself.

- Anything raised to the power of 0 is equal to 1.

- Raising a number to a negative power is the same as raising the reciprocal of the base to the power.

Roots part 1

> **Read this section if:** you don't understand why a number will have two square roots but only one cubed root or you don't know what it means to raise to the power of a fraction.

Just as subtraction is the opposite of addition, and division is the opposite of multiplication, so **a root is the opposite of a power.** If 4^2 is 16, then the square root of 16 is 4. The 3rd root (or cubed root) of 125 is 5, because 5^3 is 125, and so on.

We write it as

$$\sqrt[3]{125} = 5$$

Square roots can be written as $\sqrt[2]{36} = 6$ or just as $\sqrt{36} = 6$ (the 2 is usually omitted).

Some square roots are integers (whole numbers), but many are irrational numbers (or **surds**), which is a concept we will cover soon.

We saw earlier that when you multiply two negatives it makes a positive. So $6^2 = 36$, but so does -6^2. This means 36 has two square roots: plus 6 and minus 6. This is written as:

$$\pm\sqrt{36}$$

It doesn't work that way for numbers raised to odd powers, though:

$$
\begin{aligned}
-6^3 \ &= \ -6 \bullet -6 \bullet -6 \\
&= \ (-6 \bullet -6) \bullet -6 \\
&= \ 36 \bullet -6 \\
&= \ -216
\end{aligned}
$$

So

$$\sqrt[3]{-216} = -6$$

while

$$\sqrt[3]{216} = 6$$

The **line across the top** of a root-expression has the same effect as **brackets**; the operations beneath the line are undertaken first, then the root is calculated on the result.

For example:

$$4 \bullet \sqrt{2 + 7} = 4 \bullet \sqrt{9}$$

$$= 4 \bullet 3$$

$$= 12$$

Raising to the power of a fraction

Returning to operating with exponents, let's think about the meaning of raising something to a fraction. The convention is that raising something to a fraction is the same as finding the root, based on the denominator of the fraction. In other words:

$$27^{\frac{1}{3}} = \sqrt[3]{27}$$

$$= 3$$

Raising to a fraction doesn't result in a fraction; that happens when we raise to a negative power, as we saw earlier. This can be confusing at first; it is one of the few things in mathematics that is worth committing to memory, because it comes up a lot. To understand why it works that way, let's think about this expression:

$$(3^4)^{\frac{1}{4}} = 3^{4 \bullet \frac{1}{4}}$$

We've seen that if we raise something to a power, put it in brackets and raise everything in brackets to a power, that's the same as multiplying the two exponents, as we did above. In this case, when we multiply the exponents we are multiplying 4 by one quarter, which is 1.

$$(3^4)^{\frac{1}{4}} = 3^{4 \bullet \frac{1}{4}}$$

$$= 3^1$$

$$= 3$$

We know that $3^4 = 81$. Substituting that back into what we just looked at, we get:

$$(81)^{\frac{1}{4}} = 3$$

That only works if raising to the power of a fraction is the same as calculating that root:

$$(81)^{\frac{1}{4}} = \sqrt[4]{81}$$

$$= 3$$

When you raise a number to a fraction that has something other than 1 in the numerator, you raise it to the power of the numerator, then use the denominator to find that root. It doesn't matter what order you do this in. For example, with

$$8^{\frac{2}{3}}$$

you can either calculate 8^2, which is 64, then find the 3rd root of that, which is 4. Alternatively, you can find the 3rd root of 8, which is 2 (because $2^3 = 8$), then raise that to the 2nd power, which gets you back to 4. Either way the answer is still:

$$8^{\frac{2}{3}} = 4$$

To cement your understanding:

• If $3^2 = 9$, what is 3^3 equal to? Therefore what is $27^{\frac{1}{3}}$?

• What is $27^{\frac{2}{3}}$ equal to? Hint: you can do this the hard way or the easy way. The hard way would be to square 27 then find the cubed root of the result, and it would almost need a calculator or computer to work it out. The easy way would be to find the cubed root of 27 (which you did in the dot point above this one), then square the result: either way you get back to 9.

Summary of operations with exponents

We're about to do some algebra, using a and a b in our exponents, to indicate that we could use any numbers we liked. These numbers could represent the number 3 or the number 1,234,567; it doesn't matter.

We are showing 2 as the base, but of course we could also have any number we like there, too. This just summarises the rules (or better, the shortcuts) we have seen:

Rule	See page…
$2^a \cdot 2^b = 2^{a+b}$	54
$\dfrac{2^a}{2^b} = 2^{a-b}$	57
$(2^a)^b = 2^{a \cdot b}$ $= \left(2^b\right)^a$	54
$2^{-a} = \dfrac{1}{2^a}$	56
$2^{\frac{1}{a}} = \sqrt[a]{2}$	60
$2^{\frac{a}{b}} = \sqrt[b]{2^a}$	60
$2^1 = 2$	55
$2^0 = 1$	55

That seems like a lot to remember, but if you get stuck as we continue through the book (and we will use this a lot), come back and remind yourself of *why* these work the way they do. Once you understand that, remembering them becomes easier.

Key points:

- A root is the opposite of a power.

- Raising something to an even power creates a number with two square roots: a positive and negative one.

- The line across the top of a root-expression has the same effect as brackets.

- Raising something to the power of a fraction is the same as finding the root, based on the denominator of the fraction.

Decimal fractions

> **Read this if** you don't understand how decimals relate to fractions.

Through the ages, around the world, there have been a lot of different numbering systems, just as there are many different languages. Unlike with languages, though, one number system—the **Hindu-Arabic** one—is so superior to other ones for most applications that it is now virtually universal. It was originally developed in India, and differs from most other forms of numbering (such as Roman numerals) by using the position of numerals to determine their value.

You can think of the number two thousand, three hundred and seventy-nine (2,379) as

$$2,000 + 300 + 70 + 9$$

which is the same as:

$$2 \cdot 1,000 + 3 \cdot 100 + 7 \cdot 10 + 9 \cdot 1$$

The way we normally write numbers is actually shorthand for this.

One of the very few areas where the Hindu-Arabic system doesn't dominate is machine code for computing, which consists of ones and zeros. It still uses the same principles as the Hindu-Arabic one though; it is based on positions.

We can think of putting the numbers as belonging in their own columns:

1,000	100	10	1
2	3	7	9

With decimal fractions, instead of putting one number on top of another, we write them in a line and use the same logic we use for other numbers. We use a decimal separator for this. In this book, we will use a decimal point, but see the text box over the page for alternatives.

> The **notation for a decimal point varies** around the world. Here are some variations:
>
> Dot (.): The most commonly used decimal separator in many countries, including the United States, the United Kingdom, India, and China, is the dot. For example, one and a half is written as 1.5 This book follows that convention.
>
> Comma (,): Many European and Latin American countries use a comma as the decimal separator. For instance, in Germany, France, and Brazil, one and a half would be written as 1,5
>
> Apostrophe (') or Space: In some countries, particularly in Switzerland and Liechtenstein, an apostrophe is used as the decimal separator. In addition, a space is occasionally used to denote the decimal point in some documents to ensure clarity, especially in high-precision numbers.
>
> Other Variations: In certain contexts, such as in older texts or in specific scientific fields, different symbols might be used as decimal separators. However, these are less common in everyday use.
>
> The choice of decimal separator can affect how other numerical components are displayed. For example, in countries where a comma is used as a decimal separator, a dot or space might be used to separate thousands (e.g., 1,000.00 in the U.S. becomes 1.000,00 or 1 000,00 in parts of Europe).

So we might have the number:

$$2,379.375$$

The $.375$ on the end is a decimal fraction. We can think of it as:

$$\frac{3}{10} + \frac{7}{100} + \frac{5}{1000}$$

Using the idea of columns again, we can write this as:

1,000	100	10	1	$\frac{1}{10}$	$\frac{1}{100}$	$\frac{1}{1000}$
2	3	7	9	3	7	5

Now recall that $1000 = 10^3$ etc and also that raising to a negative power is the same as raising to the reciprocal of the power. That means we can write our table using what we call **exponential notation** (because we are using exponents):

10^3	10^2	10^1	10^0	10^{-1}	10^{-2}	10^{-3}
2	3	7	9	3	7	5

Putting a decimal place in lets us keep moving into more detail still using powers of 10.

Decimal notation lets us write numbers in great precision where it is required, and the shortcut of just using a decimal place makes it much easier to write out.

$$\frac{3}{10} + \frac{7}{100} + \frac{5}{1,000}$$

can also be expressed as

$$\frac{300}{1,000} + \frac{70}{1,000} + \frac{5}{1,000} = \frac{375}{1,000}$$

(Multiply the numerator and denominator of the first term by 100, and the numerator and denominator of the second term by 10. That gives a common denominator of 1,000.)

Decimal fractions are better in some situations than others. Some fractions, such as

$$\frac{1}{3}$$

When you were at school you may have learned long multiplication and long division. We won't be covering them in this book, but if you used to go through the steps without understanding why it worked, it might be easier to understand if you think of each number as having its own column based on powers of 10.

don't cope well with being represented as decimals: it needs to be written as 0.33333.... which is called "point 3 repeating". You could keep writing 3s forever and get closer and closer to one-third each time you do, without ever reaching it. It's better to express it as one-third if possible because that designates the exact fraction.

In other situations, using decimals is better. It allows great precision in defining the length of numbers that cannot be expressed as fractions, and there are other advantages when it comes to calculation. A calculator, for example, will use decimal fractions.

To cement your understanding:

• Write out 32,456 in columns headed by powers of 10, the way it is done above.

• Write out 6.783 the same way.

• Write out 6.783 as four fractions added together, with different powers of 10 in the denominators.

• Now write out 6.783 as 6 and a single fraction, with 1,000 in the denominator.

Key points:

• Our system of numbering depends on positions, which we can represent as columns based on powers of **10**.

• The columns of decimal fractions on the right of the decimal point are based on adding successive negative powers of **10**.

Ratios and irrational numbers

> **Read this if** you don't understand how some numbers can have an infinite number of digits after the decimal point, you think that "irrational" numbers must be less logical than other numbers or you want to learn some fun facts about the mathematics of the golden ratio.

Fractions and decimal fractions are wonderful inventions, but some numbers can't be expressed as fractions based on integers at all.

The square root of 2 is one such number, as are many square roots.

$$\sqrt{2} = 1.4142135623731 \ldots$$

This means that if you multiply 1.4142135623731 by itself you will get very, very close to 2. You won't get exactly to 2, however, unless you keep writing numbers forever (although your calculator will probably round it to 2).

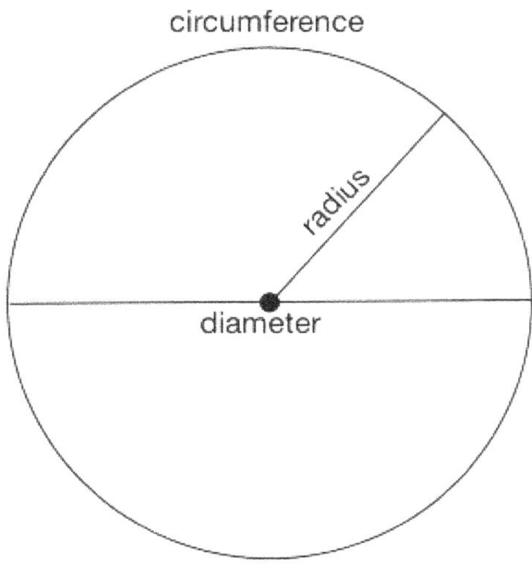

Figure 26

Numbers like this are called **irrational numbers**. They are often used because they represent an important ratio: for example, the ratio of the length of the diameter of a circle (a line passing from edge to edge through the middle) to the length of the circumference of a circle (the distance around the outside) (Figure 26).

If you multiply the length of the diameter of a circle by a number which is approximately equal to

3.14159265358979323846 2643...

you will have the length of the circumference, regardless of the size of the circle. Mathematicians call this the Greek letter pi (π), which is easier than writing out many decimal points. It doesn't matter if the circle is 1,000,000 light years across, or 1 micron; if you multiply the length of the diameter by π, you will know what the length of the circumference is. (If you are struggling with this, see the How to Love Statistics YouTube video on *Pi*.)

The problem is that to write π precisely, you would need to continue adding digits indefinitely after the decimal point, without repeating the pattern. This number can be roughly expressed as

$$3\frac{1}{7}$$

but that is just an approximation. There is no fraction that describes it accurately.

We can write this as:

$$length\ of\ diameter \bullet \pi = length\ of\ circumference$$

regardless of the length of the diameter.

Alternatively, dividing both sides by the length of the diameter, we can write:

$$\pi = \frac{length\ of\ circumference}{length\ of\ diameter}$$

regardless of the size of the circle.

The ratio of the circumference to the ratio can't be expressed as a ratio of two integers. Irrational numbers like this are also called **surds**.

Because the ratio of the circumference to a circle doesn't change, we say they keep the same proportion to one another. This brings us to another of those close relatives of the equals sign mentioned on page 28 and following. The symbol $\propto$ means "is proportional to".

$$length\ of\ diameter \propto length\ of\ circumference$$

Saying that two things are **proportional** is another way of saying that if you multiply one of them by the same fixed amount you will have the other, regardless of how big or small the first number is.

Another irrational number is 1.618033988749 ... which is given the Greek letter phi (φ).

When you divide 1 by φ and then add 1 to the result, you get back to φ again. You might want to try that on a calculator for yourself, to see just how weird

The three most-famous irrational numbers: π, $\sqrt{2}$, and another number we will explore later, e, are all involved in one of the most important equations in all of statistics: the formula for the normal distribution. (We have some way to go before we get to that.)

it is. When you multiply φ by itself then subtract 1 from the total you get back to φ yet again. It's the sort of thing you can read about, but then when you put it in the calculator and see it happen that way, there is something mind-bending about it. It's the only number for which this is true.

The ratio of phi to one (φ:1) is the "golden ratio" or "divine proportion", which was used by the ancient Greeks and by renaissance artists in painting, sculpture and architecture. (The ancient Greeks didn't use decimal notation, but were able to calculate the ratio using geometry). This proportion occurs a lot in nature, although in different ways than is sometimes understood[19].

To cement your understanding:

• Use a calculator or a spreadsheet to divide 1 by φ (using as many decimal places as possible, 1.618033988749…), then add one to the result, to confirm that you get back to φ.

• Square the result, then subtract one from that, again to confirm that you get back to 1.618033988749….

Fun fact:

Figure 27

(If what follows doesn't make sense at this stage, then ignore it until we have covered the more geometry. After that you might like to come back to this, because it really is a lot of fun.)

The ancient Greeks realised that if they drew up a "golden rectangle", with sides in the proportion of φ:1, and then removed a square from it, what was left was another golden rectangle (Figure 27). If they removed a square from that, they had another golden rectangle; and they could keep going that way. (There are six golden rectangles in this picture.) It is the only ratio for which this is true.

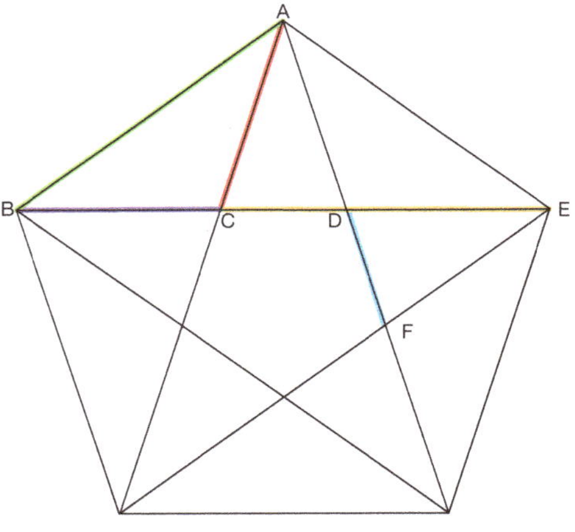

Figure 28

All of the **intervals** (in geometry a "line" goes on forever, while an "interval" is a piece of a line) in a regular five-pointed star (pentagram) inside a pentagon (five-sided figure on the outside) will have the ratio of 1:φ or φ:1, unless they are the same length. In Figure 28 for example, just some examples of many are:

AB(green):AC(red),

BC(purple):DF(blue),

EC(orange):AC(red), and

AC(red):DF(blue).

These are *all* equal to the golden ratio of φ:1; not just approximately, but with mathematical exactness.

A marvellous number is phi.
When you square it, take one, you get phi.
Take the inverse, add one, you get phi.
Add one, take the square root. It's phi.
And I'm scratching my head to know y.
A classic irrational is phi.
There's more to it than meets the i.
It's wondrous credentials
Transcend transcendentals
It's up there with googol & pi.

Howard Barnes

There is a proof of this in an appendix on page 740, but we need to cover some algebra and some geometry before it will make sense.

Key points:

- Irrational numbers can't be expressed as fractions based on integers.

- The diameter of a circle is a line from edge to edge passing through the middle, while the circumference is the distance around the outside.

- If you multiply the diameter of a circle by the irrational number π (which is approximately equal to 3.1415…) you will have the length of the circumference, regardless of the size of the circle.

- We say that the diameter of a circle is proportional to the circumference, and use this symbol to express that idea: $\propto$

- The square root of 2 is also irrational, and is approximately equal to 1.4142…

- We will explore the irrational number e (2.7182…) later in the book. $\pi, \sqrt{2} \ and \ e$ are probably the best-known irrational numbers. These will all appear in the formula for the normal distribution curve.

- Another irrational number is 1.618033988749… which is given the Greek letter phi (φ).

- When you divide 1 by the irrational number φ and then add 1 to the result, you get back to φ again.

- When you multiply φ by itself then subtract 1 from the total you get back to φ yet again.

Categories of numbers

Read this if you aren't familiar with the symbols in the left-hand column of the table below.

When you first started learning mathematics at school, the only numbers you knew were the ones you used to count. You have now met other categories of numbers. In higher mathematics, sometimes it is useful to have a language to limit and specify the types of numbers under consideration in a particular setting. You will see the following symbols used for this.

Symbol	Called	Explanation	Examples
$\mathbb{N}$	Natural, or counting numbers	The numbers we **use to count**. No surprise there, but where there *is* a surprise—in one of the few inconsistencies in mathematics—is that some people include 0 as a natural number and some don't. Normally you will be able to work out from the context whether 0 is included.	$1, 2, 3$
$\mathbb{Z}$	Integers	All **whole numbers,** including zero and negative numbers. (The z is for the German word "zahlen"). This includes the natural numbers.	$0, 1, 2, 3,$ $-57, -589$
$\mathbb{Q}$	Rational numbers	Can be described by **a fraction of integers** (the Q is for "quotient"). Includes the integers. They are called "rational" because they can be formed by a fraction of integers, not because they are sensible (although	$1, 2, 3,$ $-57, -589,$ $\dfrac{2}{3}, \dfrac{-7}{17}$

Symbol	Called	Explanation	Examples
		they are). (Note: every repeating decimal is rational.)	
$\widetilde{\mathbb{Q}}$	Irrational numbers	**Cannot be described by a fraction** of integers unless a "root" symbol is used. Includes the transcendental numbers.	$\sqrt{2}$, π, $\dfrac{1 + \sqrt{5}}{2}$, φ
$\widetilde{\mathbb{A}}$	Transcendental numbers	Cannot be formed without using a decimal fraction that continues forever without repeating the pattern.	π

To get to this point we have at least touched on numbers in each of these categories. We will see more categories as we progress.

Set notation

Read this if you aren't clear about what the symbols ∈ , ⊂, ∩ or ∪ mean.

Groups of things (and particularly, although not exclusively, groups of numbers) are called **sets**. Sets are normally indicated by **braces**. For example, the integers listed earlier can be shown as:

$$\{0, 1, 2, 3, -57, -589\}$$

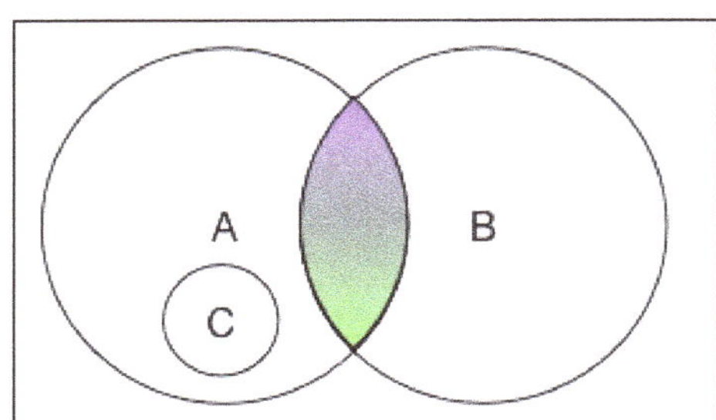

Figure 29

In this instance each of the numbers in the list is called an **element** of the set and designated by the symbol ∈ . So:

$$-57 \in \{0, 1, 2, 3, -57, -589\}$$

John Venn (1834–1923) was an English mathematician, logician, moral philosopher and Anglican minister from Cambridge University whose most important claim to fame at the time was probably his invention of a machine that bowled cricket balls, and that successfully bowled out the top-ranked batsman from a visiting Australian cricket team four times in a row. (He also contributed to what later became known as the frequentist view of probability.) Venn is now best known for Venn diagrams, as shown in Figure 29. In this instance the diagram is representing 3 sets, which we will designate A, B and C. We would say that C is a **subset** of A, because all of the elements of C are also included in A. We write this as

Figure 30
John Venn
1834–1923

$$C \subset A$$

Notice that the symbol for subset, ⊂, is similar to the symbol for element, ∈. An element is also a subset, but with only one thing in it.

The shaded area shared by both A and B is called the **intersection**, and is represented as

$$A \cap B$$

The total area of all the sets is the union of A and B, or the union of A, B and C. In this case they will be the same thing. It is written as

$$A \cup B$$

(As an aid to memory, you can think of ∪ as being the letter U, as in "Union".)

As you do more with higher mathematics, you will often see terminology like

$$\in \mathbb{Z}$$

in mathematical proofs and explanations, meaning "is an element of the set of all integers" or more simply, "is an integer".

You will also probably encounter symbols like

$$\in \{\mathbb{Z} \geq 0\}$$

This means that what you are looking at is an integer, and is also greater than or equal to zero. This is spoken as "is an element of the set of integers that are greater than or equal to zero".

We can draw the types of categories of numbers above as a Venn diagram, as in Figure 31.

In other words, a natural number will always be an integer, which will always be a rational number; but not all rational numbers will be integers. And so on.

You now have the basic building blocks of mathematical language in place!

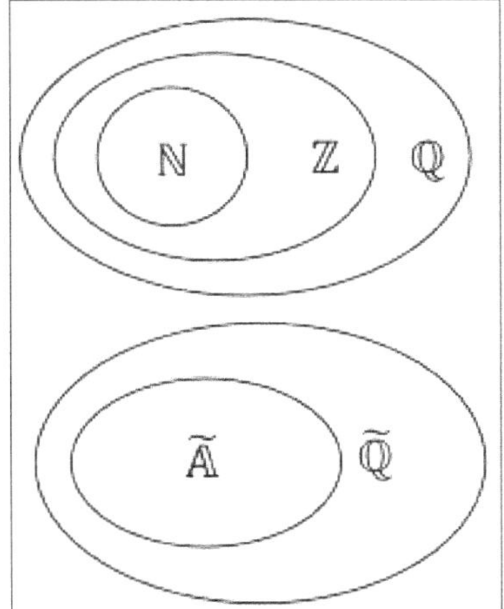

Figure 31

To cement your understanding:

• On the Venn diagram above (Figure 31), match the symbols with the descriptions on page 72

• Can you see why these types of numbers are positioned where they are? In other words, why some categories (sets) of numbers are subsets of others, while others are not?

Key points:

- Groups of things are called sets, and are normally indicated by braces.

- $\in$ means is an element of the set.

- $\subset$ means is a subset of the set.

- $\cap$ means the intersection of sets.

- $\cup$ means the union of sets.

- A Venn diagram is a way of drawing the idea of sets.

Pure mathematics, "Pascal's" triangle, and putting it all together

"When I was reading Harry Potter, I didn't find myself asking: 'When will I ever use Wingardium Leviosa? When am I going to apply the new-found knowledge I have of the rules of Quidditch?' "

Grant Sanderson[20]

Pure mathematics

Congratulations! You've made it through some of the most tedious, but also the most-challenging, mathematical ideas (either just now from reading, or earlier when you were at school). Stay with it, because, from now on, it becomes easier to enjoy.

Mathematics is, of course, an extremely practical subject, but much of it wasn't initially developed to solve practical issues: it was developed because mathematicians had fun playing with ideas. We call this **pure mathematics** as opposed to **applied mathematics**.

These are some of the ideas people developed because mathematicians were simply having fun or indulging their curiosity:

Idea	Discovery	Without this we probably wouldn't have…
Complex numbers. (We'll meet these when we look at algebra.)	Gerolamo Cardano introduced them in 1545 while solving cubic equations. Initially seen as abstract.	Electrical power grids, Wi-Fi signals, and mobile phones.

Idea	Discovery	Without this we probably wouldn't have…
Fourier analysis (We'll meet this at the end of part 1).	Joseph Fourier introduced Fourier series in 1822 while studying heat conduction.	Audio compression, image processing in cameras, and medical imaging (MRI scans).
Number theory	Studied by the ancient Greeks, Fermat, and Gauss with no clear application.	Secure online banking, messaging, and secure credit card transactions.
Imaginary exponential functions	Developed by Euler in the mid 1700s as elegant and interesting ideas.	Anything electronic.

"Pascal's" triangle

This is another of those ideas that people (this time, different people all around the world) discovered because they were just having fun with it. It turns out to be absolutely fundamental to statistics, but no one knew anything about that when people started looking at it. Before the 1600s, Indian mathematicians called this triangle the Staircase of Mount Meru, Persians called it the Khayyam triangle (after poet Omar Khayyam), Chinese called it Yang Hui's triangle, and Italians and Spanish called it Tartaglia's triangle. It is most-commonly now called **Pascal's triangle**, and that is how we'll refer to it.

We will meet Blaise Pascal (1623–1662) a few times in this book. He was a French philosopher and mathematician who was so fascinated with what was called the "arithmetic triangle" that we now associate his name with it. He died aged 39, but he packed a big contribution into a short life—including developing the understanding of air pressure and inventing the syringe.

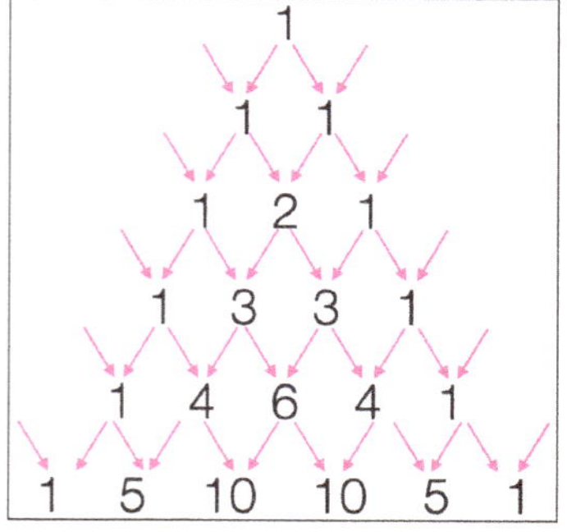

Figure 32

To draw up this triangle, take a piece of paper, and put a 1 at the top in the middle. This will be your first row. (Don't worry about why you are doing this yet; all will be revealed over time. Think of this like learning to play a game: don't start by asking why the rules are what they are, or you will never be able to enjoy playing it. Go with it for now.)

Each new row begins and ends with 1. Every other number is the sum of the numbers to the left and right directly above it. After you have drawn 6 rows, you should have something that looks like Figure 32 (without the arrows). You can see that the 2 in the middle of the third row is the sum of the 1 above it to the left and the 1 above it to the right. The three in the second place of the 4th row is the sum of the 1 above and to the left, and the 2 above and to the right. And so on. The triangle could theoretically keep going forever.

That's one form of the triangle we will be using. The **rectangular form** is made by shifting all the numbers to the left as per Figure 33. It's calculated in the same way; just displayed differently. We'll use both forms as we proceed, because the different forms are useful to illustrate different concepts.

That's the basic idea. People over the centuries have kept independently coming up with this idea because it turns out to have remarkable properties. Apart from being fundamental to statistics it will illustrate many of the ideas in this book. For now, we'll use it to revise a few of the ideas we've covered so far.

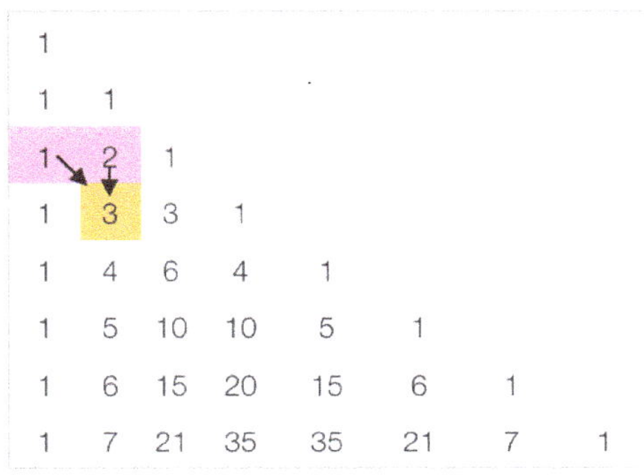

Figure 33

Positive and negative numbers revision

If you alternate the signs in Pascal's triangle (as per Figure 34, which shows positive numbers as black and negative numbers as red), then each row except the first one adds up to zero. That means, no matter how large you make the table, the total value of the whole triangle will add to 1, which is the value of the first row.

Figure 34

Prime numbers and multiples revision

In Pascal's triangle, if, and only if, any row has a prime number as the second number in the row, the rest of the numbers of that row will either be 1 or will be multiples of the prime number in the second position (Figure 35). For example, the row with 7 in the second position contains the other numbers 1, 21 and 35. Both 21 and 35 are multiples of 7.

1										
1	1									
1	2	1								
1	3	3	1							
1	4	6	4	1						
1	5	10	10	5	1					
1	6	15	20	15	6	1				
1	7	21	35	35	21	7	1			
1	8	28	56	70	56	28	8	1		
1	9	36	84	126	126	84	36	9	1	
1	10	45	120	210	252	210	120	45	10	1
1	11	55	165	330	462	462	330	165	55	11	1

Figure 35

Raising numbers to powers revision

							Sums of the numbers in the rows form powers of 2.
1							$1 = 2^0$
1	1						$1 + 1 = 2 = 2^1$
1	2	1					$1 + 2 + 1 = 4 = 2^2$
1	3	3	1				$1 + 3 + 3 + 1 = 8 = 2^3$
1	4	6	4	1			$1 + 4 + 6 + 4 + 1 = 16 = 2^4$
1	5	10	10	5	1		$1 + 5 + 10 + 10 + 5 + 1 = 32 = 2^5$
1	6	15	20	15	6	1	$1 + 6 + 15 + 20 + 15 + 6 + 1 = 64 = 2^6$
							etc

If you add up the numbers in each horizontal row of the triangle, you get a power of 2. (Remember that anything raised to the power of 0 is equal to 1, and that anything raised to the power of 1 is equal to itself.)

Likewise, when written out as a number, each row is a power of 11.

					The rows, written as numbers form powers of eleven[21]
1					$1=11^0$
1	1				$11=11^1$
1	2	1			$121=11^2$
1	3	3	1		$1331=11^3$
1	4	6	4	1	$14{,}641=11^4$

As we progress, we will see more examples of where mathematical concepts can be illustrated by this triangle. It will form something of a unifying theme on our journey.

Algebra: Telling Stories with Mathematical Language

Let's just take a moment. Breathe. If you and mathematics didn't get on very well in school, and you have glanced ahead, you are likely to be feeling somewhat intimidated at this stage. Some of the expressions suddenly get a lot longer, they start to have letters in them as well as numbers, and they are hard to read if you haven't read the fine print that explains what is going on.

It's ok!

To get to this point, you've survived the concepts of multiplying fractions and understanding decimal notation. Conceptually, nothing that follows (even into high-level statistics) is likely to be more challenging than that. It's just that you now have the tools to use mathematics for more than shopping. You are about to learn how to build whole sentences, paragraphs, and complete stories. Once you can form stories, you can construct arguments, and once you construct arguments, you can formulate ideas and move your (and other people's) understanding forward.

Algebra and algebraic notation

> **Read this if** you never really understood algebra, or if you have forgotten most of what you learned about it.

(If you are struggling with this, see the How to Love Statistics YouTube video on *Why Algebra is Hard*.)

Algebra, like fractions, is one of those topics where some people get left behind. The basic concept is easy enough.

Take this equation:

$$3 + 4 = ?$$

You probably have no difficulty recognising that the question mark could be replaced by a 7. If we used question marks in algebra, we would run out of symbols, so we use letters instead. When you see a letter, you can substitute it in your head for a question mark. You just use a letter for a number when you don't know what the number is. The most common letter is "*x*" (usually in italics).

The equation becomes:

$$3 + 4 = x$$

There are six issues that often cause people to become confused:

1. The key to solving algebraic equations is to remember that the equals sign ("=") means that two things are the same. This means what you do to one side of the equation you can do to the other, and both sides will still equal one another. (You've seen this; it's where we started this book.) To solve basic algebraic equations, you aim to isolate *x*, by applying the opposite operation of whatever is currently acting on it. The opposite of addition is subtraction, and vice versa. The opposite of

multiplication is division, and vice versa. The opposite of raising a number to a power is taking its root. Take the equation:

$$3 \bullet x^2 + 14 = 26$$

The inverse of addition is subtraction. 14 has been added on the left, so we subtract it. We also subtract it on the right, to keep the equation balanced.

$$3 \bullet x^2 + 14 = 26 - 14$$
$$3 \bullet x^2 = 12$$

The inverse of multiplication is division. We therefore divide the left side by three, and do the same to the opposite side, to keep it balanced.

$$3 \bullet x^2 = \frac{12}{3}$$
$$x^2 = 4$$

The inverse of squaring is to take the square root (remembering that there will be two of them: a positive and a negative). That leaves:

$$x = \pm\sqrt{4}$$
$$= \pm 2$$

To cement your understanding:

• Replace x in the original equation ($3 \bullet x^2 + 14 = 26$) with 2, confirm that it works out. Then replace it with -2, and check that it still works.

• Find the value of x using the same approach, in the equation $3 \bullet x^2 + 14 = 89$, using exactly the same approach. Did you get 5 and -5? If not, can you see where you went wrong?

---

2. When people are introduced to algebra, they are often introduced to the concept of equations, together with issues like double negatives, brackets and the distributive property of multiplication, all at the same time. We've already covered these ideas; what follows is just an extension of them.

3. A cross (which looks like the letter x) has, up until this point in people's school life, been used as the symbol for "multiplied by". (That's why this book has been using "•" instead.) Suddenly they see it representing something different in equations. This book doesn't use x for multiplication very much, so that shouldn't be a problem.

4. From now on the multiplication symbol mostly gets omitted. If you see two letters together, a number and letter together, or a number or letter beside a bracket, just assume that there is a multiplication symbol in there that has been omitted. If in doubt, multiply. So, for example:

$$3(4) = 3 \bullet 4$$
$$= 12$$

and

$$3y(z4) = 3 \bullet y \bullet z \bullet 4$$

To help you to get used to the idea, for a little bit longer in this book we'll use a dot to show multiplication, as in $3 \bullet y \bullet z \bullet 4$. After a little while we'll adopt standard mathematical practice and leave the dots out unless we really need them. It will be obvious from the context what is going on. This takes some getting used to, but it works fine once you get the idea. You will find you don't miss the dots.

5. You may have heard that "x is the unknown quantity", but letters can take on some subtly different nuances in algebra. Mathematicians often generalize specific problem-solving methods into a formula (plural formulae) applicable to a variety of situations. This generalisation usually involves replacing specific numbers with letters, termed variables and constants. Understanding variables and constants can initially be challenging but quickly becomes intuitive.

Take the equation:

$$3 \bullet x^2 + 14 = 26$$

Here, x^2 is a variable, and 3, 14, and 26 are constants. To create a general formula, we replace constants with letters, giving us

$$A \bullet x^2 + B = C$$

In this formula, A, B, and C are still constants (typically from the alphabet's start) representing any value, while x remains a variable. We aren't trying to find A, B or C: we are just using them as placeholders for numbers in a formula. It is still x we are trying to isolate.

Transforming the equation into a formula, we isolate x:

This is our starting point: $\qquad\qquad\qquad\qquad A \bullet x^2 + B = C$

Subtract B from both sides, as a start to getting x on $\quad A \bullet x^2 = C - B$
its own. (Once again, subtraction is the opposite of
addition, so we are undoing the addition on the left
by using subtraction, and doing the same thing to
the other side, to keep the equation balanced.)

Divide both sides by A to finish getting x on its own. (Similarly, division is the opposite of multiplication, so we are undoing the multiplication on the left by dividing, then doing the same thing on the right.)

$$x^2 = \frac{C - B}{A}$$

Then take the square root of both sides, so that x^2 becomes x.

$$x = \pm\sqrt{\frac{C - B}{A}}$$

The constants are some of the symbols that give a formula its structure.

So back to the equation:

$$3 \bullet x^2 + 14 = 26$$

We solved this earlier, easily enough, (in this case, because we've kept it simple) from first principles, but we also have a formula to solve it, and the more complex it is, the more useful it will be to have a formula.

Our formula was designed to solve an equation in the form:

$$A \bullet x^2 + B = C$$

So we can let $A = 3$, $B = 14$, and $C = 26$

Our formula for solving it is:

$$x = \pm\sqrt{\frac{C - B}{A}}$$

We can plug in numbers to this formula, and we get

$$x = \pm\sqrt{\frac{26 - 14}{3}}$$

$$= \pm\sqrt{\frac{12}{3}}$$

$$x = \pm\sqrt{4}$$

$$= \pm 2$$

which is, of course, the same answer we arrived at calculating it from first principles.

6. With **polynomials** (expressions with more than 2 terms) there is a particular approach needed to simplify them. For example, if we are simplifying:

$$x - 1 + 2x^4 + 2y - 1xy + 2 - 6xy - 2x^4$$

it might not be obvious, at first, how to approach it. The first thing is to notice that there are different variables. We need to think about x as being a different variable to x^4, which is a different variable to xy, etc. Showing the terms which share the same variables in the same colours, we have:

$$x - 1 + 2x^4 + 2y - 1xy + 2 - 6xy - 2x^4$$

We can gather those similar terms together. That gives us:

$$x + 2x^4 - 2x^4 - 1xy - 6xy + 2y + 2 - 1$$

Now we can start to simplify. We can add and subtract those terms that have the same variable. At this point, we can think of adding and subtracting different numbers of the same variables in the same way we think of adding and subtracting different numbers of anything else: company shares, elephants or mice. Four mice, minus three mice leaves one mouse. It's exactly the same principle.

We have x on its own, and can't do anything with it, so it can stay there. $2x^4$ is added then subtracted, which means they cancel one another out. We have $-1xy$ and $-6xy$, which makes $-7xy$. We have $+2y$ on its own, so we can't do anything with it. Finally we have $2 - 1$, which leaves 1.

Putting all of that together, we get:

$$x - 1 + 2x^4 + 2y - 1xy + 2 - 6xy - 2x^4 = x - 7xy + 2y + 1$$

That's as far as we can simplify this one, but we've been able to get it from 8 terms down to 4 without too much effort.

To cement your understanding:

• Solve the equation $x^3 + 30 = 3$, making sure that you can understand why $x = -3$

• Make up some examples of your own. Can you solve them all? (Don't panic if you can't; there are some tricks to equations that we haven't come to yet.)

Key points:

- What you do to one side of an equation you do to the other. Both sides will still equal one another.

- To solve basic algebraic equations, try to get x on its own, by using the opposite operation to whatever the number does to x at the moment.

- From now on the multiplication symbol mostly just gets omitted.

- Mathematicians often generalise specific problem-solving methods into a formula (plural formulae) applicable to a variety of situations.

- This generalisation usually involves replacing specific numbers with letters, termed variables and constants.

- Different powers of the same variable are different variables.

- In simplifying a polynomial, we can add and subtract those terms that have the same variable.

Recap on priority
of mathematical operations

Read this if you are at all unsure about the priority of mathematical operations, or you don't understand the difference between factorising and expanding terms.

When we simplify equations, exponents (indices) are calculated first. Multiplication and division come next, followed by addition and subtraction. You can use brackets or braces to override this order, and horizontal lines (such as those in fractions or root symbols) work the same way as brackets.

Take the complex-looking equation:

$$\frac{3 \bullet (\sqrt[4]{600 + 5^2})}{2 + 1} + x = 0$$

Start by calculating 5^2, which equals 25. Add 600 to 25 (under the horizontal line belonging to the root sign) and add 2 + 1 (under the horizonal line of the fraction). At this point we have

$$\frac{3 \bullet (\sqrt[4]{625})}{3} + x = 0$$

Take the 4th root of 625 (which is 5). (Feel free to use a computer or calculator for something like this if the answer isn't immediately obvious.)

$$\frac{3 \bullet 5}{3} + x = 0$$

Cancel out the 3s (divide the numerator and denominator by 3), leaving

$$5 + x = 0$$

Subtract 5 from both sides, to get x on its own.

$$x = -5$$

Not so bad when you take it one step at a time.

To cement your understanding:

Solve the equation $x^{1+2} + 3 \bullet 9 = 0$ using the rules above, to find x. Hint: start by adding $1 + 2$. Then multiply 3 by 9. Then subtract 27 from both sides. Then take the cubed root of both sides.

Expanding algebraic terms

The distributive property of multiplication (page 37) also applies to algebraic expressions, and it becomes more important to understand it when letters have replaced numbers, because we can't simply add them and multiply by the result.

$$(a + b) \bullet (c + d) = a \bullet c + a \bullet d + b \bullet c + b \bullet d$$

If we take the expression:

$$x \bullet (5 + x + y)$$

we can't do what is in the bracket first, because we can't simplify it more than we have. What we can do is use the distributive property of multiplication to get:

$$x \bullet 5 + x \bullet x + x \bullet y$$

This is all starting to get messy with the $\bullet$ sign included, so let's try it again without. You will see that this makes it easier to follow, and we will only use the $\bullet$ sign in future when we really need it for clarity. Trying again, we get:

$$x(5 + x + y) = x5 + xx + xy$$

The convention in algebra is that we usually put constants before variables within each individual term. Because of the commutative property of multiplication, this doesn't change the value of the term.

$$x(5 + x + y) = 5x + x^2 + xy$$

We've also simplified xx to x^2. Finally, by convention we usually place higher powers of x to the left of lower powers.

$$x(5 + x + y) = x^2 + 5x + xy$$

Factorizing

Factorizing is the opposite of expanding an algebraic term using the distributive property.

In the expression we just saw:

$$x^2 + 5x + xy$$

we can factor out an x, because x is a factor of each of the terms of the expression.

$$x^2 + 5x + xy = x(x + 5 + y)$$

We've come full circle, back to the original expression.

To cement your understanding:

• Expand out $(a + b + c)(d + e)$. Make sure you end up with 6 terms added together, each made up of two letters multiplied together.

• Factor out x from $3x + yx + x$. (Note: the last expression within the brackets will be 1, because $1x = x$)

Some techniques for factorizing

In mathematics an **identity** is an equation that holds true for all values of the variables. There are some convenient techniques involving factorizing, that use the following identities.

To cement your understanding:

Check the identities below, by expanding out the expressions on the right-hand side of each formula, to confirm that you end up with the expressions on the left-hand side.

Identify	Formula	Example
Difference of squares	$a^2 - b^2 = (a - b)(a + b)$	$x^2 - 9 = (x - 9)(x + 9)$
Sum of cubes	$a^3 + b^3 = (a + b)(a^2 - ab + b^2)$	$x^3 + 8 = x^3 + 2^3$ $= (x + 2)(x^2 - 2x + 4)$
Difference of cubes	$a^3 - b^3 = (a - b)(a^2 + ab + b^2)$	$x^3 - 27 = x^3 - 3^3$ $= (x - 3)(x^2 + 3x + 9)$

Another useful technique that sometimes works is the grouping method. Take the expression

$$x^3 + x^2 + x + 1$$

At first there is no obvious way to factorise it. We can factorise x out of the first three terms, but will still be left with a 1 on the end. We can, however, group the terms like this:

$$(x^3 + x^2) + (x + 1)$$

We can now factor out x^2 from the first set of brackets.

$$x^2(x + 1) + (x + 1)$$

And now we can now factor out $x + 1$.

That gives

$$x^3 + x^2 + x + 1 = (x + 1)(x^2 + 1)$$

We aren't going to go into diophantine equations in this book because they don't get a lot of use in statistics, but if you want to find out more about how these factoring techniques can be used, you can look up and explore that topic.

Key points:

- In equations, exponents (indices) are calculated first.

- Then multiplication and division are calculated before addition and subtraction.

- Brackets and braces are used where you want to override this rule.

- Horizontal lines (either with root symbols or fractions) have the same effect as brackets.

- The distributive property of multiplication also applies to algebraic expressions .

- Factorizing is the opposite of expanding an algebraic term using the distributive property.

Quadratic equations

Read this if you don't know how to do quadratic equations, you don't know the difference between a constant and a variable, you don't know what a coefficient is, you don't know how the quadratic formula came about, or you would like to know how to use algebra to calculate the numeric value of the golden ratio.

With the expression $(4 + 7)^2$, we add the 4 to the 7 (giving 11) then square the result (giving 121). If we write it as $(x + y)^2$ though, we think of it as

$$(x + y)(x + y)$$

You can expand this out by multiplying each of the terms in one set of brackets by each of the terms in the other, using the distributive property of multiplication. This gives us

$$(x + y)^2 = x^2 + xy + xy + y^2$$
$$= x^2 + 2xy + y^2$$

That means you have one lot of x^2, one lot of y^2 and 2 lots of xy (x multiplied by y).

To check whether that works, if we were to substitute 4 for x and 7 for y, we would have

$$\begin{aligned}
11^2 &= (4 + 7)^2 \\
&= (4 + 7)(4 + 7) \\
&= 4^2 + 4 \cdot 7 + 7 \cdot 4 + 7^2 \\
&= 4^2 + 2 \cdot 4 \cdot 7 + 7^2 \\
&= 16 + 56 + 49 \\
&= 121
\end{aligned}$$

which is the same result.

> **To cement your understanding:**
>
> Calculate $(3 + 6)^2$ in two different ways: first by adding 3 and 6 and squaring the result, then by expanding it out, as in $(3 + 6)(3 + 6)$, calculating the value of each term then adding them.

Coefficients of polynomial expressions

In the polynomial (more than two terms) expression:

$$4x^2 - 3x^1 + 8x^2 + 4x^0 + 17x^1 - 5x^0 - 2x^2$$

x is a variable. The constants in each term of the expression that are placed before the variable and that multiply it are called **coefficients**. So, the coefficients in this expression are: 4, –3, 8, 4, 17, –5, and –2.

We can simplify the expression in this way, which is similar to what we did when we looked at algebraic notation:

This is our starting expression.	$4x^2 - 3x^1 + 8x^2 + 4x^0 + 17x^1 - 5x^0 - 2x^2$
Colour each term, based on which power x is raised to, in order make what follows a bit clearer.	$4x^2 - 3x^1 + 8x^2 + 4x^0 + 17x^1 - 5x^0 - 2x^2$
Rearrange the expression to gather the different powers of x together (we can do this, because of the commutative property of addition).	$4x^2 + 8x^2 - 2x^2 - 3x^1 + 17x^1 + 4x^0 - 5x^0$
Factor out the different powers of x. (This is the bit that is different to what we did earlier; last time we just added the values. This has the same effect.)	$x^2(4 + 8 - 2) + x^1(-3 + 17) + x^0(4 - 5)$
Add the numbers in brackets, and put the numbers before the letters as per convention.	$10x^2 + 14x^1 - 1x^0$

Write x^1 as x and x^0 as 1. $\qquad\qquad\qquad\qquad 10x^2 + 14x - 1$

We cannot combine the coefficients any further than this. For example, we can't mix the coefficient of x^2 with the coefficient of x, when they are raised to different powers.

That means that in the equation

$$10x^2 + 14x - 1 = ax^2 + bx + c$$

we can safely say that $10 = a, 14 = b$ and $-1 = c$. There are no other values that they can take. They can be expressed differently; for example, we might have said $a = 4 + 8 - 2$, but the value is still 10.

Expressions where the highest power of x is x^2 are **quadratic expressions,** and they can always be reduced to this form.

To cement your understanding:

• Simplify the quadratic expression $3x^2 - 5x^1 + 11x^1 + 6x^2 + 2x^0 - 6x^0 - 1x^2$ to an expression with 3 terms in it by combining the coefficients, as we did above.

• Having done that, in the expression $3x^2 - 5x^1 + 11x^1 + 6x^2 + 2x^3 - 6x^0 - 1x^2 = ax^2 + bx + c$, what are the values of a, b and c ? (Check that you arrived at a $= 8$, b $= 6$ and c $= -4$)

Quadratic formula

It gets tricky to solve a **quadratic equation** (i.e. find what x is) such as

$$7x^2 + 5x - 378 = 0$$

because it isn't easy to get x on its own, on one side of the equation (try it if you are unsure). To do it we use what we learned about what happens when $(x + y)^2$ is expanded:

$$(x + y)^2 = x^2 + 2xy + y^2$$

In this instance it's often easier to use a general formula called the **quadratic formula.** You will use it so rarely, however, that what matters is that you understand how it is derived, so that when you see it referred to or need to look it up, you will have an idea of what it is about. Take this general equation:

$$ax^2 + bx + c = 0$$

Assume that a, b and c are constants, and we are trying to find x. Note that you aren't going to need to work something like this out for yourself if you don't want to; it can just be useful to follow along, so that you can see how others have worked it out. This sort of proof (and most of what we will cover in this book) was worked out many years ago, typically by extremely intelligent people who put a great deal of time into finding a solution.

Subtract c from both sides.

$$ax^2 + bx + \cancel{c} - \cancel{c} = -c$$

$$ax^2 + bx = -c$$

Divide both sides by a. In the first coefficient, the numerator and denominator cancel out. $\frac{a}{a} = 1$, and can be ignored.

$$\frac{\cancel{a}}{\cancel{a}}x^2 + \frac{b}{a}x = -\frac{c}{a}$$

$$x^2 + \frac{b}{a}x = -\frac{c}{a}$$

At this point, we will use a trick called a **substitution**. We will see a lot of substitutions as we progress. Think of it like this. You are doing the weekly shopping, and at first you think you will only be buying a few things. Then as your arms fill up, you realise it will be easier if you put all your shopping into a basket, so you do that. You don't plan to take the basket home; you will give it back when you leave the store. You are just using it for now, to make things easier.

Our "basket" is a completely new variable, that will carry some of our existing variables. We'll call it y. That may seem like a strange thing to do at this point; it may seem as though it is making things harder. In fact, as you will see, it will end up making it simpler. Don't worry if you can't see why this particular substitution is used; you will see soon see that it works.

Our new variable is going to be based on two of our existing letters: a and b.

$$y = \frac{b}{2a}$$

If we multiply both sides by 2 we get

$$2y = 2\frac{b}{\cancel{2}a}$$

$$2y = \frac{b}{a}$$

So now we'll return to the work we were doing, but we'll replace $\frac{b}{a}$ with $2y$.

Substitute $2y$ for $\frac{b}{a}$ in the equation

$$x^2 + 2yx = -\frac{c}{a}$$

Add y^2 to both sides. Again, that may seem strange, but stay with it. The important point at this stage is that the equation still balances. The left-hand side of this equation is now in the same form as what you get by expanding out $(x + y)^2$; that is $x^2 + 2xy + y^2$ (except that we have reversed the x and y in the 2nd term, which makes absolutely no difference).

$$x^2 + 2yx + y^2 = -\frac{c}{a} + y^2$$

We saw earlier that $(x + y)^2 = x^2 + 2xy + y^2$. That means we can replace the latter expression with the former.

$$(x + y)^2 = -\frac{c}{a} + y^2$$

Swap the positions of the terms on the right-hand side of the equation. It is more usual to put a negative term after a positive one, although, by the commutative property, it doesn't matter. But this is the way it is normally seen.

$$(x + y)^2 = y^2 - \frac{c}{a}$$

Take the square root of both sides.

$$x + y = \pm\sqrt{y^2 - \frac{c}{a}}$$

Subtract y from both sides. We have succeeded in getting x on its own.

$$x = -y \pm \sqrt{y^2 - \frac{c}{a}}$$

Substitute back $y = \frac{b}{2a}$ (This is like handing back our shopping basket.)

$$x = -\frac{b}{2a} \pm \sqrt{\left(\frac{b}{2a}\right)^2 - \frac{c}{a}}$$

Expand out the term in brackets.

$$= -\frac{b}{2a} \pm \sqrt{\frac{b^2}{4a^2} - \frac{c}{a}}$$

Multiply the numerator and denominator of $\frac{c}{a}$ by $4a$, to get a common denominator.

$$= -\frac{b}{2a} \pm \sqrt{\frac{b^2}{4a^2} - \frac{4ac}{4a^2}}$$

$$= -\frac{b}{2a} \pm \sqrt{\frac{b^2 - 4ac}{4a^2}}$$

We will see in the shortly why the square root of a fraction as whole is the same as the square root of the numerator divided by the square root of the denominator.

$$= -\frac{b}{2a} \pm \frac{\sqrt{b^2 - 4ac}}{\sqrt{4a^2}}$$

$$= -\frac{b}{2a} \pm \frac{\sqrt{b^2 - 4ac}}{2a}$$

$$x = \frac{-b \pm \sqrt{b^2 - 4ac}}{2a}$$

This is called the **quadratic formula.** To solve quadratic equations, you can manipulate them into the format of $ax^2 + bx + c = 0$, then use a calculator to find x.

Back to our original formula: $7x^2 + 5x - 378 = 0$

In this instance $a = 7$, $b = 5$ and $c = -378$.

Putting those values into the quadratic formula gives us:

$$x = \frac{-5 \pm \sqrt{25 + 10584}}{14}$$

$$= \frac{-5 \pm \sqrt{10,609}}{14}$$

$$= \frac{-5 \pm 103}{14}$$

$$= \frac{98}{14} \; or \; \frac{-108}{14} \; (two \; options, because \; we \; had \pm 103)$$

$$= 7 \; or \; -7.71428 \dots$$

Both solutions will solve the equation. You can confirm this by putting each of these numbers in turn into the original equation: $7x^2 + 5x = 378$

To cement your understanding:

Solve the equation $3x^2 - 8x = 10$ using the quadratic formula and a calculator.

Hint: 1. Subtract 10 from each side, so that you have something that is in the form of $a = 3$, $b = -8$ and $c = -10$. Use the quadratic formula $x = \frac{-b \pm \sqrt{b^2 - 4ac}}{2a}$

Put the answers back into your calculator to check that $3x^2 - 8x = 10$. Note that you probably won't get exactly 10, due to rounding.

Fun fact: We can use this approach to find a numerical value for φ, the number of the golden ratio, simply starting from 3 lines that the Greeks used to define the golden ratio. We can also use this approach to see why when we add 1 to φ we get φ^2, and when we subtract 1 from φ we get its reciprocal. The working out for this is shown in an appendix (page 733).

Key points:

- $(x + y)^2 = x^2 + 2xy + y^2$

- The constants in each term of an expression that are placed before the variable and that multiply it are called coefficients.

- Expressions where the highest power of x is x^2 are called quadratic expressions.

- Quadratic equations (when x and its powers are the only variables) can always be expressed in the form $ax^2 + bx + c = 0$.

- The quadratic formula for solving quadratic equations is

$$x = \frac{-b \pm \sqrt{b^2 - 4ac}}{2a}$$

Polynomial multiplication and convolutions

> **Read this if** you don't know how to multiply two polynomials, and/or you aren't familiar with convolutions.

Multiplying polynomials

As you might expect, when we multiply polynomials, there is a two-step process. We multiply every possible term, then combine the coefficients of equal powers of each variable. It's just like what we have already done when we looked at quadratic equations, except that there are more terms to multiply. For example:

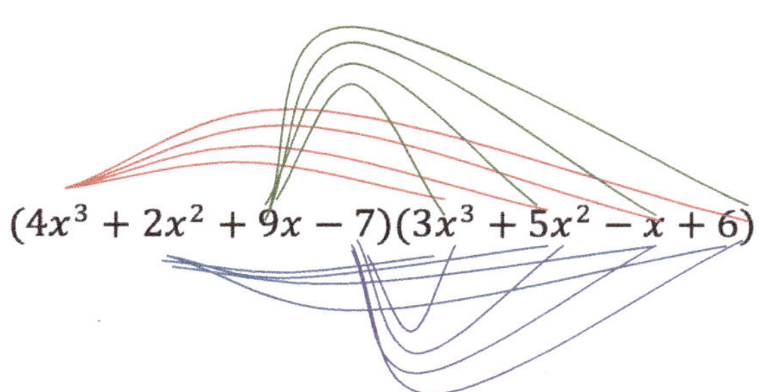

$$(4x^3 + 2x^2 + 9x - 7)(3x^3 + 5x^2 - x + 6)$$

If you ever do this, the easiest way is probably to put it into a table:

	$4x^3$	$2x^2$	$9x$	-7
$3x^3$	$12x^6$	$6x^5$	$27x^4$	$-21x^3$
$5x^2$	$20x^5$	$10x^4$	$45x^3$	$-35x^2$
$-x$	$-4x^4$	$-2x^3$	$-9x^2$	$7x$
6	$24x^3$	$12x^2$	$54x$	-42

That covers the first step, of multiplying each of the terms in the first expression by each of the terms in the second expression. We can achieve our second step, of gathering the terms that add to the same powers of x, by adding the coefficients in the diagonals (with similar colours).

This works out to:

$$12x^6 + 20x^5 + 6x^5 + -4x^4 + 10x^4 + 27x^4 + 24x^3 - 2x^3 + 45x^3 - 21x^3 + 12x^2 - 9x^2 - 35x^2 + 54x + 7x - 42 =$$

$$12x^6 + (20 + 6)x^5 + (-4 + 10 + 27)x^4 + (24 - 2 + 45 - 21)x^3$$
$$+ (12 - 9 - 35)x^2 + (54 + 7)x - 42$$

$$= 12x^6 + 26x^5 + 33x^4 + 46x^3 - 32x^2 + 61x - 42$$

We mentioned earlier that, by convention, the higher powers of the variable are put before the lower powers.

Introduction to convolutions

Convolutions are a concept—like riding a bicycle or swimming—that is difficult to explain in words and takes some getting used to, but becomes straightforward once you grasp the idea. At this stage, we are simply introducing the concept, so if anything that follows isn't immediately clear, don't worry—just let it sink in. We'll revisit it a couple more times in this book, and it should make sense by the end.

You are already familiar with operations such as addition, subtraction, multiplication, and division. There is another operation, called convolution, that you may not have encountered unless you have studied mathematics at a tertiary level. It is symbolised by an asterisk, $*$, or sometimes by $\otimes$. (You can think of the asterisk as combining the t-shaped cross of addition (+) with the x-shaped cross of multiplication (x) because convolution involves both.)

Convolution is rarely used in basic algebra, but it becomes important in mathematical statistics and various engineering disciplines. The reason for introducing it here is that it closely resembles polynomial multiplication. When multiplying polynomials, we pair terms from two sets of brackets, multiply them, and then add the results.

Similarly, when performing a convolution on two sets of numbers or functions, we multiply every possible combination and then sum the results based on a specific criterion—typically aligning terms that contribute to the same output.

Let's take the pairs of terms from the polynomials above that multiply together to give a coefficient and x^3 (that is, those from the green diagonal). These are:

$$4x^3 \bullet 6 \ = 24x^3$$

$$2x^2 \bullet -x \ = -2x^3$$

$$9x \bullet 5x^2 \ = 45x^3$$

$$-7 \bullet 3x^3 \ = -21x^3$$

$$\overline{46x^3}$$

By multiplying these pairs together and then adding the results, we have performed a convolution of them.

To cement your understanding:

Multiply out $(6x^4 + 3x^2 + 5x)(2x^3 + 7x^3 + 6x^2 + 2x + 11)$ in any way you like, to get a sense of the way that you need to multiply all the terms together, then group like terms.

Key points:

- To multiply polynomials, multiply every possible term, then combine the coefficients of equal powers of each variable.

- A convolution is symbolised by an asterisk, $*$, or sometimes by $\otimes$.

- To do a convolution of two sets of numbers or two formulae, we multiply every possible combination, then add those that give a similar result on whatever aspect we are interested in.

Roots part 2

Read this if you don't know why

$$\sqrt{a}\sqrt{b} = \sqrt{ab}$$

or why

$$\sqrt{\frac{a}{b}} = \frac{\sqrt{a}}{\sqrt{b}}$$

Read it if you don't know how to represent the square root of a negative number, you don't know what a "complex number" is, or you don't understand what the symbols $\mathbb{R}$, $\widetilde{\mathbb{R}}$ or $\mathbb{C}$ mean.

Operations with roots

Taking the square root of two numbers multiplied together is the same as multiplying their square roots. For example:

$$\sqrt{4} \bullet \sqrt{9} = 2 \bullet 3$$
$$= 6$$
$$= \sqrt{36}$$
$$= \sqrt{4 \bullet 9}$$

In general terms, we are saying that:

$$\sqrt{a} \bullet \sqrt{b} = \sqrt{ab}$$

To see why it is true, we'll use a substitution again.

$$\sqrt{a} = x \text{ and } \sqrt{b} = y$$

That means, if we square both sides:

$$a = x^2 \text{ and } b = y^2$$

Multiply a and b together.

$$ab = x^2 y^2$$

Express the right-hand side differently. $ab = xxyy$

Rearrange the terms on the right (which we can do, because of the commutative property of multiplication). $ab = xyxy$

$xyxy$ is another way of saying $(xy)^2$ $ab = (xy)^2$

Take the square root of both sides $\sqrt{ab} = xy$

Substitute back: as we originally said, $\sqrt{a} = x$ and $\sqrt{b} = y$

$$\boxed{\sqrt{ab} = \sqrt{a}\sqrt{b}}$$

To cement your understanding:

Find the square root of 5,184 without using a calculator. (Hint: 5,184 = 81 • 64. Swap the formula around so that you think of it as $\sqrt{ab} = \sqrt{a}\sqrt{b}$ Then check that you arrived at 72.

In the same way, taking the square root of a fraction is the same as taking the square root of the numerator and the square root of the denominator. (We assumed this was true in the proof of the quadratic formula.) For example:

$$\sqrt{\frac{36}{4}} = \sqrt{9}$$

$$= 3$$

$$= \frac{6}{2}$$

$$= \frac{\sqrt{36}}{\sqrt{4}}$$

To prove that more generally:

Use the same substitution as we did above: $\sqrt{a} = x$ and $\sqrt{b} = y$

That means, if we square both sides: $a = x^2$ and $b = y^2$

Divide a by b. $\dfrac{a}{b} = \dfrac{x^2}{y^2}$

<table>
<tr><td>Express the right-hand side differently.</td><td>$= \dfrac{x}{y} \dfrac{x}{y}$</td></tr>
<tr><td>Express the right-hand side differently again.</td><td>$= \left(\dfrac{x}{y}\right)^2$</td></tr>
<tr><td>Take the square root of both sides.</td><td>$\sqrt{\dfrac{a}{b}} = \dfrac{x}{y}$</td></tr>
<tr><td>Substitute back: as we originally said,

$\sqrt{a} = x \text{ and } \sqrt{b} = y$</td><td>$\boxed{\sqrt{\dfrac{a}{b}} = \dfrac{\sqrt{a}}{\sqrt{b}}}$</td></tr>
</table>

To cement your understanding:

Find the square root of 0.09 without using a calculator. (Hint: express 0.09 as a fraction with 100 in the denominator.)

Square roots of negative numbers

Figure 36

You can't have a square root of a negative number using any of the types of numbers we have seen so far, but there are times in mathematics when it would be really useful if you could. For those times, Raphael Bombelli from Bologna (1526-1572), in his textbook on algebra, came up with a solution, which he called "plus of minus". Sometime afterwards, Leonhard Euler (1707–1783) (or possibly someone else) used the letter i to symbolise $\sqrt{-1}$. (We'll meet Euler again later.)

This means that i can be used as part of the description for the square root of any negative number. For example:

$$\sqrt{-2} = \sqrt{-1 \cdot 2}$$

$$= \sqrt{-1}\sqrt{2}$$

$$= i\sqrt{2}$$

Getting a bit philosophical, i is called an 'imaginary number' or an 'unreal number', because there is nowhere to put it on an ordinary number line. However, all numbers are just useful ways of communicating ideas: they are as real and as unreal as that. You can't pick up i oranges the way you can pick up 3 oranges, but neither can you pick up –3 oranges or 0 oranges or π oranges: these are all just useful ways to communicate concepts, so the terminology is somewhat confusing.

Complex numbers

Complex numbers have both a real and an imaginary component added together, with the imaginary part coming after the real part. For example:

$$4 + i5$$

Complex numbers are useful to the point of being quite necessary in a number of highly practical disciplines, including electronics and surveying. (In electronics, for historical reasons, the letter j is usually used instead of i. In this book we'll stick with i.)

Earlier we looked at some categories of numbers. We now have some more to add to our list.

Symbol	Called	Explanation	Examples
$\mathbb{R}$	Real numbers	Any number that can be described on a normal number line.	All the categories of numbers we have seen to date.
$\widetilde{\mathbb{R}}$	Unreal, or imaginary numbers	A number that can't be placed on a normal number line. Based on the square roots of negative numbers.	$i \quad \sqrt{-347}$
$\mathbb{C}$	Complex numbers	Numbers with a real and an imaginary component.	$4+i5$

Both real and unreal numbers are subsets of complex numbers. With real numbers, the i is multiplied by 0, whereas with unreal numbers, the real part is 0.

To cement your understanding:

- Calculate $\sqrt{-3}$ the way we did above.

- Match the symbols $\mathbb{R}$, $\widetilde{\mathbb{R}}$, and $\mathbb{C}$ to the numbers: $3 + i8$, $\sqrt{-347}$ and 462.

Key points:

- Taking the square root of two numbers multiplied together is the same as multiplying their square roots.

- Taking the square root of a fraction is the same as taking the square root of the numerator and the square root of the denominator.

- $\sqrt{-1}$ is designated by the letter i.

- i is called an unreal or an imaginary number, but it is actually as valid as any other number.

- Complex numbers have both a real and an imaginary component added together, with the imaginary part coming after the real part.

- $\mathbb{R}$ symbolises real numbers, $\widetilde{\mathbb{R}}$ symbolises unreal or imaginary numbers, and $\mathbb{C}$ symbolises complex numbers.

- All numbers can be expressed as complex numbers.

Simultaneous equations

> **Read this if** you don't know what a simultaneous equation is, and/or you don't know two ways to solve them

While the equation

$$3x + 9 = 24$$

has only one solution ($x = 5$), the equation

$$3x + y = 24$$

has an unlimited number of solutions. These come in pairs, for example:

$$x = 4 \quad y = 12$$

$$x = -400 \quad y = 1{,}224$$

$$x = 27.43 \quad y = -58.29$$

It is usually impossible to find values for a *single* equation with *two* variables, such as $3x + y = 24$. If we have some additional information however, such as $4x + y = 31$, then we may have enough information to work out the values of each of the two variables.

There are two broad strategies.

The first approach is **substitution**. This is the approach we used at the start of this book, in the equations we looked at involving double negatives in addition and multiplication, so it will have some familiarity.

Using this strategy, once again our two equations are

$$3x + y = 24 \text{ and } 4x + y = 31$$

If $4x + y = 31$, then $y = 31 - 4x$ (subtracting $4x$ from both sides, to get y by itself). Substituting this into the other equation ($3x + y = 24$) gives

$$3x + (31 - 4x) = 24$$

We now have a single equation with a single variable, and once again it is straightforward to solve it to find that $x = 7$, as follows:

Remove the brackets (we don't need them here).	$3x + 31 - 4x = 24$
Add $3x$ to $-4x$	$31 - x = 24$
Subtract 31 from both sides of the equation.	$-x = -7$
Multiply both sides by -1	$x = 7$

We can replace x with 7 in either of the equations, then use algebra to find the value of y.

Using the first equation ($3x + y = 24$):	$3(7) + y = 24$
	$21 + y = 24$
Subtract 21 from both sides.	$y = 3$
Using the second equation ($4x + y = 31$):	$4(7) + y = 31$
	$28 + y = 31$
Subtract 28 from both sides.	$y = 3$

It doesn't matter which of the equations we use. We also could have used this approach to find the value of y, then substituted it back in to find the value of x.

The second approach is **elimination**. Line up two equations under one another, like an ordinary subtraction:

Equation 1:	$4x + y = 31$
Equation 2:	$3x + y = 24$
Equation 1 – Equation 2:	$x = 7$

That was a particularly simple case. More commonly, to make this work you need to multiply each term in one of the equations by a constant, so that when you subtract one equation from the other, one of the variables is eliminated.

As a general rule, to solve simultaneous equations you need at least the same number of equations that you have variables. In this case, we had two variables and so we needed two equations. If we had had three variables, we would have needed three equations, and so on.

> **To cement your understanding:**
>
> • Solve the simultaneous equations $x + 2y = 19$ and $4x + 3y = 41$ using the first strategy: isolate one of the variables (it doesn't matter which), then substitute it into the other equation. Hint: the steps are: 1) pick one of the equations; it doesn't matter which. 2) Pick one of the variables in that equation; it also doesn't matter which. 3) Get that variable on its own on one side of the equation. 4) Replace the variable in the second equation, so that you now have an equation with only one variable. 5) Solve that equation by getting the variable on its own. 6) Use this to work out the value of the other variable.
>
> • Solve the simultaneous equations $8x + 2y = 62$ and $3x + y = 24$, first by multiplying the second equation by 2, then subtracting the second equation from the first.

> **Key points:**
>
> - It is usually impossible to find values for a *single* equation with *two* variables.
>
> - We normally need the same number of equations as the number of variables.
>
> - We can combine two equations by substitution or by elimination.

Functions and variables

Read this if you don't know what $f(x)$ means, you don't understand the difference between a dependent and an independent variable, or you don't know what an inverse function is.

Function notation

Algebra as we understand it was developed in Persia by Muhammad ibn Musa al-Khwārizmī (about 780–850 AD). He wrote "The Book of Completion and Balancing". In Persian, the words "al-jabr" mean completion and balancing (as in balancing equations). He was also the person who introduced the Hindu form of letters to the Islamic world, from where it eventually came to Europe. He was an astronomer and geographer, and his name lay behind the word "algorithm".

Several hundred years later, these ideas found their way to Europe, which was, until then, still in its dark ages. These ideas were an important part of allowing Europe as a whole to progress. The proofs of more complex mathematical ideas, though (including the early developments of calculus, and much of the early work in statistics), still heavily relied on the sorts of geometrical drawings that were used by the ancient Greeks. Reading them today, it is hard to understand what they were talking about. Some improved notation was needed to simplify more complex proofs.

Swiss mathematician Leonhard Euler (1701-1783) (pronounced "oiler")—unquestionably one of the greatest mathematicians ever—developed some ideas and notation to make things simpler. We will meet Euler again several times. (As well as driving mathematics forward, he also made contributions to physics, astronomy, topology, engineering, fluid dynamics, optics, music theory...). He was certainly one of the most prolific mathematicians of all times, as well as one of the

Figure 37
Leonhard Euler
1701–1783

best: his publications when stacked on a bookshelf are measured by the metre, and not all of his works have been published yet.

We've seen that a variable (e.g. x) can be used as a place holder for a number; either a number we want to find, or just any number, so that we can create a formula. We've also seen that a collection of terms on one side of an equation is an **expression**. Another word for a term or an expression is a **function**. A function can be a single term, like

$$5x^3$$

or something more complicated, like

$$5x^3 + 19x + 746.32$$

The idea behind a function is that, once we know what the variable is, we can then calculate the value of the function. If, for example,

$$x = 1$$

it is straightforward to substitute 1 for x, and to calculate that the value of the function must be

$$5(1^3) + 19(1) + 746.32 = 770.32$$

You can think of a function as being a bit like an app or computer program: once it is created, you input something and get something else as an output.

We write a function of x as $f(x)$. The function is what you do to x to obtain a result.

Until it is defined, this notation is really saying "this will apply, no matter what function of x we are talking about".

We can use different letters to indicate that there are different functions involved. For example $f(x) + g(x) = h(x)$. This means that function h is made up by adding two other functions of x.

If, for example,

$$f(x) = x^2$$

and

$$g(x) = 2x + 7$$

then

$$h(x) = f(x) + g(x)$$
$$= x^2 + 2x + 7$$

Dependent and independent variables

Take $y = f(x)$ defined by

$$3x + 5 = y$$

In fact the two expressions $y = 3x + 5$ and $f(x) = 3x + 5$ are virtually equivalent; it's just that in some contexts it is more helpful to talk in terms of y, and in others it is more helpful to talk about $f(x)$.

We say that y is a function of x. In this case we say that x is the **independent variable**, and y is the **dependent variable**. x can take any value at the moment, while the value of y depends on the value of x.

Now if we turn it around to get x on its own:

$$3x + 5 = y$$

Subtract 5 from both sides
$$3x = y - 5$$

Divide both sides by three
$$x = \frac{y - 5}{3}$$

Now we would say that x is a function of y. We have now made y the independent variable (it can take any value) and x the dependent variable. That is, the value of x depends on the value of y.

Inverse functions

If y is a function of x, then the terminology we use is that x is the **inverse function** of y. If you think of getting dressed in the morning, putting on your shirt first and then your jacket, as a function, then the inverse function is doing the opposite: taking your jacket off, then your shirt, to return you to the situation you started with.

The notation we use for this is a little confusing at first: we write an inverse function as

$$x = f^{-1}(y)$$

That looks a lot like saying that we are raising f to the power of –1; but that isn't what it means, when it is used with function notation like this. It's just a convention to do it this way.

For example, if

$$y = x^2$$

$$= f(x)$$

then

$$x = \pm\sqrt{y}$$

$$= f^{-1}(y)$$

We say that these are **inverse functions** of one another.

Functions with more than one independent variable

We can use this notation to denote a function with more than one independent variable, in which case we write something like $f(x, y)$ meaning that it is a function of both x and y; that is, a function that has both x and y in it, that individually contribute to the result. So $y = 5x^3$ is just a function of x, not a function of x and y.

In the expression $z = x^5 + x\sqrt{y}$, however, both x and y are independent variables; the value of one does not affect the value of the other. But z is dependent on both x and y. In this instance we say "z is a function of x and y", and write it as

$$z = x^5 + x\sqrt{y}$$

$$= f(x, y)$$

To cement your understanding:

- How would you write, in mathematical notation, that y is a function of x?

- How would you write, in mathematical notation, that x is an inverse function of y?

- How would you write that z is a function of both x and y?

Key points:

- Another word for a term or an expression is a function.

- The idea behind a function is that, once we know what the variable(s) is/are, we can then calculate the value of the function.

- We write a function of x as $f(x)$. Until it is defined, this notation is really saying "this will apply, no matter what function of x we are talking about".

- We can use different letters to indicate that there are different functions involved.

- An independent variable can take any value, whereas a dependent variable can only take values produced by a function.

- An inverse function is the reverse of a function.

- If $y = f(x)$ then $x = f^{-1}(y)$

- We can use this notation to denote a function with more than one independent variable, in which case we write something like $f(x, y)$

Sequences and limits

> **Read this if** you don't know how to mathematically describe numbers in a sequence, you don't know the difference between an arithmetic progression and a geometric progression, you don't understand whether infinity is actually a number, you don't understand the concept of finding the limit as something approaches infinity, you want to learn some more fun facts about Pascal's triangle, or you want to know what Pascal's triangle has to do with the golden ratio, and what any of that has to do with a Fibonacci sequence.

(If you are struggling with this, see the How to Love Statistics YouTube video on *Sequences*.)

At this point we start to move beyond only using algebra in equations and start to use it to describe patterns of numbers.

A sequence, or progression, is a list of numbers that has a pattern we can describe. If you see a sequence like:

$$3, 7, 11,...$$

and have to guess the next number, you would probably say 15, based on the idea that the sequence is obtained by adding a 4 to the previous number. That might not be what the pattern really represents, though. It could, for example, created by adding 4 for the first two times, adding 14 for the next two, adding 24 for the two after that, and so on. It could actually be many different things.

Patterns of numbers like this can be important, and it isn't always obvious just by looking at a handful of examples what the underlying pattern is, so we need a mathematical way to clearly describe what to do to create the pattern.

If you have ever had to assemble flat-pack furniture, you will know that the instructions typically go for a number of pages, because there isn't much repetition. They might say "assemble each drawer the same way", but most of the instructions will be different for different parts of whatever you are building.

Knitting patterns tend to be more difficult for the non-knitter to follow, saying things like "*Row 1 (RS): K1, *k1, p1; rep from * to last st, k1. Rep row 1 another 32 times, ending with a RS row.*" They use more technical abbreviations, but are usually more concise than instructions for building flat-pack furniture, because there is more repetition. A couple of pages is usually all that is needed to know how to knit a whole garment.

The mathematics used to describe patterns of numbers is both more concise than this, and easier to follow, once you understand the basic terms. Just as knitters need to keep track of which row they are up to, sequences need a way of designating which item in the sequence you are up to, but the pattern rarely changes. One short instruction usually covers the entire pattern. Sequences are written like Figure 38.

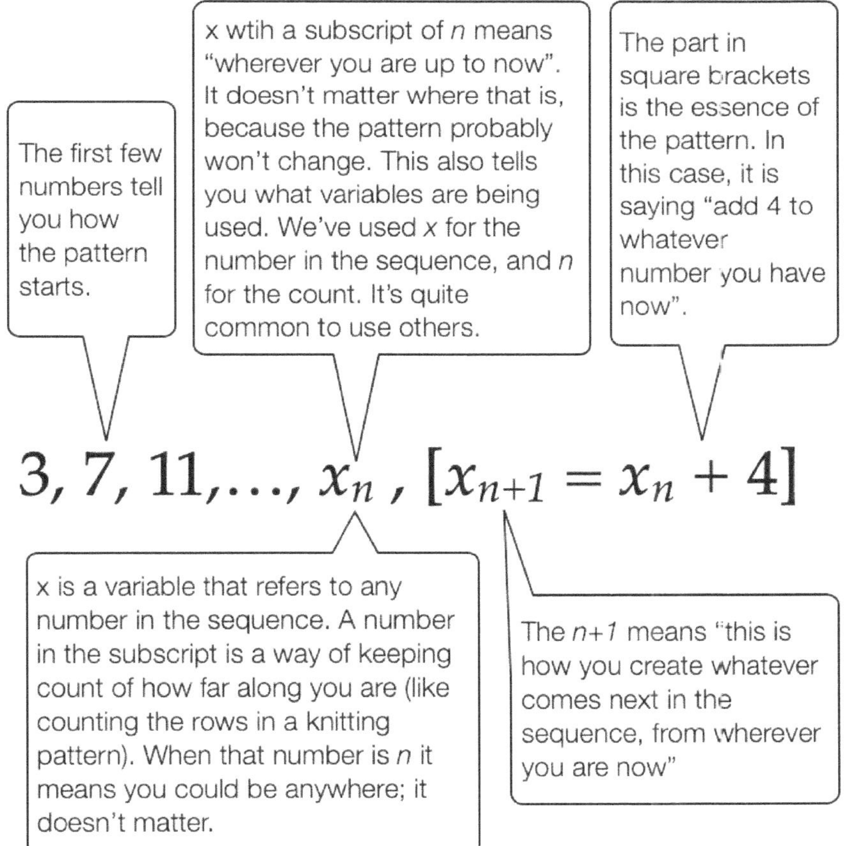

$$3, 7, 11,\ldots, x_n, [x_{n+1} = x_n + 4]$$

Figure 38

The way of describing these in mathematics is to start by giving the first few numbers in the sequence. This is like the picture on the front of flat-pack furniture instructions or a knitting pattern, telling you where you are going. After that there are some mathematical-convention tools to tell you how to keep building the pattern, based on what you already have.

To do this we need a variable that will describe any number in the sequence. For now we will use x for this, because that is the most familiar variable to us at this stage. (It is actually more common to use a with sequences, but it doesn't matter, we can use whatever we want, so we'll stay with x for now.) This is *almost*, but not quite, the same way we have used x so far. That is because, to this point, x could be any number at all, until we know more. When we are thinking about a sequence, x can be any number *in the sequence.*

The second thing we need is a way to describe the relationship that the different numbers in the sequence have to one another. The simplest way to do this is to have a way to count how far along the sequence we have come. This is like identifying where you are up to in a knitting pattern, by giving you a row number. We've seen that raising something to a power can be described by putting a number into superscript or small characters raised above the line, ^{like this} (as in x^4). To keep count of which number in a series we are looking at, mathematicians often use a subscript, or small characters dropped below the line, _{like this}. So in the sequence we are looking at, we would say that $x_1 = 3$, $x_2 = 7$, $x_3 = 11$, and so on.

The final thing we need is another variable that can stand in the place of the numbers in the subscript, to say that whatever we are talking about can apply to any instance of x in this series. The letter i is sometimes used as this variable, but we've already used that for the square root of negative 1, so we will use n, at least for now. When used like this, n will always be a natural or counting, number.

We write it like x_n. When you see that, think of it as meaning "wherever you are up to now". If you are at the first place in the sequence n will be 1. If you are the 100th place, n will be 100. But if you just see "n", think "wherever you are up to now; it doesn't matter where".

If we want to say "the number in the sequence that comes before some other number in the sequence (and it doesn't matter which one), we write it as x_{n-1} and likewise, if we want to talk about the next number in the sequence we write it as x_{n+1}.

> **To cement your understanding:**
>
> • Suppose you see this section out of a long sequence: $\dots 323, 327, 331, 335, 339, \dots$ and you are interested in looking at the number 331. You haven't actually counted every number, so you don't know how far along 331 is. So you decide to designate it as x_n
>
> $$x_n = 331$$
>
> What is x_{n-1}? What is x_{n-2}? What is x_{n+1}? What is x_{n+2}?
>
> Can you see that these numbers are 327, 323, 335 and 339 respectively? If not, can you see where you went wrong?
>
> • If we change our focus in the same series, and make 327 our x_n, what will x_{n+1} be now? Can you see that it will be 331? If so, you've got this.

So, if we want to use a formula to describe the sequence of numbers we looked at earlier, we usually give the first few numbers in the sequence, then three dots, then x_n, then a formula that we can use to calculate the next number in the sequence.

The description of our formula will be:

$$3, 7, 11, \dots, x_n, [x_{n+1} = formula\ goes\ here]$$

In this case, we want to say that we find the next number by adding 4 to the number we are currently at. In other words, $x_n + 4$ will tell us the value of x_{n+1}, regardless of how far along in the sequence we are. We can unambiguously describe our sequence as:

$$3, 7, 11, \dots, x_n, [x_{n+1} = x_n + 4]$$

> **To cement your understanding:**
>
> What would this look like if a sequence started with 2, and added 5 to each term to get the next number? What would it look like if it started with 4 and multiplied by 3 each time?
>
> Did you get:
>
> $$2, 7, 12, \dots, x_n, [x_{n+1} = x_n + 5]$$
>
> $$4, 12, 36, \dots, x_n, [x_{n+1} = 3x_n]$$
>
> If not, can you see where you went wrong?

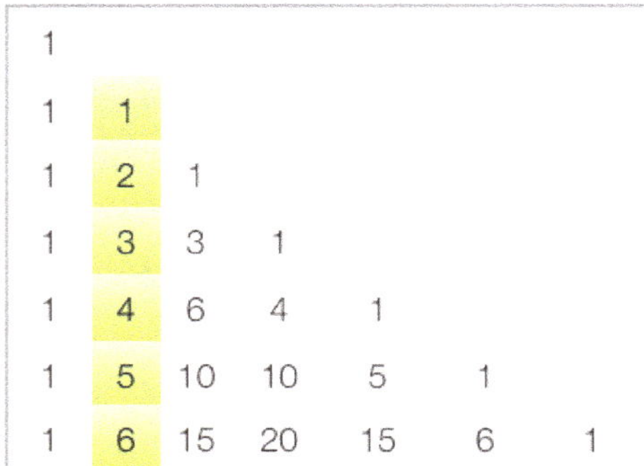

Figure 39

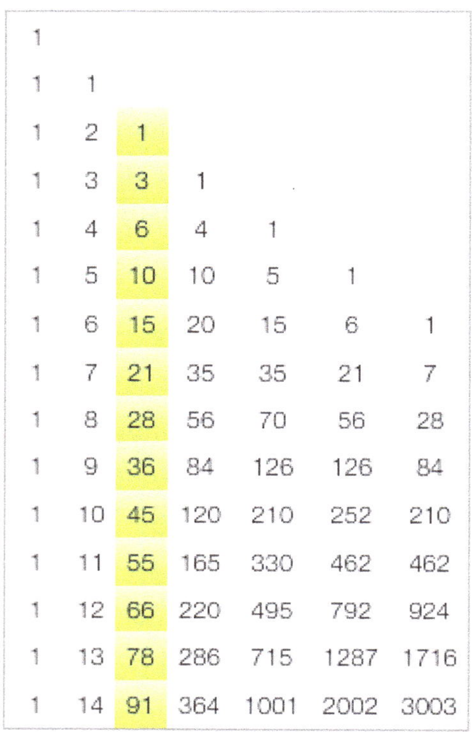

Figure 40

Pascal's triangle contains some interesting sequences.

The second column (Figure 39) has the positive integers. These are themselves a sequence. This particular list of numbers is an **arithmetic progression**, which means that each number of the sequence is made up by having the same number added to it. In the case of the list of integers in column 2, the number added each time is 1.

If a sequence is made up by multiplying each term by the same number to get the next number (e.g. $3, 6, 12, 24, 48$, which involves multiplying each number by 2 to get the next number), it is called a **geometric progression**.

The third column in Pascal's triangle (Figure 40) is a more interesting sequence, although it might not be immediately obvious what its pattern is. Two things are happening in this sequence. One is that each number is obtained by adding one more than was added the previous time. In other words, $1 + 2 = 3$, $3 + 3 = 6$, $6 + 4 = 10$, $10 + 5 = 15$, and so on. At the same time, what is happening in this sequence is that every two successive numbers add together to form the square of a number. The first two numbers, 1 and 3, add together to make 4, which is the square of 2. The next two, 3 and 6, add together to make 9, which is the square of 3. The two after that, 6 and 10, add together to make 16, which is the square of 4, and so on.

We can write the formula for this sequence as

$$1, 3, 6, \ldots, x_n, [x_{(n+1)} = (n+1)^2 - x_n]$$

If you count down 6 places, $x_n = 21$ and $n = 6$.

Putting those numbers into this formula, we have:

$$1, 3, 6, 10, 15, 21, [x_7 = (6 + 1)^2 - 21]$$

The formula is telling us that the next place in the sequence ($x_{(n+1)}$) will be equal to the square of $(6 + 1)$, which is 49, minus 21. That is, 28. In other words, $28 = 7^2 - 21$, which is another way of saying that $21 + 28 = 7^2$.

We know that, in this sequence, the sum of each two successive numbers makes the square of a number. We have changed things around, so that we can use a formula to find the next number in the sequence.

To cement your understanding:

In Figure 40, suppose we wanted to find out what is in the 9th position in the column. (Yes, you could just look it up; but follow through with this, because the important thing is to understand the way the notation is working.) Count down to position 8. Replace n with 8, and x_n with 36. So we have the number in the 8th position; now we want to find the number in the 9th position. We designate this number $x_{(n+1)}$ or in other words, x_9. You can calculate it by using this formula, using $n = 8$ and $x_n = 36$. The formula, again, is

$$x_{(n+1)} = (n + 1)^2 - x_n$$

Did you arrive at 45? If not, can you see where you went wrong?

There is an even more interesting sequence hidden in Pascal's triangle. Leonardo Bonacci (1140-1245), also known as Fibonacci, was a mathematician from Pisa in Italy who travelled within the (more scientifically advanced) Islamic world and brought ideas back to Europe. He was one of the people who introduced our current system of numbering (**Hindu-Arabic numerals**) into Europe. He also brought back a method for predicting animal population growth which involved the following sequence:

$$1, 1, 2, 3, 5...$$

In other words, each two successive numbers added together make the next number.

$$1 + 1 = 2, 1 + 2 = 3, 2 + 3 = 5, \text{ and so on.}$$

To make it crystal clear in mathematical language how to find the next number in the sequence, we can write it as:

$$1, 1, 2, 3, 5,..., x_n, [x_{(n+1)} = x_n + x_{(n-1)}]$$

This is another way of saying: find the next number by adding the current number to the previous number. The current number is x_n. The previous number is x_{n-1}. That means the next number $x_{(n+1)}$ will be $x_n + x_{(n-1)}$. This will apply to any number in the sequence.

To cement your understanding:

The next number after 5 in the Fibonacci sequence will be 8. Find the number after 8, using the formula. (Hint: In this instance we say that $x_n = 8$. Because 8 is the next number after 5, that means that $x_{(n-1)} = 5$. The number you are trying to find will be $x_{(n+1)}$)

This is now called the **Fibonacci sequence** (The sequence was probably first identified by the Indian scholar Pingala in roughly 200BC; he called it "mountain of cadence". He also identified a form of Pascal's triangle.) If you add up the diagonals in Pascal's triangle (add up each coloured diagonal), you get (drumroll) the Fibonacci sequence (Figure 41), which you can see along the top.

A sequence that has a limited number of terms is a **finite sequence**. If there is no limit to the number of terms, it is an **infinite sequence**.

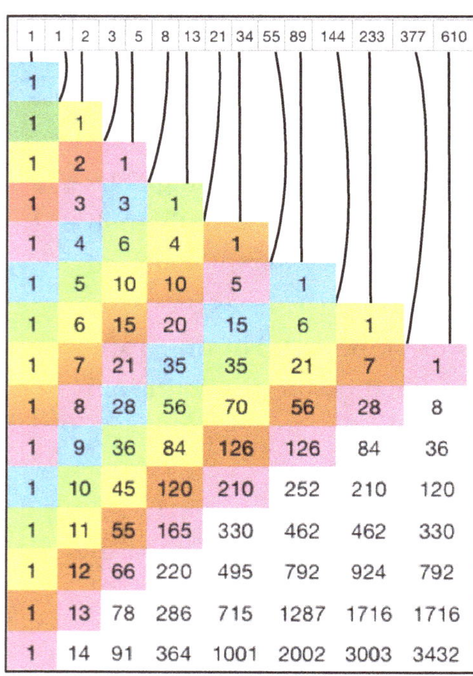

Figure 41

The concept of **infinity** in mathematics isn't actually a number: it's a way of saying that something is unbounded, or continues without ever coming to a stop. The symbol used for it looks like the number 8 on its side: ∞

Each of the sequences we have looked at so far have been infinite: there isn't any limit on the number of terms (n). In these particular series there won't be any limit on the size of the term (x). It will keep getting bigger and bigger. Sequences like this, that just keep getting bigger, are described as being **divergent**.

Some infinite sequences, though, reach a point that they will not go beyond, no matter how many terms are added. These sequences are **convergent**. We've already seen something like this, when we looked at repeating decimals.

The sequence

$$\frac{1}{1}, \frac{1}{2}, \frac{1}{3}, \ldots, x_n, \left[x_{n+1} = \frac{1}{n+1}\right]$$

will be infinite – it will have an unlimited number of terms – but it will also be convergent. The larger n gets, the smaller and smaller x will become. It will never reach 0, and it will never become negative (get smaller than 0).

You can write what happens to a convergent sequence like this ("lim" means "limit"):

$$\lim_{n \to \infty} \frac{1}{n} = 0$$

This notation means as n approaches infinity, its reciprocal will get closer and closer to zero, but never quite reach it and never get past it. So, for example,

$$\frac{1}{10,000}$$

is a very small number, while

If you are wondering about the practical relevance of all of this, Fibonacci's work was an example of mathematical breakthroughs allowing progress in other areas of life. He wrote a book called *Liber Abaci*, or *Book of Calculation* which contained many examples of financial transactions. The ideas in this book made commercial lending more feasible in Northern Italy, which in turn made commercially-motivated sea voyages possible. In turn, that created the wealth that funded the Italian renaissance, and also provided a motivation for European colonial expansion.

$$\frac{1}{1,000,000,000,000,000,000,000,000,000,000,000,000,000,000,000}$$

is a much smaller number, but still greater than 0.

Fun fact: The Fibonacci sequence is divergent, but the *ratio* between successive numbers (dividing each number by the number that immediately precedes it) is convergent (Figure 42). Do you recognise the number in the bottom of the column on the right? The limit on the ratio between successive numbers in the Fibonacci sequence, as x approaches infinity is, in fact,

Fibonacci sequence (sum of the previous two numbers)	Ratio of this number to the previous number (1:1, 2:1, 3:2, 5:3 8:5, etc)
1	
1	1
2	2
3	1.5
5	1.66666666666667
8	1.6
13	1.625
21	1.61538461538462
34	1.61904761904762
55	1.61764705882353
89	1.61818181818182
144	1.61797752808989
233	1.61805555555556
377	1.61802575107296
610	1.61803713527851
987	1.61803278688525
1597	1.61803444782168
2584	1.61803381340013
4181	1.61803405572755
6765	1.61803396316671
10946	1.6180339985218
17711	1.61803398501736
28657	1.6180339901756
46368	1.61803398820533
75025	1.6180339889579
121393	1.61803398867044
196418	1.61803398878024
317811	1.6180339887383
514229	1.61803398875432
832040	1.6180339887482
1346269	1.61803398875054
2178309	1.61803398874965
3524578	1.61803398874999
5702887	1.61803398874986
9227465	1.61803398874991
14930352	1.61803398874989
24157817	1.6180339887499
39088169	1.61803398874989

Figure 42

exactly phi (φ) ; the number of the golden ratio. And so Pascal's triangle meets φ, believe it or not.

Key points:

- A sequence is a list of numbers that has a pattern which can be described by a mathematical formula. It is sometimes also called a progression.

- Expressing a sequence mathematically requires a variable that can be used to describe any number in the sequence.

- It also requires a variable to keep count of how far along you are in a sequence.

- Subtracting 1 from the variable that keeps count means go to the previous item in the series, whereas adding one means go to the next item.

- An arithmetic progression is one where each number of the sequence is made up by having the same number added to it.

- An geometric progression is one where each number of the sequence is made up by being multiplied by the same number.

- Sequences don't have to be either arithmetic or geometric.

- If a sequence has a limited number of terms is called a finite sequence.

- If there is no limit to the number of terms, it is called an infinite sequence.

- Sequences that keep getting bigger are divergent.

- Sequences that reach a limit, even with an infinite number of terms, are convergent.

Logarithms

> **Read this if** you don't understand what a logarithm is, you thought that logarithms were originally based on exponents, you didn't know what any of that has to do with Pascal's triangle, or you aren't sure what "logarithm identities" are.

It wasn't easy to do calculations with numbers that had many digits before the invention of the calculator, and the late 1500s were definitely before the invention of the calculator. John Napier (1550-1617), the 8th Laird of Merchiston (now within Edinburgh), developed some ingenious ways to make it easier. He didn't invent decimal notation, but he was the one who popularised it.

Napier is more famous, though, for his invention of **logarithms**[22]. His idea was to find a way for people to add or subtract, rather than having to multiply or divide. His approach was, first, to make a sequence of multiples of some number, which he called the base. So in Figure 43 we are using a base of 2, and the list is multiples of 2, in a geometric progression.

Number to be multiplied or divided
2
4
8
16
32
64

Figure 43

Then he allocated an index to each number, which is created by adding the same number each time to the previous number; that is, an arithmetic progression (Figure 44).

Logarithms were used in a two-stage process. To divide, say, 2,147,483,648 by 134,217,728, the first step was to look up the index of each number, and subtract them: in this case (Figure 44) the index of 2,147,483,648 is 31, while the index of 134,217,728 is 27.

$$31 - 27 = 4$$

You then use the resulting index (4) to look up the answer: 16. So

$$\frac{2,147,483,648}{134,217,728} = 16$$

Or to multiply 16,384 by 65,536 you would look up the index of each number and add them.

$$14 + 16 = 30$$

The number with index 30 is 1,073,741,824, so

$$16,384 \cdot 65,536 = 1,073,741,824$$

Of course, the numbers in the left-hand column get further and further apart, leaving many numbers between them that the table can't be used for. To make this work practically, Napier's logarithm tables were based on multi-digit decimal fractions that were only separated by a few decimal points.

Napier and a friend developed tables, and (slightly different) logarithm tables were in used in schools until at least the late 1970s. He used .9999999 instead of 2 as the **base**, or the number you multiply each number in the first column by to get the next number. People performing large calculations could look up the logarithm to base .9999999 of the numbers they were trying to multiply, add them together, then use the tables in reverse to find the number they were seeking from the logarithm.

It's extraordinary to realise that the calculations needed to build the Eiffel Tower and Golden Gate Bridge, and to put men on the moon, were all done with tools such as these (including the slide rule, which was based on logarithms).

Number to be multiplied or divided	Index to be added or subtracted
2	1
4	2
8	3
16	4
32	5
64	6
128	7
256	8
512	9
1,024	10
2,048	11
4,096	12
8,192	13
16,384	14
32,768	15
65,536	16
131,072	17
262,144	18
524,288	19
1,048,576	20
2,097,152	21
4,194,304	22
8,388,608	23
16,777,216	24
33,554,432	25
67,108,864	26
134,217,728	27
268,435,456	28
536,870,912	29
1,073,741,824	30
2,147,483,648	31

Figure 44

To cement your understanding:

Use Figure 44 to multiply 32,768 by 16,384. Did you get 536,870,912? If not, can you see where you went wrong?

Here's the thing though. Now that we have calculators, all of that might have been of purely historical interest if it wasn't for something that was noticed *after* Napier had already prepared his logarithm tables. At the time, not many people paid much attention to the idea of raising numbers to different powers. That didn't happen until René Descartes (1596-1690) (who was 46 years younger than Napier, and who we will meet later) gave us our current notation for raising numbers to powers. Once that happened, people noticed that they could calculate the numbers they wanted to multiply or divide by raising the base to the power of the index. So in Figure 44 if you raise 2 (the base) to the power of the index (or logarithm) in the second column, it will give you the number in the first column. For example, 2 raised to the power of 10 (which you can look up in the index column) is 1,024 (which you can look up in the left-hand column.)

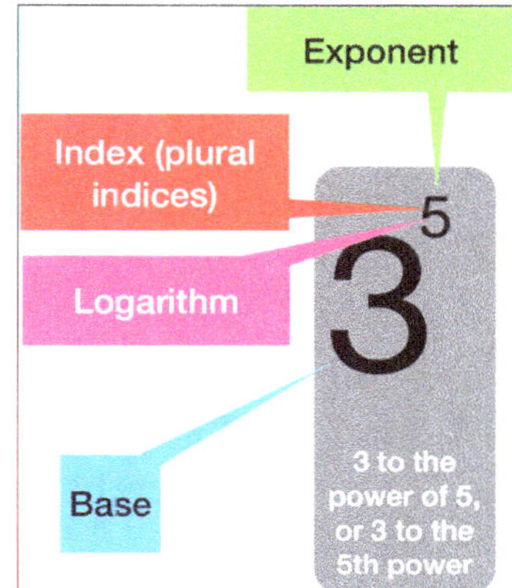

Figure 45

That is why we now use the words "index", "logarithm" and "exponent" interchangeably (Figure 45).

The way we think about it now, logarithms use the fact that you can multiply the same number raised to two different exponents by adding the exponents, e.g.

$$6^2 \bullet 6^5 = 6^{(2+5)}$$

The **logarithm** of a number is the power to which a base is raised, to equal that number. $6^2 = 36$, so we say that the logarithm of 36, base 6, is 2. Likewise $6^5 = 7,776$, so the logarithm of 7,776, base 6, is 5. We write this as

$$\log_6 7,776 = 5$$

or, more commonly,

$$\log_{10} 10,000 = 4$$

Happily, we don't need to use logarithm tables any more to undertake calculations, but logarithms are still important for some of the mathematics behind statistics. The important thing, at this stage, is to remember that "logarithm" is another word for "exponent".

To cement your understanding:

• What is the logarithm, base **10**, of **10,000**? (Can you see why it is 4?)

• Calculate **100** multiplied by **10**, using logarithms base **10**, without using a calculator. (Hint: **100** is **10** squared, **10** is **10** to the first, and you add the exponents. That means you add **2 + 1**, giving you 3. Raise **10** to that power.)

Logarithm identities

We've seen that in mathematics an **identity** is an equation that holds true for all values of the constants that are used. There are some useful ones involving logarithms, particularly when we will come to look at **exponential equations**. These are equations where the variable is part of the exponent or index. One common strategy with these is to take the logarithm of both sides, then work with that. (We will learn more about the importance of this later.)

When we looked at raising numbers to powers (pages 54 and following) we saw that:

1) When you multiply a base raised to a power by that base also raised to a power, you can just add the exponents.	$a^m a^n = a^{m+n}$
2) When you divide a base raised to a power by that base also raised to a power, you can just subtract the exponents.	$\dfrac{a^m}{a^n} = a^{m-n}$
3) If we raise something to a power, then raise the whole thing to a power, that is the same as multiplying the exponents.	$a^{m^n} = a^{mn}$

Keeping in mind that logarithms are the same thing as exponents, these translate directly to the following identities:

Name of identity	Formula
1) Logarithm addition rule	$\log_a(mn) = \log_a m + \log_a n$
2) Logarithm subtraction rule	$\log_a\left(\dfrac{m}{n}\right) = \log_a m - \log_a n$
3) Logarithm power rule	$\log_a(m^n) = n\log_a m$

The final logarithm identity we'll look at is the **change of base formula**. It doesn't flow so naturally from what we already know about working with exponents. The formula states that:

$$\log_a m = \frac{\log_x m}{\log_x a}$$

This is a particularly useful formula if, for example, your calculator gives you logarithms to base 10, and you are trying to solve an equation using logarithms to a different base. This is the proof.

Let... $y = \log_a m$

From the definition of logarithms, that means $m = a^y$

Take the logarithm of both sides, in base x. $\log_x m = \log_x a^y$

Use the logarithm power rule $= y \log_x a$

Divide both sides by $\log_x a$ $\dfrac{\log_x m}{\log_x a} = y$

We started by saying $y = \log_a m$, which means, swapping sides in the equation that

$$\log_a m = \frac{\log_x m}{\log_x a}$$

To cement your understanding:

- Find $\log_3 63 - \log_3 7$. Hint: use the logarithm subtraction rule to combine this into the logarithm of a single number. Can you see why it is $\log_3 9$? Now ask: what power do you raise 3 to, to get 9? Can you see why it is 2?

- Simplify $\log_2(8^3)$. Hint: use the power rule to write this as three times log base two of eight. Then think about what power you raise two to, to get eight. Multiply that by 3. Did you get the answer nine? If not, can you see where you went wrong?

- Evaluate $\log_7 100$, if you have a calculator that tells you that $\log_{10} 7 \approx 0.845$. (Hint: use the change of base formula, and convert to base 10.

Fun fact:

Another of Napier's inventions for simplifying calculations was a system of rods (or "bones", because that's what they were usually made of) with multiples of the numerals from 0 to 9, and the numbers divided by diagonal lines (Figure 46). At the simplest level if he wanted to multiply, say 542 by 7 he could arrange the 5, 4 and 2 rods, look down the index rod to 7 (on the left, and read off the answer (adding the numbers in the parallelograms): 3,794. In other words, the three rods with multiples of 5, 4 and 2 are placed together, then an index rod is placed to the left. Look down to number 7, and read the answer across: $3, (5 + 2), (8 + 1), 4 = 3{,}794$. The rods could

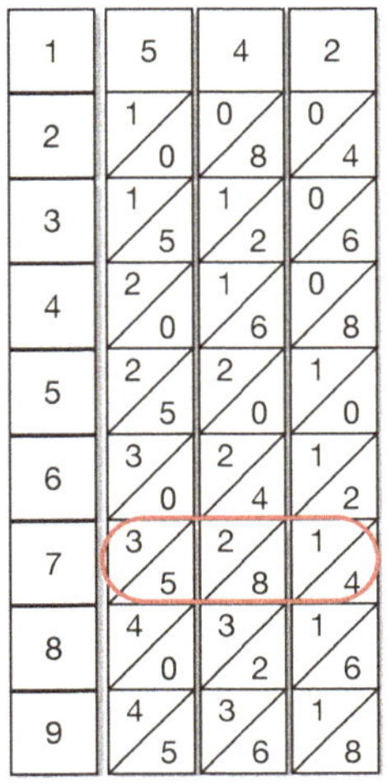

Figure 46

be used to multiply, divide and calculate the square roots of large numbers, and also convert fractions to decimal notation.

To cement your understanding (of the fun fact):

Use Figure 46 to multiply 542 by 9. (The 5, 4 and 2 "rods" are already where they need to be. Read down the index rod on the left for the number 9, then read off the result, adding the numbers in parallelograms (5 + 3 and 6 + 1). That will give you 4,878. Check that answer, either by multiplying it out using long division, or with a calculator.

Key points:

The logarithm of a number is the power to which a base is raised in order to equal that number. That means a logarithm is the same as an exponent, or an index.

$$\log_a(mn) = \log_a m + \log_a n$$

$$\log_a\left(\frac{m}{n}\right) = \log_a m - \log_a n$$

$$\log_a(m^n) = n\log_a m$$

$$\log_a m = \frac{\log_x m}{\log_x a}$$

Series

> **Read this if** you glance down the page and the symbols confuse you, you want to learn more fun facts about Pascal's triangle, you don't know what "linearity" means (and you will definitely need to know in order to understand statistics), you don't know an easy way to calculate the value of a geometric series, you don't know what the product operator is, or you are surprised to learn that starting from 1, any series of odd numbers will always equal a square number.

(If you are struggling with this, see the How to Love Statistics YouTube video on *Series*.)

A series is the sum of the numbers in a sequence; the result when you add them together. The notation can be a bit confusing and intimidating when you aren't used to looking at it, because you have to work out what to do starting at the bottom, then moving to the formula, then moving to the top. At first you might think that it seems so strange it won't be relevant, but in fact we'll use it over and over again.

Figure 47 shows that you start by assigning a numeric value to a letter (often n, as in this case, though it can be any variable) and then apply a formula to that variable.

Figure 48 introduces the capital sigma notation (Σ), another Greek letter, which signals the next step. At this stage, you take the number just calculated, add one to n (or whichever variable you are using), and repeat the process. This continues until n reaches the number specified at the top.

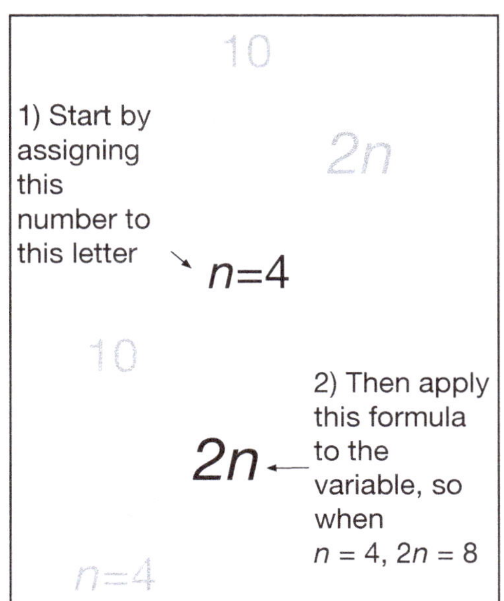

Figure 47

So, as per Figure 48, we have:

$$\sum_{n=4}^{10} 2n = 2 \bullet 4 + 2 \bullet 5 + 2 \bullet 6 + 2 \bullet 7 + 2 \bullet 8 + 2 \bullet 9 + 2 \bullet 10 = 98$$

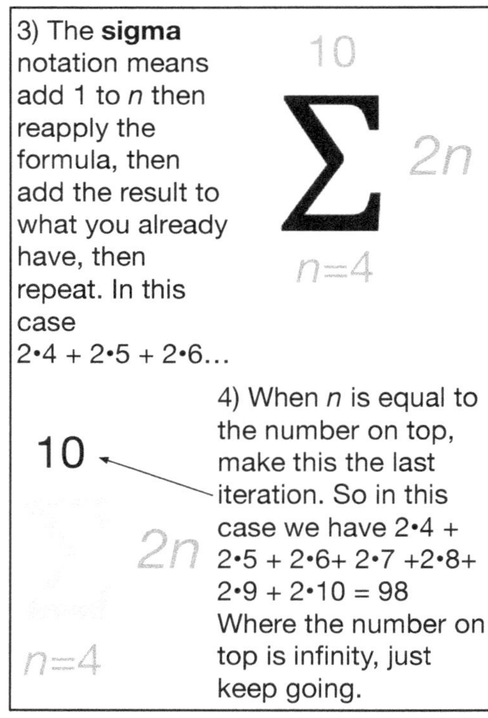

Figure 48

This is a **finite series** because there is a point where it stops. When a series has no end it is an **infinite series**, and is expressed as:

$$\sum_{n=1}^{\infty} \frac{3}{10^n}$$

It doesn't matter whether the small numbers ($n = 1$, and ∞) are above and below, or to the side of the sigma; that's just a matter of convenience.

Normally, if there is an infinite series there will be a fixed limit that the sum of terms will approach, although infinite terms are added. (Unexpected, but true.) In this instance, we have seen this expression before, but written differently.

We can expand it as:

$$\sum_{n=1}^{\infty} \frac{3}{10^n} = \frac{3}{10^1} + \frac{3}{10^2} + \frac{3}{10^3} + \frac{3}{10^4} + \frac{3}{10^5} \dots$$

In other words, it is the same as writing 0.33333..., which is the same as one-third. So we can say that:

$$\sum_{n=1}^{\infty} \frac{3}{10^n} = \frac{1}{3}$$

To cement your understanding:

• Calculate the value of the series $\sum_{n=0}^{5} 2^n$. Did you get 63? If not, can you see where you went wrong?

• Calculate the value of the series $\sum_{n=0}^{4} \frac{3}{10^n}$. Did you get 3.3333 ? If not, can you see where you went wrong?

Fun fact:

If you start at the top of any column of Pascal's triangle, make that the beginning of a series, and stop at any point you like, the value of the series will be given by the number in the following column and the next place down. So, in Figure 49, 1 + 6 + 21 = 28, and 1 + 10 + 55 + 220 = 286.

You can try this with any column of figures in the triangle, starting at the top and stopping anywhere you like; it will always work.

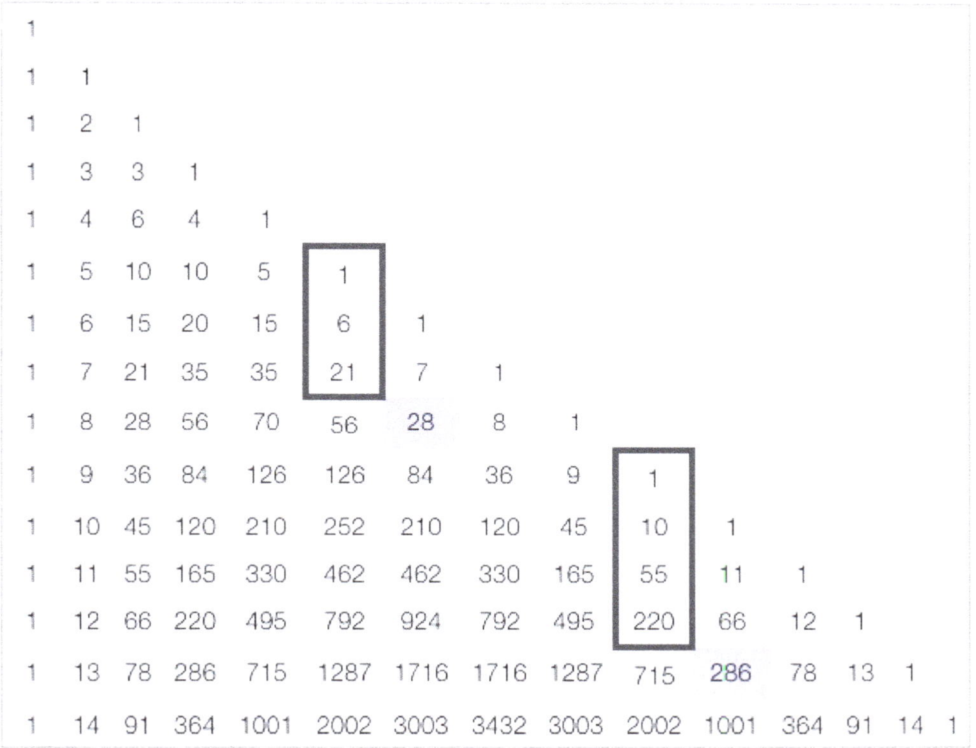

Figure 49

Linearity

Series have a property called **linearity**. That means a series made of different functions added together is the same as making a series of each of the functions and adding the results together. So, for example:

$$\sum_{x=1}^{4} (3^x + x^2) = \sum_{x=1}^{4} 3^x + \sum_{x=1}^{4} x^2$$

This follows naturally from the facts that a series is a group of terms added together, and addition is commutative. You can change the order of the terms, and can group them together in different ways, without changing the overall value of the expression.

In this example:

$$\sum_{x=1}^{4} (3^x + x^2) = (3^1 + 1^2) + (3^2 + 2^2) + (3^3 + 3^2) + (3^4 + 4^2)$$

$$= (3^1 + 3^2 + 3^3 + 3^4) + (1^2 + 2^2 + 3^2 + 4^2)$$

$$= \sum_{x=1}^{4} 3^x + \sum_{x=1}^{4} x^2$$

$$= 150$$

This can be expressed more generally, thinking of $3^x = f(x)$ and $x^2 = g(x)$:

$$\sum [f(x) + g(x)] = \sum f(x) + \sum g(x)$$

Linearity also implies that a constant can be factored out, and the sum can be multiplied by the constant instead of the individual terms; the result will be the same. This is based on the distributive property of multiplication, and is simply the same as saying, for example, from an example we saw earlier:

$$\sum_{n=4}^{10} 2n \;=\; 2 \bullet 4 + 2 \bullet 5 + 2 \bullet 6 + 2 \bullet 7 + 2 \bullet 8 + 2 \bullet 9 + 2 \bullet 10$$

$$= \; 2(4 + 5 + 6 + 7 + 8 + 9 + 10)$$

$$= \; 2 \sum_{n=4}^{10} n$$

$$= \; 98$$

If we make this more general, and change the 2 to a C representing any constant, we can express this as:

$$\sum C f(x) = C \sum f(x)$$

where C is a constant.

Series of natural numbers

As well as being fun, this is also a useful result that we'll need when we come to look at statistics. When German mathematician Karl Friedrich Gauss (whom we will learn more about as we go along) was 10 years old, a teacher asked the class to add the numbers from 1 to 100, as a way of keeping them busy. In a ridiculously short amount of time Gauss gave the answer: 5,050. This is how he worked it out.

Write out the series starting at 1:	1 +	2 +	3 +	... +	100
Then write it out backwards:	100 +	99 +	98 +	...+	1
Then add each number in the series on top to the number immediately below it in the second series:	101 +	101 +	101 +	... +	101

There are 100 terms of 101. When you multiply them together, you get 10,100. But we had to add the series to itself, so this is twice the total we need. If we halve the total, we get 5,050.

We can generalise this for any total. If we replace 100 with x, it could be any natural number. That means we write what we had above as:

Write out the series starting at 1:	1 +	2 +	3 +	... +	x
Then write it out backwards:	x	$(x - 1) +$	$(x - 2) +$	... +	1
Then add the two series together:	$(x + 1) +$	$(x + 1) +$	$(x + 1) +$	... +	$(x + 1)$

Each pair of terms will add up to $x + 1$, and there will be x pairs. The total of this will be twice the original series, so if we divide it by two, that gives the value of the series. That makes it:

$$\sum_{n=1}^{x} n = 1 + 2 + 3 + \cdots + x$$

$$= \frac{x(x + 1)}{2}$$

For example, the sum of the numbers from 1 to 100 is

$$\frac{100(100 + 1)}{2} = 5,050$$

> **To cement your understanding:**
>
> - Use this approach to add up the numbers from 1 to 300 (feel free to use a calculator). Did you get 45,150? If not, can you see where you went wrong?
>
> - Now use the approach to add up the numbers from 50 to 75. (Hint: add the numbers from 1 to 75, then add the numbers from 1 to 49, then subtract the second number from the first.) Did you get 1,625?

Alternatively, you could read the totals from the third column of Pascal's triangle, if column 2 went down to 100: pick any number in the second column, and the number to the right and one down from that will be the total of that column (Figure 50). (Refer back to page 133 if you can't remember what this is about.)

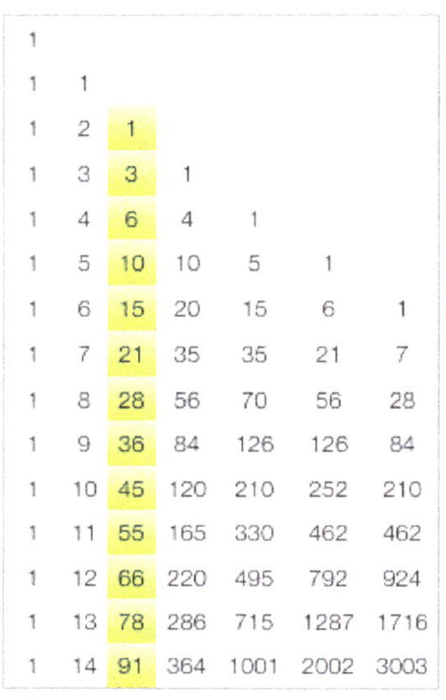

Figure 50

Fun fact:

A series of odd numbers, starting from 1, will always add up to a square number. For example:

$$1 + 3 = 4$$

$$= 2^2$$

$$1 + 3 + 5 = 9$$

$$= 3^2$$

$$1 + 3 + 5 + 7 = 16$$

$$= 4^2$$

You can keep this sequence going for as long as you like. You can see why this works when you realise that an odd number of identical squares can always be used to form a right angle, and both arms will be the same length (Figure 51). (If you do this with even numbers, the arms will be of different lengths.)

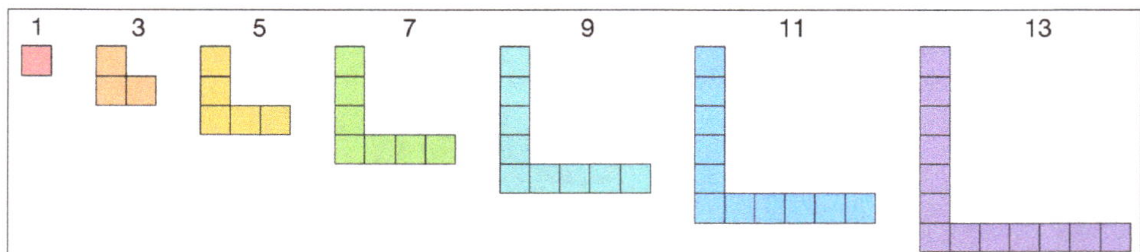

Figure 51

If you nest these right angles, you form squares (Figure 52).

In this way, every time you add an odd number, you increase the number being squared by 1.

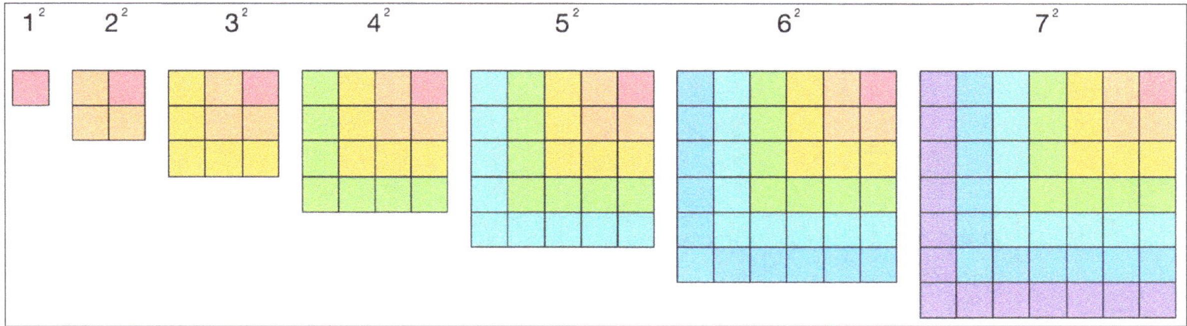

Figure 52

Key points:

- A series is the sum of the numbers in a sequence.

- Series are described using the sigma notation.

- A finite series has a limited number of terms, while an infinite series has no limit on the number of terms.

- Linearity implies that a series made of different functions added together is the same as making a series of each of the functions and adding the results together:

$$\sum [f(x) + g(x)] = \sum f(x) + \sum g(x)$$

- Linearity also implies that a constant can be factored out, and the sum can be multiplied by the constant instead of the individual terms; the result will be the same.

$$\sum Cf(x) = C \sum f(x)$$

- The sum of a series of natural numbers can be calculated by the formula

$$\sum_{n=1}^{x} n = \frac{x(x+1)}{2}$$

Geometric Series

Read this if you don't know an easy way to calculate the value of a geometric series (or if you aren't completely clear about what one is).

When we looked at sequences or progressions we saw that in an arithmetic progression we add the same amount each time, while in a geometric progression we multiply by the same amount each time. Then when we looked at logarithms we saw that they were originally developed by using a geometric progression for the numbers we are calculating, and an arithmetic progression for the index we are using. As you might expect, the same terminology applies to series: in an arithmetic series we add the same amount to each term, while with a geometric series we multiply each term by the same amount. We are still adding the terms, but each term we add is the same as the previous one, except that it is multiplied by a specified amount. (Sequences and series don't have to be either arithmetic or geometric. They can have all sorts of forms. These are just two forms that are quite common.)

Now let's explore that in a little more depth.

Thinking back further, when we looked at *multiplication* (page 28), we said that one way to think about it was as a count of the number of times we *added* something to itself, as in

$$4 + 4 + 4 = 3 \cdot 4$$

When we looked at raising numbers to a *power* (page 53), we said that raising to a power is the same as a count of the number of times we *multiplied* something by itself, as in

$$4 \cdot 4 \cdot 4 = 4^3$$

This next bit can seem a little confusing at first, but stay with it, you will get it. Earlier we looked at the series

$$2 \bullet 4 + 2 \bullet 5 + 2 \bullet 6 + 2 \bullet 7 + 2 \bullet 8 + 2 \bullet 9 + 2 \bullet 10$$
$$= 8 + 10 + 12 + 14 + 16 + 18 + 20$$

You can see that this is an arithmetic series: we add the same amount (in this case 2) to each term to make the next term, then we add all the terms together (because that's what a series involves). When we put this into series notation, however, we expressed it using multiplication:

$$\sum_{n=4}^{10} 2n$$

To cement your understanding:

You've already done this, but once again use the formula from sigma notation to write out this series, to confirm to yourself that the *multiplication* within the formula has the effect of *adding* the same thing to each term.

Can you see where this is going? When we want to describe a geometric series using the sigma notation, we usually have to express it using the notation for raising something to a power. So we can express a geometric series that starts with 1, and multiplies each term by 2, by using this notation:

$$1 + 2 + 4 + 8 + \cdots = 2^0 + 2^1 + 2^2 + 2^3 + \cdots$$

$$= \sum_{n=0}^{\infty} 2^n$$

Note that in this case we started by setting n as 0, so that the first term will equal 1.

To cement your understanding:

Write out the first four terms of $\sum_{n=0}^{\infty} 5^n$, to confirm to yourself that this notation has the same effect as multiplying each term by 5.

We've already seen a geometric series, first when we were looking at decimal fractions:

$$3.33333\ldots = \sum_{n=0}^{\infty} \frac{3}{10^n}$$

$$= \frac{3}{10^0} + \frac{3}{10^1} + \frac{3}{10^2} + \frac{3}{10^3} + \frac{3}{10^4} + \frac{3}{10^5} \cdots$$

$$= 3 + \frac{3}{10} + \frac{3}{100} + \frac{3}{1,000} + \frac{3}{10,000} + \frac{3}{100,000} \cdots$$

In this case we create each term after the first one by multiplying the previous one by one over ten.

We could just as easily have written it as:

$$3 + 3\left(\frac{1}{10}\right)^1 + 3\left(\frac{1}{10}\right)^2 + 3\left(\frac{1}{10}\right)^3 + 3\left(\frac{1}{10}\right)^4 + 3\left(\frac{1}{10}\right)^5 + \cdots$$

If we were to generalise it by replacing 3 with the letter a, and $\frac{1}{10}$ with the letter r, we would have a formula to describe any geometric series:

$$S_n = a + ar^1 + ar^2 + \cdots + ar^n$$

This is the way that mathematicians wrote about a geometric series, before the sigma notation (Σ) was developed.

As in the cases we have looked at up until now, series usually start with a constant, so by convention the initial number we start with is designated by a. (When we progress further we will see that series don't have to be based on constants; they can involve variables.)

We already know that

$$3.33333 \ldots = 3\frac{1}{3}$$

There is a way to prove this mathematically. That may not seem relevant at this stage, because we know the answer, but this way lets us easily calculate the value of any geometric series. This older notation is the easiest way to see why it works.

Think of it in this way:

Call this general expression for a geometric series equation1:

$$S_n = a + ar + ar^2 + \cdots + ar^n \text{ (equation 1)}$$

Multiply both sides of the equation by r, then distribute the r.

$$rS_n = r(a + ar + ar^2 + \cdots + ar^n)$$

$$rS_n = ra + rar + rar^2 + \cdots + rar^n$$

Rearrange each term on the right.

$$rS_n = ar + ar^2 + ar^3 + \cdots + ar^n + ar^{n+1} \text{ (equation 2)}$$

Put r on the right of a in each term and combine it with the powers of r that are already there. Call this equation 2.

Note that rar^n is the same as ar^{n+1}. (As a reminder, r^n is the same as saying "multiply r by itself n times". If you multiply that by another r, you are saying "multiply r by itself $n + 1$ times".

Also notice now that the right-hand side of equations 1 and 2 are almost the same. The only difference is that equation 1 has an a by itself at the beginning, and equation 2 has an ar^{n+1} on the end. So when we treat these as simultaneous equations (refer back to simultaneous equations on page 108 if you are unsure about this), and subtract equation 2 from equation 1, most of the terms cancel out.

Once again, this is equation 1

$$S_n = (a + ar + ar^2 + \cdots + ar^n)$$

This is equation 2

$$rS_n = (ar + ar^2 + \cdots + ar^n + ar^{n+1})$$

Subtract equation 2 from equation 1

$$S_n - rS_n = (a + ar + ar^2 + \cdots + ar^n) - (ar + ar^2 + \cdots + ar^n + ar^{n+1})$$

Distribute the negative sign in equation 2

$$= a + ar + ar^2 + \cdots + ar^n - ar - ar^2 - \cdots - ar^n - ar^{n+1}$$

Change the colour coding

$$= a + ar + ar^2 + \cdots + ar^n - ar - ar^2 - \cdots - ar^n - ar^{n+1}$$

Rearrange the terms and note that most of them cancel one another out.

$$= a + ar - ar + ar^2 - ar^2 \ldots ar^n - ar^n - ar^{n+1}$$

$$= a + \cancel{ar} - \cancel{ar} + \cancel{ar^2} - \cancel{ar^2} \ldots \cancel{ar^n} - \cancel{ar^n} - ar^{n+1}$$

This is what we are left with:

$$S_n - rS_n = a - ar^{n+1}$$

Factor out S_n on the left, and a on the right.

$$S_n(1 - r) = a(1 - r^{n+1})$$

Divide both sides by $(1 - r)$. This gives us a formula for the sum of *any* geometric series; but we haven't finished yet.

$$S_n = \frac{a(1 - r^{n+1})}{(1 - r)}$$

Now, where the absolute value of r is greater than 1, each term will get bigger and bigger. If it is an infinite geometric series (there is no limit on the number of terms) we say that the series will **diverge** to infinity: the total of the series will be infinitely large. Where the absolute value of r is less than 1, we say that an infinite geometric series will **converge**: the terms will reach a limit in their size, and the total of the series will be a finite number.

This is because where r is a fraction whose absolute value is less than 1 (e.g. $\frac{1}{2}$, $\frac{-3}{4}$), the larger n gets, the smaller r^n gets. For example:

$$\left(\frac{1}{2}\right)^{50} = \frac{1}{1,125,899,906,842,620}$$

which is an extremely small number. As n approaches infinity, r^n approaches 0. If r^n approaches infinity, then obviously r^{n+1} will, too. So, in the formula we just derived, where n approaches infinity, r^{n+1} will approach 0; it will disappear if r is between –1 and 1 or, in other words, its absolute value, symbolised by $|r|$, is between 0 and 1.

That makes the formula for the sum of an infinite geometric series::

$$S_\infty = \frac{a(1 - \cancel{r^{n+1}}0)}{(1 - r)} \; where \; 0 < |r| < 1$$

$$\boxed{S_\infty = \frac{a}{(1 - r)} \; where \; 0 < |r| < 1}$$

To use this formula to calculate the value of 3.333333... as a fraction, we can write this in one of the forms we used earlier:

$$3 + 3\left(\frac{1}{10}\right)^1 + 3\left(\frac{1}{10}\right)^2 + 3\left(\frac{1}{10}\right)^3 + 3\left(\frac{1}{10}\right)^4 + 3\left(\frac{1}{10}\right)^5 + \cdots$$

That means we can let $3 = a$, and let $\frac{1}{10} = r$

$$\frac{3}{\left(1 - \frac{1}{10}\right)} = \frac{3}{\left(\frac{9}{10}\right)}$$

Putting those values into the formula $\frac{a}{(1-r)}$ gives:

Multiply the numerator and denominator by 10

$$= \frac{30}{9}$$

Convert the improper fraction to an integer and a fraction

$$= 3\frac{1}{3}$$

So the formula works in this instance. Not that we needed it; we knew the answer. But the formula is useful to find the answer if it isn't immediately obvious.

> **To cement your understanding:**
>
> Have a guess at what this series is equal to. It involves adding an infinite number of terms, so will the total be infinity?
>
> $$\sum_{n=0}^{\infty} \frac{1}{2^n} = 1 + \frac{1}{2} + \frac{1}{4} + \frac{1}{8} + \cdots$$
>
> Note that this is a geometric series.
>
> Once you have had a guess, use the formula to see if you were right. Then follow the working out below, to see if you came to the same answer.

To find the total of the series:

$$1 + \frac{1}{2} + \frac{1}{4} + \frac{1}{8} + \cdots$$

This is our series. Using the old-style notation, we can make a (our starting number) = 1 and r (the number we are multiplying by) = $\frac{1}{2}$

$$S_{\infty} = 1 + \frac{1}{2} + \frac{1}{4} + \frac{1}{8} + \cdots$$

The absolute value of $\frac{1}{2}$ is between 0 and 1, so we can put those values into the formula $S_{\infty} = \frac{a}{(1-r)}$. That gives us:

$$= \frac{1}{\left(1 - \frac{1}{2}\right)}$$

$$= \frac{1}{\left(\frac{1}{2}\right)}$$

Multiply the numerator and denominator by 2; or just think of it as: one half goes into 1 two times.

$$= 2$$

You can see why when you realise that, with each term you add, you will be getting halfway between where you are now and the number 2 (that is, $1 + \frac{1}{2} + \frac{1}{4} + \cdots$). You will get closer and closer to 2, without ever getting past it.

Key points:

- In an arithmetic series, the same amount is added to each term. In the formula this will be shown as multiplication.

- In a geometric series, each term is multiplied by the same amount, although the terms themselves are still added. In the formula this will be shown as raising to a power.

- The formula for the sum of a geometric series is $S_n = \frac{a(1-r^{n+1})}{(1-r)}$

- The formula for the sum of an infinite geometric series where $0 < |r| < 1$ is

$$S_\infty = \frac{a}{(1-r)}$$

The product operator and factorials

Read this if you don't know what a "product operator" is, you don't know what an exclamation mark represents in mathematics, you don't know how to calculate how many ways there are to deal 5 cards from a deck with 52 cards, you don't understand why $\frac{n!}{(n-r)!} = n(n-1)(n-2)\ldots(n-r+1)$, you don't know how many terms there will be in $n(n-1)(n-2)\ldots(n-r+1)$, you don't know what $0!$ means, or you can't think of an obvious way to simplify the following expressions: $n(n-1)!$, $(n+1)n!$, or $(n-r-2)!(n-r-1)$. **Note:** all of this is highly relevant to understanding statistics, so if you aren't familiar with it and you want to really understand statistics, don't be tempted to skip it.

(If you are struggling with this, see the How to Love Statistics YouTube video on *Combinations, Permutations and n choose r.*)

The product operator

We've seen that a series involves adding a sequence of terms. Even in the case of a geometric series, where each of the terms is multiplied by the same number to get the next term, the terms themselves are still added. The **product operator**, on the other hand, represented by capital pi ($\prod$), (not to be confused with lower-case pi (π), which is the ratio of the circumference of a circle to its diameter) indicates that we should multiply a sequence of terms that follow a specific rule or formula, up to a certain value, rather than adding them.

For example:

$$\prod_{n=1}^{3} n^2 = 1^2 \bullet 2^2 \bullet 3^2$$

$$= 1 \bullet 4 \bullet 9$$

$$= 36$$

This is telling us to square each of the natural numbers from 1 through 3, and then to multiply the result. The idea of multiplying sequences of terms is important in probability, and therefore in statistics.

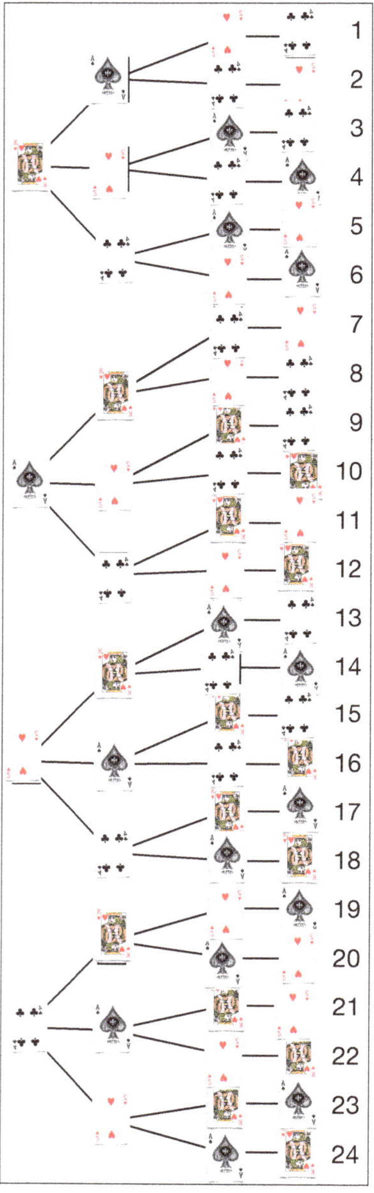

Figure 53

Probability began when people were trying to work out how to win at gambling, or "games of chance", and that's a good place for us to start. If it helps you can use a pack of playing cards, and deal them out as we go along, to get a sense of what is happening.

How many ways are there to arrange four playing cards? There will be 4 possible cards for the first position. For each of these, there will be 3 possible cards for the second position. For each of these, there will be 2 cards left that can go in the third position, and for each of these there will be one card left as the last card (Figure 53).

That means the number of possible ways to arrange the cards is:

$$(4)(3)(2)(1) = 24$$

Using the product operator, we can write this as

$$\prod_{n=1}^{4} n = (1)(2)(3)(4)$$

(Logically, we would write this as (4)(3)(2)(1), because we are starting with four cards. The only reason for reversing the order when we are using the product operator is convention, to be consistent with what we do using series notation we are starting with n is equal to 1 in the beginning, and adding 1 each time. The order in multiplication doesn't matter, of course.)

Now suppose that we only want to deal out 2 cards from a selection of 4, and we want to know how many ways there are to do that. We have 4 possible cards for the first card dealt, 3 possible cards for the second card and after that we'll stop dealing. Using the product operator, we can write this as

$$\prod_{n=3}^{4} n = (3)\,(4)$$

So there are 12 possible ways to deal 2 cards from a set of 4.

To cement your understanding:

Using this same logic, how many ways are there to deal out 3 cards from a pack of 10? Did you get 720? If not, can you see where you went wrong?

That is a neat-enough way to express it, but it isn't the form mathematicians usually use. As in any other language, within mathematics there is often more than one way to express an idea. The two ways to express this idea that follow are a little more challenging to get your head around, but they are used more frequently because, when they are included as part of longer formulae within mathematical proofs, they are easier to work with using algebra.

The two different ways to express this same concept are an older-style one which we are about to look at, and a (slightly) more recent shorthand called a factorial, which we'll look at soon.

Alternative 1 for expressing the same idea

If you wanted to write out the number of ways you could sort an entire deck of 52 cards, you could write it as:

$$(52)(51)(50) \dots (2)(1)$$

The three dots in mathematics means the same thing that it does in English: "and so on" or "keep following this pattern". In this case, keep following this pattern until you get down to 2 then 1. It's a really big number, by the way; roughly equal to:

80,658,175,170,943,900,000,000,000,000,000,000,000,000,000,000,

000,000,000,000,000

The same formula will tell us how many selections we can make from any set of distinct objects, regardless of the number of objects we have. We know that the last two numbers we will multiply together will always be 2 and 1 (although of course multiplying by 1 makes no difference), regardless of how many objects there are (provided there are at least 2), but the first number we use to start multiplying will be the number of objects in our set. We can write $(52)(51)(50) \dots (2)(1)$

as:

$$52 \bullet 51 \bullet 50 \bullet \ldots \bullet 2 \bullet 1 = 52 \bullet (52 - 1) \bullet (52 - 2) \bullet \ldots \bullet 2 \bullet 1$$

We are reducing by 1 each time, because we have already dealt that card – so it isn't available to choose from.

When we write it like that, we see that we can replace 52 with any natural number we like. (Reminder: a natural number is a number we use to count with, and is indicated with the symbol $\mathbb{N}$ - see page 72). Let's say that the number of objects in a set that we want to organise is n, and n is an element in the set of natural numbers ($n \in \mathbb{N}$). Our formula then becomes:

$$n(n - 1)(n - 2) \ldots (2)(1)$$

To cement your understanding

Replace n in the formula $n(n - 1)(n - 2) \ldots (2)(1)$ with 6, to confirm that it works in the same way as the product operator.

Now let's think about what happens if we aren't selecting the whole deck, but are only selecting a certain number from the deck. This time let's imagine we have a deck of 24 cards, and we are selecting 5 cards from it.

To make our first choice, we are selecting from 24, so we have 24 options to choose from. Having done that, we now have $24 - 1 = 23$ cards from which to make choice number 2, because we have already removed one, so we have 23 options. For the third choice we have already removed 2 cards, so we have $24 - 2 = 22$ cards to choose from. For the fourth choice we have already removed 3, so we have $24 - 3 = 21$, and for the fifth choice we have $24 - 4 = 20$ cards to pick from; we have 20 options available to us. Remember, what we are doing is multiplying the options available.

You might remember when we first looked at sequences (page 116) we used a subscript notation to indicate which term of the expression we are up to. While this terminology isn't usually used or needed in the context we are in now, we will use it to make what is happening clearer.

$$(24)_1 \bullet (24 - 1)_2 \bullet (24 - 2)_3 \bullet (24 - 3)_4 \bullet (24 - 4)_5$$

In other words, we are always reducing our starting number by one less than whichever choice we are up to. On choice number 2 we remove 1 card. On choice number 3 we remove another card, meaning we have now reduced our starting number by 2. So, if we were to make it into a formula—instead of saying we are selecting 5 cards we say we are selecting r of them (where r is a counting/natural number, i.e. $r \in \mathbb{N}$)—then the last choice

(choice number r) will be the one where we have already removed one less than r (that is, $r - 1$) cards.

If we say we are starting with n cards rather than 24 or 52 or any other number, then the last choice will be from $n - (r - 1)$ cards.

If we make $n = 24$ and $r = 5$, then that means that when we are about to make the last choice (the 5th one), it will be a choice from

$$24 - (5 - 1) = 24 - 4$$
$$= 20$$

That's how many options we have (so that works). Of course, as we saw at the start of this book, $24 - (5 - 1)$ is the same as $24 - 5 + 1$ which, of course, still adds up to 20.

Another way to think of it is: the $+1$ is there because you haven't made the last choice yet, and you won't have to. What we are looking at is the options available.

So putting all of that together, our generic formula for selecting r items, in a specific order, from n possibilities, will be:

$$n(n - 1)(n - 2) \dots (n - r + 1)$$

To cement your understanding:

Take a pack of playing cards, deal 13 cards, and count the number you are selecting from each time, to confirm for yourself that this works.

Alternative 2: factorials

Now let's look at the (slightly) more recent shorthand for the same thing, which French mathematician Christian Kramp developed in 1808. It's called factorial, and the notation is an exclamation mark, as in

$$4! = 4 \bullet 3 \bullet 2 \bullet 1$$

This is the most-commonly used way to express the concept we have been looking at. In terms of the order of operations, factorial comes before anything else. In other words, $5 \bullet 3!$ doesn't equal 15 factorial; it equals 5 multiplied by 3 factorial (that is, $5 \bullet (3 \bullet 2 \bullet 1) = 30$).

> **To cement your understanding:**
>
> • Write out 11! (11 factorial). You can calculate the value if you like, but don't worry if you don't want to. The main thing is to make sure you understand what 11! means.
>
> • Divide 5! by 3! Did you get 20 and, if not, can you see where you went wrong?

Let's think about what happens if you are playing a game with 20 cards in a deck, and you just want to select 9 cards. To find a generic formula using the factorial (exclamation mark) notation the formula needs to only take the first nine numbers in the factorial, and get rid of the rest. (This would be easy to express using the product operator notation, but it is slightly more difficult with the factorial notation.)

First of all, we can think of 20! as being the same as:

$$20 \cdot 19 \cdot 18 \cdot 17 \cdot 16 \cdot 15 \cdot 14 \cdot 13 \cdot 12 \cdot 11 \cdot 10 \cdot 9 \cdot 8 \cdot 7 \cdot 6 \cdot 5 \cdot 4 \cdot 3 \cdot 2 \cdot 1$$

However, everything in blue is the same as 11!, so we can write the same thing as:

$$20 \cdot 19 \cdot 18 \cdot 17 \cdot 16 \cdot 15 \cdot 14 \cdot 13 \cdot 12 \cdot 11!$$

We can replace everything we don't need, from 11 down, with 11!

We are only interested in multiplying the 9 numbers between 20 and 12, and we want to get rid of the 11! after that, because at that point we have stopped dealing out cards. But doing it in this way, we only want to use factorial notation. We can do this by showing it as 20 factorial divided by 11 factorial.

$$\frac{20!}{11!} = \frac{20 \cdot 19 \cdot 18 \cdot 17 \cdot 16 \cdot 15 \cdot 14 \cdot 13 \cdot 12 \cdot 11!}{11!}$$

The 11! in the numerator and denominator cancel one another out.

$$\frac{20 \cdot 19 \cdot 18 \cdot 17 \cdot 16 \cdot 15 \cdot 14 \cdot 13 \cdot 12 \cdot \cancel{11!}}{\cancel{11!}}$$

That leave us with:

$$20 \cdot 19 \cdot 18 \cdot 17 \cdot 16 \cdot 15 \cdot 14 \cdot 13 \cdot 12$$

which is what we are after.

We can also write this as

$$\frac{20!}{(20-9)!}$$

In other words, we are dividing our starting number (20) by the *difference* between our starting number and the number we want to select ($20 - 9 = 11$), to eliminate those numbers we don't require. Making this a generic formula, if n is the number of cards in the deck, and r is the number of cards we are selecting, the formula will be:

$$\frac{n!}{(n-r)!}$$

So now we have three equivalent ways of expressing the same idea. We could use the product operator (but there isn't a widely-used generic expression for any n and r using the product operator), or we could use the second way:

$$n(n-1)(n-2)\ldots(n-r+1)$$

or the third, most common way:

$$\frac{n!}{(n-r)!}$$

All of these are equivalent.

$$n(n-1)(n-2)\ldots(n-r+1) = \frac{n!}{(n-r)!}$$

To cement your understanding:

We've seen three different ways to calculate a factorial that stops before you get to 1: the product operator, $\frac{n!}{(n-r)!}$ and $n(n-1)(n-2)\ldots(n-r+1)$. Use these last two approaches to work out how many ways there are to select 5 cards from a group of 7. (Hint: Use $n = 7$ and $r = 5$, put these into each expression, and calculate the result. In both cases, you should get to 2,520.) If you are struggling, the full working is on page 153.

While the idea of a factorial began with playing cards, it has much wider use within statistics and within mathematics more generally. There is a useful convention that, at first, appears illogical: factorial zero is equal to one.

$$0! = 1$$

We will see later that there is a definition of factorial that makes this logical, but there is quite a bit of mathematics to get through first. For now, it is best to just accept it as a convention, just as all languages have conventions.

When doing algebra with factorials, it can be important to remember what the notation means.

There are several formulae that come from this notation that will be useful later.

First formula:

$$n \cdot (n-1)! = n!$$

For example, if $n = 8$

$$8 \cdot (8-1)! = 8 \cdot 7!$$
$$= 8 \cdot (7 \cdot 6 \cdot 5 \cdot 4 \cdot 3 \cdot 2 \cdot 1)$$
$$= 8!$$
$$= 40{,}320$$

Second formula:

$$(n+1)! = (n+1) \cdot n!$$

For example, if $n = 8$

$$(8+1)! = 9!$$
$$= 9 \cdot (8 \cdot 7 \cdot 6 \cdot 5 \cdot 4 \cdot 3 \cdot 2 \cdot 1)$$
$$= 9 \cdot 8!$$
$$= (8+1) \cdot 8!$$
$$= 362{,}880$$

Third formula:

$$(n-r-1) \cdot (n-r-2)! = (n-r-1)!$$

For example, if $n = 8$ and $r = 3$

$$(8-3-1) \cdot (8-3-2)! = (8-3-1)!$$
$$4 \cdot 3! = 4!$$
$$= 24$$

> To cement your understanding:
>
> • Use $n = 7$ (or any other natural number you like) to prove to yourself that $n(n-1)!=n!$
>
> • Use $n = 7$ (or any other natural number you like) to prove to yourself that $(n + 1)! = (n + 1)n!$
>
> • Use $n = 7$ and $r = 4$ (or any other natural numbers you like where $n > r$) to prove to yourself that $(n - r - 2)!\,(n - r - 1) = (n - r - 1)!$
>
> • Use $n = 10$ and $r = 5$ (or any other natural numbers you like where $n > r$) t to prove to yourself that $n(n - 1)(n - 2)\dots(n - r + 1)$ will have r things multiplied together

Fun fact:

The Earth is thought to be about 4.5 billion years old. If that is correct, it means it has been around for about 2,890,800,000,000,000,000 minutes. So if you had started shuffling a deck of cards at the beginning of the Earth, and came up with a new shuffle every minute, then by now you would hardly have made a dent on the 80,658,175,170,943,900,000,000,000, 000,000,000,000,000,000,000,000,000,000,000,000 possibilities. Even if you had asked all 8 billion or so people on the planet to help you, that would still hardly make a blip. Note that mathematically, you wouldn't be dividing by the number of minutes or the 8 billion people; you would be subtracting 8 billion times the number of minutes: 23,126,400,000,000, 000,000,000,000,000,000. That's another really big number, but if you subtract it from the number of ways a deck of cards can be shuffled, you would barely notice a change; it would be less than a rounding error. (If you doubt that, do the subtraction for yourself, either manually or with a spreadsheet.) That means that each time you shuffle a deck of cards there is virtually no chance that anyone has ever put a deck of cards into that specific order before, in the entire history of the world.

Yet deliberately sorting a deck of cards into order by suit then by card—in other words, finding one of these options—will probably take you less than 10 minutes.

Answer to the question on page 151:

First approach:

$$\frac{n!}{(n-r)!} = \frac{7!}{(7-5)!}$$

$$= \frac{7!}{2!}$$

$$= \frac{(7)(6)(5)(4)(3)\cancel{(2)}\cancel{(1)}}{\cancel{(2)}\cancel{(1)}}$$

$$= (7)(6)(5)(4)(3)$$

$$= 2{,}520$$

Second approach:

$$n(n-1)(n-2)\dots(n-r+1) = 7(7-1)(7-2)(7-3)(7-5+1)$$

$$= (7)(6)(5)(4)(3)$$

$$= 2{,}520$$

Key points:

- The product operator represented by capital pi ($\prod$) indicates that we should multiply a sequence of terms that follow a specific rule or formula, up to a certain value.

- To multiply a descending sequence of r natural numbers stopping before 1, you can use the product operator.

- The factorial of a natural number, denoted as $n!$, means the product of all natural numbers from 1 to n.

- Both of the following expressions tell you how many ways you can select r objects from a set of n of them.

$$n(n-1)(n-2)\dots(n-r+1)$$

    or

$$\frac{n!}{(n-r)!}$$

- $n(n-1)! = n!$

- $(n+1)! = (n+1)n!$

- $(n-r-2)!\,(n-r-1) = (n-r-1)!$

Combinations, permutations, and n choose r

> **Read this if** you don't know the difference between a permutation and a combination, or you don't know what $\binom{n}{r}$ means.

(If you are struggling with this, see the How to Love Statistics YouTube video on *Combinations, Permutations and n choose r*.)

In the game of poker, the highest hand is a royal flush: this consists of the ace, king, queen, jack and ten of the same suit. We are *almost* at the point of being able to predict the chances of dealing a royal flush.

When we were looking at factorials we saw that the number of ways to choose r items from a set of n (e.g. 5 specific cards from a deck of 52) is:

$$\frac{n!}{(n-r)!}$$

That means the chance of dealing the ace, king, queen, jack and ten in spades, *in that order*, is one chance in:

$$\frac{52!}{(52-5)!} = \frac{52!}{47!}$$

$$= 52 \bullet 51 \bullet 50 \bullet 49 \bullet 48$$

$$= 311{,}875{,}200$$

In most games involving playing cards, however, including poker, no one cares in what order the cards are dealt. That means J, K, 10, A, Q of spades has the same value as A, K, Q, J, 10 of spades. It will still be a royal flush.

At this point, we need some words to describe two related but different concepts. When we are talking about a hand of cards, where it is the actual cards that matter, not the order in which they were dealt, we call it a

combination of cards (or of anything else, for that matter). When we are using set notation (page 74), the order of the items in a set does not matter; all that matters is which items are included. That makes a set a combination. The items are combined, and we do not care how.

In situations where we are interested in the order in which items appear, we call it a **permutation**. In genetics, for example, the order of molecules in a short strand of DNA is important. If the order is changed, even though the same molecules are present, it can create a mutation. (The words "permutation" and "mutation" come from the same Latin root, meaning "exchange".) In words, if we say "the dog chased the cat", that has a different meaning from "the cat chased the dog". This is the same combination of words but a different permutation.

The formula

$$\frac{n!}{(n-r)!}$$

will give us the number of permutations when selecting r objects from a set of n. For a given collection of items, there will be more permutations than combinations. The number of permutations will be the same as the number of combinations multiplied by the number of ways in which the items in each combination can be arranged. That means we can find the number of possible combinations if we know the number of permutations.

When we are thinking about a hand of five cards, it can be arranged in 5! ways: there are five options for the first card, for each of these there are four options for the next card, for each of these there are three options for the next card, etc. For each combination of 5 cards, there will be 5! (that is 120) permutations.

If we are using a generic formula and we replace the number 5 with the letter r, the number of ways r objects can be arranged is $r!$

That means:

$$r! \bullet (Number\ of\ combinations) = Number\ of\ permutations$$
$$= \frac{n!}{(n-r)!}$$

If we divide both sides of this equation by $r!$ to get the number of combinations on its own, that gives us:

$$Number\ of\ combinations = \frac{n!}{r!\,(n-r)!}$$

Therefore the number of possible hands (combinations) of 5 cards from a deck of 52, where order doesn't matter is:

$$\frac{52!}{5!\,(52-5)!} = \frac{52 \bullet 51 \bullet 50 \bullet 49 \bullet 48 \bullet \cancel{(47!)}}{5! \bullet \cancel{47!}}$$

$$= \frac{311{,}875{,}200}{120}$$

$$= 2{,}598{,}960$$

You have one chance in 2,598,960 of dealing a royal flush in spades, in the game of poker.

The expression

$$\frac{n!}{r!\,(n-r)!}$$

which is the number of combinations possible when choosing r items from a set of n, comes up a lot in mathematics, and particularly in statistics. While the factorial symbol is shorthand for writing out a lot more mathematics, the expression for combinations has its own shortcut:

$$\binom{n}{r}$$

Note that this isn't a fraction: it's just a form of shorthand.

It can also sometimes be written as

$$^nC_r$$

It is pronounced as "n binomial r" or "n choose r". (We'll see shortly why it can be called "binomial".)

$$\boxed{\frac{n!}{r!\,(n-r)!} = \binom{n}{r}}$$

This is a general formula for the number of combinations of r items that can be selected from a set of n.

To cement your understanding:

• How many *permutations* of 4 playing cards are there out of a deck with 7 cards in it? Did you get that there are 840 and, if not, can you see where you went wrong?

• How many *combinations* of 4 playing cards are there out of a deck with 7 cards in it? (If you didn't get 35, can you see where you went wrong?)

• The n choose r idea is useful for all sorts of games and puzzles, not just those based on playing cards. For example, the New York Times has an on-line game called "Connections", in which you are given 16 words or phrases, and have to combine them into four groups of four, based on what they have in common. Using this approach, how many options do you have for picking your first four words? (The order you pick them in doesn't matter.) (Hint: it's much easier to work this out if you cancel out everything you can first, rather than starting with 16!, which is a big number that your calculator won't like very much.) Did you get that there are 1,820 options for the first four words? If not, can you see where you went wrong?

• Now imagine that you have guessed your first two sets of words correctly but cannot think of any meaningful connections between any four of the final eight words. You are tempted to just guess randomly, knowing that the game will give you four chances to make a mistake. Using this approach, if you were to just guess randomly, how many options are there for selecting the next four words? Did you get that there are 70 options? If not, can you see where you went wrong?

• Once you have correctly selected the next four words, you have four words left, and need to select four of them. If the order doesn't matter, logic would tell you that there is one way to do this. Check that it works out that way, using the n choose r approach. Keep in mind that $0! = 1$

Key points:

- With a combination, the order of items doesn't matter. With a permutation it does. That means, for a given collection of items, there will be more permutations than combinations.

- The formula

$$\frac{n!}{(n-r)!}$$

  gives the number of **permutations** when selecting r objects from a set of n.

- The formula

$$\frac{n!}{r!\,(n-r)!}$$

  gives the number of **combinations** when selecting r objects from a set of n.

- This can be written as

$$\binom{n}{r}$$

  and sometimes as

$$^nC_r$$

  and is pronounced "n choose r" or "n binomial r".

Binomial expansion

> **Read this if** you aren't extremely familiar with the binomial theorem, you don't know why the coefficients can be calculated with n choose r, or why both can be read from Pascal's triangle. The binomial theorem is central to statistics.

(If you are struggling with this, see the How to Love Statistics YouTube video on *The Binomial Theorem*.)

The binomial theorem is one of the fundamental concepts in statistics. Almost every statistical idea we encounter will link back to it in some way. The basic idea is that we take an expression with two terms (a binomial), raise it to a power, and then use series notation to describe the expression that results. We've already seen what happens when you raise a binomial expressions to the power of 2 (page 93): $(x + y)^2 = x^2 + 2xy + y^2$. Now we are looking at what happens we you raise it to the power of any natural number.

Just why this is such a fundamental idea will have to wait until we've covered more concepts but broadly, it will help us calculate the chances of something happening versus the chances of it not happening if we repeat an activity. (Don't worry if that doesn't make sense yet; it will when we get there.)

When we expand the binomial (using the distributive property of multiplication), the first six expansions are:

$(x + y)^0 = 1$

$(x + y)^1 = x + y$

$(x + y)^2 = x^2 + 2xy + y^2$

$(x + y)^3 = x^3 + 3x^2y + 3xy^2 + y^3$

$(x + y)^4 = x^4 + 4x^3y + 6x^2y^2 + 4xy^3 + y^4$

$$(x + y)^5 = x^5 + 5x^4y + 10x^3y^2 + 10x^2y^3 + 5xy^4 + y^5$$

$$(x + y)^6 = x^6 + 6x^5y + 15x^4y^2 + 20x^3y^3 + 15x^2y^4 + 6xy^5 + y^6$$

To cement your understanding:

Can you see any patterns in this yet? (Don't worry if you can't.)

Expand $(x + y)^3$ to check that you get the same result as the one above.

Now let's explore in more detail what happens when we raise a binomial to the fourth power.

$$(x + y)^4 = (x + y)(x + y)(x + y)(x + y)$$

$$(x + y)^4 = (x + y)(x + y)(x + y)(x + y)$$

That means the possible *combinations* are $xxxx, xxxy, xxyy, xyyy, yyyy$.

Remember that a combination means we aren't worried about the order. There are a lot more *permutations*; we could keep going and reach these combinations in different ways. In order to develop our formula, however, we only need the combinations; then we'll have to work out how many ways there are to get each combination.

We can write these combinations as x^4, x^3y, x^2y^2, xy^3, y^4. Over many years different mathematicians in different parts of the world worked on this. It was probably Blaise Pascal who came up with the following way of formalising it in the mid 1600s (although he didn't use modern notation, which hadn't been invented then; he mostly had to explain it verbally):

$$\sum_{r=0}^{4} x^{4-r}y^r = x^{4-0}y^0 + x^{4-1}y^1 + x^{4-2}y^2 + x^{4-3}y^3 + x^{4-4}y^4$$

$$= x^4 + x^3y + x^2y^2 + xy^3 + y^4$$

To cement your understanding:

Replace the 4 on the left hand side with 2 then expand out the series, to ensure that you can see why it comes to $x^2 + xy + y^2$. (Remember that anything raised to the power of 0 equals 1, and that anything raised to the power of 1 equals itself.)

That gets us part of the way there, but we still need a way to calculate how many times each term will appear (that is, the coefficients of the terms; how many versions of each combination we need to add together).

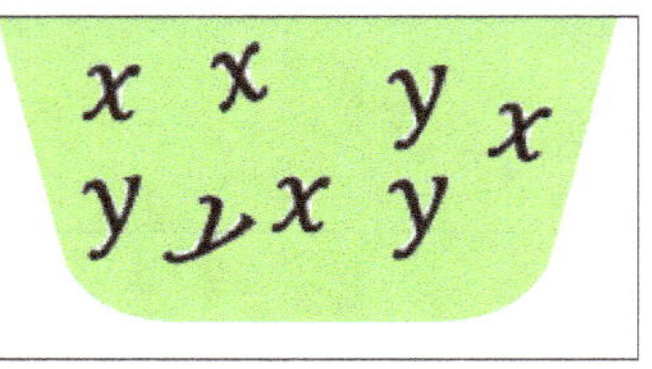

Figure 54

It may not be obvious at first, but this is actually an n choose r problem. It might be easier to imagine why this is if you just imagine having a basket full of x's and y's. You will be dipping your hand in four times, and selecting one letter each time. You can nominate the number of y's you will end up with, and you win a prize if you get exactly that number. So the question becomes: how many ways can you achieve a given number of y's? If you hadn't read the material on combinations and permutations, you could work it out for each option. There will be only one way to get no y's; every time you dip in, you get an x. There are four ways to get one y: you could pick one out on the first, second, third or fourth choice, but everything else would have to be an x. There are six ways to get two y's. You could select y (1) on the first and second, (2) first and third, (3) first and fourth, (4) second and third, (5) second and fourth or (6) third and fourth selections. You could keep going, or you could just use n choose r: how many ways are there to select two y's out of four choices?

$$\frac{4!}{2!\,(4-2)!} = \frac{4!}{2!\,2!}$$

$$= \frac{4 \cdot 3 \cdot \cancel{2} \cdot \cancel{1}}{2 \cdot 1 \cdot \cancel{2} \cdot \cancel{1}}$$

$$= \frac{12}{2}$$

$$= 6$$

To calculate the chances of getting any other number of y's, we can just replace r with the number we are looking for.

$$\frac{4!}{r!\,(4-r)!} = \binom{4}{r}$$

That way r indicates which term of the expression we are up to, starting with 0. It tells us how many y's there will be in the term.

In other words:

$$\sum_{r=0}^{4} \frac{4!}{r!\,(4-r)!} x^{4-r}y^r = \frac{4!}{0!\,(4-0)!} x^4 + \frac{4!}{1!\,(4-1)!} x^3 y + \frac{4!}{2!\,(4-2)!} x^2 y^2$$

$$+ \frac{4!}{3!\,(4-3)!} xy^3 + \frac{4!}{4!\,(4-4)!} y^4$$

$$= \quad \frac{4!}{1(4!)}x^4 + \frac{4!}{1(3)!}x^3y + \frac{4!}{2!\,2!}x^2y^2 + \frac{4!}{3!\,1!}xy^3 + \frac{4!}{4!\,1}y^4$$

$$= \quad \frac{4!}{1(4!)}x^4 + \frac{4\bullet\cancel{(3!)}}{1\bullet\cancel{(3!)}}x^3y + \frac{4\bullet3\bullet\cancel{(2!)}}{2!\,2!}x^2y^2 + \frac{4\bullet\cancel{(3!)}}{3!\,1!}xy^3$$
$$+ \frac{4!}{4!\,1}y^4$$

$$= \quad x^4 + 4x^3y + 6x^2y^2 + 4xy^3 + y^4$$

To cement your understanding:

Replace the 4 on the left hand side with 3 then expand out the series, to ensure that you can see why it comes to $x^3 + 3x^2y + 3xy^2 + y^3$.

Notice that, as you would expect, this is symmetrical. There is the same coefficient for a given power of y as there is for that same power of x. It doesn't matter whether we are calculating the chances of getting one y or one x; the chances will be the same.

This formula will work for any expansion to a counting number, so we can replace the exponent by n. (Isaac Newton worked out how to use this with non-integers, but we don't need to go there.)

That give us a formula for a **binomial expansion**, or a series that describes what happens to each term when a binomial is raised to a power, then expanded out.

$$(x + y)^n = \sum_{r=0}^{n} \frac{n!}{r!\,(n-r)!}x^{n-r}y^r$$

$$= \sum_{r=0}^{n} \binom{n}{r}x^{n-r}y^r$$

This is called the **binomial theorem**, and we will see a lot of it in Part 2.

Believe it or not, the coefficients of binomial expansions form Pascal's triangle.

You can see why that is by looking carefully at Figure 55. It shows the way that successive binomial expansions can be built up, by multiplying each term by x and each term by y.

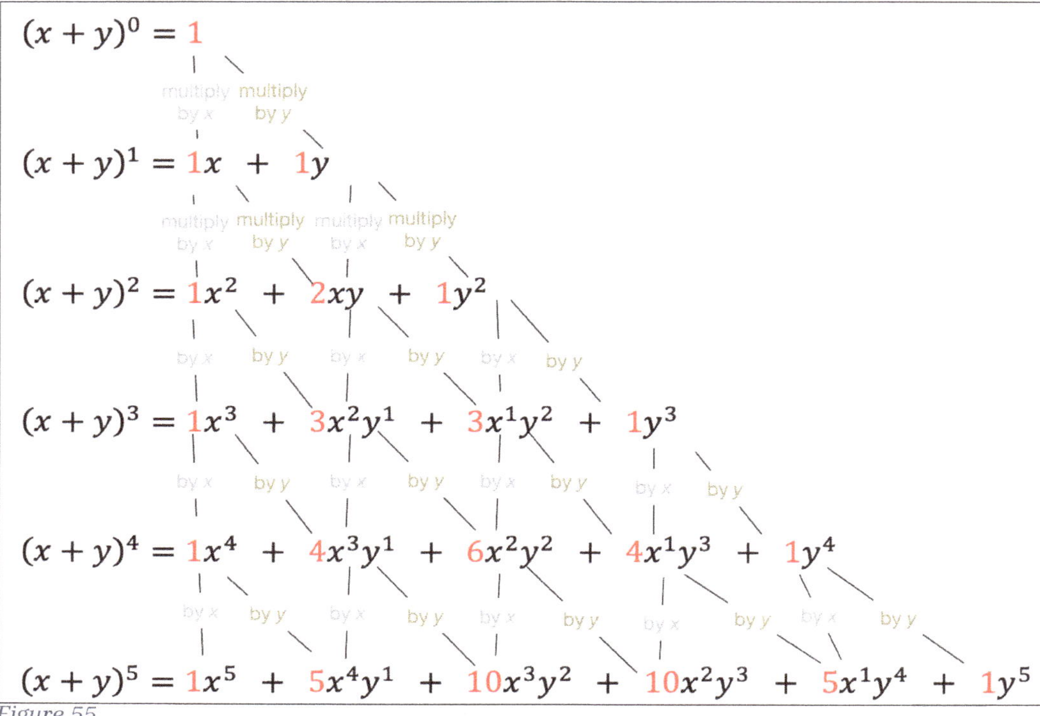

Figure 55

That in turn means that you can use n choose r to build up Pascal's triangle (not that you would want to; it makes the calculations much more difficult). It is fairly awesome that it works that way, though. (If you want a more rigorous proof of why the numbers of Pascal's triangle can be calculated by

$$\binom{n}{r}$$

there is one in an appendix on page 736. It's a bit confusing and doesn't take us forward particularly (which is why it is relegated to an appendix); it's just there for those who would really like to know.)

There is a related but different way to think of it. Imagine you have a device like Figure 56. (Something very similar was designed by Francis Galton, who we will meet again.) You put a ball bearing in at the top, and each time it hits the top of a grey triangle it could go to the left or the right. The arrows show different paths that a ball bearing can take. If it goes to the left, it accumulates an x, and if it goes to the right it accumulates a y.

At the top triangle there is only one way to get an x, and one way to get a get a y. At the second row, it can have 2 x's if it rolls down the left-hand side, an x and a y if it ends up in the middle position, and 2 y's if it rolls down the right-hand side. There is only one path that will get to either one x or one y,

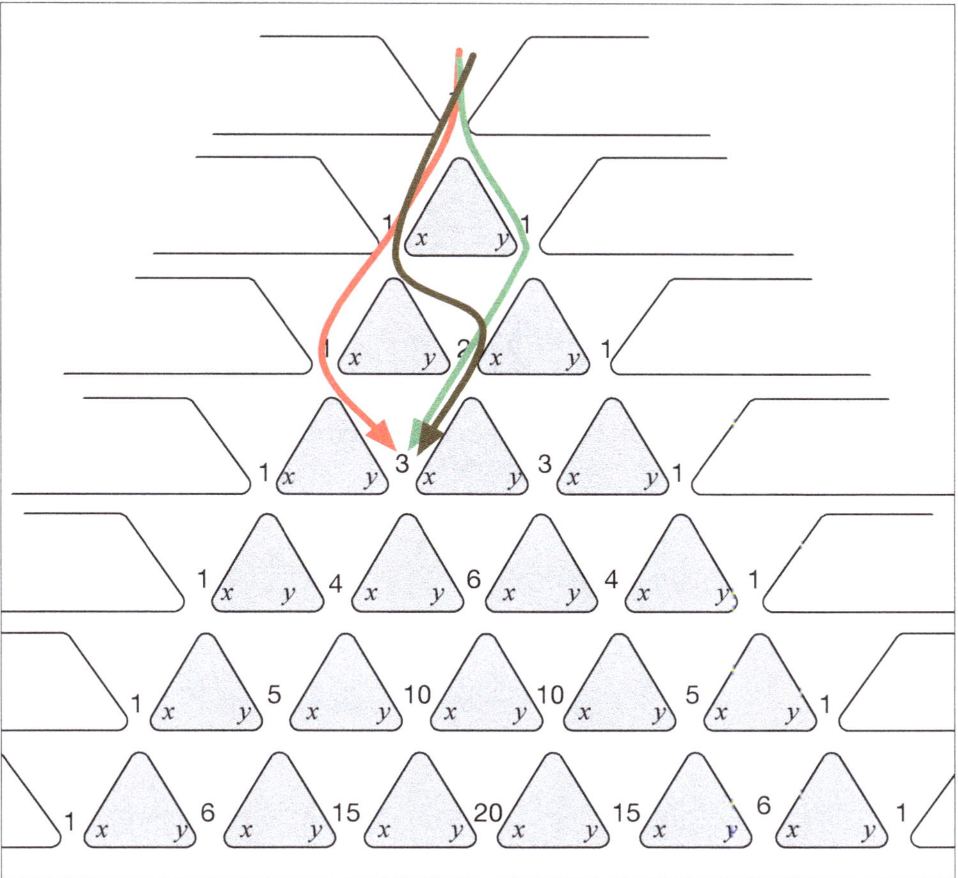

Figure 56

but two paths will lead to x and a y. The numbers shown on the figure indicate the number of paths that will get you to that point; and if you follow them through you will be able to see why the numbers are those of Pascal's triangle. In other words, the numbers are telling you how many possible ways there are to achieve that combination.

At the position where the arrows meet, it will have accumulated 2 x's and 1 y, but there are three ways it can have done so. That means that three different permutations will lead to the same combination.

Key points:

- The binomial theorem is:

$$(x + y)^n = \sum_{r=0}^{n} \frac{n!}{r!\,(n-r)!} x^{n-r} y^r$$

$$= \sum_{r=0}^{n} \binom{n}{r} x^{n-r} y^r$$

- The coefficients of binomial expansions form Pascal's triangle

Geometry: mathematical language in pictures

A picture may tell a thousand words, but it can never replace them. In mathematics, too, numbers, symbols, and pictures each have their own role to play. Getting geometry—mathematics in pictures—to work smoothly with the rest of mathematics, however, proved more difficult through the ages than you might expect. Arguably, that difficulty kept the world in what we would now call a backward state until relatively recently, especially when compared with how long humans have been around.

The geometry of the ancient world was already highly sophisticated. It made possible the Great Wall of China, the Taj Mahal, the pyramids, the Parthenon, and the Roman aqueducts. What lagged behind were numbers and symbols, which lacked the sophistication to push human achievement beyond those wonders.

It took three great inventions to propel mathematics—and with it, the world—forward: India gave us our modern number system, Persia gave us algebra, and France gave us the Cartesian plane (which we will cover in this section). When those three innovations joined forces with geometry, human knowledge exploded, and we are still riding the wave of that ever-growing surge.

Before we get there, though, let's revisit some basic geometry, so that what follows will fall into place.

Fun fact:

Pictures are processed in a different part of our brains to numbers. When you have spent a couple of hours working hard to understand new mathematical concepts, you might find that an area slightly in front of and above your ear feels warmer[23]. When you study geometry you might find that as well as this area, and an area slightly behind and above your ear warms up.

This warming is because your system shunts blood to different parts of the brain where it is needed. In order for you to be able to feel increased warmth through your hands, the heat has to travel through: a layer of thin membrane, the clear fluid that bathes your brain fluid, two more layers of much thicker membrane, bone (which has its own layers of membrane on each side), connective tissue, muscle, layers of skin and hair. In other words, the change of blood flow within your brain is dramatic, so stretching your brain in a new way, even if it feels uncomfortable (which hopefully it won't) will still be good for your cognitive capacity. It will also, of course, take you further along the road towards being able to understand statistics in depth.

Basic Geometric Concepts

> **Read this if** you struggled with geometry at school, you never studied it at all, or if you glance through the headings and diagrams and see something you aren't familiar with.

Einstein said: "*never memorise something you can look up.*" There are a lot of words listed here, but no need to memorise them. If any of this is new to you, however it, is helpful to understand the concepts.

Points, lines, planes and spaces

Dimensions are the minimum number of directions you can move in to always be able to get from any point to any other within a given shape. If you are on a straight path and can go forward and backwards, that is one dimension.

In geometry a **point** refers to a location, but has no size. It is drawn with a dot, but that is just a representation, because a dot has size; a point doesn't. In language we use a word to describe some aspect of something. If you talk about going "home" you don't need to describe what it is made of; the word itself probably conveys all that the person you are speaking to needs to know. If they need more information, you can elaborate further. Similarly, when we speak about a point all we are describing is a specific location.

A point will often be indicated by a dot with a letter beside it.

A point has **no dimension**: no movement is possible, while staying within that point. As soon as you move anywhere you are at another point.

An **interval**, or **segment**, is the shortest distance between two points. It is drawn on a page with a line but it actually has no width. A **straight line** is an interval that continues in an infinite direction past the two points, without curving. A **ray** is an interval that extends past one of the points in an infinite direction, but does not extend past the other point (Figure 57).

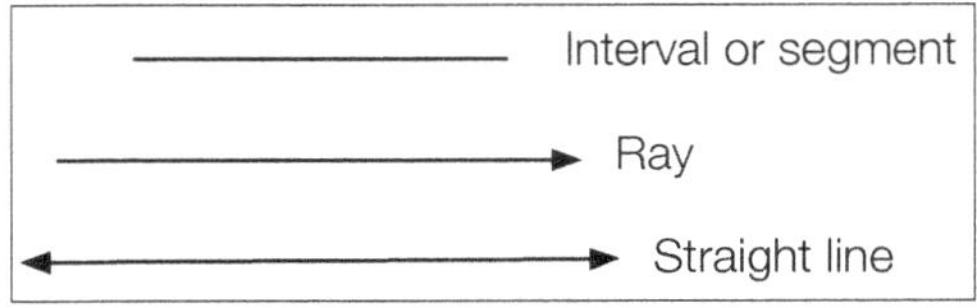

Figure 57

Intervals, rays and lines have **one dimension**: you can move in either direction, and could eventually come to every point. TO move elsewhere you would need to change direction. This in turn means that any two points will fix a line. There is one, and only one, line that will pass through any two points.

Lines can often be indicated with two letters (representing two points) with a bar on top of them:

$$\overline{AB}$$

If you are on a sporting field, and can go forward or backward, or left or right, that's two dimensions. You can get to any point on the field by going forward or backward, left or right. That won't be the quickest way to get there, but it can be done.

To cement your understanding:

Select any two spots on this page (or screen, if that is how you are viewing it). Move your finger from one to the other in two distinct movements: first to the left or right until you are directly above or below the other one, then up or down until you arrive at it. Note the way you could get from any point on the page to any other just with these two movements.

A **flat plane** is a surface like the page or screen you are reading this on, if it had no curvature, no thickness, and extended an infinite distance in every direction.

If you are in a swimming pool you can move in three dimensions. Any space-occupying objects, like a cube, a sphere, or anything else that has existence in the real world, has **three dimensions** which, for convenience, we could call length, width and height.

In this book we are mainly concerned with thinking in two dimensions. In higher level statistics, where we start combining multiple variables we need to think in terms of three or more dimensions. We won't be looking at that much in this book, though. (Once we speak of "more" than three dimensions things stop making sense geometrically, but can still make sense with algebra. It means adding more variables.)

Angles

No one knows who first came up with the idea of dividing a circle into 360 degrees; it might have been the Babylonians. (It was probably chosen because it has so many factors.) The ancient Greeks were aware of the idea of 360° (and were also possibly the inventors of it) but didn't do much with it, because their numbering system was too cumbersome.

An **angle** is a measure of rotation or the amount of turn between two straight lines, intervals or rays that intersect. We call the two lines the **arms** of the angle. A **right angle** is one where the angle forms one quarter of a full rotation. Angles are usually measured in degrees, and marked by this ° symbol. By definition a right angle is one quarter of a full revolution, is 90°, and is usually drawn with a small square where the lines meet (Figure 58).

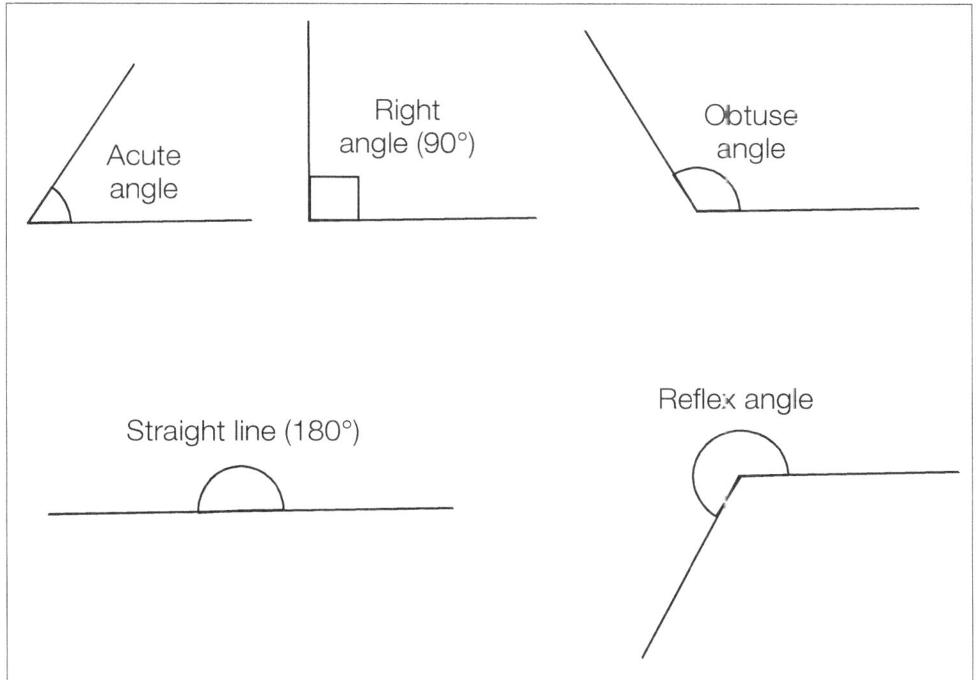

Figure 58

An **acute angle** is less than 90°, while an **obtuse angle** is between 90° and 180°. At 180° the two angles form a straight line. A **reflex angle** is one that is greater than 180° and less than 360°.

There are some mathematical applications (which we will get to) where it is helpful to think of angles of greater than 360°. You go around the circle, and just keep going. So an angle of 405° is the same as an angle of 45°. It can also sometimes be helpful to think of a negative angle. An angle of –90°

is the same as an angle of 270°. (Figure 59). The usefulness of these ideas will make more sense when we get to trigonometry.

Figure 59

While we don't normally worry about this, the "inside" of any angle will also be matched with a reflex angle on the "outside". The inside and the outside of an angle will add up to 360°.

Angles may be indicated with an angle symbol and three letters, for three points: the middle letter will be the point where the lines form an angle. This is called the **vertex** (Figure 60):

$$\angle ABC$$

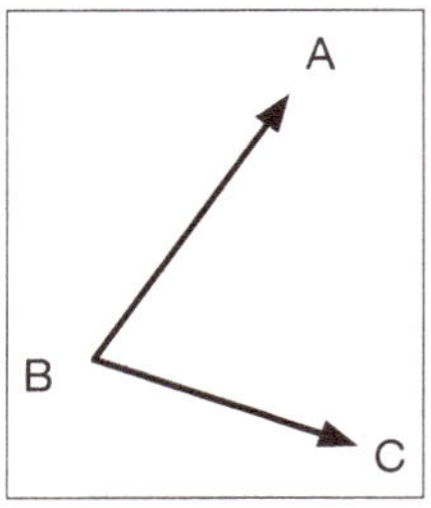

Figure 60

Parallel Lines

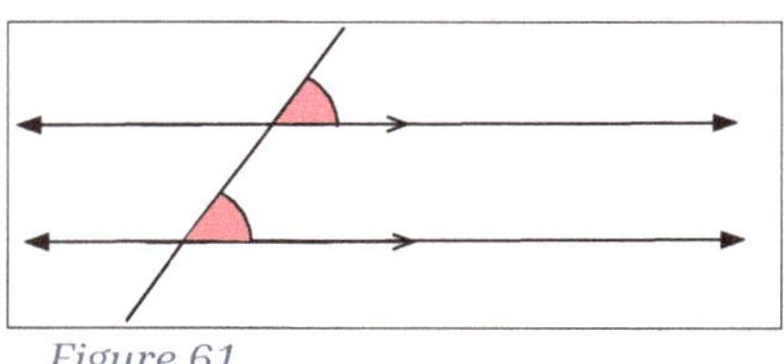

Two lines are **parallel** if they lie in the same plane but never intersect, no matter how far they extend in either direction. (This is sometimes indicated by placing arrows within them, as in Figure 61.) If you make a copy of a line and move it to a different position in the same plane

Figure 61

without twisting it in any way, you will have two parallel lines. The two lines will have the same orientation.

That means that if a third line passes through them—this is called a **transversal**—it will form the same angle with each of them.

Straight-sided, two-dimensional figures

When two straight-sided figures have the same angles and the same lengths of their sides, even if they may be in a different position with a different orientation, we say they are **congruent**, and use the symbol $\equiv$ to represent them, or sometimes the symbol $\cong$. If you put one on top of the other you couldn't tell the difference, as in Figure 62.

Figure 62

When two straight-sided figures have the same angles, but sides of different lengths, we say they are **similar**. Even though the lengths of the sides will be different, the ratios between the sides of similar shapes will be constant. In other words, if you take one of two congruent shapes, and expand it or contract it by a fixed amount, it will still be similar to the other shape: the angles won't

Figure 63

change (Figure 63). The symbol used for this is ~ . The best way to think of this for our purposes is probably **has the same shape as**, because we'll revisit this symbol in the Part 2. In the statistical sense it means "is distributed according to the pattern..." (Don't worry, that will make sense when we get there.)

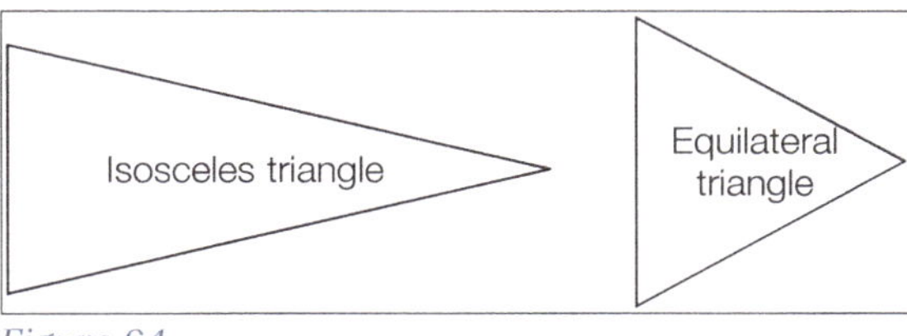

Figure 64

A shape formed by three intervals that meet at their endpoints is a **triangle**. A triangle with two sides the same length is an **isosceles** triangle, while a triangle with three sides of the same length is an **equilateral** triangle (Figure 64).

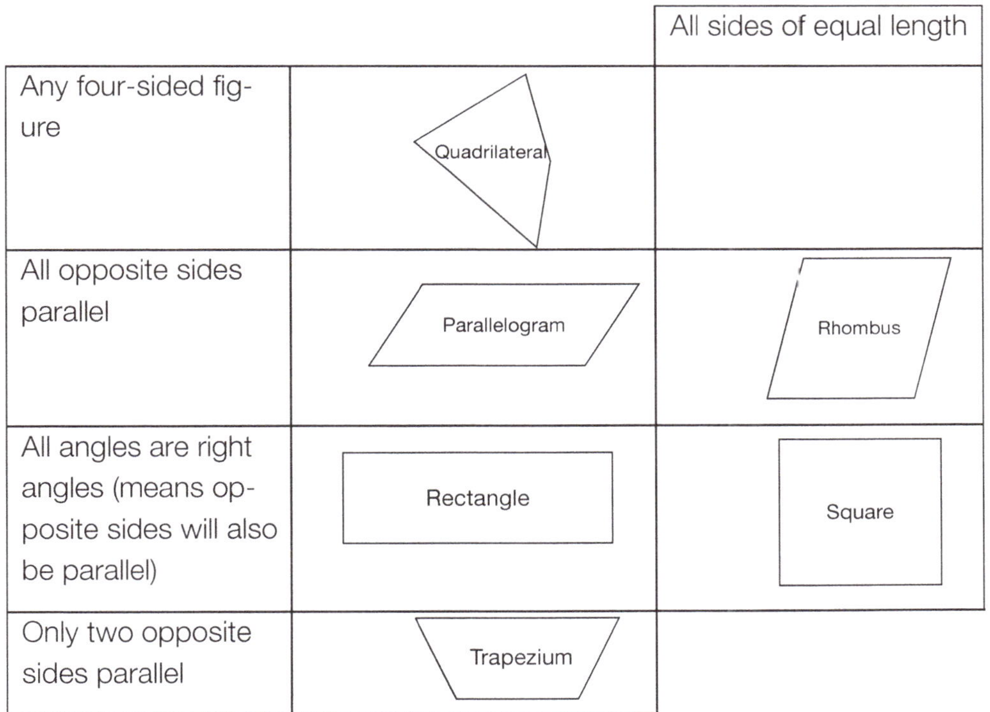

		All sides of equal length
Any four-sided figure	Quadrilateral	
All opposite sides parallel	Parallelogram	Rhombus
All angles are right angles (means opposite sides will also be parallel)	Rectangle	Square
Only two opposite sides parallel	Trapezium	

A two-dimensional figure formed by four intervals that meet at their end-points is a **quadrilateral**. A quadrilateral where the intervals meet at right angles is a **rectangle**. The opposite sides of a rectangle are parallel, and also have the same length. A rectangle where all sides are the same length is a **square**. A **parallelogram** is a quadrilateral where all opposite sides are parallel and equal in length, but the angles where they meet are not necessarily right angles. A **rhombus** is a parallelogram where all sides are equal in length. A **trapezium** is a quadrilateral with only one pair of opposite sides that are parallel.

Circles

(If you are struggling with this, see the How to Love Statistics YouTube video on *Pi*.)

A **circle** is a two-dimensional shape where all the points are the same distance from the centre. This distance is the **radius**. A **chord** is an interval that goes from one side of a circle to the other. If a **chord** passes through the centre of a circle it is the **diameter**. The diameter will be twice the length of the radius.

The distance around the perimeter of the circle is called the **circumference**. The **ratio** of the circumference to the diameter of a circle is always the same ($\pi = 3.1415\dots$), regardless of the size of the circle.

$$diameter \bullet \pi = circumference$$

Dividing both sides by the diameter

$$\frac{\cancel{diameter}}{\cancel{diameter}} \bullet \pi = \frac{circumference}{diameter}$$

$$\pi = \frac{circumference}{diameter}$$

A line that only touches a circle (or any other curve) at one point is a **tangent** to the circle (Figure 65). The idea of a tangent becomes particularly important when we come to look at calculus.

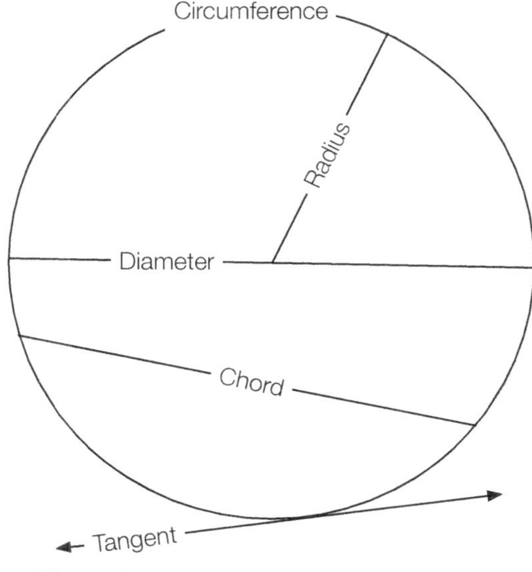

Figure 65

Key points:

- This material looked at dimensions, points, intervals and straight lines and planes.

- It defines **angles**, with **arms** and **vertices** (the plural of vertex).

- It looks at right angles, acute angles, obtuse angles, reflex angles, negative angles and angles of greater than 360°.

- It defines parallel lines and transversals.

- A **transversal** will form the same angles with both parallel lines.

- When two straight-sided figures have the same angles and the same lengths of their sides they are **congruent**. The symbols for this are ≅ or ≡.

- When two straight-sided figures have the same angles, but sides of different lengths, we say they are **similar**. The symbol used for this is ~ . The best way to think of this for our purposes is probably **has the same shape as**, because we'll revisit this symbol in Part 2

- A triangle with two sides the same length is an **isosceles** triangle, while a triangle with three sides of the same length is an **equilateral** triangle.

- Wed defied: quadrilateral, rectangle, square, rhombus and trapezium.

- **A circle** is a two-dimensional shape where all of the points are the same distance from the centre. This distance is called the **radius**.

- A **chord** is an interval that goes from one side of a circle to the other. If a **chord** passes through the centre of a circle it is called the **diameter**. The diameter will be twice the length of the radius.

- The distance around the perimeter of the circle is called the **circumference**. The ratio of the circumference to the diameter of a circle is always the same (π), regardless of the size of the circle.

- A line that only touches a circle (or any other curve) at one point is called a **tangent** to the circle.

Areas of straight-sided figures

Read this if you don't know how to calculate the area of any straight-sided figure.

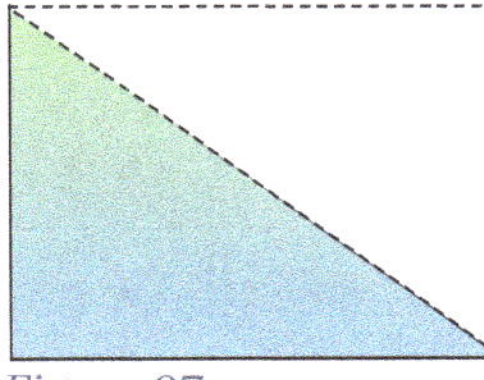

Figure 66

Thinking back to multiplication, we can see that the area of a rectangle will be the height multiplied by the width, in square units. If Figure 66 was measured in metres, it would be 4 metres high by 3 metres wide, making its area 12 square metres. If we changed the unit to miles, furlongs, cubits or microns, the result would be the same: it would be 12 squares, each based on whatever units we were using. That's really all you need to know, to calculate the area of any straight-sided figure. Let's look at why.

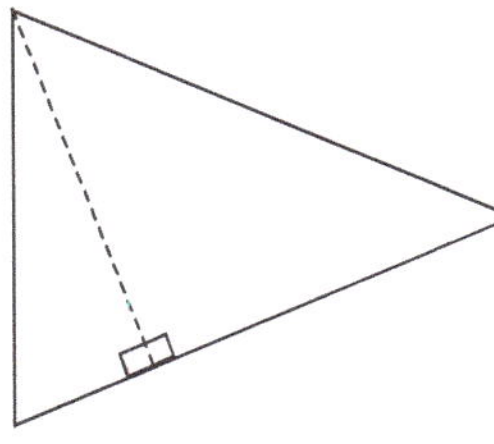

Figure 67

A right-angled triangle has half the area of a rectangle (Figure 67). That means we can find the area by multiplying the length of two sides of a rectangle, then halving the result. (The longest side, which is opposite the right angle, is the **hypotenuse**. We don't use it to calculate the area.)

Any triangle can be divided to make two right-angled triangles (Figure 68). Any other straight-sided figure can be divided into triangles by drawing lines from one corner to another (Figure 69). Sometimes you can also use rectangles within other straight-sided figures, to make the calculation of area even more straightforward.

Figure 68

To find the area of a parallelogram, think of it as a rectangle with a triangle cut off and moved to the other end. When you do, you can see that its area is given by the length multiplied by the height (Figure 70).

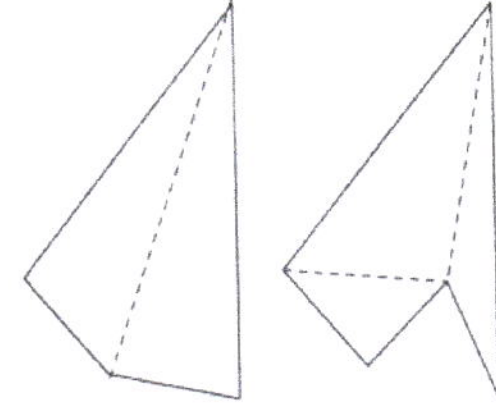

Figure 69

Figure 70

To cement your understanding:

• Draw a right-angled triangle, measure the two shortest edges (i.e. the two that aren't the hypotenuse), then calculate the area by multiplying the lengths together and halving the result.

• Draw any triangle, and draw another line from one of the angles directly to the opposite side, to make two right-angled triangles.

• Draw a six-sided figure, divide it into triangles by drawing lines between relevant corners (without crossing any lines), then divide the triangles so that each forms a right-angled triangle.

• Confirm that you can tell why the area of a parallelogram is what it is.

Key points:

- The **area of a rectangle** will be the height multiplied by the width, in square units.

- A **right-angled triangle** has half the area of a rectangle. That means we can find the area by multiplying the lengths of two sides of a rectangle, then halving the result.

- The longest side of a right-angled triangle (which will be opposite the right angle) is called the **hypotenuse**.

- Any triangle can be divided to make two right-angled triangles.

- Any other straight-sided figure can be divided into triangles by drawing lines from one corner to another.

Pythagoras' theorem

> **Read this if** you would like a proof of Pythagoras' theorem. (The idea of summing squares comes up a lot in statistics.)

(If you are struggling with this, see the How to Love Statistics YouTube video on *Pythagoras Theorem*.)

Pythagoras' theorem was known long before Pythagoras (who seems to have been more of an ancient-Greek cult leader than anything else, although not much is known about him). You may remember this from school: the square of the hypotenuse (the longest side of a right-angled triangle) is equal to the sum of the squares of the other two sides. (The word comes from the idea of stretched, as in tension, and under, as in hypoglycaemic.

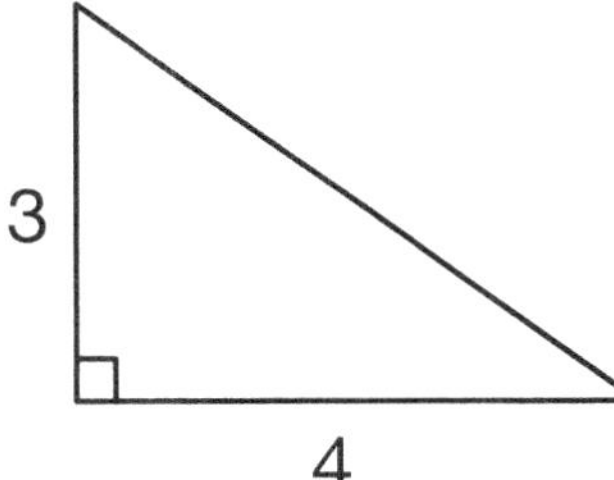

Figure 71

The idea is that the side is stretched under the right angle.) That means we can find the length of the hypotenuse (in Figure 71), by calculating:

$$3^2 + 4^2 = 9 + 16$$

$$= 25$$

So 25 is the square of the longest side; that means the length will be the square root of 25, which is 5.

In trying to generalise this result the ancients didn't have algebra, so they thought about it in terms of drawing actual squares on the sides, as in Figure 72. If the ancients *had* had algebra, they could have expressed it as: if a is the length of one side, b is the length of the second and h is the length of the hypotenuse, then:

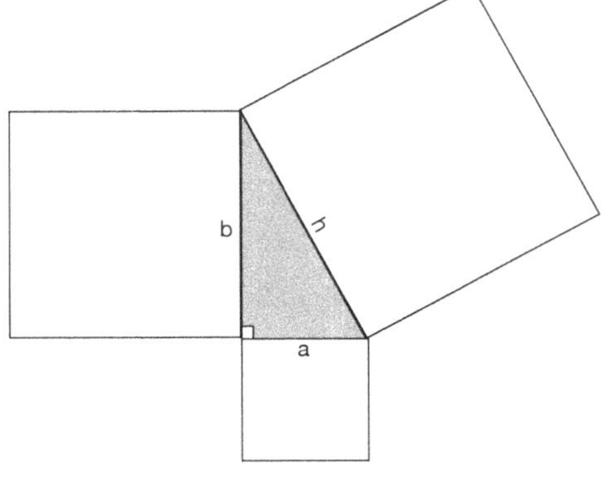

Figure 72

$$h^2 = a^2 + b^2$$

We can therefore take the square root of each side, and we are left with:

$$h = \sqrt{a^2 + b^2}$$

This is definitely NOT the same as the square root of a^2 plus the square root of b^2:

$$\sqrt{a^2} + \sqrt{b^2}$$

For example, using the lengths of Figure 71,

$$\sqrt{a^2 + b^2} = \sqrt{3^2 + 4^2}$$

$$= 5$$

This is the correct answer.

If we'd squared each side, then taken the square root again (in other words, just reversing what we had done) and added the results together we would have 7, which is definitely the wrong answer.

$$\sqrt{a^2} + \sqrt{b^2} = \sqrt{3^2} + \sqrt{4^2}$$

$$= 3 + 4$$

$$= 7$$

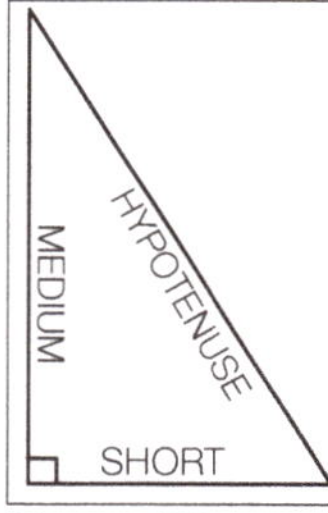

Figure 73

To prove this theorem, from Figure 72, what the Greeks were saying was that the area of the square with an h in it is the same as the area of the square with a in it, plus the area of the square with b in it.

This idea of the square root of sums of squares comes up a lot in statistics (it forms part of the definition of **standard deviation**), and historically it first appeared here, and so it is worth understanding how it can be proven. Here's one way to prove it.

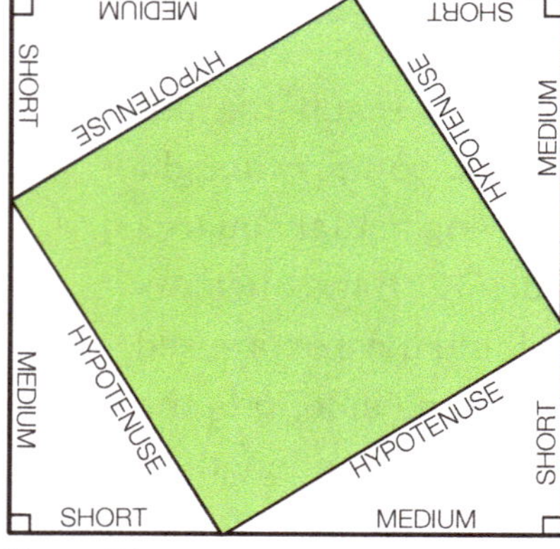

Figure 74

Take a right-angled triangle, of which you know the length of the short and the medium sides, but not the length of the hypotenuse (Figure 73). Now make three more identical copies of this triangle and arrange them into a square as in Figure 74. The area we are interested in is the green area: the square on the hypotenuse. Now if we don't change the area of the outer square, and rearrange the triangles without letting them overlap, then the green area can't change. It will always be

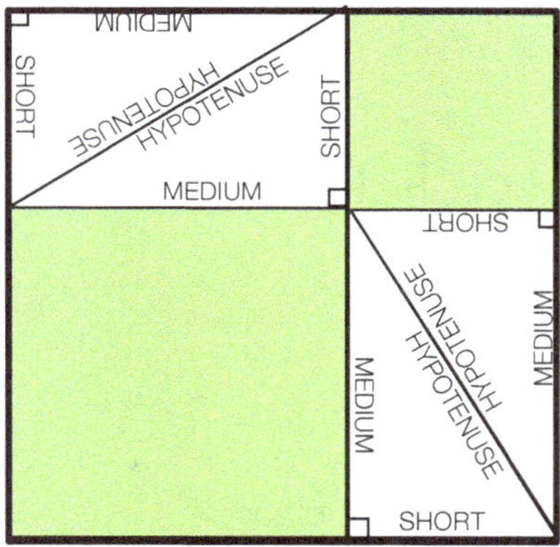

Figure 75

the area of the whole outside square, minus the areas of the four triangles. If we rearrange the triangles as in Figure 75, we can see that it is now equal to two squares, based on the short and medium sides of the triangle we started with, proving that the area of the square on the hypotenuse is equal to the total areas of the squares on the other two sides: which, in turn, proves that if you square the length of the short side, square the length of the medium side and add them together, you will have the square of the length of the hypotenuse.

Using this theorem, the Greeks recognised that if you draw a right-angled triangle with each of the shorter sides having the length of 1, then it was impossible to write the length of the hypotenuse as a fraction based on two different integers. We now know that it will be the square root of two ($\sqrt{2}$ = 1.4142135623731 …) and it was probably the first irrational number discovered. (We covered irrational numbers on page 67.) We will see this again later, in the formula for the **normal distribution curve**.

(For interest, you might also want to view the How to Love Statistics YouTube video on *The Problem with Counting Pythagorean Triples.*)

<table>
<tr><td>

Key points:

- Pythagoras' theorem states that in a right-angled triangle, the length of the hypotenuse squared is equal to the sum of the squared lengths of the other two sides.

- The idea of sums of squares, and the square root of a sum of squares, frequently arises in statistics. This is where it began.

- There is an easy way to prove Pythagoras' theorem.

- A right-angled triangle with two sides of **1** unit will have a hypotenuse of the square root of two, which was probably the first irrational number discovered.

</td></tr>
</table>

Angles, degrees and radians

> **Read this if** you don't understand why the angles of a triangle add up to a straight line, you don't know what that has do to with the angles of other straight-sided figures, or if you aren't familiar with radians.

Angles of a triangle

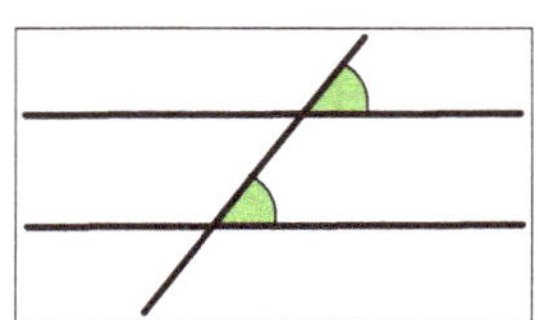

Figure 76

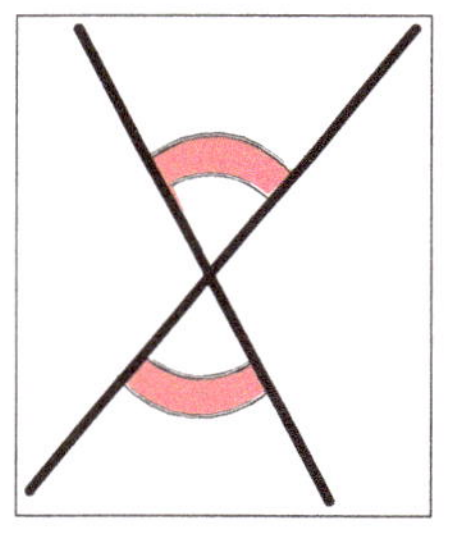

Figure 77

If you take two parallel lines and a transverse which crosses them (Figure 76), then, as we saw on page 172, the angles where the new line meets each of the parallel lines will be the same, because the orientations of parallel lines are the same. If you cross two straight lines, then the angles on either side of the crossing point will also be the same (Figure 77). (You can see this because the difference between each angle and 180°, or a straight line, will be the same for both of them.)

In Figure 78, the two dotted lines are parallel. You can see that, along the top line, the angles represented by A, B and C together add up to the angle of a straight line, which we saw earlier (page 171) will always be 180°. You can also see that, based on what we saw in Figure 76, the angle in Figure 78 at A will be the same as the angle at a, and the angle at C will be the same as the angle at c. Based on what we saw in Figure 77, the angle in Figure 78 at B will be the same as the angle at b.

Putting all of that together, we can see that the total of the angles A, B and C, which is 180°, will be the same as the total of the angles a, b and c. This means that the angles of any triangle will always add up to the same as the angles of a straight line, or 180°.

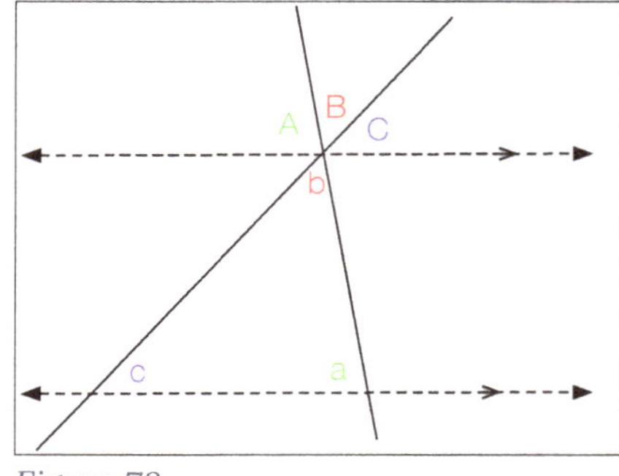

Figure 78

Angles of straight-sided figures

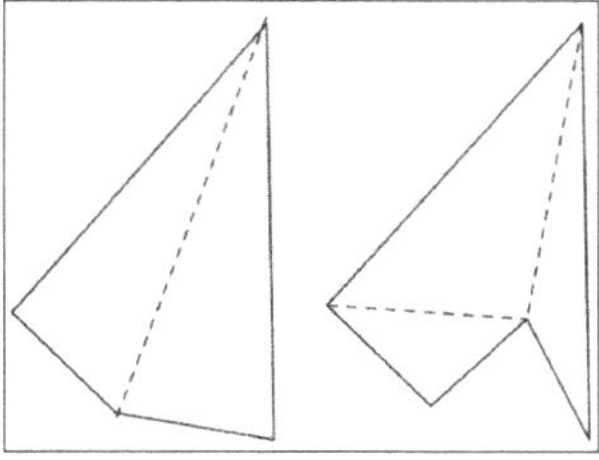

Figure 79

The angles of any figure with four straight edges will add up to 360°, because you can always draw an interval between two opposite corners and divide it into two triangles, and the angles of any figure with five straight sides will add to 540°, because you can always draw two intervals between three corners and divide it into three triangles (Figure 79). This is the same as the number of degrees if you rotated all the way around in a circle, then turned half-way around again.

To cement your understanding:

• Draw something similar to Figure 78, to confirm for yourself that the angles of a straight line will be the same as the angles of a triangle.

• Draw a 6-sided figure, and divide it into triangles by drawing lines between some of the corners (without crossing any lines) to work out how many degrees it will have. Did you get 720? If not, can you see where you went wrong?

You now have enough information to understand the proof that the ratios of the sides of a pentagram (a five sided star) within a pentagon (a five-sided figure) are all in exactly the golden ratio. There is an appendix showing this proof, on page 740.

Radians

(If you are struggling with this, see the How to Love Statistics YouTube video on *Degrees and Radians*.)

Roger Cotes (1682-1716) had an even shorter life than Blaise Pascal. As well as being a Cambridge University professor of astronomy and experimental philosophy, he was heavily involved in editing Isaac Newton's *Principia Mathematica*. Cotes realised that there was another way of measuring angles that could sometimes be more useful than using degrees.

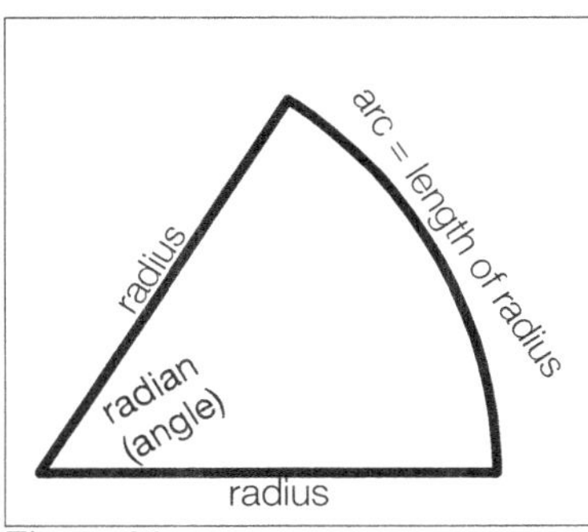

Figure 80

The measure he developed later became known as the **radian** (not to be confused with radius). A radian is the angle formed when the length of an **arc** (a section of the circumference of a circle) is the same as the radius (Figure 80). It is roughly equal to 57°. It doesn't matter how large the radius is: it could be

1cm or 4 light years. Provided that the length of the arc is the same as the length of the radius, the angle formed in the way shown in Figure 80 will always be the same: one radian.

We saw earlier that pi (π) (3.14159...) is an irrational number that gives the ratio of the circumference of a circle to the diameter, regardless of the size of the circle. We also saw that another way to write that is

$$diameter \bullet \pi = circumference$$

Because the diameter of a circle is twice the radius, we can also say

$$radius \bullet 2\pi = circumference$$

This is another way of saying that the length a radius will go 2π (6.28318...) times around the circumference of the circle, regardless of the size of the circle (Figure 81).

One particularly useful thing about working in radians when we use them for mathematical proofs is that the length of an arc is equal to the size of the angle (when it is measured in radians) multiplied by the length of the radius. If the radius is 3 units long, then the length of an arc where the angle is 1 radian will be 3 units. The length of an arc where the radius is 2 radians will be 6 units, and so on (Figure 82).

Because 1 radian is the angle created by 1 length-of-the-radius on the circumference of the circle, that means that there will be 2π radians in an angle of 360 degrees (going all of the way around the circle). There are therefore π radians in straight line (180°), and $\frac{\pi}{2}$ radians in a right angle, or 90°.

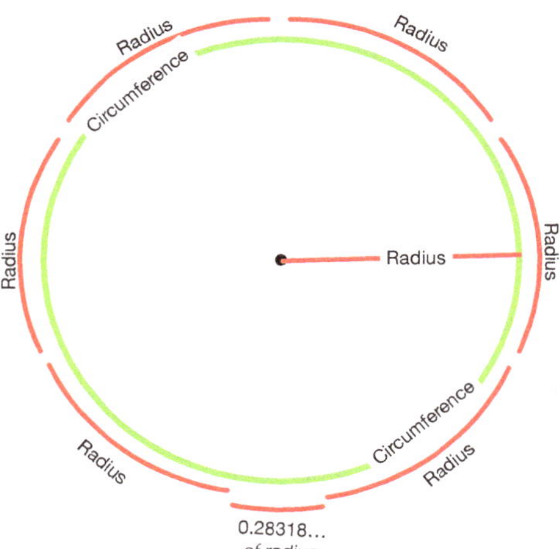

Figure 81

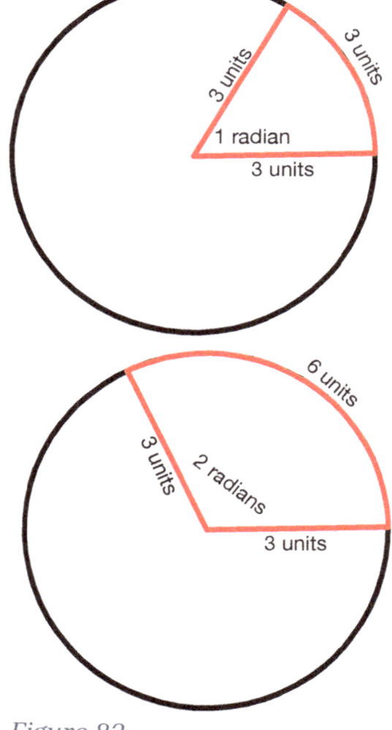

Figure 82

To cement your understanding:

• How many radians will there be in the angles of a triangle? Can you see that is the same number as the ratio of the circumference of a circle to its diameter?

• How many radians will there be in the angles of figures that have 4, 5, 6 and 7 straight sides?

• If the length of the radius of a circle is 3 units, how long will the arc be if the angle is 3 radians?

When using degrees and radian for statistics you aren't ever likely to have to convert them (you are learning about them primarily because it will help you to understand other concepts) and if you ever need to in practice you will probably use a calculator. Just so you know, though, we've seen that

$$180° = \pi \; radians$$

That means that

$$1 \; degree = \frac{\pi}{180} \; radians$$

$$\approx 0.0174532925199433...$$

and

$$1 \; radian = \frac{180}{\pi} \; degrees$$

$$\approx 57.2957795130823...$$

Fun geometric fact:

If you draw a circle with any number of points on it and draw lines between all of them, then if you go to the row of Pascal's triangle that is one more than the number of points you have drawn, the second number is the number of points on the circle, the third number is the number of lines between them, the fourth number is the number of triangles that have all of their corners on the circumference, the fifth

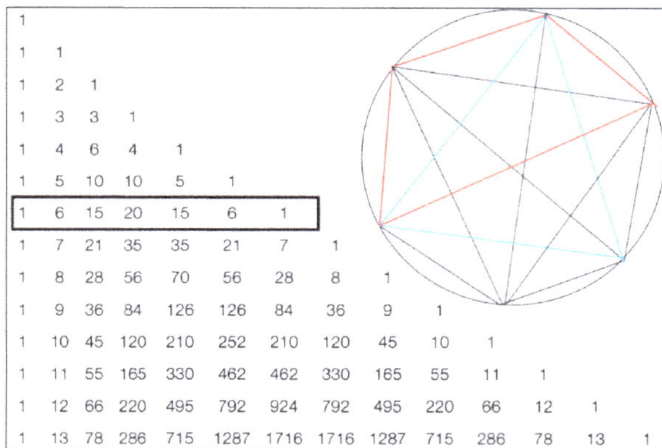

1													
1	1												
1	2	1											
1	3	3	1										
1	4	6	4	1									
1	5	10	10	5	1								
1	6	15	20	15	6	1							
1	7	21	35	35	21	7	1						
1	8	28	56	70	56	28	8	1					
1	9	36	84	126	126	84	36	9	1				
1	10	45	120	210	252	210	120	45	10	1			
1	11	55	165	330	462	462	330	165	55	11	1		
1	12	66	220	495	792	924	792	495	220	66	12	1	
1	13	78	286	715	1287	1716	1716	1287	715	286	78	13	1

Figure 83

on the circumference, the fifth

number is the number of four-sided figures that have all of their corners on the circumference, and so on. So, for example, in Figure 83, the circle has 6 points on the outside of it. If we go to the 7th row of the triangle the second number, 6, is the same as the number of points on the circle. The third number, 15, is the same as the number of lines between the points, the 4th number, 20, is the number of triangles that have all of their corners on the circumference, the 5th number, 15, is the same as the number of four-sided figures that have all of their corners on the circumference, the 6th number, 6, is the same as the number of five-sided figures, and the last number, 1, is the same as the number of six-sided figures.

Key points:

- The angles of a triangle add up to the angles in a straight line, or **180°**.

- If you cross two straight lines, then the angles on either side of the crossing point will be the same.

- The angles of any figure with four straight edges will add up to **360°**.

- A radian is the angle formed when the length of an arc (a section of the circumference of a circle) is the same as the radius. It is roughly equal to **57°**.

- The length of an arc is equal to the size of the angle (when it is measured in radians) multiplied by the length of the radius.

- There are 2π radians in an angle of **360°** (going all of the way around the circle).

- There are π radians in straight line (**180°**), and $\frac{\pi}{2}$ radians in a right angle, or **90°**.

Area of a circle and a segment

Read this if you don't know *why* the area of a circle is πr^2, or you don't know how to calculate the area of a segment.

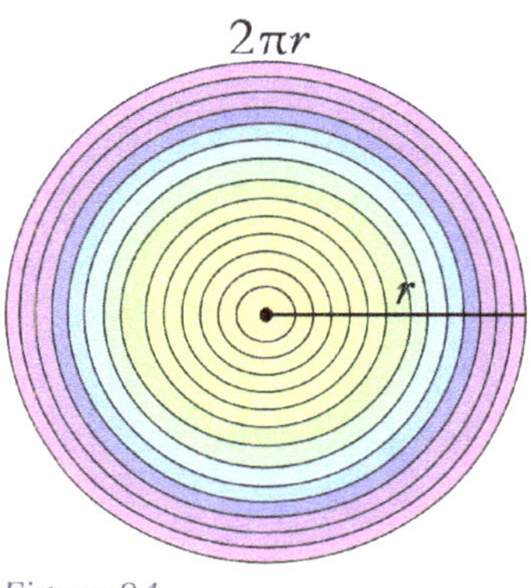

Figure 84

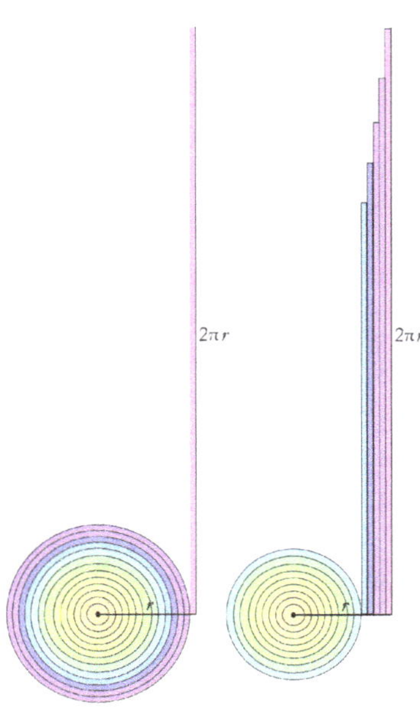

Figure 85

So far we've learned how to find the area of any straight-sided figure. Finding the area of a shape with curves in it, such as a circle, is a bit more challenging, but Archimedes of Syracuse in Sicily (287-212 BC) worked out how to do it using a method that was a forerunner of what we now call integral calculus (which we'll get to).

Imagine a circle made of extremely thin, but not elastic, threads, each circled around the centre (Figure 84).

Then imagine straightening out the first thread and placing it at right angles to the radius. Note that it will have the same thickness, and it will still be $2\pi r$ in length; its area won't have changed, just its shape. Do that with each successive thread and note that as you go closer to the centre of the circle, the threads get shorter (Figure 85).

Now imagine that the threads become as thin as you can imagine, and there are as many of them as you can imagine. We now have a right-angled triangle, with width r and height $2\pi r$ (Figure 86). We've changed a shape with a curved edge to one whose shape we know how to calculate.

If this was a rectangle with width r and height $2\pi r$ we could multiply the dimensions together, to get $2\pi r^2$. As it is a right-angled triangle, the area is half of that; just πr^2. So that is the area of the circle.

$$Area\ of\ a\ circle = \pi r^2$$

where r is the length of the radius.

Area of a segment

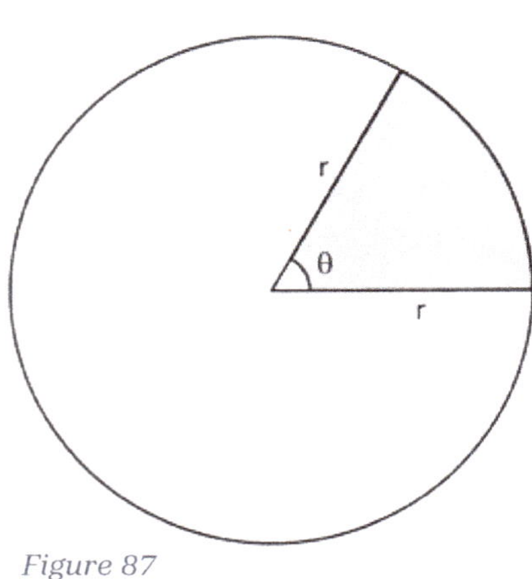

$2\pi r$

r

Figure 86

A segment of a circle is like a slice cut out of a round cake (Figure 87). To find its area, we need to know the angle made by the two edges of the segment near the centre of the circle. In this case let's say we have an angle of 45°. We also need to know the length of the radius of the circle; let's make that 4 units.

We already know that the area of the whole circle will be πr^2; that is, $\pi 4^2$.

The strategy we can use is to work out what proportion of the *angle* of the whole circle is made by the *angle* 45°. This proportion will be the same as the proportion of the *area* of the segment to the *area* of the circle.

In other words:

$$\frac{Area\ of\ segment}{Area\ of\ whole\ circle} = \frac{Angle\ of\ segment}{Angle\ of\ whole\ circle}$$

We know three of these numbers. The whole area of the circle is $\pi 4^2$. The angle of the segment is 45°. The size of the angle of the whole circle is 360°. (We'll start by calculating this in degrees, because that is likely to be more familiar to you, but it gets even simpler in radians.)

Figure 87

Substituting those numbers in, we get:

$$\frac{Area\ of\ segment}{\pi 4^2} = \frac{45°}{360°}$$

Multiplying both sides of the equation by $\pi 4^2$ gives us:

$$Area\ of\ segment = \frac{45\pi 4^2}{360}$$

$$\approx 6.283 \ldots square\ units$$

To turn that into a general formula, we'll let our angle equal θ (the Greek letter theta), which is often used for angles (the way we have been using x) and let our radius equal r. That makes the formula:

$$Area\ of\ segment = \frac{\theta \pi r^2}{360}$$

That assumes we are measuring the angle θ in degrees.

Now let's assume we are measuring it in radians, instead. We would need to convert our angle to radians.

Instead of saying that the size of the angle of the whole circle is $360°$, we can say that it is 2π radians. Substituting those values into the equation we started with, we get:

$$Area\ of\ segment = \frac{\theta \pi r^2}{2\pi}$$

The π in the numerator and denominator cancel one another out, giving us:

$$Area\ of\ segment = \frac{\theta r^2}{2}$$

when the angle is measured in radians.

And yes, circle geometry is also relevant to statistics.

To cement your understanding:

• Check through that you understand how we came to the formula using degrees.

• Find the area of a segment of a circle of radius 2 units, and angle cf 2 radians. Can you see why it is 4 square units?

Key points:

- Area of a circle $= \pi r^2$ where r is the length of the radius.

- Area of segment (when the angle θ is measured in degrees) $= \frac{\theta \pi r^2}{360}$

- Area of segment (when the angle θ is measured in radians) $= \frac{\theta r^2}{2}$

Trigonometry

> **Read this if** you don't know the difference between a sine, a cosine and a tangent, you don't know what any of these have to do with a unit circle, you can't understand how they could be graphed or why that would be of any practical use to anybody, or you don't know the difference between cos A squared and cos squared A.

(If you are struggling with this, see the How to Love Statistics YouTube video on *Trigonometry*. The second part of the video requires an understanding of the Cartesian plane, so you can just watch the first half for now.)

The ability to calculate unknown geometrical measurements from known ones was essential for ancient civilisations. People used it for sophisticated astronomy, which underpinned calendars and religious observance. It was crucial for surveying and land management, which helped resolve disputes about property rights—especially if markers had been removed (by floods or deliberately). It was helpful for navigation (particularly beyond the horizon), which enabled greater trade (and warfare), and it was vital for constructing large buildings. In other words, it was something that allowed people to live in cities rather than isolated communities, which is what "civilisation" originally meant.

Trigonometry (in some form) was central to all of those applications, and it still is. Without it, our world would be far more fragmented. For reasons we'll come to later (page 214), trigonometry is also now essential for any application that uses mathematics to describe recurring waveforms. That includes physics, computer graphics, signal processing, biology, aspects of medicine and some statistics.

The basic shape of trigonometry is the right-angled triangle. Any straight-sided shape can be broken into triangles, and any triangle can be divided into right-angled ones (Figure 88). We've covered enough so far to calculate some unknown from known information, using a right-angled triangle. Using Pythagoras' theorem, if we know the length of two sides we can find the

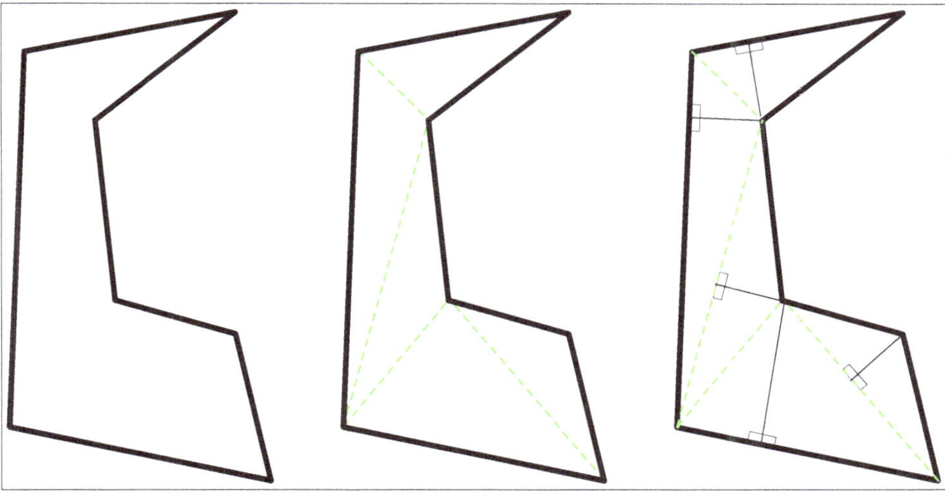

Figure 88

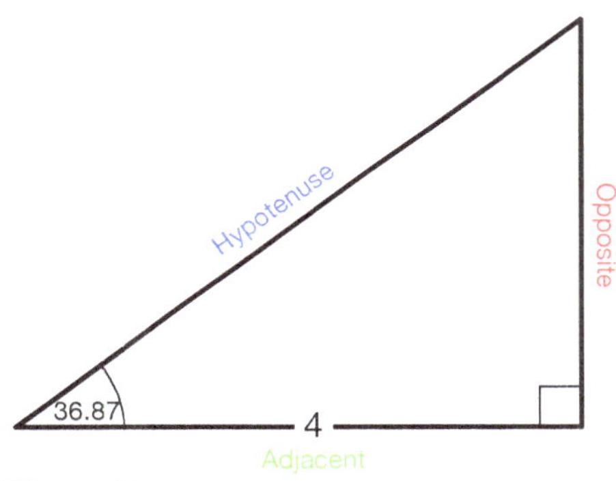

Figure 89

third. Using the fact the angles of a triangle add to 180°, if we know two angles we can find the third. We need trigonometry, though, to relate the length of sides to the size of angles.

In Figure 89 we will call the side that is adjacent to the angle that we know (36.87°) the **adjacent**, and the side that is opposite to the angle we know the **opposite**. We saw when we looked at Pythagoras' theorem that the longest side of a right-angled triangle is the **hypotenuse**. Suppose you want to calculate the length of the opposite side and the hypotenuse. This can be quite a frustrating problem, because you can see intuitively there can only be one answer.

If you change the length of the adjacent side you would have to change the opposite and hypotenuse, in the same proportion. For example, if you increased the adjacent to 5 metres, increasing its

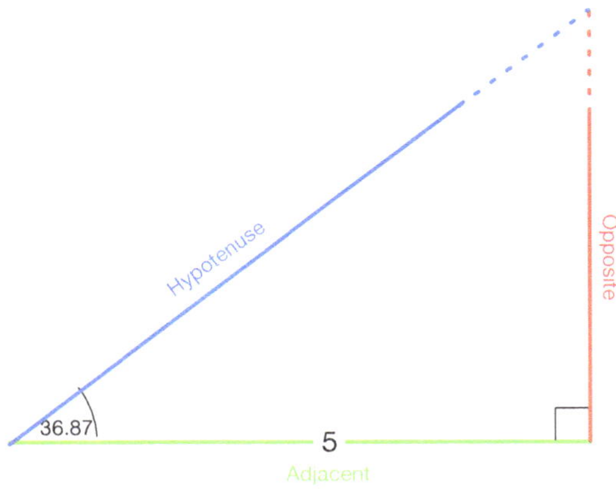

Figure 90

length by one quarter, then you would have to increase the length of the opposite and the hypotenuse by one quarter, so that they would still meet (Figure 90). If you didn't change the length of the sides, you would have to change one of the angles. Once you know the size of one angle in addition to the right angle, and the length of one side, there can only be one triangle.

> **To cement your understanding:**
>
> Play with Figure 90, either in your head or on paper, to confirm that you can't change one of the angles without changing the sides, and you have to change all of the sides proportionately if you don't want to change the angle.

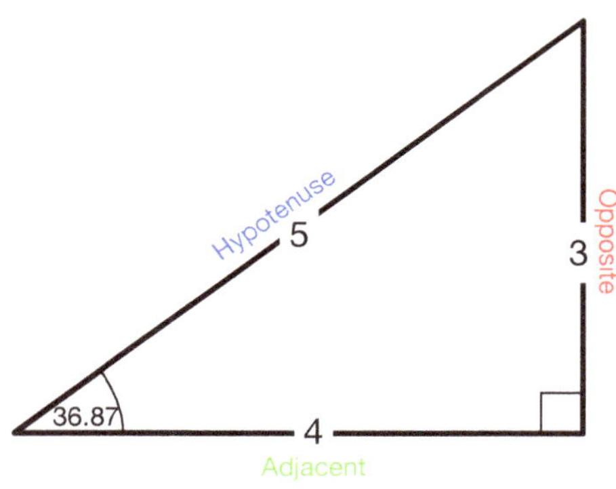

Figure 91

The solution to the problem is actually simple. Just draw up the triangle using a ruler and protractor, using centimetres or inches, and measure them. If you do, you get the results in Figure 91. Obviously to do that you would have to be painstaking with your drawings and measurements, because a slight error in your drawing could turn into a large error when a building is constructed and a larger error in navigation. But if you drew it up in centimetres or inches, you could write the same measurement in any other units.

If you want the adjacent side to be 4.5 metres, but the angle of inclination to stay the same, you don't need to remeasure. You can solve this quite easily, without doing more measurement, based on what you have learned about **similar triangles** (page 172). If you keep the angle the same, then the *ratios* of the sides will be constant.

> **To cement your understanding:**
>
> Before reading on, based on what you know now, can you think of a way to calculate the length of the opposite and the hypotenuse if the adjacent side is 4.5 metres long?

For an angle of 36.87°:

$$\frac{adjacent}{hypotenuse} = \frac{4}{5}$$

This ratio will *always* be four fifths with an angle of 36.87°, regardless of the size of the triangle. That is what allows trigonometry to work.

You can therefore work out the length that the hypotenuse needs to be based on some algebra.

Let's call the new hypotenuse NH. We know that the new adjacent is 4.5. And we know what the ratio of the adjacent to the hypotenuse will be.

This is the equation you need to solve:

$$\frac{4.5}{NH} = \frac{4}{5}$$

Multiply both sides by NH.

$$4.5 = \frac{4}{5}NH$$

Multiply both sides by $\frac{5}{4}$ which is the reciprocal of $\frac{4}{5}$. Remember that if you multiply a fraction by its reciprocal, it equals 1.

$$4.5\frac{5}{4} = NH$$

Multiply that out to give you a single decimal fraction.

$$5.625 = NH$$

You could now calculate the length of the opposite side in a couple of different ways. You could use Pythagoras' theorem to work it out, or you could use a similar approach to the one we did above, based on the fact that you know that

$$\frac{opposite}{adjacent} = \frac{3}{4}$$

Once again, this ratio will stay the same regardless of the size of the triangle. Let's call the new opposite NO. We know that the new adjacent is 4.5.

This is the equation you need to solve:

$$\frac{NO}{4.5} = \frac{3}{4}$$

Multiply both sides by 4.5

$$NO = \frac{3}{4}4.5$$

Multiply that out to give you a single decimal fraction.

$$NO = 3.375$$

Ancient people in different parts of the world measured the ratios of sides for different angles, and used the ratios in a range of different situations. The words we now use for them are:

$$sine = \frac{opposite}{hypotenuse}$$

$$cosine = \frac{adjacent}{hypotenuse}$$

$$tangent = \frac{opposite}{adjacent}$$

The words "sine", "cosine" and "tangent" came from different places. "Sine" was the result of a Latin translation of an Arabic translation of a Sanskrit word. "Cosine" probably came from Latin, indicating something that was related to sine. And the use of the word "tangent" will make more sense when we see how it relates to rise-over-run, then later when we see the way rise-over-run relates to calculus.

You can remember this by the mnemonic "SOHCAHTOA" (**S**ine = **O**pposite over **H**ypotenuse, **C**osine = **A**djacent over **H**ypotenuse, **T**angent = **O**pposite over **A**djacent). (SOHCAHTOA becomes more memorable when you say it out loud.)

Sine, cosine and tangent are usually abbreviated to sin, cos and tan.

These ratios used to be read from printed tables. Now, of course, we look them up on a spreadsheet or some calculators. (When we come to look at the Taylor series later in this book, you will see that there is a way to calculate them without having to measure anything, which is what makes it possible for computers to work them out.)

To cement your understanding:

• Use a spreadsheet or a calculator to look up the sine, cosine and tangent of an angle of 30° (Make sure your device is set to degrees, not radians.)

• Using the sine (which works out neatly to 0.5), if the hypotenuse is 1 unit long, how long will the opposite side be? Did you get that it is 0.5 and, if not, can you see where you went wrong?

• Use a spreadsheet or a calculator to look up the sine, cosine and tangent of an angle of 60°. Which of these is the same as the sine of 30°?

The slope of a straight line

(If you are struggling with this, see the How to Love Statistics YouTube video on *The Slope of a Line*.)

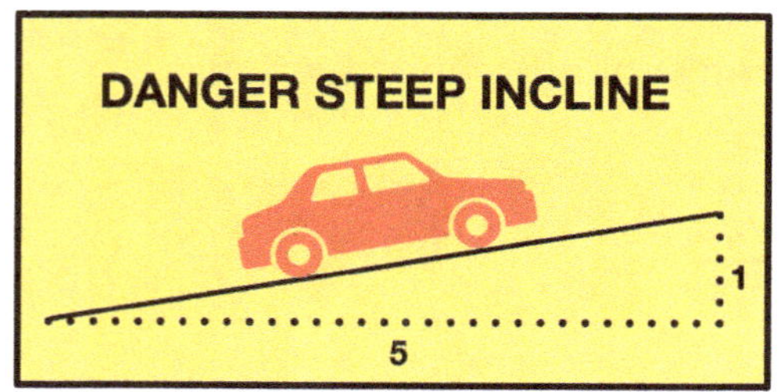

Figure 92 (Based on Pythagoras' theorem you will actually need to travel a distance of $\sqrt{5^2 + 1^2}$ units on the road to travel 5 horizontally and 1 vertically, but that doesn't interest us; we are only looking at the slope.)

There are two ways to describe the slope, or gradient, of a line, compared to the horizontal. One way is to use the angle from the horizontal We might say that something is pitched with a 45° slope.

Another way is to think about how far a line rises vertically compared to how far it moves horizontally. This is called **rise-over-run.**

We might say that the roof of a house or a (very steep) road has a **gradient** of 1 to 5 (Figure 92). That is, for every 1 units that you increase in elevation, you will travel 5 units along the road in a horizontal direction. It doesn't matter what units you use; it isn't relevant. It doesn't matter if you travel 10 kilometers horizontally and 2 vertically. What matters is the *ratio* of how far you move up compared to how far you move along. In this case we would say the slope, or **gradient**, is $1:5$. That means we can describe this slope as

$$\frac{1}{5}$$

Trigonometry allows us to link these two different ways of thinking about sloping lines. The rise-over-run gradient is the same as the tangent of the angle of elevation.

Inverse trigonometric functions

In Figure 93 we know the tangent (1 over 5, or 0.2), but don't know the size of the angle it relates to: the angle of inclination. We saw that for each angle, there can only be one value for the sine, one value for the co-sine and one value for the tangent. The same works in reverse (or it does for angles between 0° and 90°, which is all that we've looked at so far).

Figure 93

That means that once you have a table that gives you the tangents of different angles, then you can look it up in reverse: look up the tangent, and read across to see what angle it relates to. This is an example of something we talked about when we looked at functions (page 113): it's an **inverse function**. We reverse the function. Instead of using an angle to find a ratio, we use the ratio to find the angle.

An inverse trigonometric function is like any other inverse function, and is written with a superscript of $^{-1}$. It means the angle that will give a particular trigonometric ratio.

If we have an angle of theta (θ) and its tangent is a number represented by A:

$$\tan\theta = A, then$$

$$\tan^{-1} A = \theta$$

This is known as the **arc tangent** (or **arc cosine** or **arc sine**).

It's worth remembering that $^{-1}$ in the superscript in this instance doesn't mean the reciprocal of the trigonometric ratio, it means doing the opposite

Angle (degrees)	Angle (radians)	Sine	Cosine	Tangent
10.25	0.1789	0.1779	0.984	0.1808
10.5	0.1833	0.1823	0.9832	0.1854
10.75	0.1876	0.1865	0.9825	0.1898
11	0.192	0.1908	0.9816	0.1944
11.25	0.1963	0.195	0.9808	0.1989
11.5	0.2007	0.1994	0.9799	0.2034
11.75	0.2051	0.2037	0.979	0.208
12	0.2094	0.2079	0.9782	0.2125
12.25	0.2138	0.2122	0.9772	0.2171

Figure 94

of the function. When you see arc sine, arc cosine or arc tangent, remember that you are actually looking at an angle.

In Figure 93, the angle of inclination will be

$$\tan^{-1}\left(\frac{1}{5}\right) \approx 0.1974 \ldots radians$$

$$\approx 11.31 \ldots degrees$$

To find the arc tangent, you just use a calculator, spreadsheet or another program that has inverse trigonometric functions.

Reciprocal trigonometric ratios

The reciprocals of trigonometric functions have their own names. (This is probably one of those things that is easier to look up when you need it, rather than feeling you need to memorise it.)

sine	**cosecant**
$\sin\theta = \dfrac{opposite}{hypotenuse}$	$\csc\theta = \dfrac{1}{\sin\theta} = \dfrac{hypotenuse}{opposite}$
cosine	**secant**
$\cos\theta = \dfrac{adjacent}{hypotenuse}$	$\sec\theta = \dfrac{1}{\cos\theta} = \dfrac{hypotenuse}{adjacent}$

tangent	cotangent
$\tan\theta = \dfrac{opposite}{adjacent}$	$\cot\theta = \dfrac{1}{\tan\theta} = \dfrac{adjacent}{opposite}$

If the difference between an arc sin and a cosecant is starting to slip from your mind, then feel free to forget it for now: just remember, if you ever need it, that it was in this book, so that you can look it up. (It comes up later, but not very much.)

Describing the squares of trigonometric functions

If you say "cosine of A squared" it is ambiguous: does it mean that you square A and then calculate the cosine, or that you square the cosine of A?

The cosine of $60°$ is 0.5. So cos squared $60°$ is

$$\cos^2 60° = (\cos 60°)^2$$
$$= (0.5)^2$$
$$= 0.25$$

Cos of $60°$ squared, on the other hand, is

$$\cos(60°^2) = \cos(3{,}600°)$$
$$= 1$$

To make it clearer, we use the expression $\cos^2 A$. It is pronounced "cos squared A" and is the same as $(\cos A)^2$. It means find the cosine of angle A, then square the result. This is different from $\cos(A^2)$, which we pronounce "cos of A squared". This involves first squaring the angle A, then finding the cosine of the result.

The same applies, of course, to other trigonometric ratios.

Key points:

- For any acute angle in a right-angled triangle, the **sine** is the ratio of the opposite side to the hypotenuse, the **cosine** is the ratio of the adjacent side to the hypotenuse and the **tangent** is the ratio of the opposite to the adjacent.

- These are usually shortened to sin, cos and tan.

- These can be remembered by saying **SOHCAHTOA**.

- These ratios **do not change** with different triangle sizes: each angle will only ever have one sine, one cosine and one tangent.

- The **inverse function** of a trigonometric ratio is called the arc sine, arc cosine or arc tangent. It is the angle that belongs to a given ratio. These are written as $\sin^{-1} A$, $\cos^{-1} A$ and $\tan^{-1} A$ where A is a ratio.

- The **reciprocal** of sine is cosecant, the reciprocal of cosine is secant, and the reciprocal of tangent is cotangent.

- The expression $\cos^2 A$ is pronounced "cos squared A", and is the same as $(\cos A)^2$.

- The expression $\cos(A^2)$ is pronounced "cos of A squared".

The Cartesian plane

> **Read this if** you aren't familiar with the Cartesian plane, or you don't know the sorts of formulae you would use to draw a straight horizontal or vertical line.

(If you are struggling with this, see the How to Love Statistics YouTube video on *The Cartesian Plane*.)

Figure 95
René Descartes
1596-1690

René Descartes (1596-1690) was another French philosopher (like Blaise Pascal, but 27 years older). He's the man who gave us "I think, therefore I am", and he clearly did. As we mentioned earlier, he also gave us the notation of using superscript (smaller font raised above the line $^{\text{like this}}$) to describe raising something to a power (as in 4^5). That was an important contribution to mathematics, but it was nothing compared to his invention of the Cartesian plane. Descartes thought up an invention that is to mathematics a bit like the wheel is to transportation, in that no one had ever invented it previously (when it could have been incredibly helpful), once someone did it was revolutionary, and when you see it you are tempted to think "that's so simple and obvious I could have invented it myself!"

The **Cartesian plane** is based on the idea of using two number lines and turning one at right angles to the other, in a way that makes it easy to describe geometric shapes and their positions using numbers and algebra (Figure 96). Descartes also assigned a letter to each of the two number lines (axes). The horizontal axis (which is often called the x axis) starts at 0 (also known as the **origin**) and goes to the right for positive numbers and to the left for negative numbers, just like the number line we saw early in this book. The vertical axis (which is often called the y axis) also starts at the origin and goes upward for positive numbers and downward for negative ones.

His invention allowed any point on an infinitely large plane to be described with two numbers: a horizontal coordinate and a vertical coordinate, giving the horizontal and vertical distances from zero. So, for example, the point on Figure 96 that is indicated by a pink triangle, and that is 4 units to the left

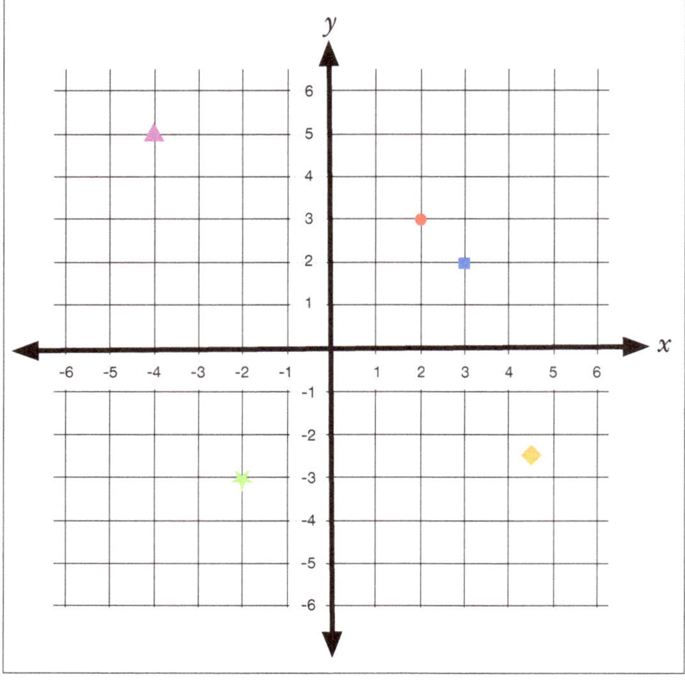

Figure 96

of the origin and 5 units above it would be called (–4,5). The first number always represents the horizontal distance, while the second one represents the vertical distance.

To cement your understanding:

• Using Figure 96, find the points $(3,2), (4.5, -2.5), (-2, -3)$, and $(2,3)$. Did you get that they were the locations of the blue square, the orange diamond, the green star, and the red dot (in that order)?

• Draw up a Cartesian plane on a piece of paper similar to Figure 96 (or just use Figure 96). Plot the positions $(1,1), (-1,1), (-1,-1)$ and $(1,-1)$. Note that you can join these dots to form a square, with the origin at the centre.

• Looking now at Figure 97, what will all of the x values on the purple line be equal to? What will all the x values on the orange line be? What will all of the y values on the light blue line be equal to? What will all of the y values on the green line be equal to? Did you get $-4, 3, 5$ and -3? If not, can you see where you went wrong?

• What point will you arrive at if you subtract 6 from the x value of the point $(3,3)$, without changing the y value? Work this out using the numbers, and then by looking at Figure 97. In both cases, did you get that you will arrive at the point (−3,3)?

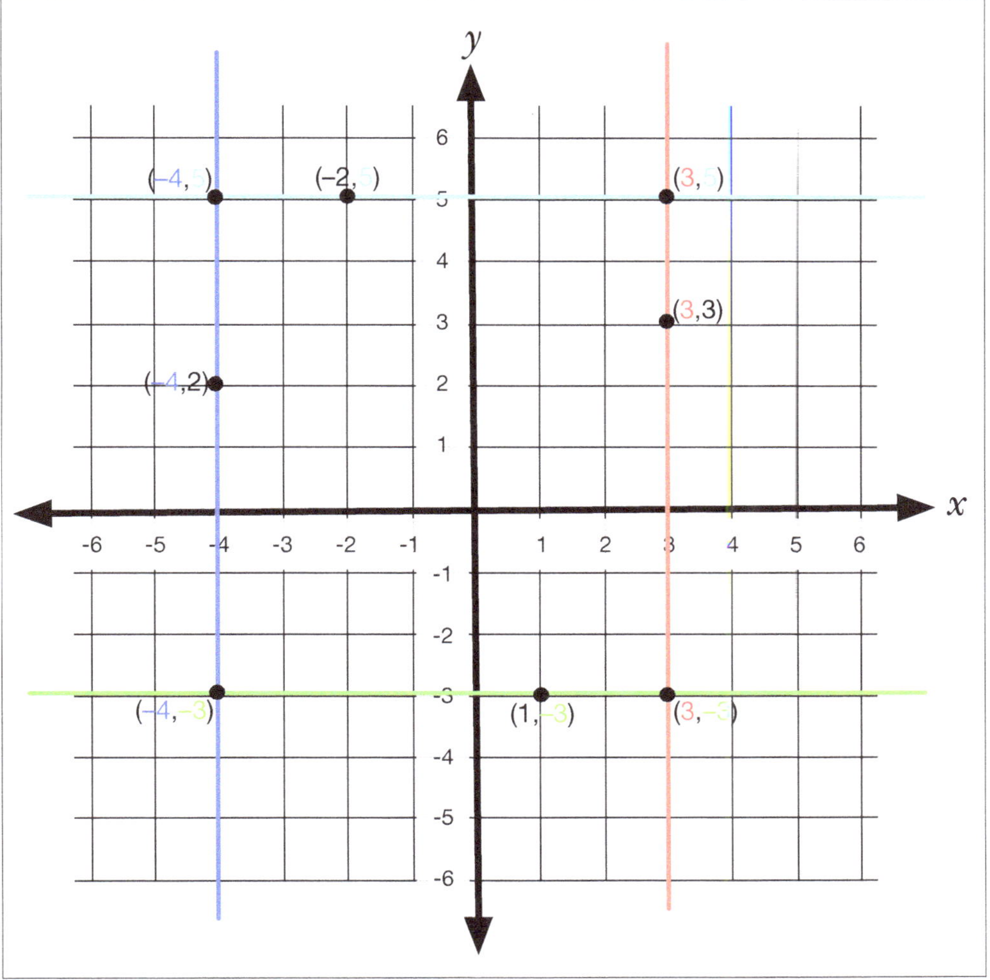

Figure 97

From that point on, mathematicians could combine algebra with geometry more effectively. What made it particularly interesting was that it was now possible to draw with equations. This opened the way for the explosion of mathematical understanding that underpins the technological advances we now enjoy.

This was a huge conceptual leap for the world, even though people had been using sophisticated geometry for thousands of years, and sophisticated algebra (at least in the Islamic world) for centuries. It can also be a conceptual leap for us if we aren't used to it. (Once you become familiar with it, it will seem like second nature.) It can be an effort to realise that when we are talking about an equation we are talking about algebra, and we may also be talking about a line drawn on a page; at the same time.

In Figure 97 the purple line will have the equation:

$$x = -4$$

It doesn't matter what value you give y, x will always equal -4 if the point is on this line. This will be true for every point that is on this line, and it won't be true for any points that aren't on this line.

The light blue line will have the equation

$$y = 5$$

> **To cement your understanding:**
>
> In Figure 97, what will be the equations of the green and the orange lines? Can you see that they are $y = -3$ and $x = 3$?

> **Key points:**
>
> - The Cartesian plane is based on the idea of using two number lines and turning one at right angles to the other.
>
> - The horizontal axis (which is often referred to as the x axis) starts at 0 (also known as the origin), and goes to the right for positive numbers and to the left for negative numbers.
>
> - The vertical axis (which is often referred to as the y axis) also starts at the origin, and goes upward for positive numbers and downward for negative ones.
>
> - Any point on an infinitely large plane can be described with two numbers: a horizontal coordinate and a vertical coordinate, giving the horizontal and vertical distances from zero.
>
> - This approach makes it possible to draw with equations.
>
> - A line where x is equal to a constant will be vertical, and a line where y is equal to a constant will be horizontal.

Sloping and curved lines

Read this if you don't know how to determine the slope of a straight line when you are given its equation, you don't understand the concept of rise-over-run, you don't understand the effect of adding a constant to the end of an equation for a straight line, you don't know what Δx means or you don't know what equations would draw a circle, a parabola, an ellipse or a hyperbola.

(If you are struggling with this, see the How to Love Statistics YouTube video on *The Slope of a Line*.)

The equations in the previous section on the Cartesian plane had only one variable, which is unusual on a Cartesian plane. It only applies to straight lines which are either perfectly horizontal, or perfectly vertical. Lines that slope and lines that curve will always have 2 variables, and the equation will describe the relationship between them.

By convention an equation using the Cartesian plane will usually have an x as the independent variable, and y as the dependent one. That is, the value of y depends on the value of x. (We covered dependent and independent variables on page 113.) In other words, pick any real value you like for x, and the equation will tell you what value to use for y. All of the (x, y) points, and only the (x, y) points, that make the equation true will be on the line drawn by the equation.

The equation $y = x$ draws a straight line, sloping at 45° from the horizontal (Figure 98). In this instance, the x and y values will be the same. This means you can select any point you like on the x (horizontal) axis, and

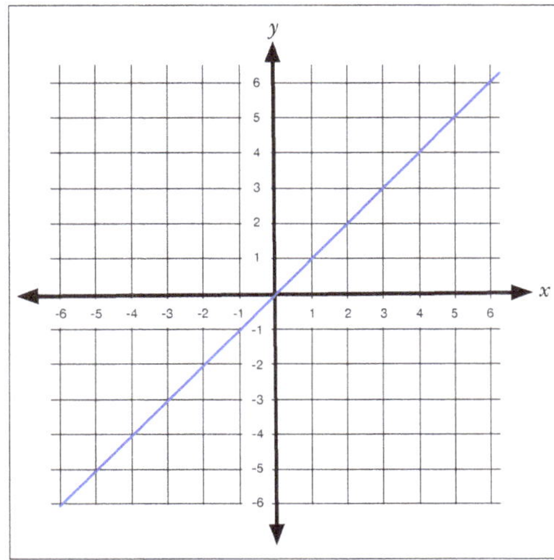

Figure 98

replace the x in the equation with that value. That will give you the value of y. Move y units vertically from your point on the x axis and you will be on the blue line.

To cement your understanding:

Using Figure 98, put your finger on any position you like on the x axis. Use that as your value of x, and calculate the y value using the equation $y = x$. Now move vertically that number of units, to check that you have landed on the blue line.

We looked at measuring the slope of a line in the context of trigonometry (page 194). We can move this diagram onto the Cartesian plane, which will allow us to derive an equation for the sloping line that represents the road (Figure 99). Now we need to work out what equation to use. As a starting point, neither x nor y will be raised to a power. Or, more accurately, they will both be raised to the power of 1. (This form is called a **linear equation**, because it draws a straight line.) We know this, because equations with x or y raised to powers other than 1 always draw curved lines, and our line is straight.

We also saw earlier (page 93) that **quadratic** expressions (i.e. those where the highest power of x is x^2) can always be simplified to the form:

$$Ax^2 + Bx + C$$

where A, B and C are constants (any of which may be zero).

In the same way it is always possible to reduce linear equations on a Cartesian plane to the form:

$$y = Ax + B$$

Where A and B are constants.

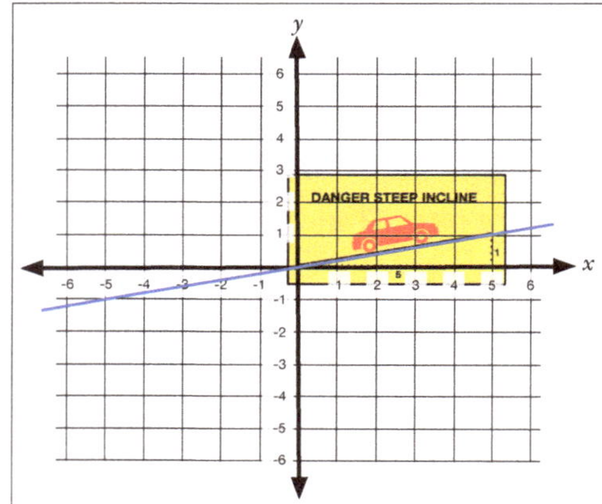

Figure 99

To cement your understanding:

Put the equation $6 + 2x + 3 = -3y + 11x + 40 - 7$ into the form $y = Ax + B$. Hint: Use algebra to get all of the constants and terms with x in them onto the right-hand side of the equation, and all of the terms with y in them onto the left-hand side. Then gather the terms with the same variable, and simplify. Then divide both sides by the coefficient of y, to get y on its own as the dependent variable. Did you get it to the form $y = 3x + 8$?

So $y = Ax + B$ is a general formula that we can use for any straight line. With that as our starting point, we can work out our formula. Now we are looking for the specific formula for the line in Figure 99, which means we need to find numeric values for A and B. (Up until now we have usually started with a specific example, then built that into a general formula. In this situation we are doing the reverse of that. We have a general formula. Now we are going to make it specific, so that we can describe one particular line from Figure 99.)

Because A and B are constants, they will be the same for any values of x and y on this line. The line passes through the point (0,0), or the origin, because that's how we drew it. Substituting those values in for x and y gives the equation

$$0 = A \bullet 0 + B$$

That can only be true if $B = 0$, so we can ignore it. It will equal 0 for all values of x and y: that's why we call it a constant. That makes our equation:

$$y = Ax + 0$$

$$= Ax$$

Dividing both side by x gives us

$$\frac{y}{x} = A$$

Now we also know that our line in Figure 99 passes through the point (5,1). That means we can also substitute these values, because whatever A is equal to with these values of x and y, it will be equal to for all values. So we have

$$\frac{1}{5} = A$$

That means we can draw a line with this slope on a Cartesian plane, using the equation (substituting the value we just calculated for A):

$$y = \frac{1}{5}x$$

At the point where $x = 5$, y will equal 1 and, in fact, at any point on the line the y value will be one fifth of the x value.

We're talking about a straight line, which means that no matter where abouts on the line you are, the slope won't change; it will keep heading in the same direction.

There is something else to notice about this. We said that A was equal to y divided by x. But y divided by x also gives us the slope of our line: it's the rise (the y value) over the run (the x value).

In other words, the constant we multiply x by (the **coefficient** as discussed on page 94) always determines the slope of a straight line when it is written in the format

$$y = Ax$$

To cement your understanding:

• Figure 100 shows a steeper straight line than Figure 99. Can you work out its formula, either just by looking at it, or using the same approach that we did above? Hints: 1) It's still a straight line, so it will have the same general formula: $\qquad y = Ax + B$
2) It passes through the points $(0,0)$ and $(1,2)$.

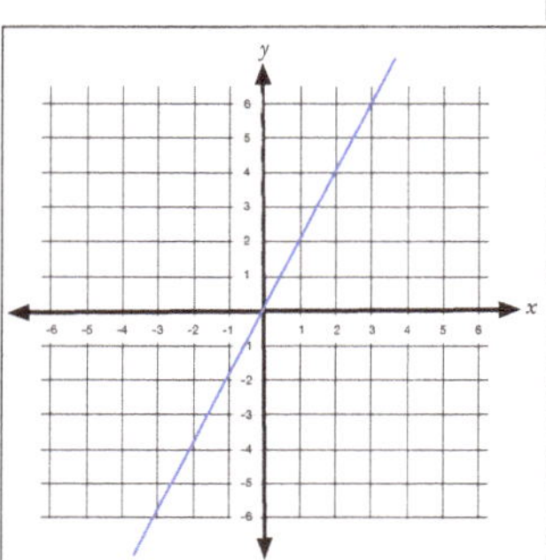

Figure 100

• What is the slope of this line? Can you see why it is 2?

• Pick any other point that the line passes through, such as $(-3, -6)$ or $(2,4)$. Confirm that they can be used to solve the equation.

Adding a constant to the equation for a straight line

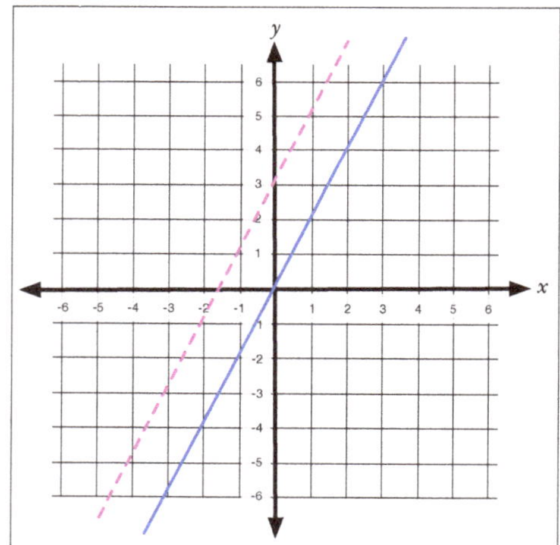

Figure 101

The equation $y = 2x + 3$ also draws a straight line (the pink, dotted one in Figure 101). The effect of adding the 3 to the x increases every y value by 3. You can see that when $x = 0$, $y = 3$. It doesn't change the slope of the line in any other way.

To cement your understanding:

• Using Figure 101 or a copy of it, draw the line $y = 2x - 4$. To do this, pick at least 2 values for x, calculate the values for y for each of them, and mark both the (x, y)

points; then draw a line through them. You should have drawn a line which is parallel to the two already shown on the diagram, but that crosses the y axis (that is, the point where $x = 0$) at the point where $y = -4$.

• Check some other values of x and y on the line you have just drawn, to make sure that each point has values for x and y that solve the equation $y = 2x - 4$.

The idea of delta

When we worked out the formula for a straight line, we drew it up (back at Figure 99) going through the origin. We didn't have to; we just chose to do it that way to keep the algebra as simple as possible. In the real world it isn't always that simple, and there is another way to do it if the line doesn't pass through (0,0). It's worth thinking this through at this stage, because the idea we are about to look at is a key concept in calculus, and also because the idea of not having much information and using it to

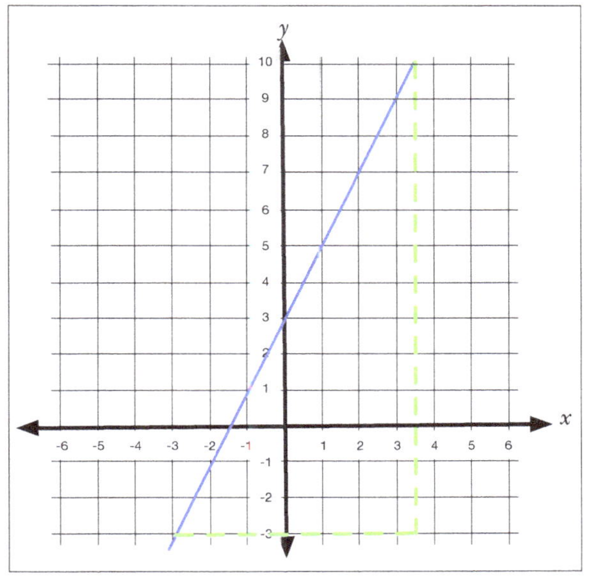

Figure 102

develop a useful formula is fundamental to statistics.

Supposing the only line we had had to work with was the blue one in Figure 102, and we wanted to find its equation. It's a straight line, so we know it will be able to be reduced to the form

$$y = Ax + B$$

It passes through the point (0,3), so where $x = 0$ (where it crosses the y axis) B must equal 3. But what will A, the slope be? Suppose we only have two points to work with: $(-3, -3)$ and $(3.5, 10)$. What we need to calculate is the vertical rise over the horizontal run. We can work this out, first of all, by subtracting the lower y value from the higher one. This will give us the vertical rise, or the length of the vertical green dotted line in Figure 102. (You could, of course, just count the squares on the vertical green dotted line to get the same answer.) We can then subtract the lower x value from the higher one. This will give us the horizontal run, or the length of the horizontal green dotted line in Figure 102. (Again, we could just look at the diagram to get the length of the horizontal green dotted line.) Finally we can divide the first number by the second, to get the rise over the run.

So, if A is the slope of the line, that give us:

$$A = \frac{10 - (-3)}{3.5 - (-3)}$$

$$= \frac{13}{6.5}$$

$$= 2$$

That makes the formula for the line

$$y = 2x + 3$$

which is, of course, exactly the same line we looked at earlier (except that then it was dotted and coloured pink).

We use the capital of the Greek letter delta (Δ) to describe the differences between the y values (delta y) and the difference between the x values (delta x). That means the slope of the line—the rise-over-run—is given by

$$\frac{\Delta y}{\Delta x}$$

That's all you need to understand about the idea of delta at this stage; the significance will become clearer a little later.

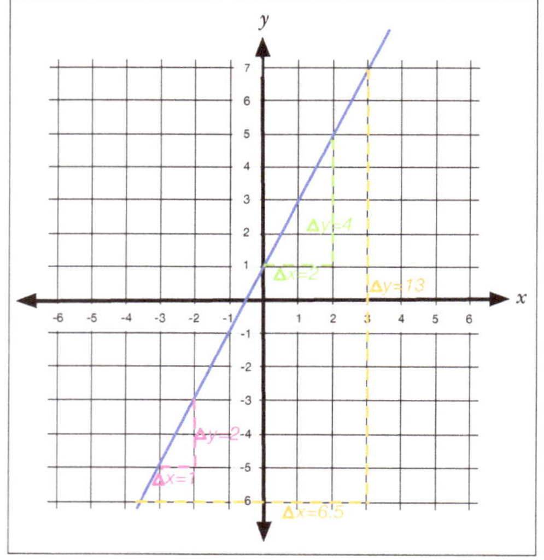

Figure 103

To cement your understanding:

Using Figure 103, compare the delta values given by the orange, green and pink lines. Can you see why, regardless of the starting point or the size of Δx , the ratio remains constant because the slope of the line doesn't change?

So far we've been looking at straight lines, which hasn't involved raising x to a power (or, of you prefer to think of it this way, we have x raised to the power of 1). As mentioned earlier, once we start raising x to a power other than 1, the lines start being curved.

The circle

The equation $x^2 + y^2 = 1$ draws a circle, with a radius of 1 (Figure 104). If we want to show this with y as the dependent variable, the way we have with our other equations so far, we can do some simple algebra to get there.

This is our starting equation.

$$x^2 + y^2 = 1$$

Subtract x^2 from both sides.

$$y^2 = 1 - x^2$$

Take the square root of both sides, noting that there will be positive and negative values for the square root.

$$y = \pm\sqrt{1 - x^2}$$

A circle with a radius of 1 is called **unit circle**, and we'll meet it again soon.

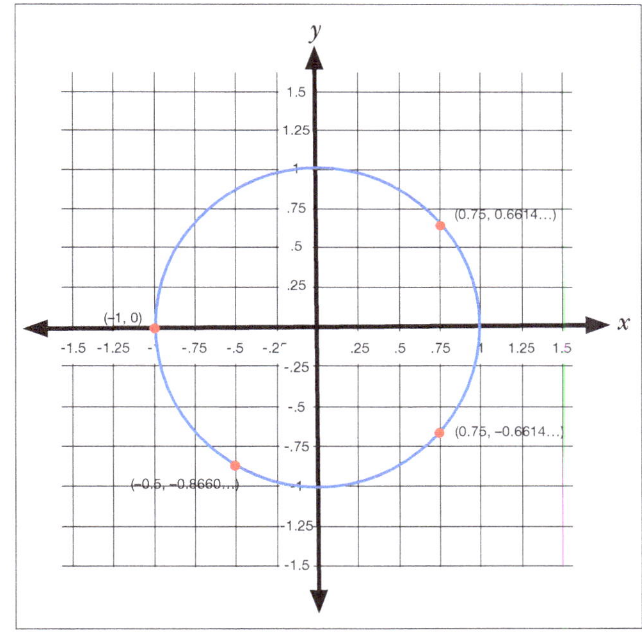

Figure 104

To cement your understanding

Select at least one of the red points on the circle in Figure 104. Square both coordinates and add the result, to reassure yourself that the result will be 1. (There will probably be a rounding error, so your calculator may not show it as exactly 1, but it will be very close.)

The parabola

The equation

$$y = x^2$$

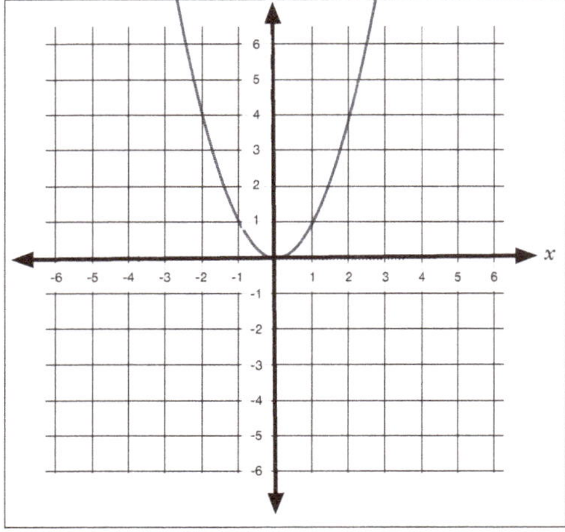

Figure 105

draws a **parabola** (Figure 105). A parabola is a u-shaped curve where the sides go on forever and get steeper and steeper. We know that multiplying two negative numbers makes a positive, and so any real negative number squared will be positive. For negative values of x, the value of x^2, and therefore the value of y, is still positive.

An appendix on page 742 has a stricter definition of a parabola, and some interesting information about why the shape is used so much in devices like torches and antennas.

> **To cement your understanding:**
>
> • Using the equation $y = x^2$ and Figure 105, if x is -2, what will y be? Check that the point with these (x, y) coordinates is on the line.
>
> • What will the value of y be if x is one half? What about if it is one-third? What will it be if x is 0? What about if x is -100? Can you see why the curve never has a negative value for y, and get's steeper and steeper the further you move from the origin?

The ellipse

The equation

$$\frac{x^2}{a^2} + \frac{y^2}{b^2} = 1$$

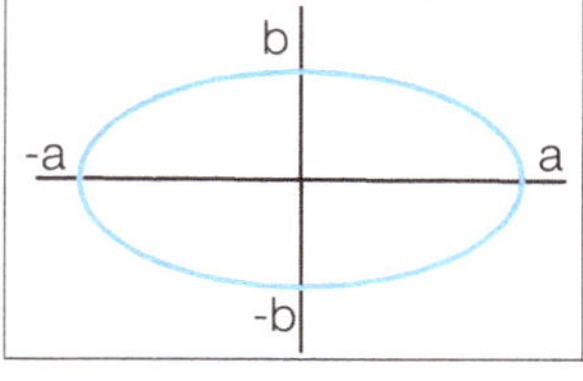

Figure 106

draws an **ellipse** (Figure 106), or oval-shape. You may note that the formula for an ellipse is very similar to the formula for a circle. That is because a circle is actually a type of ellipse. There is also a stricter definition of an ellipse in an appendix on page 742.

We will briefly meet ellipses again in Part 2, because planetary orbits are elliptical (we'll see why that is relevant when we get there), and the ellipse is one of the fundamental shapes of randomness. For example, if you throw a handful of pebbles at an angle at a pool of water, the pattern they make when they first strike the water will be roughly elliptical.

> **To cement your understanding:**
>
> If you are confused about ellipses, put a piece of paper on a piece of cardboard or similar, and put two thumb tacks in. Tie a piece of string, with plenty of slack in it, to both tacks. Then with a pencil draw the shape you get by going around the pins with the string stretched to its limit. That should give you an ellipse.

The hyperbola

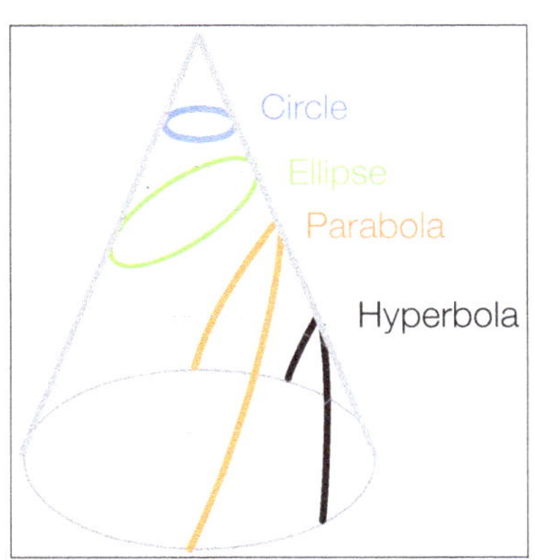

Figure 107

The equation

$$y = \frac{1}{x}$$

draws a **hyperbola**, in which the larger x becomes the closer y gets to 0 without ever touching it, and vice versa (Figure 107). A situation in which two lines get closer and closer without ever touching is called an **asymptote.** A hyperbola is actually two curves.

The appendix on page 742 also has a stricter definition of a hyperbola.

> **To cement your understanding:**
>
> Pick some negative, and positive, high and low values for x, and calculate the value of y for each of them. Can you see why it has the shape it has?

Fun fact: Why they are called conic sections

Figure 108

The circle, the ellipse, the hyperbola and the parabola are known as **conic sections**, because each of them can be formed by cutting through a cone at a different angle (Figure 108). (In the case of the parabola and the hyperbola the cone would have to be of infinite length, and for the hyperbola, in order to form the second curve, the cone has to extend above the point, so that it is like two cones placed point-to-point.)

Of course, while we have looked at the slope of a straight line, we can't talk about the slope of a line like a parabola, because it keeps changing, getting steeper and steeper. Or can we? We can, in fact, and we will, when we get to look at differential calculus. Don't panic about that; we have already covered the key concepts. Now we just have to put them together. We will cover a few more concepts in geometry before we get there, though.

Key points:

- By convention an equation using the Cartesian plane will usually have an x as the **independent** variable, and y as the **dependent** one.

- A linear equation is one where x is raised to the power of 1. It draws a straight line, and can always be expressed in the form: $y = Ax + B$

- In this form A gives the rise-over-run slope of the line, and B gives the elevation of the line.

- That means the rise-over-run slope of a line is given by $\frac{\Delta y}{\Delta x}$ where Δx is an arbitrary increase in the value of x, and Δy is the corresponding value which, when added to y, will keep an equation balanced when Δx is added to x.

- The equation $x^2 + y^2 = 1$ draws a **circle**, with a radius of 1. This is called a **unit circle**.

- The equation $y = x^2$ draws a **parabola.**

- The equation $\frac{x^2}{a^2} + \frac{y^2}{b^2} = 1$ draws an **ellipse.**

- The equation $y = \frac{1}{x}$ draws a **hyperbola.**

Trigonometry on the Cartesian plane

Read this if you don't know what a unit circle has to do with trigonometry, you don't know how you would graph trigonometric ratios on a Cartiesan plane, or you don't know why you would bother to.

(This is covered in the second part of the How to Love Statistics YouTube video on *Trigonometry*.)

The unit-circle definition of trigonometric ratios

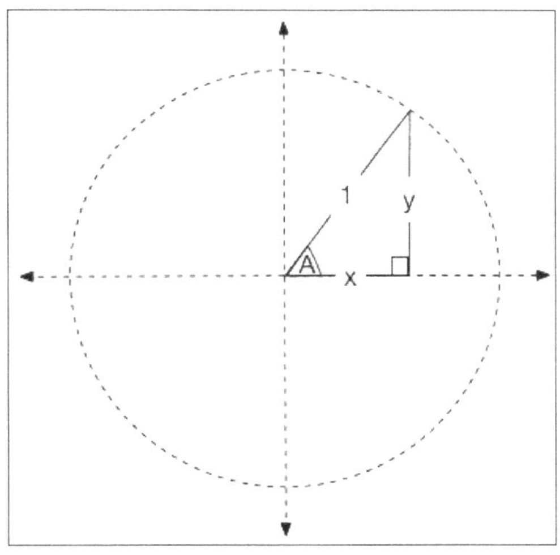

Figure 109

The approach to trigonometry based on a right-angled triangle works wonderfully for acute angles (those of less than 90°), but it stops making sense after that, because a right-angled triangle always has one angle of 90° and two acute angles (i.e. of less than 90°). The development of the Cartesian plane made it possible, among other things, to change the definition of trigonometric ratios (which had been around for a long time prior to that), so that it could cope with angles of more than 90°.

Using this approach, angles are defined by looking at the amount of counter-clockwise/anti-clockwise rotation, starting from the positive x axis. (The axis that goes horizontally to the right from the origin.)

In Figure 109, if you draw a circle with a radius of 1 (which is called a **unit circle**) you can see that the length of the opposite side of the triangle to angle A will be the same value as the y coordinate, and that sine A will be this value over 1 or, in other words, the same as the y value. Similarly cosine A will be the x value. The tangent of A will be

$$\frac{y}{x}$$

This definition allows the use of trigonometry for angles of any size, including obtuse and reflex angles (Figure 110).

Trigonometric ratios can even be assigned to negative angles (those which are formed by a clockwise rotation from the positive x axis) and to angles of greater than 360° using this approach.

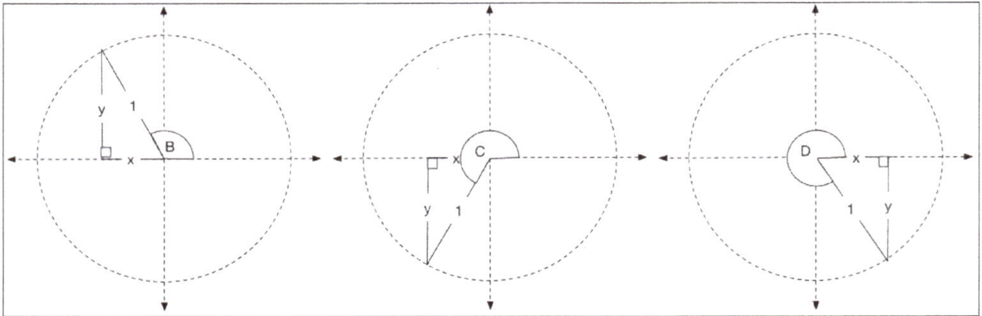

Figure 110

To cement your understanding:

• You can see from Figure 109 that the sine and cosine of angle *A* will both be positive, and therefore that the tangent will be positive. What will happen to the positivity/negativity of the sine, cosine and tangent for: angles greater than 90° but less than 180°, angles greater than 180° but less than 270°, and angles greater than 270° but less than 360°?

• Can you see why the sine, cosine and tangent of an angle of 380° will be the same as those for an angle of 20°?

• What will the sine and cosine of an angle of 0° be?

• What will the sine and cosine of an angle of 90° be?

Graphing the sine and cosine functions

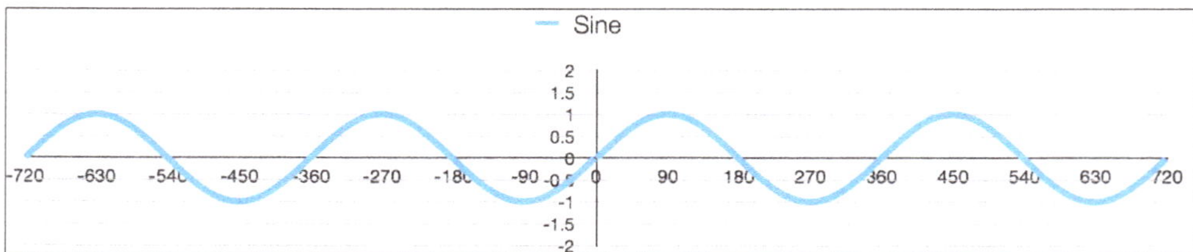

Figure 111

In another of those magical mathematical wonders, it turns out that when you graph the equation $y = \sin x$, it creates an oscillating wave form

(Figure 111). At this point it is easy to get confused by the fact that we have been thinking of the angle as being an angle on the graph. This graph works differently. It's such a useful graph that it is worth the mental gymnastics of thinking it through.

The *x* values in Figure 111 represent the values of the angles in degrees. That's the confusing bit. Once you've got that, the rest is easier. The *y* values represent the sine of the angle, as it did when we looked at the unit circle. What's happening here is we've sort of rolled out the angles around a circle and flattened them onto the *x* axis.

As this happens, the wave represents what the *y* value would be doing if the graph were still a circle. In other words, the sine value gives us a way to take a cyclical (literally "moving in a circle") phenomenon and graph it out lengthwise in a wave pattern. This wave form, after suitable mathematical manipulation, can be the basis for a description of *any* naturally occurring periodic wave form, from sound to light to electrical current to human neurology, and more. The graphs might not look like this: they might have sharp edges or peaks at various points. But if you see a graph for any periodic phenomenon, it will probably have some trigonometry in the mathematical formula, to create a recurring wave.

Now imagine what happens to the adjacent side/cosine while this is going on. As the opposite side increases, the adjacent size side gets smaller, and vice versa. When you graph them, the cosine wave look exactly the same as the sine wave, except they are out of sync.

To cement your understanding:

• Using Figure 110 or something similar, calculate the values of cosine (the *x* value) for angles of: − 270° (rotate clockwise for negative angles), −180°, −90°, 0°, 90°, 180°, 270°, 360°, 450°, and 540°. Mark these points on the graph of Figure 112. Can you see why a cosine wave is the same as a sine wave, except that it is shifted along the *x* axis?

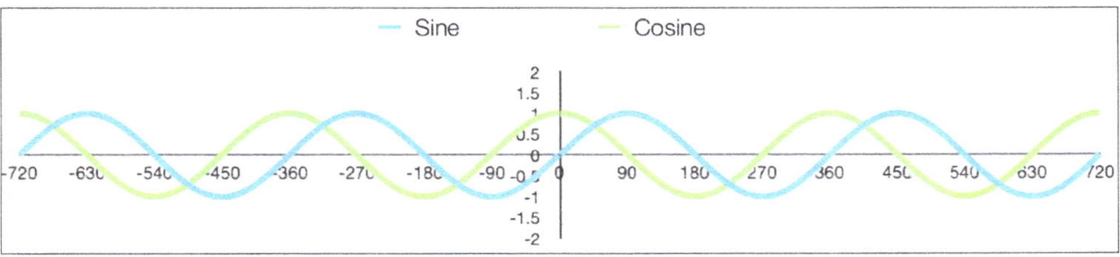

Figure 112

Key points:

- Using a unit circle, the sine is the y value and the cosine is the x value.

- The unit circle definition allows trigonometry to be used for angles of any size, including negative angles and angles of more than 360°.

- When graphed, sine and cosine form smooth undulating curves, which would be identical except for being displaced horizontally.

- Sine and cosine can be used within the mathematical formula to describe any periodic wave form.

Trigonometric identities

> **You can probably afford to skip this** if you aren't interested in knowing where trigonometric identities come from. The main identities, together with their proofs, are here. (Except for some proofs which are a bit complicated and are in an appendix.) If you are happy to just believe that these identities are right, and you are short of time or patience, then skip to the material polar coordinates. But don't skip that; it's only about a page long, and the understanding behind it becomes important.

There are a number of formulae (or **identities**) that apply to trigonometric functions. This becomes important in using calculus with trigonometry. They aren't the sort of thing you are likely to want to memorise, but if you know where they come from and where to look them up, they will be useful later.

Trigonometric ratios of negative angles

From the unit circle (Figure 113), because the cosine represents the x value, you can see that for any angle A the cosine of a negative angle (one that starts by moving clockwise rather than anti-clockwise) will be the same as the cosine of the positive angle.

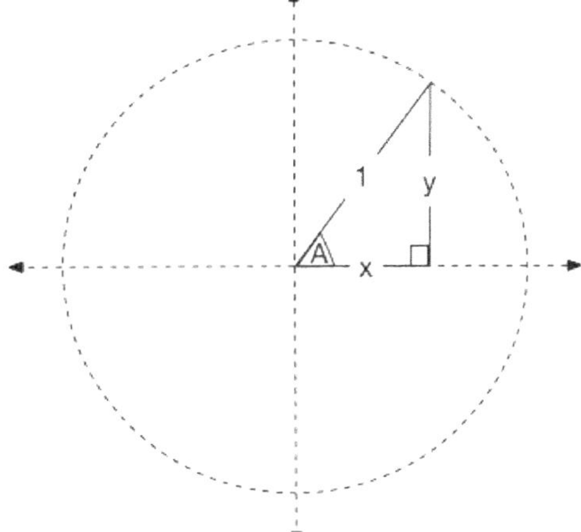

Figure 113

$$\cos(-A) = \cos A$$

In other words, whether the angle is positive or negative, the x value will be the same.

Because the sine is the y value, the sine of a negative angle will be the negative sine of the positive angle.

$$\sin(-A) = -\sin A$$

In other words, if an angle is negative, the *y* value will be negative but will have the same absolute value as it would if the angle were positive.

Pythagoras' theorem and the unit circle

Using Pythagoras' theorem and the unit circle:

$$\sin^2 A + \cos^2 A = 1$$

Trigonometry and the alternative angle

There are π radians in the angles of a triangle, and $\frac{\pi}{2}$ in a right angle. That means:

If one angle of a triangle is a right angle and another is designated θ, the third angle will be

$$\frac{\pi}{2} - \theta$$

(We are using A and θ interchangeably here, not to confuse you, but to help you to keep the flexibility of thinking that different symbols can replace others. If you just see one symbol used consistently in a text, you might be completely thrown when you see things written differently somewhere else. Better to start off being flexible.)

In a right-angled triangle, the side opposite of one angle (other than the right angle) will be the side adjacent to the other (Figure 114).

All of that means:

$$\sin(90° - \theta) = \cos \theta$$

and

$$\cos(90° - \theta) = \sin \theta$$

This still applies where θ is greater than $90°$.

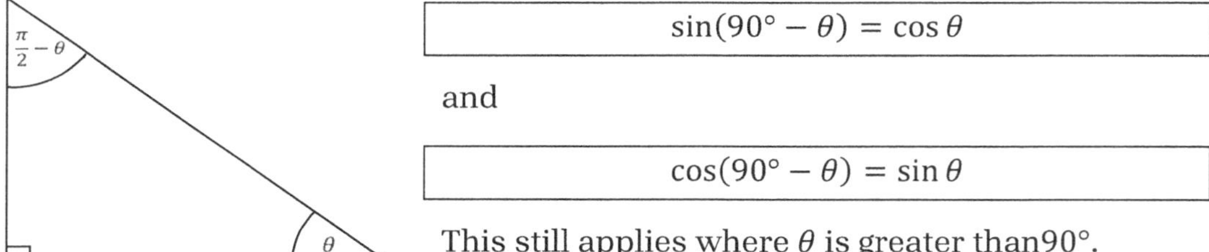

Figure 114

To cement your understanding:

If any of that wasn't clear, draw up your own diagrams similar to those above, and use it to work out the three identities above for yourself.

Sine and cosine of the sum of two angles

The next two identities have a bit more complexity in the working out. The working out is therefore in the appendix (page 742) in case you want to check it out. The sine of the sum of two angles: in this case θ and ζ (the Greek letters theta and zeta) is given by.

$$\sin(\theta + \zeta) = \sin \theta \cos \zeta + \cos \theta \sin \zeta$$

Which is very neat, and very useful, when it comes to calculus and other areas of higher mathematics. (And yes, we will use it when we look at statistics.)

Using a similar proof in the same appendix, the cosine of the sum of two angles: $\cos(\theta + \zeta)$

$$\cos(\theta + \zeta) = \cos \theta \cos \zeta - \sin \theta \sin \zeta$$

Other identities linking sine and cosine

Other formulae can be derived from these. For example:

$$\cos(2\theta) = \cos(\theta + \theta)$$

$$= \cos \theta \cos \theta - \sin \theta \sin \theta$$

$$\cos(2\theta) = \cos^2 \theta - \sin^2 \theta$$

We saw earlier that:

$$\sin^2 \theta + \cos^2 \theta = 1$$

Subtract $\cos^2\theta$ from each side $\qquad \sin^2\theta = 1 - \cos^2\theta$

Use the identify for $cos\ 2\theta$ we worked out immediately above. $\qquad \cos 2\theta = \cos^2 \theta - (1 - \cos^2\theta)$

Simplify $\qquad \cos 2\theta = 2\cos^2 \theta - 1$

Add 1 to both sides and divide both sides by 2

$$\frac{\cos 2\theta + 1}{2} = \cos^2 \theta$$

$$\sin^2\theta + \cos^2\theta = 1$$

Subtract $\cos^2\theta$ from each side $\qquad \sin^2\theta = 1 - \cos^2\theta$

Substitute with the identity we found immediately above

$$sin^2\,\theta = 1 - \frac{\cos 2\theta + 1}{2}$$

Write 1 as $\frac{2}{2}$

$$= \frac{2}{2} - \left(\frac{\cos 2\theta + 1}{2}\right)$$

$$= \frac{2 - \cos 2\theta - 1}{2}$$

$$\boxed{sin^2\,\theta = \frac{1 - \cos 2\theta}{2}}$$

Let $\theta = \frac{x}{2}$ Multiply both sides by 2

$$2sin^2\,\frac{x}{2} = 1 - \cos x$$

The law of cosines

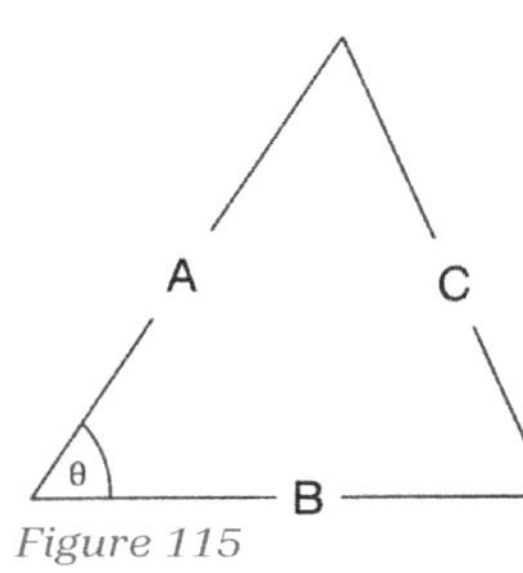

Figure 115

Now suppose we have a non-right-angled triangle where we know two of the sides and one of the angles, and we want to find the third side. In Figure 115, assume that we know θ, A and B, but not C.

First, draw an interval from the angle formed by A and C, to form a right angle where it intersects B. Designate this line E. Call the length from this intersection to θ D (Figure 116).

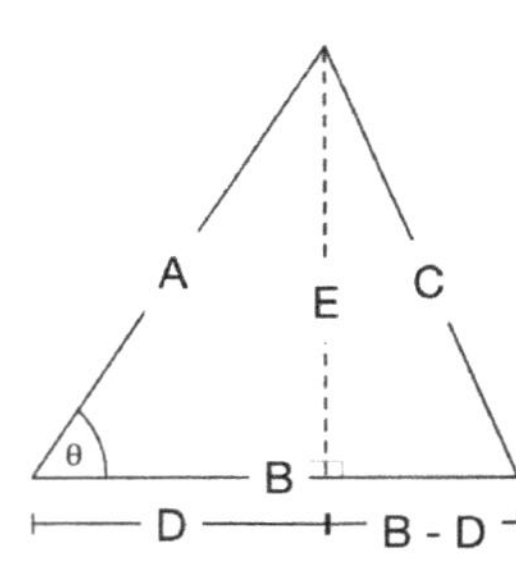

Figure 116

Use the definition of cosine, then multiply both sides by A.

$$\frac{D}{A} = \cos \theta$$

$$D = A \cos \theta$$

Use the definition of sine, then multiply both sides by A.

$$\frac{E}{A} = \sin \theta$$

$$E = A\sin \theta$$

Use Pythagoras' theorem.

$$C^2 = E^2 + (B - D)^2$$

Substitute $A \cos \theta$ for D, and $A\sin \theta$ for E.

$$C^2 = A^2 \sin^2 \theta + (B - A \cos \theta)^2$$

Expand the expression in brackets.

$$= A^2 \sin^2 \theta + B^2 - 2AB \cos \theta + A^2 \cos^2 \theta$$

Gather the terms and factor out A^2.

$$= A^2(\sin^2 \theta + \cos^2 \theta) + B^2 - 2AB \cos \theta$$

Recall that we saw earlier that $\sin^2 \theta + \cos^2 \theta = 1$ (based on the unit-circle definitions of sine and cosine, and Pythagoras' theorem).

This is known as the law of cosines.

$$C^2 = A^2 + B^2 - 2AB \cos \theta$$
$$\textit{where } C \textit{ is the side opposite angle } \theta$$

The law of cosines isn't just used to find C, of course. If you know three sides you can use it to find any angle, (and therefore all three angles) and with any two sides and one angle you can use it to find the third side. Where θ is a right angle, cosine $\theta = 0$. In that instance, the law of cosines will give Pythagoras' theorem. (It doesn't prove Pythagoras' theorem, because we used the theorem in the proof of the law of cosines.)

To cement your understanding:

• If any of that wasn't clear, work out the identities listed above for yourself. Reason out as much of it as you can on your own, before referring back to the text to check your thinking, or if you get stuck.

• Put some real angles into some drawings and use a calculator or spreadsheet to check that the formulae work when numbers are substituted for algebraic letters.

Key points:

- $\cos(-\theta) = \cos\theta$

- $\sin(-\theta) = -\sin\theta$

- $\sin^2\theta + \cos^2\theta = 1$

- If one angle of a triangle is a right angle and another is designated θ, the third angle will be $\frac{\pi}{2} - \theta$

- $\sin(90° - \theta) = \cos\theta$

- $\cos(90° - \theta) = \sin\theta$

- $\sin(\theta + \zeta) = \sin\theta\cos\zeta + \cos\theta\sin\zeta$

- $\cos(\theta + \zeta) = \cos\theta\cos\zeta - \sin\theta\sin\zeta$

- $\cos 2\theta = \cos^2\theta - \sin^2\theta$

- $\cos 2\theta = 2\cos^2\theta - 1$

- $\frac{\cos 2\theta + 1}{2} = \cos^2\theta$

- $\sin^2\theta = \frac{1 - \cos 2\theta}{2}$

- $2\sin^2\frac{\theta}{2} = 1 - \cos\theta$

- The law of cosines states that for any triangle with sides of length A, B and C:

  $$C^2 = A^2 + B^2 - 2AB\cos\theta \text{ where C is the side opposite angle } \theta$$

Polar coordinates

Read this if you aren't familiar with polar coordinates. Without understanding this you won't understand, for example, how the formula for the normal distribution curve is derived.

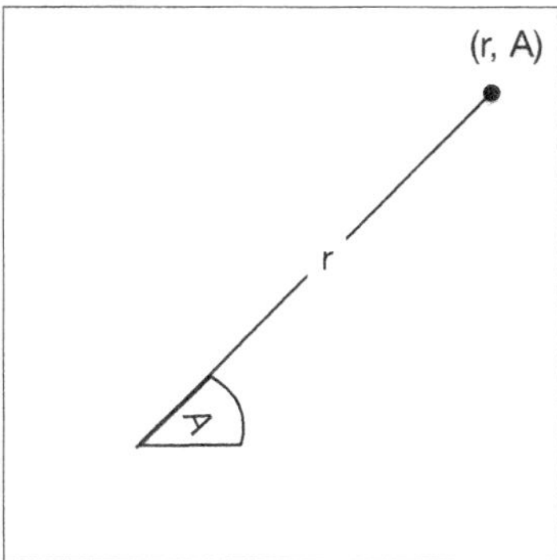

Figure 117

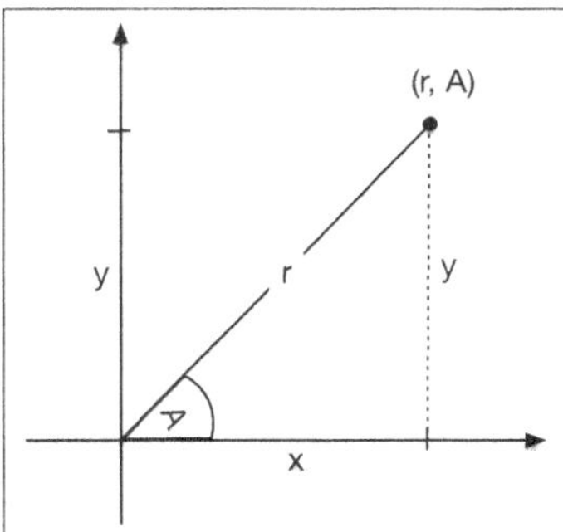

Figure 118

The Cartesian "rectangular" coordinates in an x-y plane are incredibly useful, but there are other ways to describe points on a plane which can be more useful in particular circumstances. **Polar coordinates** use an angle starting from the horizontal axis on the right at the origin, and the length of a radius. We can describe any point on an infinite plane this way (Figure 117).

To convert x-y coordinates to polar coordinates (Figure 118), the sine of the angle will be the y value over r, while the cosine of the angle will be the x value over r. This means that

$$y = r \sin A$$

and

$$x = r \cos A$$

(multiplying both sides of each of these equations by r). Using Pythagoras' theorem we can also see that

$$r^2 = x^2 + y^2$$

The formula for a circle with its centre at the origin and a radius of 1 units, using Cartesian coordinates is:

$$x^2 + y^2 = 1$$

whereas the formula for the same circle in polar coordinates is simply

$$r = 1$$

To cement your understanding:

Think of the point $(3,4)$ in Cartesian coordinates. What will be the length of r if you convert it to polar coordinates? (Hint: Pythagoras' theorem).

The complex plane

Read this if you don't know what the complex plane is.

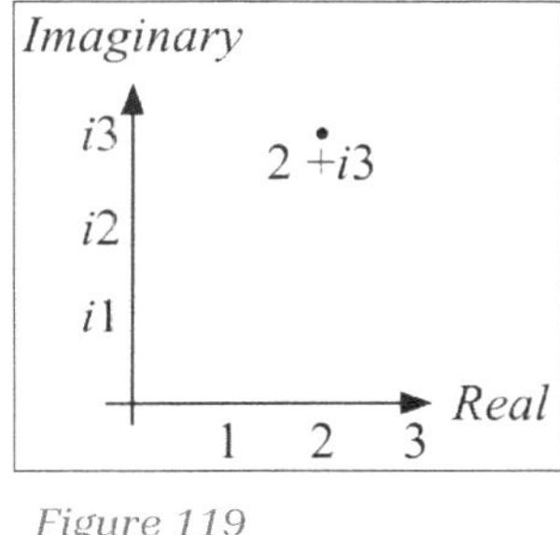

Figure 119

The use of *i* to represent the square root of minus one, as discussed earlier (page 105), arises from the fact that multiplying two negative numbers produces a positive result. This means that the square root of a negative number cannot exist within the real number system— unless a new type of number is introduced. Because of this, some people view *i* as not being "real" since it cannot be placed on a traditional number line.

There are some real-world applications, such as surveying, where that presents a problem. Danish (Norwegian born) surveyor and cartographer Caspar Wessel (1745-1818) solved this in 1797. He needed to be able to graph *i* for his cartography and couldn't do it, so he took a leaf from Descartes book, and created a plane where the "real" number line was the horizontal axis, while the "imaginary" number line (i.e. each number was multiplied by *i*) was the vertical axis. Coordinates on the plane were described by adding the real and imaginary component (Figure 119). This means that, unlike the Cartesian plane in which each point is represented by 2 numbers, in the complex plane each point is represented by a *single* number that has *two parts*.

Descarte's invention of the x–y plane revolutionised mathematics because it made it easier to link geometry and algebra. This meant that some algebraic issues could be visualised, and geometric problems could be addressed with algebra. There was a limit, though. With geometric shapes it is easy to move them to any size or position: just draw them differently. On the rectangular plane it is easy enough to change the size of an object or to

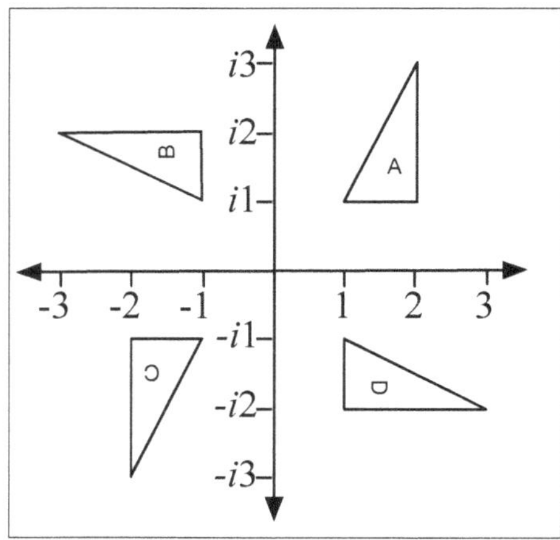

Figure 120

move it left, right, up or down through algebraic manipulation. But rotating objects is more challenging.

On the complex plane, however, simply multiplying points of an object by i has the effect of rotating them anticlockwise though $\frac{\pi}{2}$ radians (90°), about the zero point, or the origin. For example, in Figure 120 triangle A is bounded by the points: $(1 + i1), (2 + i1)$ and $(2 + i3)$. If you multiply each of these points by i you get the points bounding figure B: $(-1 + i1), (-1 + i2)$ and $(-3 + i2)$. When you do this, you will see that, because i multiplied by i is real (-1), while the real part multiplied by i becomes imaginary, the real and imaginary components of a complex number exchange positions. Similarly, if you multiply the points of triangle B by i you get triangle C, which is bounded by: $(-1 - i1), (-2 - i1)$ and $(-2 - i3)$. So when you multiply the points in triangle A by i then do it again, it has the same effect as multiplying the points in triangle A by -1, which is what you would expect.

Some other aspects of working in the complex plane—complex conjugates, dividing complex numbers and taking the square roots of complex numbers—aren't particularly relevant to the statistics we'll be covering, and so can be found in an appendix on pages 745 and following.

To cement your understanding:

• Draw the point $(-2 + i3)$. Now multiply the number by $i2$. (This means i multiplied by 2. When we use i multiplied by a number, the convention is to put the i first). Note that when you do, the real part, (-2), becomes imaginary, and the imaginary part becomes real (because multiplying i by i gives -1). Did you get $(-6 - i4)$? If not, can you see where you went wrong?

• Can you see where this would go on a diagram?

Calculus: the Mathematics of Predictable Change

Calculus is the mathematics of rates of change and of accumulation: of direction, of speed, of volume under a curved surface or, in fact, of anything which changes smoothly. That makes it useful for everything from astronomy to sub-atomic particles. Engineers of every description, economic analysts, environmental scientists and many other people trying to understand and improve the world are dependent on its insights.

For our every-day, human-scale of existence, however, thinking in terms of "predictable change" can lead to disappointment. Calculus, as a description of the way the world really works, has its limitations.

> *"Since all models are wrong the scientist must be alert to what is importantly wrong."*[24]
>
> George Box

At the human scale, the world isn't usually all that predictable; or at least not in the way that calculus works. To come to terms with uncertainty we need statistics; but understanding the mathematics of predictable change is a vital step towards improving our understanding of uncertainty.

Differential calculus

Read this if you aren't familiar with differential calculus.

(If you are struggling with this, see the How to Love Statistics YouTube video on *Differential Calculus*.)

The Cartesian plane was such a leap forward that shortly after its invention, English astronomer/physicist Isaac Newton (1642-1726) and German diplomat/public administrator/scientist Gottfried Leibniz (1646-1716) both (probably) independently used it to come up with the idea of calculus.

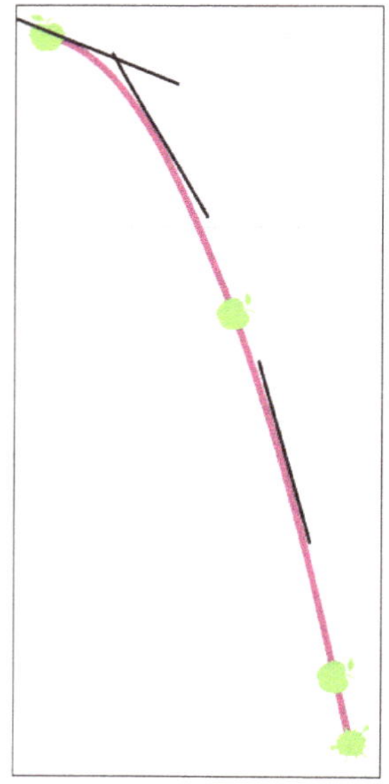

Figure 121

It's just possible that Newton (who also described gravity and used a prism to create a rainbow) was actually hit on the head by an apple as the old story goes, and was thereby encouraged to think about gravity. He may therefore have wanted to know how fast the apple was falling when it hit his head. It was clear that the higher it fell from, the faster it would fall. Falling objects accelerate, they don't start out at a constant speed. Assuming you could find an equation that described the descent, how could you work out the ever-changing speed at any instant (Figure 121)? That is one example of many situations where it becomes helpful to thinking about the slope of a curve.

We saw on page 194 that the slope of a straight line stays constant for its entire length, and that we can calculate it by looking at the vertical rise over the horizontal run. In the case of a curve, however, the slope keeps changing, just as the speed of Newton's falling apple kept changing.

To keep the mathematics simple, we'll stop thinking about a falling apple at this point, and just think about what it means to calculate the slope of a curve, and how we go about it. Remember that a tangent to a circle (page 174) is a straight line which just touches the circle at a single point. To calculate the slope of a curve at any point, we actually calculate the slope of a tangent to the curve at that point, using the rise over the run.

This will be different for every point on the curve. So what we need is an equation, in terms of x and y, that will describe the slope of a tangent to a curve *for a given value of x*. That way, whatever value of x we are looking at, we should be able to tell the slope of the tangent at that point.

We'll approach the problem as a three-stage process.

We'll start with a fairly simple curve in mathematical terms. Imagine a parabola with the equation of $y = x^2$ (Figure 122). First, we'll find the slope of a line that crosses the curve at two distinct points, so isn't a tangent. Second, we'll use algebra to come up with a formula to describe the slope of such a line between any two points. Third, we'll shrink the distance from the first point to the second one, until there is no distance left.

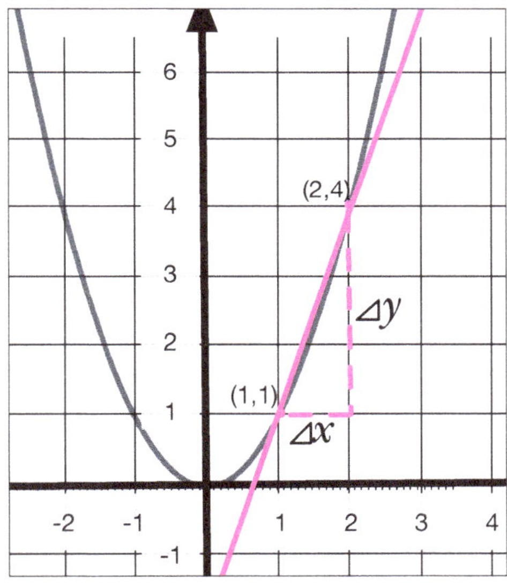

Figure 122

Stage 1: As our starting point, let's make $x = 1$, which means that y will also equal 1 (because $1^2 = 1$). We'll find our second point by adding 1 to x. That will make x equal 2, at which point y will equal 4 (because $2^2 = 4$). So we could draw a straight line through these two points (1,1) and (2,4) then work out its slope. As x has increased by 1, y has increased by 3, so the slope of this line (the vertical rise over the horizontal run) is 3 : 1, which is the same as just 3. That is how we can do it for two specific points.

Stage 2: Now let's come up with something more general, so that we can do it for any values we choose to give x and y. We can do this by adding something to x (as we've seen on page 207, we usually use the capital of the Greek letter delta, Δ, for this) then adding something corresponding to y, to keep the equation balanced. We'll call these Δx and Δy (delta x and delta y). The equation becomes:

$$y + \Delta y = (x + \Delta x)^2$$

We want to know what the ratio of this change in y is to the change in x, for any value we choose to give to x: that will give us the vertical rise over the horizontal run. That means we want to use algebra to come up with an expression for Δy over Δx.

This is our starting point. Call this equation 1.

$$y = x^2$$

Add something to x, which we'll call Δx, and a corresponding amount to y, which we'll call Δy, so that the equation stays balanced.

$$y + \Delta y = (x + \Delta x)^2$$

Expand out the binomial expression on the right. Call this equation 2.

$$y + \Delta y = x^2 + 2x\Delta x + (\Delta x)^2$$

Subtract equation 1 from equation 2 (see pages 108 and following on simultaneous equations).

$$\cancel{y} + \Delta y = \cancel{x^2} + 2x\Delta x + (\Delta x)^2$$

$$\Delta y = 2x\Delta x + (\Delta x)^2$$

Now divide both sides by Δx.

$$\frac{\Delta y}{\Delta x} = 2x + \Delta x$$

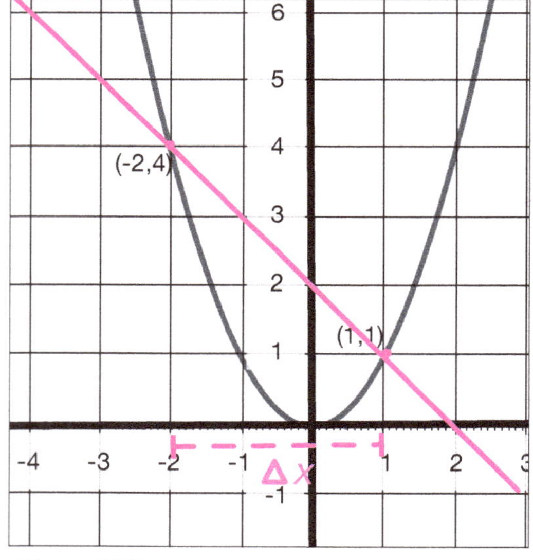

Figure 123

We've now solved part of the problem: we have a formula that will tell us the slope of the line in terms of x and whatever amount we decide to add to x (i.e. Δx). We started with $x = 1$, and Δx also equal to 1 (we increased x by 1 to find the second point). Plugging those numbers into the formula $2x + \Delta x$ gives us $2 \bullet 1 + 1 = 3$, which is what we worked out earlier as the slope between points (1,1) and (2,4). So that part seems to be working.

This formula works so well that it doesn't matter which point you start with, out of any two points where the line crosses the curve. In Figure 123 for example, finding the slope between the points $(-2,4)$ and $(1,1)$, we can start with the point $(-2,4)$. Our starting x value will be -2, and our Δx will be 3 (we have increased the x value from -2 to 1, which is a difference of 3). Using the formula

$$\frac{\Delta y}{\Delta x} = 2x + \Delta x$$

give us

$$\frac{\Delta y}{\Delta x} = 2(-2) + 3$$

$$= -1$$

Alternatively, if we make the starting point (1,1), then our starting x value will be 1, and our Δx will be -3 (because we are moving to the left on the x axis). Using the formula

$$\frac{\Delta y}{\Delta x} = 2x + \Delta x$$

gives us

$$\frac{\Delta y}{\Delta x} = 2(1) - 3$$

$$= -1$$

We get the same answer: the line slopes backwards, and each forward movement of x is matched by an identical fall in y.

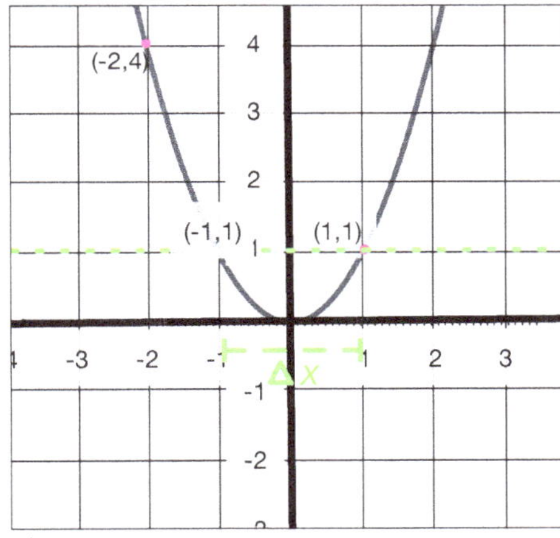

Figure 124

Stage 3: The next issue for us to resolve is that we don't want the slope of a line crossing a curve at two points; we want it to just touch the curve at one point. In Figure 123 the pink line doesn't remotely resemble a tangent to the curve. In Figure 124 if we take point (1,1) as the one we are interested in, and reduce Δx from -3 to -2, the line now passes through the points $(-1,1)$ and (1,1). It has flattened out to have a slope of 0, and, while still not close to being a tangent to the curve, it is a slight improvement. In Figure 125 if we make Δx smaller still, reducing it down to -1, the line passes through the points (0,0) and (1,1). It now has a positive slope, and is starting to vaguely resemble a tangent to the curve. Then in Figure 126 if we reduce Δx to $-\frac{1}{2}$, the line passes through the points $\left(\frac{1}{2},\frac{1}{4}\right)$ and (1,1).

As Δx gets smaller, the distance between the two points decreases, and the line gets closer and closer to being a tangent to the curve. That means we could *almost* just make $\Delta x = 0$, and have the equation we are looking for.

Almost, but not quite. The problem is what happens on the left-hand side of the equation. We are trying to find the ratio of

$$\frac{\Delta y}{\Delta x}$$

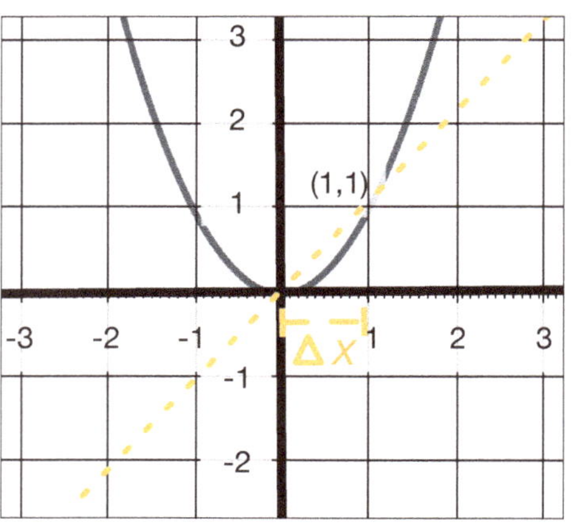

Figure 125

If $\Delta x = 0$, this becomes meaningless. When Δx is getting smaller and smaller, provided is stays slightly above 0, we don't have a problem. Δy will change in proportion to Δx, because it is the amount that y has to change when x changes, in order to keep the equation balanced: in other words, in order to keep the points on the curve. It doesn't matter if Δy and Δx are as small as we can imagine because, as we saw when we looked at the slope of a straight line and first looked at the idea of delta (page 207), if the line is straight (which it is when we are thinking of the tangent, not the curve) then the ratio between Δy and Δx won't change, no matter what size Δx is.

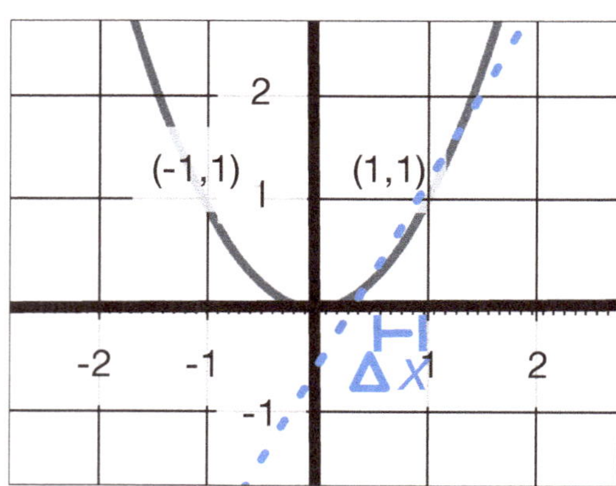

Figure 126

That means we can make Δx as close as we like to 0, as long as it doesn't actually become 0. The way around this is to use the same concept that we first saw when we looked at converging sequences (page 116): the idea of a limit. That is, a number gets smaller and smaller, getting closer and closer to another number, never quite reaching it but never able to get past it.

Back to our equation for a tangent to a curve, the Δx on its own on the right-hand side gets smaller and smaller until it can be ignored. We are left with:

$$for\ the\ curve\ y = x^2, \quad \lim_{\Delta x \to 0} \frac{\Delta y}{\Delta x} = 2x$$

That means that the slope of a line just touching the curve of a parabola with this equation, for *any* value of x, will be $2x$. If you think about the parabola with this equation (Figure 122) you can see that this seems to describe it well. The bigger the value of x, the steeper the slope of a line which touches the parabola will be. As x gets close to zero the slope of the line will become flatter until it is horizontal, and then as x becomes negative the line will slope backwards.

It was Leibniz who used delta to indicate the changes, and he also came up with the convention we use for the writing the limit as delta x and delta y approach zero: the lower case delta, or δ. These days we usually just use the letter d. So

$$\lim_{\Delta x \to 0} \frac{\Delta y}{\Delta x} = \frac{dy}{dx}$$

That's pretty much what there is to differential calculus (and you thought it was scary). You can use this approach to find the slope of virtually any curve described by an equation with x and y. This process is called **differentiating** the equation or finding the **differential**. The equation that results is known as the **derivative**.

Leibniz vs prime notation

The notation we have seen so far is called the Leibniz's notation. Sometimes it is more helpful to use function notation (page 111) to describe derivatives, using something close to an apostrophe, which is called **prime**.

If

$$y = f(x)$$

$$\frac{dy}{dx} = f'(x)$$

Using this notation, we sometimes describe the derivative as *f* **prime of** *x*.

One of the situations where this is useful is when we come to take the derivative *of* the derivative, or subsequent derivatives after that.

These are written as:

Function	$f(x)$
First derivative	$f'(x)$
Second derivative (i.e. derivative of the derivative)	$f''(x)$
Third derivative	$f'''(x)$ or $f^{(3)}(x)$
Fourth derivative	$f^{(4)}(x)$
nth derivative	$f^{(n)}(x)$

Key points:

- **Differential calculus** involves identifying formulae that will describe the **slope of a tangent** to a curve, for any value of x.

- This provides a way of assessing the **rate of change** of a curve at any point.

- The process to achieve this is to add a small amount (Δx) to the x value(s) in the formula, add a corresponding amount to the y(s) (Δy), using algebra to find a formula in terms of $\frac{\Delta y}{\Delta x}$, then taking the limit as Δx approaches 0.

- $\lim\limits_{\Delta x \to 0} \frac{\Delta y}{\Delta x} = \frac{dy}{dx}$

- This process is called **differentiating** the equation or finding the **differential**. The resultant formula is known as the **derivative**.

- **Prime notation** is sometimes used when differentiation is described using function notation. One of the situations where this is useful is in using higher derivatives.

Some basic rules of differential calculus

> **Read this if** you don't know how to use the power rule, or you don't know how to find: the derivative of: a polynomial expression, 1 over x, the square root of x, or $(x + A)^2$. Read it if you can't calculate the minimum value of a quadratic equation (and if you don't know that, you won't understand least squares in statistics), or the closest point of a curve to a line. Finally, read if it you don't understand what $f'(x) = \lim\limits_{dx \to 0} \frac{f(x+dx)-f(x)}{dx}$ means.

The derivative of a polynomial expression, and the addition rule

Now we'll look more generally about how this approach can be applied to any polynomial. Let's start by looking at one where the highest power of x is 3.

Let equation 1 be:
$$y = Ax^3 + Bx^2 + Cx + D$$

Add Δy to y and Δx to each x
$$y + \Delta y = A(x + \Delta x)^3 + B(x + \Delta x)^2 + C(x + \Delta x) + D$$

Expand out the binomial terms
$$y + \Delta y = Ax^3 + 3Ax^2\Delta x + 3Ax\Delta x^2 + A\Delta x^3 + Bx^2 + 2Bx(\Delta x) + B(\Delta x)^2$$
$$+ Cx + C\Delta x + D$$

Subtract equation one from this
$$\cancel{y} + \Delta y = \cancel{Ax^3} + 3Ax^2(\Delta x) + 3Ax\Delta x^2 + A\Delta x^3 + \cancel{Bx^2} + 2Bx\Delta x + B(\Delta x)^2$$
$$+ \cancel{Cx} + C\Delta x + \cancel{D}$$

$$\Delta y = 3Ax^2\Delta x + 3Ax\Delta x^2 + A\Delta x^3 + 2Bx\Delta x + B\Delta x^2 + C\Delta x$$

Divide both sides by Δx

$$\frac{\Delta y}{\Delta x} = 3Ax^2 + 3Ax\Delta x + A\Delta x^2 + 2Bx + B\Delta x + C$$

As Δx becomes smaller and smaller we write it as dx, and all terms that are multiplied by dx also approach zero, so we can ignore them.

$$\frac{\Delta y}{\Delta x} = 3Ax^2 + \cancel{3Ax\Delta x} + \cancel{A\Delta x^2} + 2Bx + \cancel{B\Delta x} + C$$

$$\frac{dy}{dx} = 3Ax^2 + 2Bx + C$$

To cement your understanding:

If this all looks confusing then get a pen and paper and write it out for yourself.

The addition rule

You might have noticed when we were working that through that we would have arrived at the same place if we had differentiated each of the terms on the right individually, then added them together. This is an example of the addition rule. To find the derivative of a function that includes two or more terms that are **added or subtracted**, you can just take the derivatives of the individual terms and add them. It can be expressed more generally as follows.

Where g and h are two functions that have x in them (rather than what we have seen in the past, which is using letters to represent individual values) then

$$f'(g + h) = f'(g) + f'(h)$$

This is known as the **addition rule**.

To get a more visual sense of what is happening, you can think about a function with an addition included in it, as just drawing a curve of each of the separate terms, then putting one on top of the other. So if you have a curve of $y = x^3$ and then add another curve of, say, $y = 6x^2$ then to find y at any point just take the value of x^3, then add (or subtract if it is negative) another $6x^2$ in order to reach the value of $x^3 + 6x^2$.

Then when you increase the x value by dx, the change in the dy value of the whole curve will increase by the dy of the first curve plus the dy of the second curve.

The power rule

(If you are struggling with this, see the How to Love Statistics YouTube video on *Why the Power Rule for Differentiation Works.*)

You might start to see a pattern here. The long way to do these equations is what we did: add Δy to each y and Δx to each x, expand out the resulting binomial expressions, subtract the first equation, divide both sides by Δx then treat any terms on the right that still have Δx in them as being equal to zero.

The shortcut for finding the derivative of a polynomial expression is to multiply each term by the power of x and reduce the power of x by 1. This is called the **power rule.**

If we write out equation one from page 235 as:

$$y = Ax^3 + Bx^2 + Cx^1 + Dx^0$$

it makes it easier to see that if we multiply each term by its exponent and then reduce the exponent by 1 we get the same answer we calculated earlier:

$$\frac{dy}{dx} = 3Ax^{(3-1)} + 2Bx^{(2-1)} + 1Cx^{(1-1)} + 0Dx^{(0-1)}$$

$$\frac{dy}{dx} = 3Ax^2 + 2Bx + C$$

To cement your understanding

When we started looking at differentiation on page 228 we found the derivative of the curve $y = x^2$. Use the power rule to find the derivative, and confirm that you get the same answer that we did when we worked it out the long way.

The derivative of a reciprocal

$$y = \frac{1}{x}$$

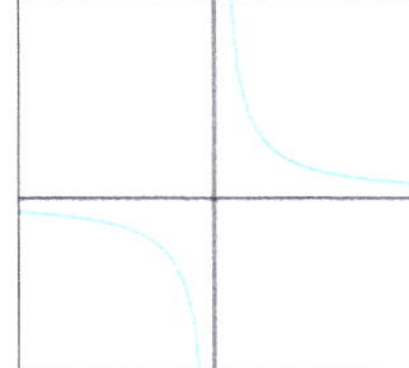

Figure 127

The graph of this expression forms a double hyperbola (Figure 127). (See the appendix on page 740 if you want to know more about what a hyperbola is.)

$$y = \frac{1}{x}$$

Multiply both sides by x.

$$xy = 1$$

(From now on we'll dispense with the capital delta, and just use the lower case.)

$$(y + dy)(x + dx) = 1$$

Expand the terms.

$$yx + xdy + ydx + dydx = 1$$

Substitute $\frac{1}{x}$ for y (because they are the same thing).

$$\frac{1}{x}x + xdy + \frac{1}{x}dx + dydx = 1$$

Clean up the equation.

$$1 + xdy + \frac{dx}{x} + dydx = 1$$

Subtract 1 from both sides.

$$xdy + \frac{dx}{x} + dydx = 0$$

Divide both sides by dx.

$$x\frac{dy}{dx} + \frac{dx}{xdx} + \frac{dydx}{dx} = 0$$

$$x\frac{dy}{dx} + \frac{1}{x} + dy = 0$$

Divide both sides by x

$$\frac{dy}{dx} + \frac{1}{x^2} + \frac{dy}{x} = 0$$

Subtract $\frac{1}{x^2}$ from both sides.

$$\frac{dy}{dx} + \frac{dy}{x} = -\frac{1}{x^2}$$

As dx approaches zero then $\frac{dy}{x}$ approaches zero, and can be ignored.

$$\frac{dy}{dx} = -\frac{1}{x^2}$$

We could have written the initial equation as

$$y = x^{-1}$$

Using the power rule: applying the shortcut of multiplying the term by the exponent then reducing the exponent by 1 would have given us

$$\frac{dy}{dx} = -x^{-2}$$

which is another way of writing the answer we arrived at, and would have been much quicker! So the power rule also applies in this situation.

The derivative of a square root

$$y = \sqrt{x}$$

Square both sides.

$$y^2 = x$$

Use the same analysis that we used above for $y = x^2$ but exchange the letters.

$$\frac{dx}{dy} = 2y$$

Take the reciprocal of each side (turning the numerator into the denominator and the denominator into the numerator).

$$\frac{dy}{dx} = \frac{1}{2y}$$

Because $y = \sqrt{x}$.

$$\frac{dy}{dx} = \frac{1}{2\sqrt{x}}$$

Now, we could have written $y = \sqrt{x}$ as

$$y = x^{\frac{1}{2}}$$

Then we could have just applied the same shortcut that worked in the equations above: multiplying the term by the exponent then reducing the exponent by one. That gives us

$$\frac{dy}{dx} = \frac{1}{2}x^{\frac{-1}{2}}$$

Which is another way of writing the answer we worked out. Once again, the power rule works.

The derivative of a variable plus a constant all squared

A is a constant

$$y = (x + A)^2$$

Expand the terms.

$$y = x^2 + 2Ax + A^2$$

Add dy and dx and expand the terms again.

$$y + dy = (x + dx)^2 + 2A(x + dx) + A^2$$
$$= x^2 + 2xdx + dx^2 + 2Ax + 2Adx + A^2$$

Subtract the initial equation.

$$\delta y = \cancel{x^2} + 2xdx + dx^2 + \cancel{2Ax} + 2Adx + \cancel{A^2}$$
$$= 2xdx + dx^2 + 2Adx$$

Divide both sides by dx.

$$\frac{dy}{dx} = 2x + dx + 2A$$

Rearrange the terms and factor out the 2.

$$\frac{dy}{dx} = 2(x + A) + dx$$

$$\lim_{\delta x \to 0} \frac{dy}{dx} = 2(x + A)$$

In other words, the power rule also applies to the equation

$$y = (x + A)^2$$

There is no need to expand the terms applying it. Which makes sense, because if x can be equal to any value, then it can be equal to another value of x with a constant added, and so it should work the same way.

To cement your understanding:

Work through the calculations above for yourself, writing them out if that is helpful, to make sure that you understand the reasoning.

The minimum value of a quadratic equation

Statistics has a lot to do with making the best guess we can with the available information (such as when we know there are errors but don't know where they are or how big they are). Differential calculus plays a role in helping us to know when the values in one equation are as close as they can be to the values expressed by another equation, if they are never exactly there. This is a conceptual leap in mathematics because the subject is usually about precision: when things are exactly equal. Statistics can involve the idea of "how close can we get?"

Take the generalised form of a quadratic equation:

$$y = Ax^2 + Bx + C$$

If the values of the constants are positive, this equation will form a parabolic shape. Figure 128 graphs the equation

$$y = 3x^2 + 4x + 5$$

To cement your understanding:

• Before reading on, if you were to try to work out the minimum value of the equation $y = 3x^2 + 4x + 5$, think about what the slope of the curve would be at the minimum point.

• What equation would you use to find this slope?

• Can you work out the answer before reading ahead?

We can use calculus to find the x value that will give the lowest possible value of y, by finding the point where the slope of the tangent to the line is horizontal. In other words, the slope will equal zero. If $y = Ax^2 + Bx + C$, then

$$\frac{dy}{dx} = 2Ax + B \text{ (applying the power rule)}$$

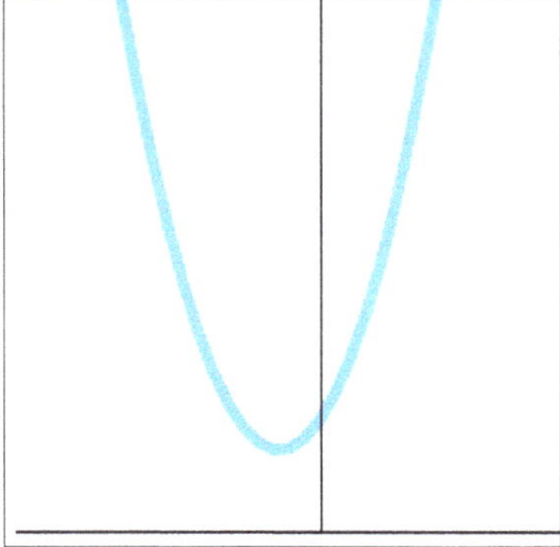

Figure 128

The lowest point of the curve will be found where this expression is equal to zero: the tangent to the curve will be flat.

$$2Ax + B = 0$$

$$2Ax = -B$$

$$x = \frac{-B}{2A}$$

This gives us a general formula for the minimum value of a quadratic equation. Using those values for the equation of our graph, $y = 3x^2 + 4x + 5$, tells us that the minimum value of the curve will occur where

$$x = \frac{-4}{6}$$

$$= \frac{-2}{3}$$

For this value of x, the lowest value of y will be

$$y = 3\left(\frac{-2}{3}\right)^2 + 4\left(\frac{-2}{3}\right) + 5$$

$$= \frac{11}{3}$$

The closest point of a curve to a line

Using a similar approach to the way we found the lowest point on a parabola, we can use calculus to find the point at which a curve that never meets a line comes closest to it. Figure 129 shows a small portion of the graphs of the line $y = 5x - 2$ (in green) and the curve $y = x^2 + x + 3$ (in blue).

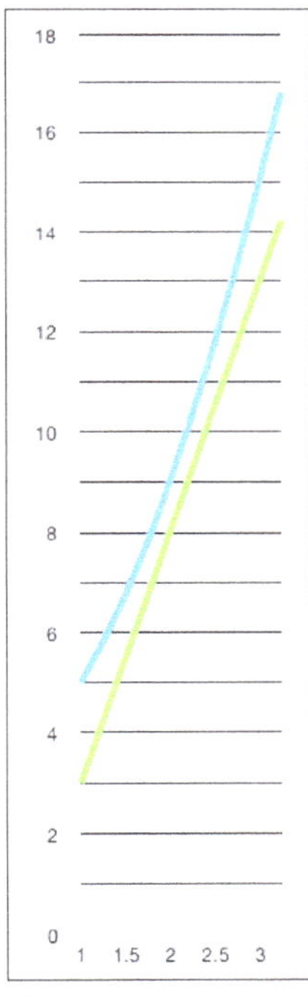

Figure 129

These two equations will never meet. There are no values of x that will produce the same value for y.

To cement your understanding:

Can you work out the how to go about this before reading ahead?

We can find the closest point between them though, by finding the point where the tangent to the blue curve is parallel to the green line. The slope of the green line is 5:1, or just 5. In other words, for each change in x, y will change 5 times. (The -2 in the equation doesn't change the slope; it drops the line down two units. In other words the line will cross the y axis—where x is equal to zero—at the point where $y = -2$.) This means we need to find the value of x where the tangent to the blue curve also has a slope of 5.

$$y = x^2 + x + 3$$

Use the power rule to differentiate.

$$\frac{dy}{dx} = 2x + 1$$

We need $\frac{dy}{dx}$ to equal 5

$$2x + 1 = 5$$

Subtract 1 from each side, then divide each side by 2.

$$x = 2$$

The point where the x value of the blue curve is equal to 2 is the point where the blue curve comes closest, in a direct line, to the green line.

Generalised form

This is a generalised form of what we have been doing.

$$f'(x) = \lim_{dx \to 0} \frac{f(x + dx) - f(x)}{dx}$$

In other words, add a small amount to each x within the function, subtract the original function, and divide the result by the same small amount, then take the limit as that amount approaches zero.

Key points:

- The addition rule is used to differentiate a sum of functions:

$$f'(g + h) = f'(g) + f'(h)$$

- This rule generalizes to sums of any number of functions.

- The power rule is used to differentiate functions of the form $f(x) = Ax^n$

$$\frac{d(Ax^n)}{dx} = nAx^{n-1}$$

- The derivative of the reciprocal of x, the square root of x and x plus a constant all squared follow the power rule: it doesn't only apply when the power of x is an integer.

- The minimum value of a quadratic equation in the form $y = Ax^2 + Bx + C$ is the point where

$$x = \frac{-B}{2A}$$

- The closest point on a curve to a line is found where the derivative of the curve equals the slope of the line.

- A generalised form of the formula for differentiation is given by

$$f'(x) = \lim_{dx \to 0} \frac{f(x + dx) - f(x)}{dx}$$

Differential calculus with trigonometry part 1

> **Read this if** you don't know what the derivates of the sine and cosine functions are.

Derivative of the sine function

There are different proofs of the derivatives of the sine and cosine functions. This is one of them.

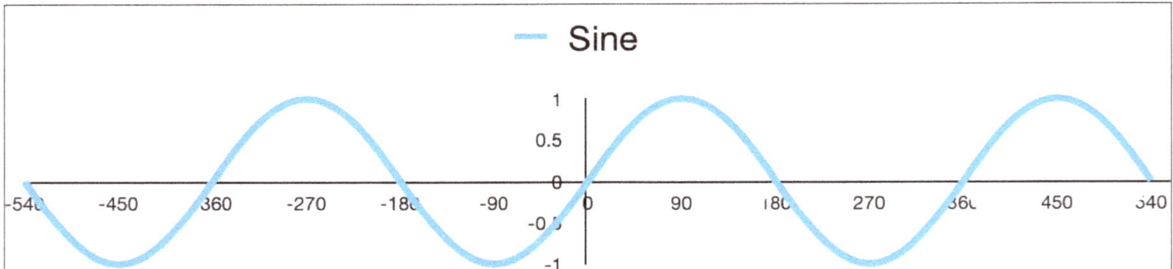

Figure 130
x axis is in degrees

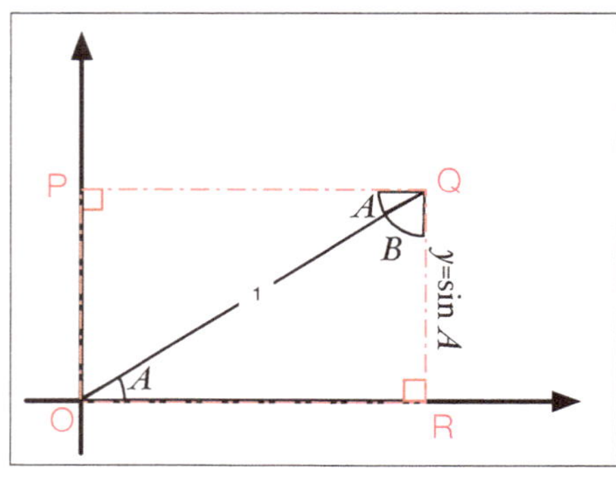

Figure 131

As we saw when looked at trigonometry and the Cartesian plane, a sine wave maps the size of an angle along the x axis (if an angle can go all the way around a circle through 2π radians and then keep going), and the value of the sine of that angle along the y axis . To find the derivative of the sine wave, we need to work out the ratio of small changes in the angle to small changes in its sine, and find the limit of that ratio when those changes approach zero. Unfortunately the power and addition rules don't help us.

Figure 131 shows a rectangle with the bottom left corner at the origin, and a diagonal of 1 unit. Interval QR is the sine of angle A. It is easiest if we use radians for this proof. The angles of the triangle $\triangle QOR$ will be the same as the angle measurement of a straight line; that is, π radians. The right angle will be $\frac{\pi}{2}$ radians, which means that angle A plus angle B must add to $\frac{\pi}{2}$ radians, or one right angle. That in turn means that angle $\angle OQP$ must be the same as angle $\angle QOR$; that is, A radians.

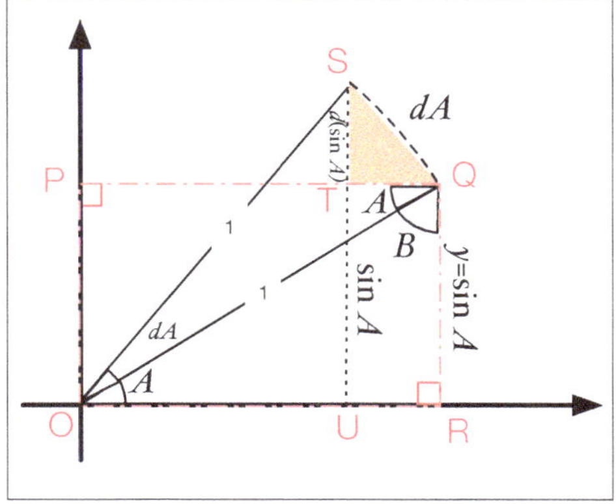

Figure 132

Now we'll create a new point on the unit circle, S. This will increase the angle of A by a small amount, which we'll call dA (Figure 132). We are using radians and so dA, as well as being the amount that we have increased the angle A, is also the length of the arc between points Q and S. (If you can't remember why this is, refer back to page 184.) The diagram also now includes a perpendicular line from point S through point T to point U.

The sine of the angle $\angle SOU$ is the length of the interval $\overline{SU}$. This interval is made of two intervals: $\overline{TU}$ which is the same length as $\overline{QR}$; that is, $\sin A$; and $\overline{ST}$, which is the corresponding amount that $\sin A$ has to increase when angle A is increased by dA. We'll designate this $d(\sin A)$. (It's more usual in mathematics to call it $\sin dA$ but that just gets confusing, so we aren't following that convention here.)

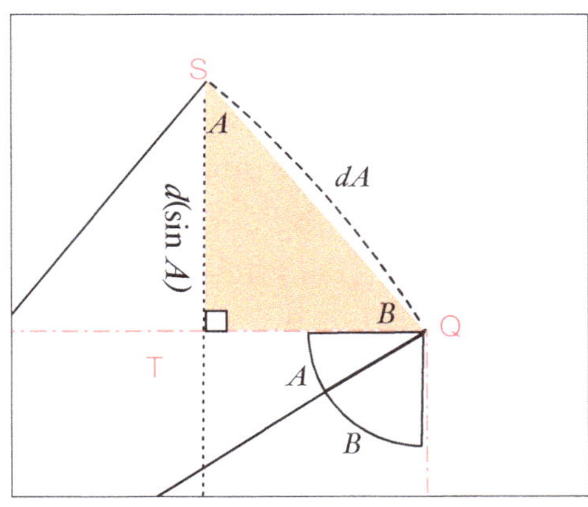

Figure 133

We're interested in the ratio of $d(\sin A)$ to dA, as dA gets smaller and smaller and approaches zero, and so we'll focus in on what is happening around the orange triangle (Figure 133).

As dA gets closer and closer to zero, the arc of length dA will get closer and closer to the hypotenuse of the orange triangle; so much so that we can treat them as being the same length. As this happens the angle marked with an A on a white background and a B on an orange background will get closer and closer to being a right angle. That's what tells us that the bottom right angle of the orange triangle is, in fact, equal to B. If that angle is equal to B, then the angle at the top of the orange triangle must equal A. We can see from this that the cosine of A (adjacent over hypotenuse) is what we have been looking for: the ratio of

$$\frac{d(\sin A)}{dA} = \cos A$$

So the derivative of the sine-wave graph is equal to the cosine.

> **To cement your understanding:**
>
> Draw up these figures for yourself, making sure that you can see how all of the values written on the diagram were calculated.

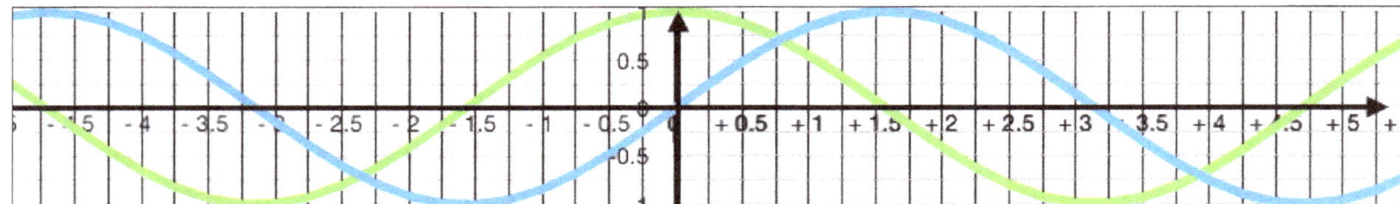

Figure 134
x axis is in radians

Figure 134 graphs a sine wave (in blue) and a cosine wave (in green). You can see that the value of cosine at any value of x does, indeed, match the slope of the sine wave. For example, at the points where the sine wave flattens out, or has zero slope—its maximum and minimum values—the cosine wave passes through zero on the y axis. Where the sine wave has a negative slope—it slopes backwards—the value of cosine is negative, and so on.

Derivative of the cosine function

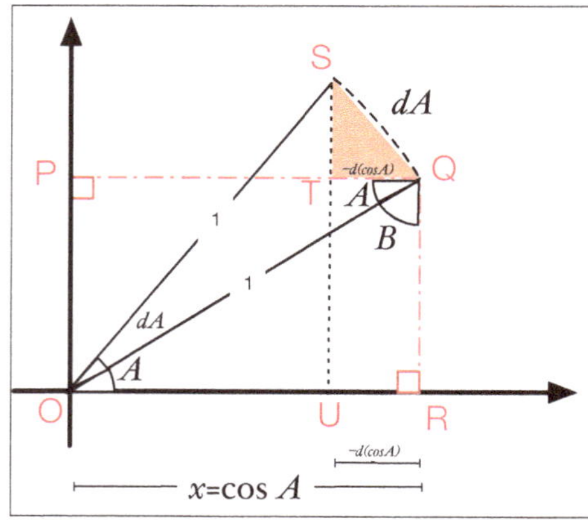

Figure 135

Using the same diagrams, starting with the cosine of A being the x axis of our first picture (Figure 135), you can use the angles we have now established to find the derivative of the cosine of A. As the angle of A near the origin increases, the length of the adjacent side decreases which means that we need to think in terms of a negative direction. The amount we need to add to the cosine of A is negative: $-d \cos A$. This is the length $\overline{\text{UR}}$ and is also the length $\overline{\text{TQ}}$.

When we focus in on the orange triangle we can see that, as dA approaches 0

$$\frac{-d(\cos A)}{dA} = \sin A$$

What we are actually after is:

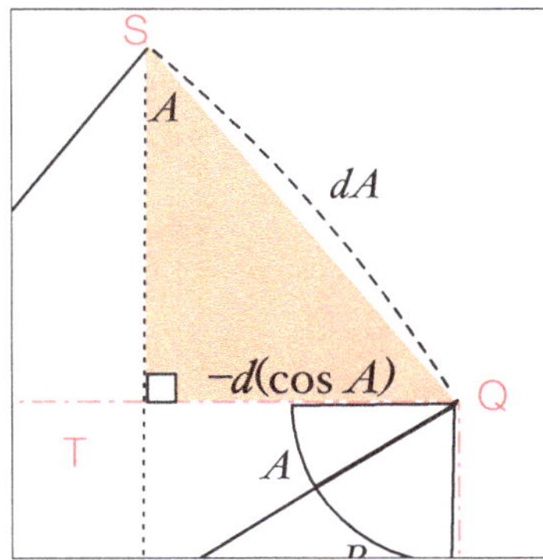

Figure 136

$$\lim_{dA \to 0} \frac{d(\cos A)}{dA}$$

That is, the ratio between a small change in the cosine to a small change in the angle, as the small change approaches 0. That gives us the derivative of the cosine function.

We can find this by multiplying both sides by -1.

$$\frac{d(\cos A)}{dA} = -\sin A$$

The derivative of cos A is therefore $-\sin A$.

To cement your understanding

Work this through for yourself, to check that it make sense.

Once again, you can use Figure 134 to trace the pattern of the slope of the cosine wave and see that it is the negative of the value of the sine wave.

Key points:

- The derivative of sine is cosine.

- The derivative of cosine is negative sine.

More rules for differentiating combined functions

> **Read this if** you aren't familiar with the product rule, the quotient rule and/or the chain rule.

We've already seen the addition rule and the power rule. There are three additional rules that are useful when differentiating more complex equations:

1. the **product rule** for functions that are multiplied together;

2. the **chain rule**, for functions within functions; and

3. the **quotient rule**, for dividing one function by another.

The product rule

Where functions are **multiplied**, you can't just take the derivatives of each and multiply them. There is a way to differentiate them, though. Let's start with two functions that are multiplied. We *could* call our two functions of x: $g(x)$ and $h(x)$. The brackets with x, however, would make the algebra that follows look unnecessarily confusing, so we will simplify, and just call them g and h. For the next section the functions are coloured, to make it a bit simpler to follow what is happening. We'll call dg the change in the output of function g when we increase x by dx, and dh the change in the output of function h when we increase x by dx. Remember that we are using g and h to represent whole functions of x, rather than just numeric values. Now we can just solve this the way we have determined derivatives previously.

This is our starting equation.
$$y = gh$$

Add a delta amount to each function, as we have been doing.

$$y + dy = (g + dg)(h + dh)$$

Expand out the terms.

$$y + dy = gh + hdg + gdh + dgdh$$

Subtract the first equation.

$$\cancel{y} + dy = \cancel{gh} + hdg + gdh + dgdh$$

$$dy = hdg + gdh + dgdh$$

Divide both sides by dx. As dx approaches zero, dh will approach zero, so the last term in this equation will approach zero and can be ignored.

$$\frac{dy}{dx} = h\frac{dg}{dx} + g\frac{dh}{dx} + \frac{dg}{dx}dh$$

$$\boxed{\frac{dy}{dx} = h\frac{dg}{dx} + g\frac{dh}{dx}}$$

That means that the derivative of two functions multiplied together will be the first function multiplied by the derivative of the second, added to the second function multiplied by the derivative of the first.

Putting this into function notation, we can say that for $f(x) = g(x)h(x)$ (where $g(x)$ and $h(x)$ are both functions of x), then

$$\boxed{f'(gh) = f'(g)f(h) + f(g)f'(h)}$$

If there were more than two terms multiplied together, we could have just repeated the process. In each case the term will involve one derivative, with the remaining functions in the term left as they were.

In other words, for $f(x) = f(g)f(h)f(i)$

$$\boxed{f'(x) = f(g)f(h)f'(i) + f(g)f'(h)f(i) + f'(g)f(h)f(i)}$$

This is known as the **product rule**.

To cement your understanding:

Work out the derivative of $y = 4x^2 \sin x$ using the product rule. (Hint: let $4x^2 = f(g)$ and $\sin x = f(h)$)

The chain rule

For **functions of other functions**, it is best to think of them as two different functions, with the "inside" having an effect on the "outside" one. Thinking about them as different graphs on different axes can help with this.

Suppose our function is $y = sin(x^3)$. Let's break that down into two different functions. We'll make $x^3 = w$. Therefore $y = \sin w$.

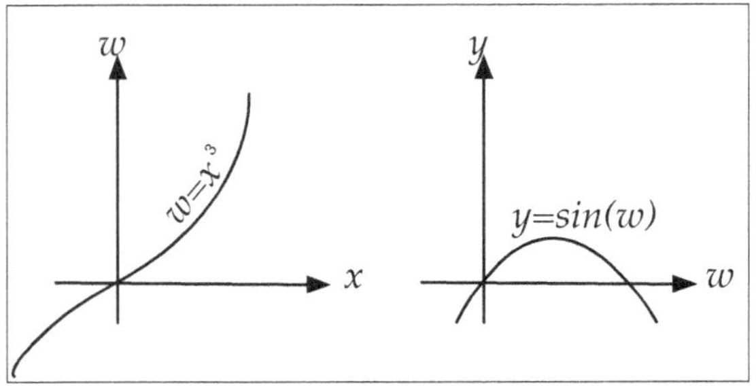

Figure 137

Now imagine two graphs (Figure 137). To get what is happening clear: as x changes, w changes (left diagram). Then as w changes, y changes (right diagram). What we want to work out is, first, what happens to y in response to a small change in x? Second, what happens to the ratio between these small changes as the change in x approaches zero?

So on the left diagram, we can say that

$$\frac{dw}{dx} = 3x^2 \text{ (using the power rule on } w = x^3)$$

On the right diagram:

$$\frac{dy}{dw} = cos\ w \text{ (because the derivative of sine is cosine)}$$

Neither of those are what we want to know, though. What we want to find is $\frac{dy}{dx}$.

We can get that by multiplying $\frac{dw}{dx}$ with $\frac{dy}{dw}$.

$$\frac{dy}{dx} = \frac{dw}{dx}\frac{dy}{dw}$$

If we do that the dw in both expressions will cancel out. So the product of the derivative of both of our equations will give us our derivative of the whole equation. That is:

$$\text{For } y = \sin x^3$$

$$\frac{dy}{dx} = 3x^2 cos(x^3)$$

More generally

$$\lim_{dx \to 0} (g(f(x))) = \big(g'(g(x))\big)\big(f'(x)\big)$$

In other words, we multiply the derivative of the "outer function" by the derivative of the "inner function".

This is called the **chain rule**.

> **To cement your understanding:**
>
> Work out the derivative of $\cos(x^4)$ using the chain rule. (Recall that we saw earlier that the derivate of the cosine of an angle is the negative of the sine of the angle.)

The quotient rule

The quotient rule applies when you are finding the differential of one function divided by another. Its proof uses the chain rule and the product rule. We'll go back to the notation we used for the product rule for this, except that we'll use the alternative way of expressing derivatives, because that is (hopefully) a simpler way to view it in this case.

Let g , h and y be functions of x (or any other variable you like).

$$y = \frac{g}{h}$$

$$= gh^{-1}$$

h is a function, so h^{-1} is a function within a function. That means we will need to use the chain rule to differentiate it. Once we know what its derivative is, we can use the product rule to find the derivative of y.

Introduce a new function of x, and designate it w. (This is just here to simplify the working. We'll get rid of it again before we finish.)

$$w = \frac{1}{h}$$

$$= h^{-1}$$

By the chain rule, multiply the derivatives of the inner and outer functions. Use the power rule to find the derivative of h^{-1}.

$$w' = -h^{-2}h'$$

$$= -\frac{h'}{h^2}$$

$y = gw$, so we can use the product rule to differentiate it.

$$y' = wg' + gw'$$

Substitute back to replace w.

$$y' = \frac{1}{h}g' + g\left(-\frac{h'}{h^2}\right)$$

$$= \frac{g'}{h} - \frac{gh'}{h^2}$$

Multiply the numerator and denominator of the left-hand term on the right side of the equation by h, to find a common denominator.

$$y' = \frac{hg'}{h^2} - \frac{gh'}{h^2}$$

$$\boxed{y' = \frac{hg' - gh'}{h^2}}$$

So to differentiate one function divided by another, multiply the function in the denominator by the derivative of the function in the numerator, subtract from that the product of the function in the numerator and of the derivative of the function in the denominator, and divide all of that by the square of the function in the denominator.

Simplifying polynomial fractions

Calculus can get particularly tricky when it comes to dealing with one polynomial divided by another, even when we use the quotient rule. There are two techniques that can help to simplify expressions of this nature: polynomial long division, and partial-fraction decomposition. If you are using mathematics for statistics you probably won't need to use these much, whereas for some disciplines using mathematics you are likely to use them all the time. There are two appendices showing how to use each of these techniques, if you are interested, on pages 750 and 753.

Key points:

- The **product rule** says that the derivative of two functions multiplied together will be the first function multiplied by the derivative of the second, added to the second function multiplied by the derivative of the first.

- The **chain rule** says that, to find the derivative of a function within a function, multiply the derivative of the 'inner function' by the derivative of the 'outer function'.

- The **quotient rule** says that the derivative of one function (the numerator) divided by another (the denominator) is the denominator function multiplied by the derivative of the numerator function, less the derivative of the denominator function multiplied by the numerator function, all divided by the square of the denominator function.

Differential calculus with trigonometry part 2

> **Read this if** you don't know what the derivatives of the tangent function or the arc tangent functions are.

Derivative of the tangent function

We saw when we looked first looked at trigonometric functions that

$$Tangent = \frac{Sine}{Cosine}$$

Use the quotient rule (page 251) to find the derivative. Remember that $(\cos x)^2 = \cos^2 x$ (page 197), that the derivative of sine is cosine (page 244), and the derivative of cosine is –sine (page 246)

$$\frac{d(\tan x)}{dx} = \frac{\cos x \cos x - - \sin x \sin x}{\cos^2 x}$$

We saw when we looked at trigonometric identities that $\cos^2 x + \sin^2 x = 1$ (page 218)

$$= \frac{\cos^2 x + \sin^2 x}{\cos^2 x}$$

Recall that the reciprocal of cosine is the secant (page 194)

$$\frac{d(\tan x)}{dx} = \frac{1}{\cos^2 x}$$
$$= \sec^2 x$$

Derivative of the arc tangent function

We haven't looked at the derivatives of other inverse functions, but this result is a useful one with integral calculus, and we will use it when we look at statistics.

Recall that this is how we can express the inverse tangent function, which is also known as the arc tangent. The y is an angle, which has x as its tangent (page 194)

$$y = \tan^{-1}x$$

$$x = \tan y$$

Based on the previous set of calculations, but with x and y interchanged.

$$\frac{dx}{dy} = \frac{1}{\cos^2 y}$$

Take the reciprocal of both sides. This is the answer, except that we want this as a function of x, not of y.

$$\frac{dy}{dx} = \cos^2 y$$

$\cos^2 y + \sin^2 y = 1$, so dividing by it doesn't change anything.

$$= \frac{\cos^2 y}{\cos^2 y + \sin^2 y}$$

Divide the numerator and denominator by $\cos^2 y$ which, once again, doesn't change anything.

$$= \frac{1}{1 + \dfrac{\sin^2 y}{\cos^2 y}}$$

This is another way of writing the same thing.

$$= \frac{1}{1 + \left(\dfrac{\sin y}{\cos y}\right)^2}$$

$$\frac{\sin y}{\cos y} = \tan y$$

$$= \frac{1}{1 + (\tan y)^2}$$

We started by saying that $x = \tan y$ because $y = \tan^{-1}x$

$$\boxed{\frac{d(\tan^{-1}x)}{dx} = \frac{1}{1 + x^2}}$$

Key points:

- The derivative of the tangent of x is secant squared x.

- The derivative of the arc tangent of x is one over one plus x squared.

l'Hôpital's rule

> **Read this if** you aren't familiar with l'Hôpital's rule, and when to use it.

This rule was first developed by Johann Bernoulli 1666–1748), one of eight mathematicians named Bernoulli, when he was a student of Guillaume-François-Antoine Marquis de l'Hôpital (1661-1704), but l'Hôpital (sometimes written as l'Hôspital) paid him in order to be able to take the credit, and his name has lived on in perpetuity as a result. (Money well spent if eternal fame is important, because l'Hôpital isn't remembered for anything else. He did, in fact, write an early textbook on calculus, but he isn't remembered for it.)

l'Hôpital's rule is useful when we want to understand what is happening when you divide one function by another, at a particular point where trying to divide them stops making sense.

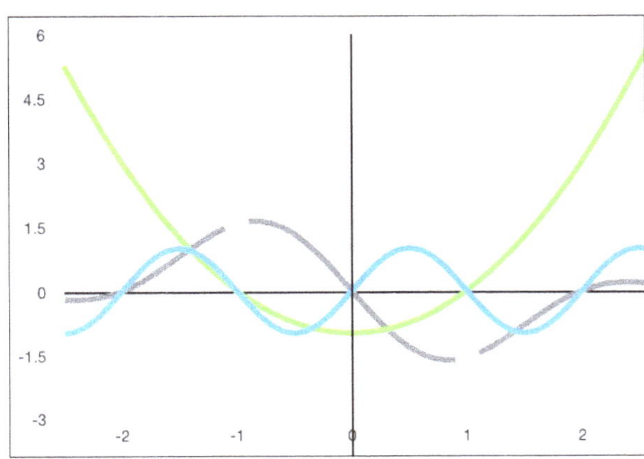

Figure 138

For example Figure 138 shows a graph of three different equations. The green parabola shows the equation

$$y = x^2 - 1$$

The blue wavy line shows the equation

$$y = \sin(\pi x)$$

The grey line shows the ratio between these two equations:

$$y = \frac{\sin(\pi x)}{x^2 - 1}$$

You can see that at the points where $x = -1$ and $x = 1$, both the blue curve and the green curve have a y value of zero. At that point the grey line has a gap. The computer that generated it can't provide a value, because it is trying to divide zero by zero. Yet we can see from the graph that it should be possible to give the ratio a value; or certainly at values where these two equations get extremely close to zero. In fact, when x is equal to 0.9999999999, y is equal to –1.57079521790074.... When x is equal to 1.0000000001, y is equal to –1.57079621354292...

So it is reasonable to assume that the value of y as x approaches 1 will be something very close to –1.57079... , but the formula doesn't tell us that. It tells us that we have zero divided by zero, and that the amount cannot therefore be calculated. What we really need is to be able to calculate the ratio as the numerator and denominator approach zero, rather than having to calculate it when they are both equal to zero.

l'Hôpital's rule was designed for this sort of situation. Instead of using differentiation to find the slope of a curve, it uses the slope of the curve to find the curve itself. It works like this.

You will recall (page 242) that we derived an equation for differentiation:

$$f'(x) = \lim_{dx \to 0} \frac{f(x + dx) - f(x)}{dx}$$

This was based on taking a point on a curve then adding a small increment to it. Suppose we chose to nominate the point where we want to work out the slope of the tangent, the value of some number x_1 (doesn't matter what that is), as in Figure 139.

So we can say that

$$x_1 + dx = x$$

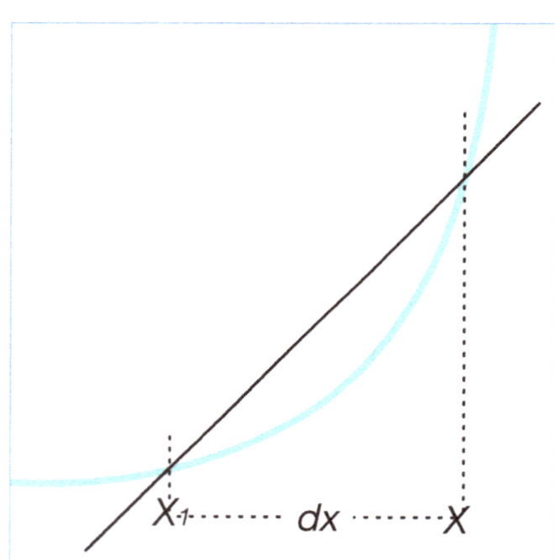

Figure 139

$$dx = x - x_1$$

Substitute the values we have defined into our equation for differentiation.

$$\frac{f(x + dx) - f(x)}{dx} = \frac{f(x) - f(x_1)}{x - x_1}$$

As dx approaches zero, x approaches x_1.

$$\lim_{dx \to 0} \frac{f(x + dx) - f(x)}{dx} = \lim_{x \to x_1} \frac{f(x) - f(x_1)}{x - x_1}$$

Now suppose we have two different functions that have x in them (it doesn't matter what they are). We'll call one $f(x)$ and the other $g(x)$.

$$f'(x) = \lim_{x \to x_1} \frac{f(x) - f(x_1)}{x - x_1}$$

By the same reasoning we used above.

$$g'(x) = \lim_{x \to x_1} \frac{g(x) - g(x_1)}{x - x_1}$$

Divide the derivative of *f(x)* by the derivative of *g(x)*.

$$\frac{f'(x)}{g'(x)} = \frac{\lim\limits_{x \to x_1} \dfrac{f(x) - f(x_1)}{x - x_1}}{\lim\limits_{x \to x_1} \dfrac{g(x) - g(x_1)}{x - x_1}}$$

Multiply the numerator and the denominator by $x - x_1$

$$\frac{f'(x)}{g'(x)} = \lim_{x \to x_1} \frac{f(x) - f(x_1)}{g(x) - g(x_1)}$$

This becomes useful where $f(x_1)$ and $g(x_1)$ both equal zero, and where they both approach infinity. Where $f(x_1)$ and $g(x_1)$ both equal zero the expression becomes:

$$\frac{f'(x)}{g'(x)} = \frac{f(x)}{g(x)}$$

where $f(x)$ and $g(x)$ approach 0.

This is l'Hôpital's rule, and it is possible to prove that it is true for any value of x for functions that don't have breaks in them, and where $g'(x) \neq 0$.

The proof is more complex than we have shown here when we are looking at the values of the function other than zero, but for our purposes in this book we only need it for when $f(x)$ and $g(x)$ do in fact both equal zero.

Back to our example in Figure 138:

$$y = \frac{\sin(\pi x)}{x^2 - 1}$$

For

$$y = \sin(\pi x)$$

$$\frac{dy}{dx} = \pi \cos(\pi x)$$

(using the chain rule).

For

$$y = x^2 - 1$$

$$\frac{dy}{dx} = 2x$$

(using the power rule).

The points where our graph broke were at plus and minus 1. $\cos \pi = -1$, so, where $x = 1$

$$\frac{\pi \cos(\pi x)}{2x} = \frac{\pi \cos(\pi \bullet 1)}{2 \bullet 1}$$

$$= \frac{\pi(-1) \bullet \cancel{1}}{2 \bullet \cancel{1}}$$

$$= \frac{-\pi}{2}$$

The cosine of $-\pi$ also equals -1, so where $x = -1$

$$\frac{\pi \cos(\pi x)}{2x} = \frac{\pi \cos(-\pi)}{2(-1)}$$

$$= \frac{\pi \cancel{(-1)}}{2 \cancel{(-1)}}$$

$$= \frac{\pi}{2}$$

These are the limits as x approaches ± 1. Putting these numbers into our graph would enable the curve in Figure 138 to be drawn continuously.

To cement your understanding:

• Calculate $\frac{x^3 - 8}{x^2 - 4}$ where $x = 2$, to confirm that it equals $\frac{0}{0}$.

• Now use l'Hôpital's rule to find the limit as x approaches 2. Confirm that your equation is now $\frac{3x^2}{2x}$. What is the value of this equation when $x = 2$?

Key point from this section:

According to l'Hôpital's rule, if $\lim\limits_{x \to c} f(x) = 0$ and $\lim\limits_{x \to c} g(x) = 0$, then

$$\lim_{x \to c} \frac{f(x)}{g(x)} = \lim_{x \to c} \frac{f'(x)}{g'(x)}$$

Taylor series

> **Read this if** you aren't familiar with Taylor series, or you want to know how computers calculate trigonometric ratios etc.

(If you are struggling with this, see the How to Love Statistics YouTube video on *Taylor/McLaurin/Power series.*)

Taylor series are used to find polynomial equations which at least approximate, if not completely replicate, non-polynomial functions. This turns out to be extremely useful in some disciplines, including physics, engineering, and statistics. Apart from anything else, a Taylor series is how a calculator works out many things, such as trigonometric functions. Considering that Brook Taylor, who came up with it, lived between 1685 and 1731, that's quite something. We will see a lot of Taylor series when we look at statistics.

In mathematics, many of the most practical ideas came from **pure mathematics**: where people basically just played with mathematical concepts, to see what they could come up with, rather than trying to solve practical problems. The practical applications came afterwards. The Taylor series probably started out like this.

With the polynomials we have differentiated so far, we have had the higher powers of x on the left of the polynomial, and the lower powers on the right. Taylor decided to write his polynomials the other way around (why not?). He then divided each term by the factorial of whatever power x was raised to.

Taylor's basic polynomial function can be written as:

$$f(x) = A + Bx + C\frac{x^2}{2!} + D\frac{x^3}{3!} + E\frac{x^4}{4!} \cdots$$

$$= A\frac{x^0}{0!} + B\frac{x^1}{1!} + C\frac{x^2}{2!} + D\frac{x^3}{3!} + E\frac{x^4}{4!} \cdots$$

To cement your understanding:

Before reading on, have a go at differentiating this for yourself, to see what happens.

The interesting thing is what happens when you differentiate it:

$$f(x) = A + Bx + C\frac{x^2}{2!} + D\frac{x^3}{3!} + E\frac{x^4}{4!} \cdots$$

$$= A\frac{x^0}{0!} + B\frac{x^1}{1!} + C\frac{x^2}{2!} + D\frac{x^3}{3!} + E\frac{x^4}{4!} \cdots$$

$$f'(x) = 0A\frac{x^{0-1}}{0!} + 1B\frac{x^0}{1!} + 2C\frac{x^{2-1}}{(2)(1!)} + 3D\frac{x^{3-1}}{(3)(2!)} + 4E\frac{x^{4-1}}{(4)(3!)} \cdots$$

$$= \cancel{0A\frac{x^{0-1}}{0!}} + \cancel{1}B\frac{x^0}{\cancel{1}!} + \cancel{2}C\frac{x^{2-1}}{\cancel{(2)}(1!)} + \cancel{3}D\frac{x^{3-1}}{\cancel{(3)}(2!)} + \cancel{4}E\frac{x^{4-1}}{\cancel{(4)}(3!)} \cdots$$

$$= B\frac{x^0}{0!} + C\frac{x^1}{1!} + D\frac{x^2}{2!} + E\frac{x^3}{3!} + \cdots$$

$$= B + Cx + D\frac{x^2}{2!} + E\frac{x^3}{3!} + F\frac{x^4}{4!} \cdots$$

The power of x multiplies the term (by the power rule). But it cancels out the first number of the factorial. It disappears, and factorial is reduced by 1.

Using the same approach, if you differentiate it again you get:

$$f''(x) = C + Dx + E\frac{x^2}{2!} + F\frac{x^3}{3!} + G\frac{x^4}{4!} \cdots$$

In other words, the polynomial looks almost identical each time you differentiate it, except that each time the coefficients shunt one space to the left along the line. That might have stayed a small mathematical curiosity, except that Taylor noticed that it could be used to find a polynomial that would approximate a non-polynomial function.

Let's take the example of finding a polynomial to approximate a sine wave: $y = \sin x$. Imagine (back then) you were trying to develop a table of values for the value of sine for different angles, or (now) you are trying to program a computer with the sine function. With what we have seen so far, you would need to draw up angles using a protractor then use a ruler to measure the distances. Once we convert it to a polynomial based on x however, it becomes much more straightforward to calculate the values. This is the type of benefit that a Taylor series can provide.

To start, find a single point where the curve drawn by the polynomial you are creating and the curve drawn by the existing function agree. It's mostly easiest to make this the value of both functions where $x = 0$. (Although to be technical, a Taylor series can be based around any value, and where $x = 0$ it is actually called a Maclaurin series. But Brook Taylor (1685-1731) came first and did the heavy intellectual lifting, so let's give him the credit. Colin Maclaurin, 1698-1746, who held the record for being the youngest professor of any discipline in any university in the world until 2008, was the one who popularised the series. He came up with his own specific, simplified case of it, but he gave Brook Taylor the credit for discovering it.)

If $x = 0$, then $\sin x = 0$.

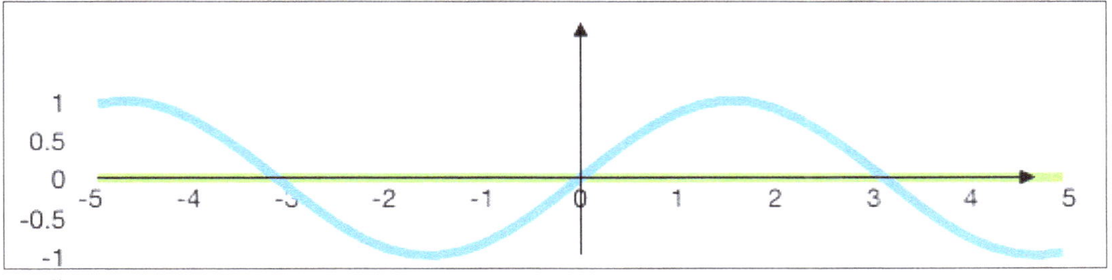

Figure 140

That's easy, so we can approximate the value of $y = \sin x$ (blue line) by the graph of $y = 0$ (green line) (Figure 140).

That gives us an approximation for the curve, but not a very good one. They really only agree at specific points.

The next step is to take the derivatives of both, and make sure that the rate of change of both also agrees, where $x = 0$.

The derivative of $y = \sin x$ is $\cos x$. Where $x = 0, \cos x = 1$. So we need to modify the expression $y = 0$ (green line), so that it still equals 0 when x is equal to 0, but when we take the derivative and insert the value of $x = 0$, the value of the function will be 1.

That may sound difficult, but Taylor was able to use his series to achieve this. If you make $A = 0$ and $B = 1$, then the value of the series where $x = 0$

is 0, which is the same as the value of $y = \sin x$ where $x = 0$. Every other term disappears, because it is multiplied by zero.

$$f(x) = 0 + 1x + C\frac{x^2}{2!} + D\frac{x^3}{3!} + E\frac{x^4}{4!}\ ...$$

$$= 0$$

Then when we take the first derivative, the one that multiplies x moves to the front of the queue. That makes the formula evaluate that when $x = 0$, we get

$$f'(x) = 1 + Cx + D\frac{x^2}{2!} + E\frac{x^3}{3!} + F\frac{x^4}{4!}\ ...$$

When we evaluate it where $x = 0$, we get 1.

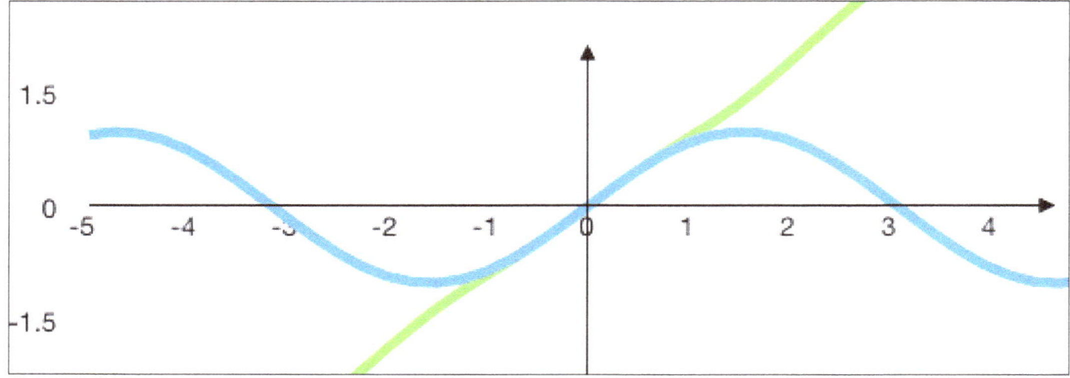

Figure 141

The graph of these two curves now looks like Figure 141.

This is a much better approximation, for values of x between -1 and 1. You can see that the slopes of the curve are very close over that distance. It's not really an approximation outside of that, so now we need to change our equation again so that the equation and the derivative both keep these values when $x = 0$, but the derivative of this derivative—that is, the rate of change *of* the rate of change—also matches.

We call this the "second derivative" and will change our notation to $f''(x)$ because that gets easier from this point.

The second derivate of $\sin x$ is the first derivative of $\cos x$, which, as we have seen, is $-\sin x$. The derivative of that is $-\cos x$, and the derivative of that is $\sin x$. If we keep taking further derivatives we keep repeating that pattern.

Derivative level	Function	Value when $x = 0$
Function: $f(x)$	$\sin x$	0
First derivative (derivative of $\sin x$): $f'(x)$	$\cos x$	1
Second derivative (derivative of $\cos x$): $f''(x)$	$-\sin x$	0
Third derivative (derivative of $-\sin x$): $f^{(3)}(x)$	$-\cos x$	-1
Fourth derivative (derivative of $-\cos x$): $f^{(4)}(x)$	$\sin x$	0

Taylor's basic polynomial function can be written as:

$$f(x) \approx \Box\frac{x^0}{0!} + \Box\frac{x^1}{1!} + \Box\frac{x^2}{2!} + \Box\frac{x^3}{3!} + \Box\frac{x^4}{4!} + \Box\frac{x^5}{5!} + \Box\frac{x^6}{6!} + \Box\frac{x^7}{7!} + \cdots$$

where the boxes represent the coefficients, which are also the values of the function, when evaluated where $x = 0$, with successive levels of differentiation. If we input the values of successive derivatives of the sine function, evaluated where $x = 0$, we get

$$\sin x = 0\frac{x^0}{0!} + 1\frac{x^1}{1!} + 0\frac{x^2}{2!} - 1\frac{x^3}{3!} + 0\frac{x^4}{4!} + 1\frac{x^5}{5!} + 0\frac{x^6}{6!} - 1\frac{x^7}{7!} + 0\frac{x^8}{8!} + \cdots$$

$$= x - \frac{x^3}{3!} + \frac{x^5}{5!} - \frac{x^7}{7!} + \frac{x^9}{9!} - \cdots$$

When we graph $y = \sin x$ and $y = x - \frac{x^3}{3!} + \frac{x^5}{5!} - \frac{x^7}{7!}$ we get Figure 142. The two lines match very closely when they are near zero, then diverge. That often happens with a Taylor series. Sometimes, though (and $y = \sin x$ is one of those times) if you were able to keep the polynomial going forever, both curves would match perfectly, for all values, meaning that the two expressions are identical.

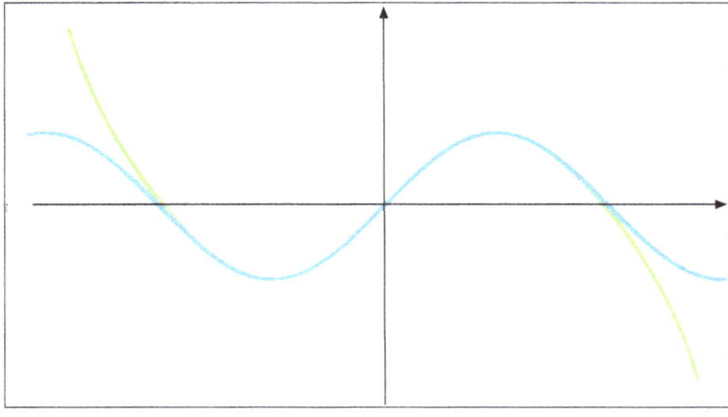

Figure 142

To cement your understanding:

- Differentiate $f(x) = 104 + 22x + 37.5\frac{x^2}{2!} + 19.33\frac{x^3}{3!} + 7\frac{x^4}{4!} + 2{,}000\frac{x^5}{5!}$

- Now differentiate the answer you got last time. Now differentiate that. Can you see that, each time, the polynomial looks almost the same, except that there is one term less in it, and the constants have all shifted one place to the left each time it is differentiated?

- Now calculate the value of each of the polynomials you have drawn, when $x = 0$. Can you see that what you have left each time is just the coefficient of one of the terms you started with?

We can therefore say that

$$\sin x = x - \frac{x^3}{3!} + \frac{x^5}{5!} - \frac{x^7}{7!} + \cdots$$

$$= \sum_{n=0}^{\infty} (-1)^n \frac{x^{2n+1}}{(2n+1)!}$$

If you were trying to develop a table of values for sine, or to program a computer with a sine function, this polynomial gives you a straightforward way to do it, to whatever level of accuracy you like, without needing a protractor and a ruler.

To cement your understanding:

- Work through the example above for yourself, calculating the derivative, the derivative of that, etc, for both sides of the equation, and checking to make sure that the values are the same when $x = 0$.

- Use the same approach to find a polynomial expression for $\cos x$.

- Confirm that $\sum_{n=0}^{\infty}(-1)^n \frac{x^{2n+1}}{(2n+1)!}$ generates the series, by putting the first few values of n in.

Making this more general, we can say that you can create a Taylor series for a function of x, evaluated at the point where $x = a$, where a is some constant (i.e. not necessarily where $x = 0$), by the formula:

$$f(x) = \frac{f(a)(x-a)^0}{0!} + \frac{f'(a)(x-a)^1}{1!} + \frac{f''(a)(x-a)^2}{2!} + \cdots + \frac{f^n(a)(x-a)^n}{n!}$$

If you replace x in this function with a, you will see that every term except the first one will disappear, because $0^0 = 1$ whereas zero raised to any other power equals zero. Each time you differentiate, the first term (which is actually a constant) will disappear, and the next term will move to the front of the queue. We can write this as:

$$f(x) \approx \sum_{k=0}^{n} \frac{f^k(a)(x-a)^k}{k!} \; for \; values \; of \; x \; close \; to \; a$$

To cement your understanding:

Expand out the formula immediately above, to confirm that it creates a Taylor series.

Key points:

- The Taylor/McLauren polynomial function can be written as:

$$f(x) \approx \Box\frac{x^0}{0!} + \Box\frac{x^1}{1!} + \Box\frac{x^2}{2!} + \Box\frac{x^3}{3!} + \Box\frac{x^4}{4!} + \Box\frac{x^5}{5!} + \Box\frac{x^6}{6!} + \Box\frac{x^7}{7!} + \Box\frac{x^8}{8!} + \cdots$$

  where the boxes represent the coefficients, which are also the values of the function, when evaluated where $x = 0$, with successive levels of differentiation.

- This can be used to create polynomial functions that approximate, or in some cases equal, non-polynomial ones. For example:

$$\sin x = x - \frac{x^3}{3!} + \frac{x^5}{5!} - \frac{x^7}{7!} + \cdots$$

$$= \sum_{n=0}^{\infty} (-1)^n \frac{x^{2n+1}}{(2n+1)!}$$

- In a true Taylor series (as opposed to a McLauren series) the result can be evaluated for any value of x, and not just 0.

Exponential equations and Euler's constant

Read this if you aren't familiar with the irrational number *e* and some different ways to calculate it. It keeps recurring, to a remarkable degree, when we enter the world of statistics.

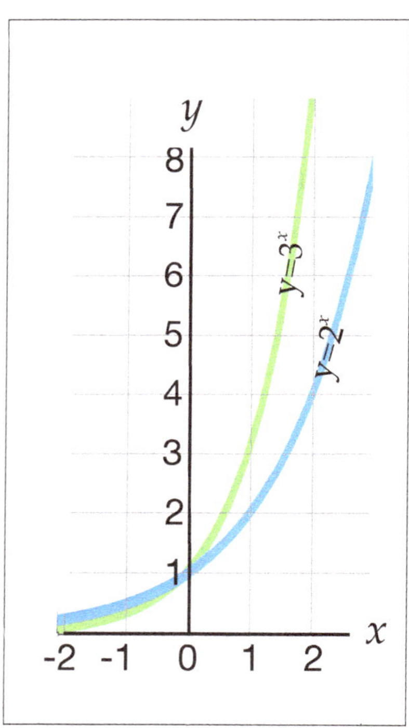

Figure 143

Up until now most of the equations we have looked at have been in the format of $y = x^2$: x is the base, which is raised to a power. Equations in the form of $y = 2^x$, where x is the index, have graphs which look like Figure 143.

When you hear about something increasing "exponentially", this is what it means.

When x becomes negative the y value is a fraction of 1. When it is 0, $y = 1$. When x is 1, y equals whatever it is you are raising to the power of x. The fractional/decimal numbers in between (which use the numerator as meaning the number of times you multiply the base by itself, and the denominator as that root of the base) form the whole equation into a smooth curve. (Refer back to the material on Powers at page 53 if you are unclear about this.)

To find the derivative:

$$y = 2^x$$

$$y + dy = 2^{x+dx}$$

When you add the exponents of a base, it is the same as multiplying the base raised to each of the exponents.

$$y + dy = 2^x 2^{dx}$$

Subtract y from the left and 2^x from the right (which we can do because they are equal).

$$dy = 2^x 2^{dx} - 2^x$$

Factor out 2^x.

$$dy = 2^x \left(2^{dx} - 1\right)$$

Divide both sides by dx.

$$\frac{dy}{dx} = 2^x \frac{\left(2^{dx} - 1\right)}{dx}$$

And at that point we get temporarily stuck. We need to work out

$$\lim_{dx \to 0} \frac{2^{dx} - 1}{dx}$$

and there isn't an easy way to do that using algebra.

Fortunately, this expression is independent of x. In other words, it doesn't matter what value of x you pick, this expression won't change. It is a convergent sequence. The smaller dx gets, the more closely the value approaches the constant value of $0.6931472...$

What that tells us is that the slope of a tangent to the curve $y = 2^x$ is proportional to itself. If we write $0.6931472...$ as C, it means:

$$\frac{dy}{dx} = 2^x C$$

The 2^x is multiplied by a constant. This makes sense when you look at Figure 143. The lower x gets, the flatter the curve, and the higher x gets, the steeper the curve.

If we had looked at the differential of $y = 3^x$ we would have found that

$$\frac{dy}{dx} = 3^x \frac{\left(3^{dx} - 1\right)}{\delta x}$$

$$\lim_{dx \to 0} \frac{3^{dx} - 1}{dx} = 1.0986123\,...$$

Which means that the slope of tangents to this curve are also proportional to itself. We could have done this with any base and achieved the same result, so that, more generally:

$$\frac{dy}{dx} = n^x \frac{\left(n^{dx} - 1\right)}{dx}$$

$$\frac{dy}{dx} = n^x C$$

Where *n* is any number, and C is a specific constant unique to that number.

Euler's constant

We've looked at the irrational numbers marked by φ and π. There is another one, and it is at least as interesting. This one is *e*. It was (probably) first identified by Leonhard Euler (1701-1783) (pronounced "oiler"), whom we first met in relation to function notation.

Figure 144
Leonhard Euler
1701–1783

Euler was a friend and pupil of at least one member of the Bernoulli family of mathematicians. One of Bernoulli's interests was compound interest. To simplify the problem Euler and/or Bernoulli thought about it in these terms:

Assume you have an account at a Swiss bank and are paid a generous rate of 100% interest each year. You start with 1 Swiss franc. After a year you will have gained 1 franc, so will have 2 francs altogether. What happens, though, if you are paid half as much interest, twice each year? That way you get compound interest: interest on the interest you have accumulated. Then the amount of money you receive back at the end of six months can be expressed as

$$1 + \frac{1}{2}$$

and then then the same calculation is applied a second time to this amount. In other words:

$$\left(1 + \frac{1}{2}\right)\left(1 + \frac{1}{2}\right) = \left(1 + \frac{1}{2}\right)^2$$

$$= \frac{9}{4}$$

$$= 2\frac{1}{4} \text{ francs}$$

That's clearly better than 2 francs. So what if you are paid one twelfth of the interest every month?

$$\left(1 + \frac{1}{12}\right)^{12} = 2.6130 \ldots \text{ francs}$$

Better still. There is a limit, however.

If interest is paid continuously, it can be expressed like this:

$$\lim_{n \to \infty} \left(1 + \frac{1}{n}\right)^n$$

As the period becomes smaller and smaller, in other words n approaches infinity, Bernoulli worked out that the limit was somewhere between 2 and 3 francs. Euler worked out that this was a binomial expansion.

As a reminder, the binomial theorem says that:

$$(x + y)^n = \sum_{r=0}^{n} \frac{n!}{r!\,(n - r)!} x^{n-r} y^r$$

He realised that if you expand that out, you let $x = 1$ and $y = \frac{1}{n}$ then use large numbers for n, (say, 100,000) it looks like this.

The first term, when $r = 0$, is:

$$\frac{100{,}000!}{0!\,(100{,}000)!} 1^{100{,}000} \left(\frac{1}{100{,}000}\right)^0 = 1$$

The second term is:

$$\frac{100{,}000!}{1!\,99{,}999!} 1^{99{,}000} \left(\frac{1}{100{,}000}\right)^1 = \frac{100{,}000}{100{,}000}$$

$$= 1$$

The third term is:

$$\frac{100{,}000!}{2!\,99{,}998!} 1^{98{,}000} \left(\frac{1}{100{,}000}\right)^2 = \frac{1}{2!} \frac{100{,}000 \cdot 99{,}999}{100{,}000 \cdot 100{,}000}$$

$$\approx \frac{1}{2!}$$

and the larger n gets, the closer this expression gets to $\frac{1}{2!}$

To cement your understanding:

• Using this approach, calculate the 4th term of the expansion (where $r = 3$).

If you keep going, this infinite *sequence* on the left approaches the infinite, convergent *series* on the right:

$$\lim_{n \to \infty} \left(1 + \frac{1}{n}\right)^n = 1 + 1 + \frac{1}{2!} + \frac{1}{3!} + \frac{1}{4!} + \frac{1}{5!} + \frac{1}{6!} + \frac{1}{7!} + \cdots$$

$$= \sum_{n=0}^{\infty} \frac{1}{n!}$$

$$= 2.718281828459045 \ldots$$

He said that

$$2.718281828459045 \ldots = e$$

Exponential equations with e

Euler also worked out that if

$$e = \lim_{n \to \infty} \left(1 + \frac{1}{n}\right)^n$$

then

$$e^x = \lim_{n \to \infty} \left(1 + \frac{1}{n}\right)^{nx} \quad \text{(raising both sides to the power of } x\text{)}$$

$$= \lim_{n \to \infty} \left(1 + \frac{x}{nx}\right)^{nx}$$

(multiplying the numerator and denominator of the second term in brackets by x).

Now as n approaches infinity so does nx, becoming indistinguishable from n.

Therefore

$$e^x = \lim_{n \to \infty} \left(1 + \frac{x}{n}\right)^n$$

Expanding this out as a binomial expression gives

$$e^x = \frac{x^0}{0!} + \frac{x^1}{1!} + \frac{x^2}{2!} + \frac{x^3}{3!} + \frac{x^4}{4!} + \frac{x^5}{5!} + \frac{x^6}{6!} + \frac{x^7}{7!} + \frac{x^8}{8!} + \cdots$$

$$= \sum_{n=o}^{\infty} \frac{x^n}{n!}$$

Euler realised that this looks like a Taylor series with no added constants, and that it doesn't matter how many times you differentiate the function on the right, it will never change.

> **To cement your understanding:**
>
> Try it for yourself if you are unsure of this.

That means that e^x must be its own derivative. If you graph the curve it falls between the curves of $y = 2^x$ and $y = 3^x$ (see Figure 143), and for any value of x, the slope of the curve will be the same as the value of y, as will the area under the curve (which we will see when we get to integral calculus).

We saw earlier that for $y = n^x$

$$\lim_{dx \to \infty} \frac{dy}{dx} = n^x \frac{\left(n^{dx} - 1\right)}{dx}$$

and that $\frac{\left(n^{\delta x} - 1\right)}{\delta x}$ is a constant. If the derivative of e^x is its own derivative, this is the same as saying for $y = e^x$

$$\frac{dy}{dx} = e^x \frac{\left(e^{dx} - 1\right)}{dx}$$

$$= e^x$$

and therefore that:

$$\lim_{dx \to 0} \frac{e^{dx} - 1}{dx} = 1$$

All of that might just appear, at this stage, to be some mathematics that happens to work out neatly. It is, also, though, absolutely fundamental to modern statistics, as well as to most other fields that involve higher mathematics. We'll see why as we progress.

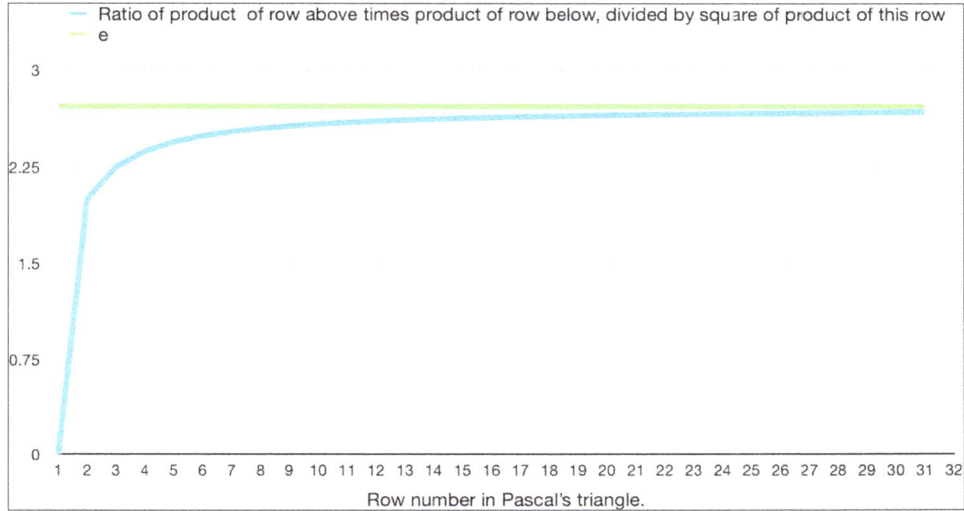

Figure 145

Fun fact: Euler's constant also makes an appearance in Pascal's triangle. If you calculate the product of each row of Pascal's triangle (multiplying all of the terms together), then, for any row, multiply the product of the row above with the product of the row below, then divide all of that by the square of the row you are on, the limit of that result as the row number approaches infinity is, incredibly, e!

Key points:

- $e = \lim\limits_{n \to \infty} \left(1 + \frac{1}{n}\right)^n$

$$= \frac{1}{0!} + \frac{1}{1!} + \frac{1}{2!} + \frac{1}{3!} + \frac{1}{4!} + \frac{1}{5!} + \frac{1}{6!} + \frac{1}{7!} + \frac{1}{8!} + \cdots$$

$$= \sum_{n=0}^{\infty} \frac{1}{n!}$$

$$= 2.718281828459045\ldots$$

- $e^x = \lim\limits_{n \to \infty} \left(1 + \frac{x}{n}\right)^n$

$$= \frac{x^0}{0!} + \frac{x^1}{1!} + \frac{x^2}{2!} + \frac{x^3}{3!} + \frac{x^4}{4!} + \frac{x^5}{5!} + \frac{x^6}{6!} + \frac{x^7}{7!} + \frac{x^8}{8!} + \cdots$$

$$= \sum_{n=0}^{\infty} \frac{x^n}{n!}$$

- Euler's constant (e) is a Taylor series where the coefficient is always 1.

- For $y = n^x$

$$\lim_{dx \to \infty} \frac{dy}{dx} = n^x \frac{(n^{dx} - 1)}{dx}$$

- For $y = e^{xx}$

$$\frac{dy}{dx} = e^x \frac{(e^{dx} - 1)}{dx}$$

$$= e^x$$

- $\lim\limits_{\delta x \to 0} \dfrac{e^{dx} - 1}{dx} = 1$

Integral calculus

Read this if you aren't familiar with integral calculus, and the difference between a definite and an indefinite integral.

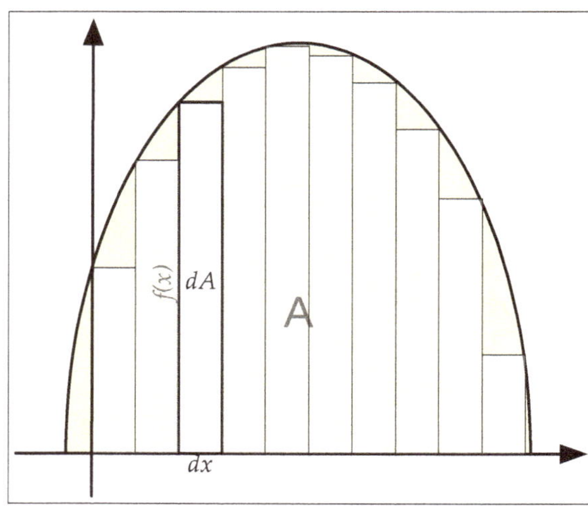

Figure 146

The same type of thinking that we used to work out a formula for the slope of the tangent to a curve at any point can also be used to calculate the area underneath a curve. (Integral calculus was actually developed before differential calculus. This type of thinking was around in the ancient world, and was used for such things as finding the area of a circle, which we covered earlier. What Newton and Leibniz did was take earlier thinking and generalise it so that it could be applied more broadly.

Differential calculus, which we have already covered, did come later, but it's usually taught in the order we are covering it in this book, because it is easier to understand this way.)

Imagine we have a curve, with area A, which is drawn by some equation

$$y = f(x)$$

(Figure 146). We can estimate the area by drawing rectangles inside the curve. Each of those will have the width of a very small change in x, which we'll call (no surprise) dx. The height of each rectangle will be the value of y, which is the same as $f(x)$ at that point. We'll call each of these minute changes in the area of the curve dA.

This tells us that the area of each small rectangle, *dA,* will be equal to *dx* multiplied by $f(x)$ (width multiplied by height). Together, they don't exactly fit the area of the curve, but the smaller and smaller *dx* gets, and the more rectangles there are to fill the space, the closer this group of rectangles will be to the whole area.

$$dA = dx f(x)$$

$$\frac{dA}{dx} = f(x)$$

In other words, the ratio of the change in the area to the change in x is equal to the formula of the curve itself. The rate of change in the area of the curve is given by the function for the curve.

Because of what we have learned about finding derivatives, this gives us a way of finding a formula for the area. We need to find the **anti-derivative**. That is, the equation whose derivative is $f(x)$. That will give us A; the area under the curve. It's quite amazing that it works so simply, but it really does.

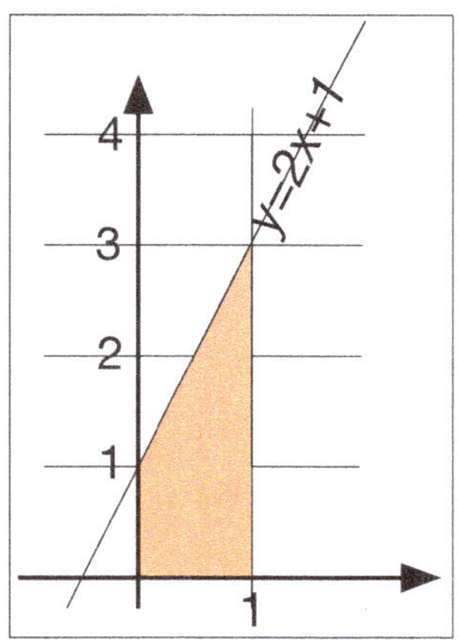

Figure 147

Let's take a simple example to see how this works in practice: finding the area under the line described by the formula of $y = 2x + 1$, between the values of 0 and 1 (Figure 147). We can work out the area of this particular angle without using calculus: there is a triangle on top of a square. The area of the square is 1 square unit, while the area of the triangle is half of the area of one unit by two units, which means it is also one square unit. Together, the orange area is 2 square units.

Finding the same area using calculus, we need to find the anti-derivative of $2x + 1$. This is the same as $2x^1 + 1x^0$. The power rule says that we should reduce the exponent by multiplying the term by the exponent then decrease the exponent by 1. Reversing that, we can increase the exponents on these terms by one, then divide each term by that amount. That leaves us with $x^2 + x^1$, or just $x^2 + x$. The value of x at the outside border of the orange figure is 1, so substituting 1 for x leaves us with the value of the area being $1^2 + 1$, which is equal to 2. Which is, of course, the right answer.

Integrative calculus is that neat, and almost that simple. It doesn't just work for straight-line equations, which we could work out without calculus. It works for curves where there is no other easy way to determine the area.

One way to think of what is happening in Figure 146 is that, starting from $x = 0$ and moving across to the right, the area under a curve is determined by the cumulative value of its rates of change at any point, and its rates of change are determined by what the curve is doing at that point. So using the equation of the curve as the expression of the rate of change of the area, you can work back to find the area for any value of the area. (You might want to read that sentence again slowly to make sure that you have got it.)

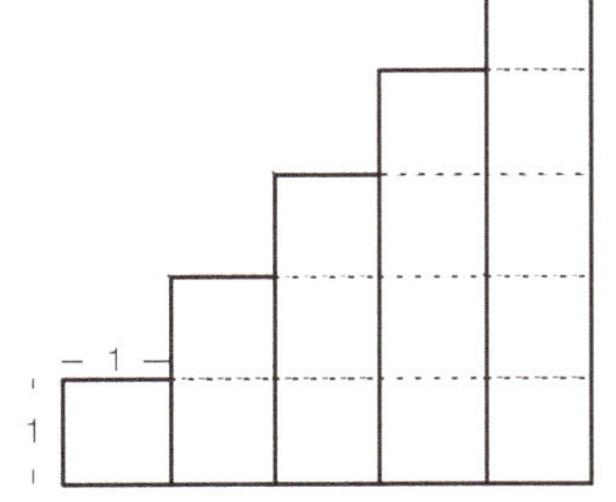

To cement your understanding:

• In Figure 148 calculate the area under the stairs by calculating the area of each rectangle (width of each stair by height of each stair), and adding the heights together.

• Can you see how the line of the stair way represents the rate at which the area is changing?

Figure 148

There is a slight complication, though. The anti-derivative of $2x + 1$ is $x^2 + x$, but it is also $x^2 + x + 1{,}275$ or, for that matter, any other number added to the equation. That is because this can be thought of as $x^2 + x^1 + 1{,}275x^0$.

When we multiply each term by the exponent, the 1,275 is multiplied by 0, and so disappears, which means that as we work backwards we can't tell if it was there or not. So the actual anti-derivative of $2x + 1$ is $x^2 + x + C$, where C is some constant (i.e. a number that doesn't vary with changes to x). Fortunately it turns out that this isn't really a practical problem in most instances.

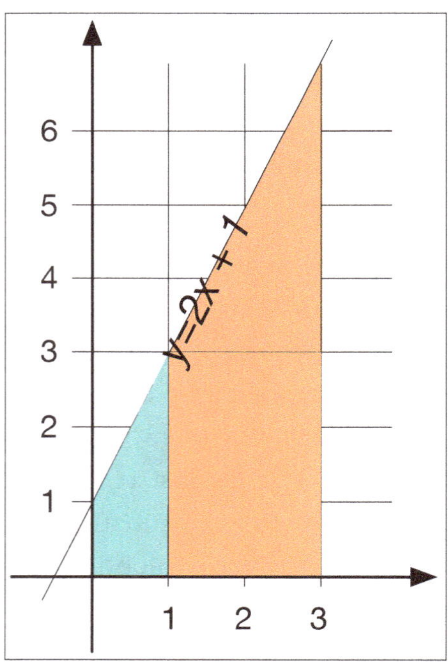

Figure 149

In Figure 149, if we want to find the area of the orange shape with x values between 1 and 3, we can find the value of the whole shape with x values between 0 and 3, then subtract the value of the blue shape with x values between 0 and 1. This gives us:

$(3^2 + 3 + C) - (1^2 + 1 + C)$, which is equal to $9 + 3 + C - 1 - 1 - C$. The Cs cancel one another out, so we are left with an area of 10.

The notation used for this is calculation is:

$$\int_{1}^{3} (2x + 1)\, dx = 10$$

The 3 and the 1 are called the **limits of integration**. The similarity of this symbol to the sigma notation (Σ) for a series is not accidental. The idea is that you are summing up a large number (approaching infinity) of the values of the anti-derivative of the function, multiplied by dx as it approaches zero, to slice up the shape. This is called a **definite integral** with respect to dx because there is a definite answer to it. The **indefinite integral**, or the more general equation used to find the area under the curve of $y = 2x + 1$ is just written as:

$$\int (2x + 1)\, dx = x^2 + x + C$$

For indefinite integrals you need to put a C at the end, but for definite integrals it isn't required.

Using function notation, the integral of a function of x written as $f(x)$ is written using a capital: $F(x)$.

To cement your understanding:

• Calculate the indefinite integral $\int (2x + 3)\, dx$, not forgetting to add a C on the end. Differentiate your answer, to confirm that it gets you back to the original expression.

• Calculate the definite integral $\int_{2}^{4} (2x + 3)\, dx$, by working out the antiderivative of $2x + 3$, calculating the value of the result where $x = 4$ and subtracting the value of the result where $x = 2$.

Key points:

- The same type of thinking that we used to work out a formula for the slope of the tangent to a curve at any point can also be used to calculate the area underneath a curve.

- The rate of change in the area of the curve is given by the function for the curve.

- That means that the anti-derivative gives the area under the curve.

- The x values that give the start and finish of the area under consideration are called the limits of integration.

- A definite integral applies to a specific area; and indefinite integral is a general formula giving the anti-derivative of a curve.

- Indefinite integrals include an added constant.

- The integral of some function of x written as $f(x)$ is written using a capital: $F(x)$.

Advanced integration

> **Read this if** you aren't familiar with linearity in integral calculus, the reverse power rule, integration by parts or integration by substitution.

The reverse power rule

The power rule for differentiating says that you multiply the term by the exponent, then subtract one from the exponent. When integrating individual terms, reverse this by adding one to the exponent, and then dividing the term by the new exponent.

$$\int x^n \, dx = \frac{x^{n+1}}{n+1} + C$$

> To cement your understanding:
>
> • Apply the reverse power rule to $\int x \, dx$.
>
> • Then differentiate your answer, to make sure that it gets you back to the starting point. Did you get $\frac{1}{2}x^2$?

Multiplying by a constant

When a term is multiplied by a constant, the constant is not affected by the integration process, and so it can be taken outside of the integration.

In other words:

$$\int Cf(x) \, dx = C \int f(x) \, dx$$

> **To cement your understanding:**
>
> Calculate $\int 3x\, dx$ in two different ways: first by leaving the 3 inside the integration, then taking it outside of the integration. Confirm that it gives you exactly the same answer.

The addition rule

The addition rule for differentiation (page 236) also works for integration: different terms added together can be integrated separately, and the results added together.

In other words:

$$\int [f(x) + g(x)]\, dx = \int f(x)\, dx + \int g(x)\, dx$$

> **To cement your understanding:**
>
> • Calculate $\int x + x^2\, dx$ in two different ways: first by applying the reverse power rule to each term, then by writing it as two separate integrations and doing exactly the same thing. Confirm that it gives you exactly the same answer.
>
> • Differentiate your answer using the power rule and the addition rule for differentiation, to confirm that it gets you back to where you started $(x + x^2)$.

Linearity of integral calculus

The way that integration behaves when multiplying by a constant, and when adding two terms together, means that integral calculus has the characteristic of linearity, in the same way that using a series does. This makes sense, because integration really is a form of series, as the size of each term that is added shrinks to approach zero while the number of terms that are added approaches infinity.

Integration by parts

Integration by parts was another clever idea from Brook Taylor, who gave us the Taylor series. It is (sort of) the reverse of the multiplication rule for differentiation.

This is the formula for the multiplication rule, where u and v are functions of x.

$$\frac{d(uv)}{dx} = u\frac{dv}{dx} + v\frac{du}{dx}$$

Subtract $v\frac{\delta u}{\delta x}$ from both sides and swap the two sides.

$$u\frac{dv}{dx} = \frac{d(uv)}{dx} - v\frac{du}{dx}$$

Integrate both sides with respect to x. The integral of $\frac{\delta(uv)}{\delta x}$ is, of course, just uv.

$$\int u\frac{dv}{dx}\,dx = uv - \int v\frac{du}{dx}\,dx$$

Now if we are trying to integrate the product of two functions of x, we can call one of them (preferably the one that is hard or impossible to integrate) u, and the other function (preferably one that we do know how to integrate) $\frac{\delta v}{\delta x}$. We find the derivative of the function we have called u and the integral of the function we have called $\frac{\delta v}{\delta x}$, which will be v, and plug these into the formula:

$$\int u\frac{dv}{dx}\,dx = uv - \int v\frac{du}{dx}\,dx$$

This may look like over-complicating the situation, but the big advantage of this approach comes when one of the two formulae that are multiplied together doesn't have a straightforward integral, or when we can't see how to integrate them in combination.

Let's take an example:

Nothing we have covered so far tells us how to integrate this. But using integration by parts, we can let:

$$\int xe^x\,dx$$

$x = u$ and $e^x = \frac{dv}{dx}$

Differentiating $u = x$

$$\frac{du}{dx} = 1$$

e^x is its own integral, because it is its own derivative

$$v = e^x$$

Using the formula for integration by parts.

$$\int xe^x\,dx = xe^x - \int e^x\,dx$$

$$= xe^x - e^x + C$$

Just checking that that really worked, we'll find the derivative of that. Let

$$y = xe^x - e^x + C$$

Finding the derivative using the addition rule and the multiplication rule, with e^x as its own derivative, and the derivative of $x = 1$

$$\frac{dy}{dx} = e^x + xe^x - e^x$$

$$= xe^x$$

It works!

To cement your understanding:

Calculate $\int x \cos x \, dx$ using integration by parts. Hint: we've seen that the derivative of $\sin x = \cos x$, which means that the integral of $\cos x = \sin x$. We also know that the derivative of x with respect to x is 1. So start by letting $u = x$ and $\frac{dv}{dx} = \cos x$. Write out the formula using these substitutions, and finish it from there.

Did you get $x \sin x + \cos x + C$? If not, can you see where you went wrong?

Integration by parts can get a bit complicated, but it gets used a lot the further we progress. So if the idea is a bit hazy at this stage don't panic; we will see plenty of examples as we go along.

Integration by substitution

It can often be helpful to substitute one variable for another. This is called integration by substitution.

One situation in which this is useful is when a term is made up of a function within a function, multiplied by the derivative of that function (or something that can be transformed into the same).

Take the following integral:

$$\int xe^{x^2} \, dx$$

We can approach this, firstly, by putting a 2 within the integral and multiplying it by $\frac{1}{2}$ outside the integral, so that the two terms will multiply out to 1, and would not change the final answer.

$$\int xe^{x^2} \, dx = \frac{1}{2} \int 2xe^{x^2} \, dx$$

You can see from this expression that the term we are integrating now looks like something that results from the application of the chain rule. If it is more complicated than something you can work out in your head, or if you want to be able to show your working, you can use a substitution.

Now let:	$u = x^2$
Find the derivative.	$\dfrac{du}{dx} = 2x$
Multiply both sides by dx. You can do this, because du and dx are incredibly small, but never quite get to zero.	$du = 2x\,dx$
Divide both sides by $2x$	$\dfrac{1}{2x}du = dx$

The integral we are working with is

$$\frac{1}{2}\int 2xe^{x^2}\,dx$$

We can substitute u for x^2 and multiply the whole term by du instead of $2x\,dx$. That gives us:

$$\frac{1}{2}\int \cancel{2xe^u}\,\frac{1}{\cancel{2x}}\,du = \frac{1}{2}\int e^u\,du$$

$$= \frac{1}{2}e^u + C$$

The integral of e^u is e^u.

$u = x^2$, so substituting back into our original equation we have:

$$\int xe^{x^2}\,dx = \frac{1}{2}e^{x^2} + C$$

You can check this result by differentiating $\frac{1}{2}e^{x^2} + C$ using the chain rule, and it will give you xe^{x^2}.

If this were a definite integral the u would disappear, and there would be no need to substitute anything back.

To cement your understanding:

We'll see a lot of integration by substitution as we move more deeply into things, so there is probably no need to practice it. It should become very clear soon, if it isn't now.

Key points:

- The reverse power rule for integration says that when integrating individual terms, add one to the exponent, and then divide the term by the new exponent.

- Integration follows linearity. That means when a term is multiplied by a constant, the constant is not affected by the integration process, and so it can be taken outside of the integration.

- The addition rule for differentiation also works for integration.

- The formula for integration by parts is:

$$\int u\frac{dv}{dx}\,dx = uv - \int v\frac{du}{dx}\,dx$$

- It can often be helpful to substitute one variable for another. This is called integration by substitution.

Multivariable calculus

> **Read this if** you find the whole idea of doing calculus with more than one variable daunting.

Partial derivatives

In the real world, not many functions are driven by a single variable. There are mostly a number of factors involved, and it is often necessary to understand how these variables are changing in response to one another. It sounds as though calculus with more than one variable would be incredibly complicated, but it's actually reasonably straightforward. Calculus with more than two variables works by dealing with one variable at a time.

The differentiation of equations with more than two variables uses **partial differentiation**.

Take the example of the equation:

$$z = 3yx^2 + 7xy^3$$

(An equation like this can be used to describe a three-dimensional shape in a space with x, y and z coordinates, as in Figure 150. More than three variables can be involved, however. If there are more than three dimensions the equation can no-longer be represented as a shape; it's just an equation.)

Partial differentiation of this equation involves finding the rate of change of z in respect to x, treating y as a constant. In other words, it is the same as normal differentiation of the shadow or projection that a three-dimensional shape would make on the x, z plain. The result is usually written with ∂x

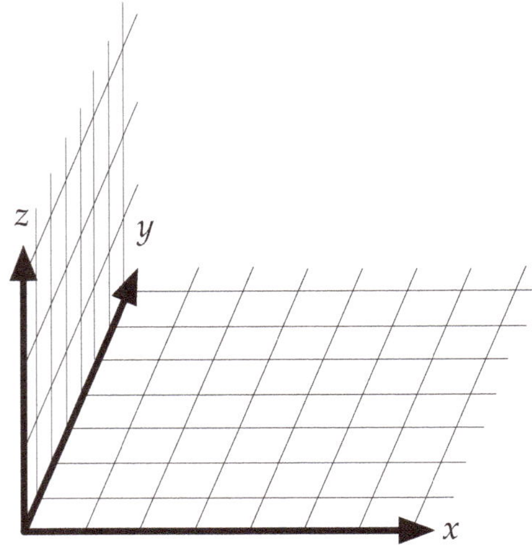

Figure 150

rather than dx. So if we find the partial derivative of $z = 3yx^2 + 7xy^3$ with respect to x, treating y as a constant, we get:

$$\frac{\partial z}{\partial x} = 6yx + 7y^3$$

Another way to think of this is to completely remove any powers of y from the equation, differentiate in the normal way, then add the powers of y back in at the end.

In the same way, the rate of change of z in respect to y can be calculated for the same equation $z = 3yx^2 + 7xy^3$, leaving powers of x as a constant. That results in:

$$\frac{\partial z}{\partial y} = 3x^2 + 21xy^2$$

To cement your understanding:

Calculate the partial derivative of $z = 5wyx^3 + w^4y^5 + x^2$ with respect to x. Did you get $\frac{\partial x}{\partial z} = 15wyx^2 + w^4y^5 + 2x$? If not, can you see where you went wrong?

Now do the same thing, but this time find the partial derivative with respect to w.

Double integration

To find *derivatives* of functions with multiple variables, it isn't necessary to end up with a single equation. However, when you are *integrating* functions with multiple variables it is usually necessary to come up with a single formula and, in fact, for definite intervals we are usually seeking a single value. In three-dimensional space this will give the volume of a space described by the equation you are integrating.

The principle works the same way: deal with one variable at a time. We write an indefinite double integral like this:

$$\iint x^2 \, y^3 \, dx \, dy$$

We deal with this in two parts: integrating the x component first (treating the power of y as a constant), then integrating the y component. To make it easier, we could have written this as:

$$\int \left(\int x^2 y^3 \, dx \right) dy$$

So in this example stage 1 gets us to:

$$\int \frac{x^3}{3} y^3 \, dy$$

Stage 2 then gets us to:

$$\frac{x^3 y^4}{12}$$

If this was a definite integral, it would be processed as though the middle integration was in brackets (although the brackets are not normally shown):

$$\int_2^4 \left(\int_1^3 x^2 y^3 \, dx \right) dy$$

Step one gives us, for the section in brackets:

$$\frac{27}{3} y^3 - \frac{1}{3} y^3 = \frac{26}{3} y^3$$

Step two gives us:

$$\int_2^4 \frac{26}{3} y^3 \, dy = \left(\frac{26}{3} \cdot \frac{256}{4} \right) - \left(\frac{26}{3} \cdot \frac{16}{4} \right)$$

$$= 520$$

The order of integration doesn't matter: putting the integration with respect to y in the brackets and doing that first would have given the same answer.

In some instances, including this one, we could also have written it by multiplying the two integrations together, as in:

$$\left(\int_1^3 x^2 \, dx \right) \left(\int_2^4 y^3 \, dy \right)$$

In this instance you will get the same answer, although not all double integrals can be written in this way.

Key points:

- To perform multivariable differential calculus, work with one variable at a time, treating the others as constants each time you do.

- This is called partial differentiation.

- To find derivatives other than a multivariable function, you don't need a single equation, but for double integration you do.

- The principle works the same way: one variable at a time.

- Some, but not all, double integrations can be the same as multiplying two separate integrations.

Changing to polar coordinates

> **Read this if** you aren't familiar with the role of the Jacobian determinant in changing to polar coordinates. This is another thing you need to understand, to see how the formula for the normal distribution curve was derived.

Sometimes integrations, and particularly double integrations, can work better with a change to polar coordinates.

For example:

$$\iint x^2 + y^2 \, dy \, dx$$

As a reminder, in polar coordinates $x^2 + y^2 = r^2$.

That would *almost* make this:

$$\iint r^2 \, dr \, d\theta$$

This obviously appears simpler to integrate. *But it isn't quite correct.*

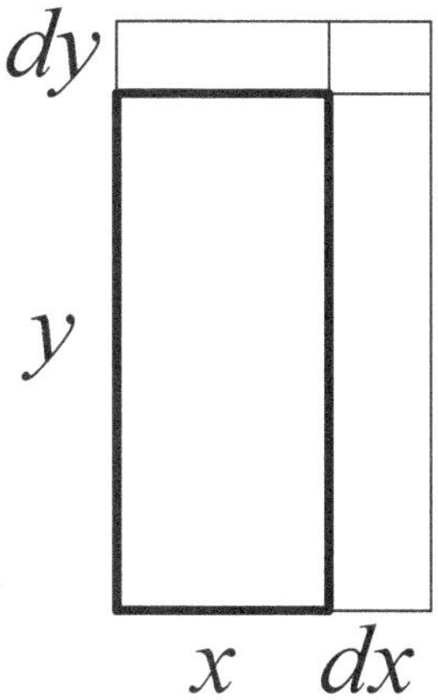

Figure 151

Jacobian

When we looked at using a substitution to make an integration easier, we couldn't always simply swap du for dx. We needed to work out what du was in terms of dx first, and that sometimes involved including a scaling factor, so that we were comparing like with like when using the different variables for integration. Now we have a situation where there are two variables, and we need an equivalent but different approach.

Changing from rectangular coordinates to polar coordinates also requires a scaling factor when changing $dy \, dx$ to $dr \, d\theta$. We can think of using $dy \, dx$ in a double integration using rectangular coordinates as being like making small changes to a rectangle (Figure 151). (We can increase the width first then the height, or the other way around; the result is the same, which is why the order of integration in a double integration doesn't matter.)

When performing a double integration with polar coordinates we can think of it as making a small change to both the angle and the radius. The arc that this creates will have a length of $rd\theta$ radians. The increase in area created by adding a small amount to the radius and a small amount to the angle will be the orange area shown on Figure 153. As dr and $d\theta$ approach zero, the orange area will approach a rectangle in shape, with height $rd\theta$ and width dr. That means that when we change a double integration that uses $dy\,dx$ into polar coordinates, we need to replace these amounts by

$$rd\theta dr.$$

Figure 152
Carl Gustav Jacob Jacobi
1804–1851

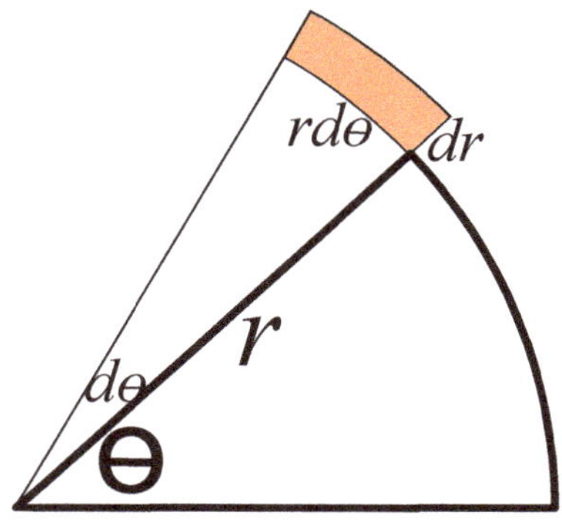

Figure 153

We need to add an extra r. This is called a **Jacobian**, named after Carl Gustav Jacob Jacobi (1804–1851). There are other Jacobian determinates used for changes to other coordinate systems. It is sometimes called the Jacobian **determinate**.

To finish our integration:

$$\iint x^2 + y^2 \, dy\,dx = \iint r^3 \, dr\,d\theta$$

$$= \int \frac{r^4}{4} \, d\theta$$

$$= \frac{r^4\theta}{4} + C$$

In linear algebra (which this book doesn't cover) a determinate tells us, among other things, the proportion of change to volume from one shape to another after it goes through a transformation. Linear algebra was actually developed after Jacobi's time. He died of smallpox aged 47 in 1851, and linear algebra first made its appearance in the middle of the 1850s. So we will just call what he came up with a **Jacobian**.

One other aspect of changing to polar coordinates is that, for definite integrals, you need to re-think the **limits of integration** (the numbers at the top and the bottom of the integral sign, showing what you subtract from what when you calculate the value). It may be necessary to sketch out the shape drawn by the x-y values, to work out roughly the area under consideration, in order to match it with equivalent polar values of r and θ.

For example, a definite integral covering all possible values using rectangular coordinates will go from plus infinity to negative infinity. The equivalent in polar coordinates is for the radius, r, to go from zero to infinity (no radius to an infinitely long one), and for the angle, θ, to go from 0 to 2π radians (from no angle to all the way around the circle).

Key points:

- When changing from rectangular to polar coordinates in an integration, it is necessary to include a Jacobian.

- This is essentially a scaling factor. In the case of converting to polar coordinates, the Jacobian is an additional r that multiplies the result.

- The limits of integration are the numbers at the top and bottom of the integral sign, showing what you subtract from what when calculating a definite integral.

- When converting an integral to polar coordinates, you also have to rethink the limits of integration.

Natural logarithms

Read this if you aren't familiar with natural logarithms.

When we looked at logarithms we learned that they were initially designed to index a geometric series with an arithmetic one, and that the understanding that they also represented an exponent of the base came considerably later. Logarithms where e is the base are called **natural logarithms**, and written as **ln**. Like other logarithms, natural logarithms were discovered/invented long before anyone thought of using them as an index, and long before Euler identified the number e. They were actually noticed first as a pattern that recurred when using a base for logarithms that was only very slightly more than one: 1.000001. They were next noticed as ways to calculate the area under a hyperbola using rectangles (in a precursor to integral calculus). It was only later Euler realised that natural logarithms were indexes with a base of e.

Natural logarithms can provide a helpful shorthand in dealing with equations that involve calculus. Take the equation $2 = e^{\ln 2}$. That's not the clever bit. It's just a definition of $\ln 2$. That is, it's what you raise e to in order to get 2. One thing it means though is that $2^x = e^{\ln 2x}$ (raising both sides to the power of x). Defining it in this way, in order to find the derivative of 2^x we use the chain rule on the right-hand side. We can think of the inner function as $w = \ln 2x$, and the outer function as e^w. The chain rule says that we differentiate each of these individually then multiply the results together to get the differential of the whole function. $\ln 2$ is a constant, so the differential of $\ln 2x$ is just $\ln 2$ (multiply the whole power by the exponent of x and reduce the exponent by 1), while the derivative of e^w is e^w (it is its own derivative).

So for the function $y = 2^x$

$$\frac{dy}{dx} = \ln2 \, e^{\ln 2x}$$

We already know that $2^x = e^{\ln 2x}$, so that means that

$$\frac{dy}{dx} = \ln2(2^x)$$

We saw earlier (page 270) that if $y = 2^x$

$$\frac{dy}{dx} = 2^x \frac{\left(2^{dx} - 1\right)}{dx}$$

This means that

$$\frac{\left(2^{dx} - 1\right)}{dx} = \ln2$$

There was nothing special about using a 2 as the base here; we could have used any number.

For $y = A^x$

$$\frac{dy}{dx} = (\ln A)A^x$$

The derivative of a natural logarithm

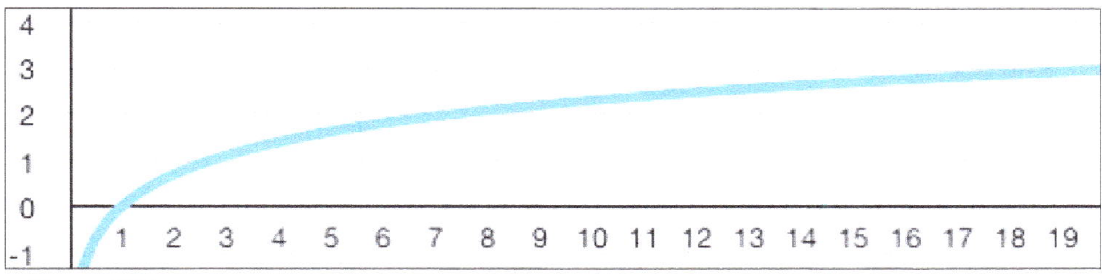

Figure 154

To find the derivative of $y = \ln x$ (the curve is shown at Figure 154), start by raising e to both sides of the equation.

$$y = \ln x$$

Raise e to both sides of the equation. $\qquad e^y = e^{\ln x}$

The expression $e^{\ln x}$ is, of course, equal to x by definition. $\qquad e^y = x$

Using x in the way we usually use y and vice versa and remembering that e^x is its own derivative. $\qquad \dfrac{dx}{dy} = e^y$

$$= x$$

Take the reciprocal of both sides.

$$\frac{dy}{dx} = \frac{1}{x}$$

From Figure 154 you can see visually that this describes the slope of the curve. When x is less than 1, the curve gets increasingly steeper. As x gets larger, the curve gets flatter, but never completely flattens.

The power rule for integration says that you add one to the exponent then divide the whole term by the new exponent. That doesn't work with x^{-1} though. You would end up trying to divide the term by zero. Instead, to integrate, based on our workings above:

$$\int \frac{1}{x} \, dx = \ln x + C$$

In other words, the area under the hyperbola defined by

$$y = \frac{1}{x}$$

is $\ln x$. That's yet another of those mathematical wonders; particularly so, since the natural logarithm as a defining idea for finding the area under a hyperbola was around for some time before Euler discovered e.

> **Key points:**
>
> - Logarithms with base e are called natural logarithms and are written as ln.
>
> - If $y = A^x$, then $\frac{dy}{dx} = \ln A \, e^{\ln Ax}$
>
> - If $y = \ln x$, then $\frac{dy}{dx} = \frac{1}{x}$
>
> - $\int \frac{1}{x} dx = \ln x + C$

Complex numbers in trigonometry and exponentials

> **You might not strictly** need what follows to understand statistics unless you plan to do much with characteristic functions (which we only touch on in Part 2), but what Euler found was so mind-blowing that you will probably be glad you have read it, even if you never use it.

We saw earlier that, using a Taylor series, we can prove that

$$\sin x = x - \frac{x^3}{3!} + \frac{x^5}{5!} - \frac{x^7}{7!} + \cdots$$

Now let's look at what happens with cos x:

Derivative level	Function	Value when $x = 0$
Function: $f(x)$	cos x	1
First derivative (derivative of cos x): $f'(x)$	$-$ sin x	0
Second derivative (derivative of –sin x): $f''(x)$	$-$ cos x	-1
Third derivative (derivative of –cos x): $f'''(x)$	sin x	0
Fourth derivative (derivative of sin x) : $f^{(4)}(x)$	cos x	1
etc		

Putting these values into a Taylor series, we get:

$$\cos x = 1\frac{x^0}{0!} + 0\frac{x^1}{1!} - 1\frac{x^2}{2!} - 0\frac{x^3}{3!} + 1\frac{x^4}{4!} + 0\frac{x^5}{5!} - 1\frac{x^6}{6!} + 0\frac{x^7}{7!} + 1\frac{x^8}{8!} + \cdots$$

$$\cos(x) = 1 - \frac{x^2}{2!} + \frac{x^4}{4!} - \frac{x^6}{6!} + \frac{x^8}{8!} - \cdots$$

We also saw that

$$e^x = \frac{x^0}{0!} + \frac{x^1}{1!} + \frac{x^2}{2!} + \frac{x^3}{3!} + \frac{x^4}{4!} + \frac{x^5}{5!} + \frac{x^6}{6!} + \frac{x^7}{7!} + \frac{x^8}{8!} + \cdots$$

Let's return to the imaginary but extremely useful number i, which is equal to the square root of –1. When we raise i to successive powers, it looks like this:

$i^0 = 1$ (everything, even i, raised to the power of 0 equals 1)

$i^1 = i$ (everything raised to the power of 1 equals itself)

$i^2 = -1$ (by definition)

$i^3 = i^2\,i = -1i$

$i^4 = i^2\,i^2 = (-1)(-1) = 1$

$i^5 = i^4 i = 1i = i$

$i^6 = i^5 i = i^2 = -1$

And so the sequence continues, in a repeating cycle

$$1,\ i,\ -1,\ -i;\ \ 1,\ i,\ -1,\ -i;\ \ 1,\ i,\ -1,\ -i;\ \ldots$$

Now if we change the x in our Taylor polynomial expansion to ix but treat it the way we treated x, we have

$$e^{ix} = \frac{(ix)^0}{0!} + \frac{(ix)^1}{1!} + \frac{(ix)^2}{2!} + \frac{(ix)^3}{3!} + \frac{(ix)^4}{4!} + \frac{(ix)^5}{5!} + \frac{(ix)^6}{6!} + \frac{(ix)^7}{7!} + \cdots$$

Use what happens when i is raised to a power:

$$e^{ix} = 1 + ix - \frac{x^2}{2!} - i\frac{x^3}{3!} + \frac{x^4}{4!} + i\frac{x^5}{5!} - \frac{x^6}{6!} - i\frac{x^7}{7!} + \frac{x^8}{8!} + \cdots$$

Separate out the terms with i:

$$e^{ix} = \left[1 - \frac{x^2}{2!} + \frac{x^4}{4!} - \frac{x^6}{6!} + \frac{x^8}{8!} + \cdots\right] + \left[ix - i\frac{x^3}{3!} + i\frac{x^5}{5!} - i\frac{x^7}{7!} + \cdots\right]$$

Factor out i:

$$e^{ix} = \left[1 - \frac{x^2}{2!} + \frac{x^4}{4!} - \frac{x^6}{6!} + \frac{x^8}{8!} + \cdots\right] + i\left[x - \frac{x^3}{3!} + \frac{x^5}{5!} - \frac{x^7}{7!} + \cdots\right]$$

The series in the first set of braces is the Taylor series expression of $\cos x$, and the series in the second set of braces is $\sin x$.

$$e^{ix} = \cos x + i \sin x$$

This is known as Euler's formula. The calculation of derivatives for trigonometric functions was based on radians, and radians operate with π (a radian being the angle formed by an arc of the length of the radius).

Fun fact:

The cosine of π is –1, and the sine of π is 0. That means, incredibly, that:

$$e^{\pi i} = -1$$

Given that e is an irrational number (with endless decimal places) based on compound interest and exponential growth, π is a different irrational number based on the ratio of the diameter of a circle to its radius, and i is an unreal number, this is some mathematics to induce a sense of awe and wonder.

φ is yet another irrational number. We already learned that $\varphi^2 - \varphi = 1$ (which is extraordinary enough), so we can also say that

$$\varphi^2 + e^{i\pi} = \varphi$$

That's just extraordinary, but as we have proven, it is also true.

There is more material in an appendix on page 745, for those who want to learn more about working with trigonometry in the complex plane.

Overall, one of the effects of moving trigonometric and exponential functions from the Cartesian plane to the complex plane is that functions which extend forward and backward infinitely in the Cartesian plane rotate around the origin in the complex plane.

Key points:

- $\cos(x) = 1 - \dfrac{x^2}{2!} + \dfrac{x^4}{4!} - \dfrac{x^6}{6!} + \dfrac{x^8}{8!} - \cdots$

- $e^{ix} = \cos x + i \sin x$

- $e^{\pi i} = -1$

Differential equations

Read this if you aren't familiar with differential equations, or you would like to start to understand why the normal distribution describes everything from people's IQs to their shoe sizes to the probability that your parcel will arrive when it is supposed to.

In real-world applications, people often have some limited information about a phenomenon and need to find the equation that explains it and predicts it. There might be information about a rate of change and what is happening at one or two points. When this happens the starting equation might include some differentials of various levels.

These can sometimes be difficult or impossible to solve using normal mathematical processes. Sometimes it is necessary to put a range of test values into a computer and just see what comes up. Sometimes it is necessary to input test values, calculate the slope that a tangent to a curve will have at that point, and draw enough of them to estimate what the curve is likely to be doing.

There are, however, some techniques that make it possible to solve some differential equations using little more than the algebra and calculus that we have covered.

Separable differential equations

Take this equation:

$$\frac{dy}{dx} = -Axy$$

where A is a positive constant (unknown fixed number) and x and y are variables (holders that can take any number). This expression is telling us that the rate of change of a function is both proportional to the value of the function itself (y) and to whatever value x has; but it isn't telling us what the

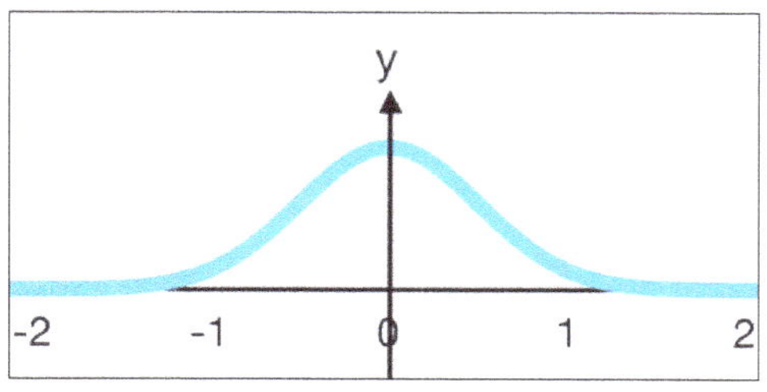

Figure 155

function itself is. In order to find the equation, first we imagine that instead of representing a limit, dy and dx are extremely small changes to x and y, but are still finite numbers. In this instance we can now divide both sides by y, and multiply both sides by dx. This gives us:

$$\frac{1}{y}dy = -Axdx$$

This is called **separating the terms**, and this is called a **separable differential equation**. Because the expressions are identical, we can integrate both sides of the equation. When we do this, we can integrate the left side with respect to y, and the right side with respect to x.

$$\int \frac{1}{y}dy = \int -Ax\,dx$$

Integrate both sides, using the reverse power rule on the right, and the fact (as we saw on page 293) that the derivative of $\ln y$ is $\frac{1}{y}$

$$\ln y + C_1 = \frac{-Ax^2}{2} + C_2$$

The constants could be anything, so we can just subtract C_1 of them from each side and call the remaining constant $(C_2 - C_1) = C_3$.

$$\ln y = \frac{-Ax^2}{2} + C_3$$

Natural logarithms are constants, and so to make the step that follows easier, we can replace C_3 with $\ln C$, because what we still have is an arbitrary constant.

$$\ln y = \frac{-Ax^2}{2} + \ln C$$

Raise e to the power of both sides. y, by definition of logarithm, is what you get when you raise e to $\ln y$.

$$y = e^{\left(\frac{-Ax^2}{2} + \ln C\right)}$$

Raising a base to two exponents added together is the same as multiplying the base raised to each exponent.

$$y = e^{\left(\frac{-Ax^2}{2}\right)} e^{\ln c}$$

$e^{\ln c} = C$ (because that's what logarithm means).

$$\boxed{y = Ce^{\left(\frac{-Ax^2}{2}\right)}}$$

Figure 155 shows the graph of this function where C and A are both given a value of 1. It is a "bell-shaped curve" which we will see a lot of in statistics. Getting back to our starting point:

$$\frac{dy}{dx} = -Axy$$

We can see that this derivation describes the varying slope of this curve. When y approaches zero (in Figure 155 from slightly more than $x = 1$ and slightly less $x = -1$) the slope of the curve approaches zero; it is flat. The same thing happens when x approaches zero. When x and y are both slightly positive the slope of the curve is negative; a line that is a tangent to it slopes backwards. And when x is negative but y is positive, the slope of the curve is positive; it slopes forward.

We will see in Part 2 that this curve, derived in this way, forms the basis of the **normal distribution curve**, and therefore of much of our understanding of statistics.

Key points:

- In a **differential equation**, you know something about the rate of change but don't have the equation and need to find it.

- This is sometimes possible using a **separable differential equation**. This involves getting the y values on one side, multiplied by dy, and the x values on the other side, multiplied by $dx.$

- An important differential equation is $\frac{dy}{dx} = -Axy$. The equation for this curve with this differential is $y = Ce^{\left(\frac{-Ax^2}{2}\right)}$. This forms the basis of the **normal distribution curve**.

Some important integrals: the gamma and beta functions, and the Gaussian integral

> **Read this if** you aren't familiar with the Eulerian integrals, the incomplete gamma function or the Gaussian integral. Or if you would like to know what π factorial equals. Or you don't know what a "dummy variable" is. Or if you would like to see the square root of π emerge from a totally unexpected source. All of this is, believe it or not, highly relevant to statistics.

Dummy variables and Eulerian integrals

When we do an indefinite integral, the result is a function, but when we do a definite integral we are usually left with a number, because the variable we are integrating with respect to disappears. A **dummy variable** is one that exists to make a function work, but disappears when the function is calculated. We'll see this technique used more as we progress. One important reason for using dummy variables in integration is that they allow us to do definite integrals, but still have a function as the output.

The following integrals seem quite weird when you first look at them, and raise the question: where did *that* come from? The answer seems to be that Euler developed them. He liked playing around with integrals, particularly with e involved (remembering that the integral of e^x is e^x), and introducing dummy variables which disappeared when the integral was calculated, apparently just to see what he could come up with when he did.

Adrien-Marie Legendre (1752-1833) was a French professor of mathematics who, among other things, wrote a textbook (*Éléments de géométrie*) which wasn't improved on for 100 years. He was the first one to publicise these integrals, but he gave credit to Euler for inventing them. These are called the Eulerian integral of the first kind, also known as **the beta function**, and the Eulerian integral of the second kind, **the gamma function.** (Beta and gamma, β and Γ, are the 2nd and 3rd letters of the Greek alphabet).

We will look at them in reverse order (because it makes more sense this way): the gamma function first, then the beta function. We will add in some closely related work (possibly, but not definitely, by Gauss).

Stay with it if these don't seem to make much sense when we get into them: they will soon enough, and become quite important when we come to look at statistics (particularly with probability distributions).

The Gamma Function

The **gamma function**, is defined as:

$$\Gamma(x) = \int_0^\infty t^{x-1}e^{-t}dt \quad (for\ any\ real\ x > 0)$$

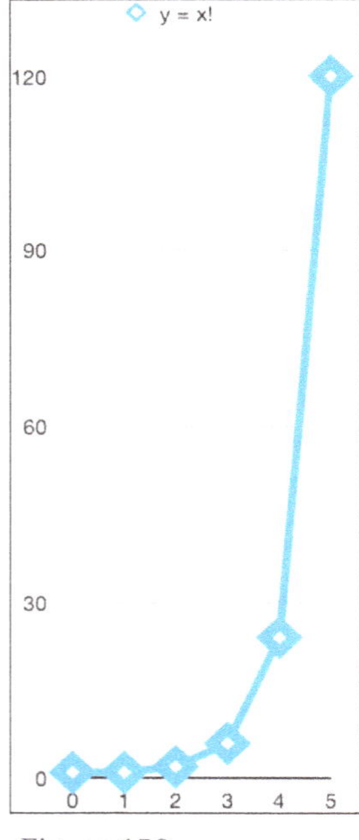

Figure 156

The t is a dummy variable. It only exists to allow the function to work. Because it is a definite integral, the t disappears.

There are a number of useful features of the gamma function, and it forms a part of some of the most important functions in statistics. One of its uses is that it enables the idea of a factorial to be extended.

Factorials involve multiplying positive integers. Based on what we have learned so far, the expression, for example, π! would not have any meaning. The graph in Figure 156 shows the factorials from 0 to 5, and as you can see, drawing a line between them makes it appear as though it should be possible to draw a smooth curve between them. The gamma function allows this to be done.

If we make $x = 1$, then the equation becomes:

$$\Gamma(1) = \int_0^\infty t^0 e^{-t}dt$$

$$= \int_0^\infty e^{-t}dt$$

The integral of e^{-t} is $-e^{-t}$. (If you are unsure about that, differentiate $-e^{-t}$ using the chain rule.) This needs to be evaluated between zero and infinity.

$$= -e^{-t}\big|_0^\infty$$

$$-e^{-\infty} = \frac{1}{-e^\infty} = 0, \text{ while } e^{-0} = e^0 = 1.$$

$$= -e^{-\infty} - -e^{-0}$$

$$--1 = 1$$

$$= 1$$

We say "gamma of one equals one", and write it as

$$\Gamma(1) = 1$$

Now if we look at $\Gamma(x + 1)$ we get:

Use integration by parts.

$$\Gamma(x + 1) = \int_0^\infty t^{(x+1-1)} e^{-t} dt$$

As a reminder, the formula for integration by parts with respect to t is: $$\int u \frac{\delta v}{\delta t} \delta t = uv - \int v \frac{\delta u}{\delta t} \delta t$$	Let $t^x = u$ and $e^{-t} = \frac{\delta v}{\delta t}$ That means... $\frac{\delta u}{\delta t} = xt^{x-1}$ and $v = -e^{-t}$

$$\Gamma(x + 1) = t^x \left(\frac{-1}{e^t}\right)\Big|_0^\infty - \int_0^\infty xt^{x-1}(-e^{-t}) dt$$

The expression $\frac{\infty^x}{e^\infty}$ evaluates to 0, because $\lim_{t\to\infty} \frac{t^x}{e^t}$ approaches 0. (Note that in this instance x is the constant, and t is the variable). You would need to use l'Hôpital's rule a number of successive times to prove this. Each time you take the derivative the numerator will be multiplied by an integer which is 1 less than the previous time (x is an integer), while the value of the denominator won't change. Eventually the numerator will be multiplied by 0.

$$= -\frac{\infty^x}{e^\infty} - 0 - \int_0^\infty xt^{x-1}(-e^{-t}) dt$$

Take the –1 outside the integration. This cancels out the –1 which was already there.

$$\Gamma(x + 1) = \int_0^\infty xt^{x-1}(e^{-t}) dt$$

x is a constant with respect to t, and so it can be taken outside the

$$\Gamma(x + 1) = x\int_0^\infty t^{x-1}(e^{-t}) dt$$

integration. $\int_0^\infty t^{x-1}(e^{-t})dt = \Gamma(x)$, by definition.

$$\Gamma(x + 1) = x\Gamma(x), where \; x > 0$$

That means if $\Gamma(1) = 1$, and each time we add 1 to x and take the gamma function, it is the same as multiplying the gamma function by x.

So:

Value of x	$\Gamma(x + 1)$	$x\Gamma(x)$	Result	Value	!
0	$\Gamma(1)$	(x is not > 0, but this result is true from our calculation above)		1	0!
1	$\Gamma(1 + 1) = \Gamma(2)$	$1\Gamma(1)$	$1 \bullet 0!$	1	1!
2	$\Gamma(2 + 1) = \Gamma(3)$	$2\Gamma(2)$	$2 \bullet 1!$	2	2!
3	$\Gamma(3 + 1) = \Gamma(4)$	$3\Gamma(3)$	$3 \bullet 2!$	6	3!
4	$\Gamma(4 + 1) = \Gamma(5)$	$4\Gamma(4)$	$4 \bullet 3!$	24	4!
etc.					

This means that $\Gamma(x + 1) = x!$, where x is a non-negative integer.

In other words:

$$\Gamma(n) = (n - 1)! \; where \; n \in \mathbb{Z} > 0$$

So we can use the gamma function as a way to calculate factorials of positive integers, but it also gives us a way to calculate something similar for numbers that aren't positive integers. As a hint to where this is heading when we explore statistics, we've seen how important the idea of factorial is for combinations, permutations, n choose r and therefore the binomial theorem. The gamma function will enable us to take some of these ideas further.

Incomplete gamma functions

An **incomplete gamma function** is important in some areas of statistics. A gamma function is defined as an integral from 0 to ∞ . A **lower incomplete gamma function** is defined from 0 to some number less than ∞, while an

upper incomplete gamma function is defined from some number greater than 0 to ∞.

A lower incomplete gamma function is given by the formula:

$$\gamma(x, a) = \int_0^a t^{x-1} e^{-t} dt$$

γ is the lower-case version of the Greek letter gamma, while Γ is the upper case.

An upper incomplete gamma function has the formula:

$$\Gamma(x, a) = \int_a^\infty t^{x-1} e^{-t} dt$$

The Gaussian Integral and gamma of one half

Now let's look at

$$\Gamma\left(\frac{1}{2}\right)$$

The working of this is quite complex, takes several pages, and uses most of the techniques we used for advanced integration plus more. There are three reasons for including it here. The first is that the approach contains a difficult integral that we will meet later when we come to the formula for the normal distribution curve, so this is an introduction to one of the techniques that is used to solve it. The second is that we will need the result when we come to look at the chi square distribution. The third is simply that the final result is surprising and delightful.

Use the definition of the gamma function.

$$\Gamma\left(\tfrac{1}{2}\right) = \int_0^\infty t^{-\frac{1}{2}} e^{-t} dt$$

Multiply outside the integration by 2, and inside by $\frac{1}{2}$, to keep everything the same.

$$= 2\int_0^\infty \tfrac{1}{2} t^{-\frac{1}{2}} e^{-t} dt$$

Use integration by substitution.

$$\text{Let } u = t^{\frac{1}{2}}$$

Use the power rule to differentiate.

$$\frac{du}{dt} = \frac{1}{2} t^{-\frac{1}{2}}$$

Multiply both sides by dt.

$$\delta u = \frac{1}{2} t^{-\frac{1}{2}} \delta t$$

If $u = t^{\frac{1}{2}}$, it also means, squaring both sides...

$$t = u^2$$

Substitute these values into the equation. Even after the substitution, unfortunately there is not easy way to integrate this; but an answer can still be found.

$$\Gamma\left(\tfrac{1}{2}\right) = 2\int_0^\infty e^{-u^2}\,du$$

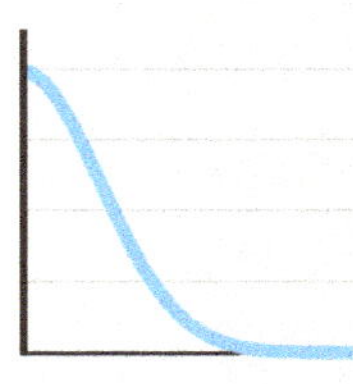

When we graph $f(u) = e^{-u^2}$, using u as the horizontal axis, for values of u greater than 0, it looks like the graph on the left,

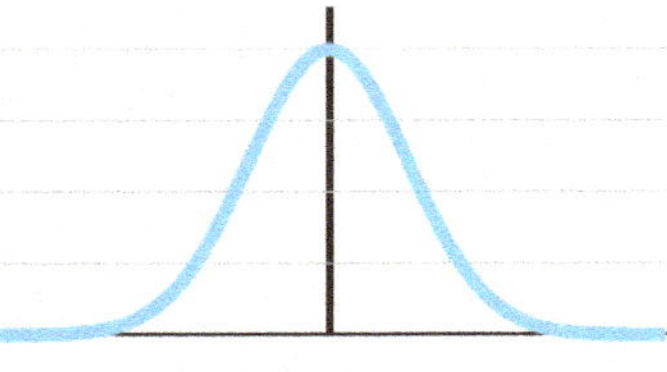

whereas when we graph it including the negative numbers, it looks like the graph on the right. It is symmetrical about 0. Because u is squared, any negative values of u will give the same values on the vertical axis as any positive values. This means that twice the integral from 0 to infinity is the same as the integral from negative infinity to infinity.

$$\Gamma\left(\tfrac{1}{2}\right) = 2\int_0^\infty e^{-u^2}\,du$$

$$= \int_{-\infty}^\infty e^{-u^2}\,du$$

Figure 157
Carl Friedrich Gauss
1777-1855

This integral is known as the **Gaussian integral**. There is no easy way to do it, but solutions have been found. Carl Friedrich Gauss (1777-1855) gets the credit for the one that follows, and may have been the person to solve it, but it uses the work of a younger German mathematician, Carl Gustav Jacob Jacobi, who came up with the Jacobian which we saw when we looked at changing to polar coordinates. Jacobi was prolific, and some of his papers have still not been published. Gauss was a powerful figure in German mathematics, and was very concerned about his reputation. A story says that he refused to allow one of his sons to study mathematics, telling him that he wasn't good enough and would therefore dilute the family name. He also didn't like to show his working, considering it a waste of time. It therefore seems more likely that it was actually Jacobi who solved this integral, but we will probably never know for sure.

To solve the Gaussian integral:

Square this, then take the square root, so that nothing changes.

$$= \sqrt{\left(\int_{-\infty}^{\infty} e^{-u^2}\,du \right)^2}$$

$$= \sqrt{\left(\int_{-\infty}^{\infty} e^{-u^2}\,du \right)\left(\int_{-\infty}^{\infty} e^{-u^2}\,du \right)}$$

It's worth remembering at this stage that when we solve a definite integral we don't end up with an equation, we end up with a number: the area under the curve. It's also worth remembering that the variable t and then the variable u are just variables to make the function work; they disappear once the integral is completed. These two facts mean it doesn't matter what variables we put into the integrals; it won't change the outcome. We can replace u in one integration with x, and the other with y.

$$= \sqrt{\left(\int_{-\infty}^{\infty} e^{-x^2}\,dx \right)\left(\int_{-\infty}^{\infty} e^{-y^2}\,dy \right)}$$

The x here bears no relation to the x in our original equation. We can think of it as a coordinate on a different plane altogether. Or actually, because there is an x and a y on the same side of the equation, it's more accurate to think of it as a three-dimensional shape in a three-dimensional space. Or just think of it as a variable in an equation if you prefer.

If this was a double integral, we could expand it out to look like the equation above (see page 286). We can therefore write the equation above as this double integral.

$$= \sqrt{\iint\limits_{-\infty}^{\infty} e^{-x^2} e^{-y^2}\,dx\,dy}$$

Use the property of adding exponents, and factor out negative 1.

$$= \sqrt{\iint\limits_{-\infty}^{\infty} e^{-(x^2+y^2)}\,dx\,dy}$$

We can change this to polar coordinates, not forgetting Jacobian determinant which creates an extra r (page 289). It is the extra r which makes it an equation that we can integrate. Remember that, in changing to polar coordinates, $x^2 + y^2 = r^2$, based on Pythagoras' theorem.

$$= \sqrt{\int_{0}^{2\pi}\!\!\int_{0}^{\infty} r\,e^{-r^2}\,dr\,d\theta}$$

Note that in the change to polar coordinates, we must rethink the limits of integration. Each of our double integrals went from minus infinity to plus infinity; on a Cartesian plain they would have covered the whole area, going on forever. In polar coordinate terms, the equivalent is for r to go from 0 (a radius of no length at all) to infinity, while the angle θ can go from 0 radians to 2π radians, or around the full circle. Finally, thanks to that extra r, we have an equation that can be integrated.

Multiply the r by 2, and the constant outside the integration by $\frac{1}{2}$, so that they will cancel one another out.

$$= \sqrt{\frac{1}{2} \int_0^{2\pi}\!\!\int_0^{\infty} 2re^{-r^2}\,dr\,d\theta}$$

We could have used integration by substitution to calculate that, but it would have made this proof even longer, and yet another substitution would have made it more confusing. You can confirm it yourself by applying the chain rule to differentiate $-e^{-r^2}$.

$$\text{Now } \int 2re^{-r^2}\,dr = -e^{-r^2}$$

$$\Gamma\left(\tfrac{1}{2}\right) = \sqrt{\frac{1}{2} \int_0^{2\pi} -e^{-r^2}\Big|_0^{\infty} \, d\theta}$$

Substitute the values into the inner integral. $-e^{-\infty^2}$ goes to zero, and $--e^0$ equals 1.

$$= \sqrt{\frac{1}{2} \int_0^{2\pi} \left\{ -e^{-\infty^2} - -e^0 \right\} d\theta}$$

The integral of 1 with respect to θ is θ. So we need to evaluate it at $\theta = 2\pi$, and subtract $\theta = 0$

$$= \sqrt{\frac{1}{2} \int_0^{2\pi} 1 \, d\theta}$$

So this integral evaluates to $2\pi - 0 = 2\pi$.

$$\Gamma\left(\tfrac{1}{2}\right) = \sqrt{\frac{2\pi}{2}}$$

$$= \sqrt{\pi}$$

Another of those awe-inspiring mathematical facts:

$$\Gamma\left(\tfrac{1}{2}\right) = \sqrt{\pi}$$

This means that $\sqrt{\pi}$ is the answer we were looking for, and is also the solution to the Gaussian integral.

Remember that we first met π as the ratio of the diameter of a circle to its circumference. Like e, it just keeps showing up in the unlikeliest of contexts.

(There is actually a way to do a Gaussian integral without using polar coordinates or a Jacobian. It was developed by Laplace who we will meet soon, and as Laplace was 55 years older than Jacobi, he almost certainly arrived at a solution first. It's in an appendix on page 758.)

We saw earlier that $\Gamma(1 + x) = x\Gamma(x), where\ x > 0$

Following from this:

$$\Gamma\left(1\tfrac{1}{2}\right) = \Gamma\left(\tfrac{3}{2}\right) = \Gamma\left(1 + \tfrac{1}{2}\right) = \tfrac{1}{2}\Gamma\left(\tfrac{1}{2}\right) = \tfrac{1}{2}\sqrt{\pi}$$

$$\Gamma\left(2\tfrac{1}{2}\right) = \Gamma\left(\tfrac{5}{2}\right) = \Gamma\left(1 + \tfrac{3}{2}\right) = \tfrac{3}{2}\Gamma\left(\tfrac{3}{2}\right) = \left(\tfrac{3}{2}\right)\left(\tfrac{1}{2}\right)\sqrt{\pi} = \frac{3\sqrt{\pi}}{4}$$

$$\Gamma\left(3\tfrac{1}{2}\right) = \Gamma\left(\tfrac{7}{2}\right) = \Gamma\left(1 + \tfrac{5}{2}\right) = \tfrac{5}{2}\Gamma\left(\tfrac{5}{2}\right) = \left(\tfrac{5}{2}\right)\left(\tfrac{3}{2}\right)\left(\tfrac{1}{2}\right)\sqrt{\pi} = \frac{15\sqrt{\pi}}{8}$$

$$\Gamma\left(4\tfrac{1}{2}\right) = \Gamma\left(\tfrac{9}{2}\right) = \Gamma\left(1 + \tfrac{7}{2}\right) = \tfrac{7}{2}\Gamma\left(\tfrac{7}{2}\right) = \left(\tfrac{7}{2}\right)\left(\tfrac{5}{2}\right)\left(\tfrac{3}{2}\right)\left(\tfrac{1}{2}\right)\sqrt{\pi} = \frac{105\sqrt{\pi}}{16}$$

We can generalise this as:

$$\Gamma\left(n + \frac{1}{2}\right) = \frac{(2n - 1)(2n - 3) \dots (3)(1)}{2^n}\sqrt{\pi}$$

($(2n - 1)(2n - 3) \dots (3)(1)$ is sometimes written as $(2n - 1)!!$ and called a **double factorial** or a **double decremental factorial**. However, in one of the instances in mathematics where one symbol is given more than one meaning, it can also mean **factorial of factorial**, so we'll avoid using it.)

To cement your understanding:

Calculate $\Gamma\left(5\tfrac{1}{2}\right)$ etc, as we did above, until you can see where the generalised equation for $\Gamma\left(n + \tfrac{1}{2}\right)$ comes from.

All of that is about as extraordinary as what we will see when we come to the normal distribution.

The Beta Function

The Eulerian integral of the first kind, **the beta function,** is another function that will be very useful when it comes to statistics. It is defined as:

$$\beta(x, y) = \frac{\Gamma(x)\Gamma(y)}{\Gamma(x + y)}$$

$$= \int_0^1 t^{x-1}(1 - t)^{y-1}dt$$

(In case you want to see the calculations showing that this integral is what you get when you combine gamma functions in the way shown here, there is an appendix covering it on page 761.)

This in turn means that, when x and y are positive integers,

$$\beta(x, y) = \frac{(x - 1)!\,(y - 1)!}{(x + y - 1)!}$$

If you think this seems vaguely reminiscent of the formula for coefficients of the binomial expansion, you would be right. This becomes relevant when we look at this formula in the context of statistics. For now, note that we can express

$$\binom{n}{r}$$

in terms of the beta function, as follows:

Use the definition of the beta function as above, substituting $x = r + 1$ and $y = n - r + 1$

$$\beta(r + 1, n - r + 1) = \frac{\Gamma(r + 1)\Gamma(n - r + 1)}{\Gamma(r + 1 + n - r + 1)}$$

Clean up the denominator

$$= \frac{\Gamma(r + 1)\Gamma(n - r + 1)}{\Gamma(n + 2)}$$

Take the reciprocal of both sides.

$$\frac{1}{\beta(r + 1, n - r + 1)} = \frac{\Gamma(n + 2)}{\Gamma(r + 1)\Gamma(n - r + 1)}$$

Multiply both sides by $\frac{1}{(n+1)}$

$$\frac{1}{\beta(r + 1, n - r + 1)} \frac{1}{(n + 1)} = \frac{\Gamma(n + 2)}{\Gamma(r + 1)\Gamma(n - r + 1)} \frac{1}{(n + 1)}$$

Note that $\frac{\Gamma(n+2)}{(n+1)} = \Gamma(n + 1)$. If you think about this in terms of factorials, you will see why this is. You can also see it if you go back to the

$$= \frac{\Gamma(n + 1)}{\Gamma(r + 1)\Gamma(n - r + 1)}$$

way we saw that the gamma function relates to the factorial function.

Using the relationship between the gamma function and factorials.

$$= \frac{n!}{r!\,(n-r)!} = \binom{n}{r}$$

$$\binom{n}{r} = \frac{1}{\beta(r+1, n-r+1)} \frac{1}{(n+1)}$$

Key points:

- **Dummy variables** can be used in integration to allow us to do definite integrals, but still have a function as the output.

- The **gamma function** is defined as

$$\Gamma(x) = \int_0^\infty t^{x-1} e^{-t} dt \quad (for\ any\ real\ x > 0)$$

- If x is a natural number, $\Gamma(x) = (x-1)!$ If x isn't a natural number, this function can be used to extend the idea of factorials.

- A **lower incomplete gamma** function is given by calculating the integral from 0 to some constant, while an **upper incomplete gamma** is given by calculating from the constant to infinity.

- The following can be calculated, in part, by using the **Gaussian integral**:

$$\Gamma\left(\frac{1}{2}\right) = \sqrt{\pi}$$

- The Gaussian integral is

$$\int_{-\infty}^{\infty} e^{-x^2}\, dx = \sqrt{\pi}$$

  Integrating it involves some complex integration, including a change to polar coordinates.

- The **beta function** is defined

$$\beta(x, y) = \frac{\Gamma(x)\Gamma(y)}{\Gamma(x+y)} = \int_0^1 t^{x-1}(1-t)^{y-1} dt$$

- The beta function can be used to express n choose r:

$$\binom{n}{r} = \frac{1}{\beta(r+1, n-r+1)} \frac{1}{(n+1)}$$

Convolution integrals

Read this if you have never studied convolution integrals, or you have studied them in engineering but aren't sure how that relates to statistics, or you have studied them but they just seemed too convoluted.

You may have thought that the integrals we have just seen looked convoluted, but convolution integrals are different again. We've looked at convolution in the context of polynomial multiplication (page 101), and saw that convolution involves taking two sets of terms or formulae, multiplying every possible combination, then adding together the results that meet a certain criteria.

When used in practice, convolution usually involves an integral of two continuous functions, and it is defined in terms of the integral. Remember that an integral is, itself, a group of terms added together, as the limit on the number of terms approaches infinity.

Our particular interest is in the application of convolution to statistics. This is largely about a way of combining two probability distributions, and we have more material to cover before we get there (including looking at what a probability distribution is).

In essence, though, the situation is that you have two variables with two separate functions, which we'll call $f(x)$ and $g(y)$, and they add to make a third variable

Convolution integrals are used in physics and engineering disciplines to show the effect that one function has on another, as their effects start to overlap. For example, in acoustical engineering convolution integrals can be used to model the overlapping effect of an initial sound source as one function, with the reverberation of the sound as another function. In that case they often use the lower case of the Greek letter tau (τ). This probably originates from the idea that, in engineering and physics, they are often looking at the effect of one function on another over a period of **time**.

Intuitive explanations for the use of convolutions in physics and engineering are a bit different from those for its use in statistics, because in engineering the use is often in combining different functions of a single variable, whereas in statistics it can be in combining functions with two variables that are added.

with its own function, which we'll call $h(w)$. The variable of this function is the sum of the other two variables:

$$x + y = w$$

We want to know what happens when we multiply every possible combination of $f(x)$ and $g(y)$. We don't want any more variables in the equation than we need to have, and we know that $x + y = w$, so we can say that $y = w - x$. That gives us $g(w - x)$. We multiply these together, then add every instance. To add every instance we integrate, with respect to x, over all possible values of x.

That means our formula for convolutions is:

$$h(w) = \int_{-\infty}^{\infty} f(x)g(w - x)dx$$

If all of that is confusing at this point, don't worry about it too much. It will fall into place later in the book; for now, all you need is a vague sense of what a convolution is, and to recognise that this is the formula for a convolution integral.

There is a piece of terminology to understand here. We saw when we first looked at convolutions (page 101) that an asterisk (*) is used as the sign for a convolution. Where two functions of x, which we'll call $f[x]$ and $g[x]$ are convolved together into a new function which we'll call $h[x]$, this is written as:

$$h[x] = (f * g)[x]$$

This notation can take some getting used to. It isn't saying that the convolution of f and g are multiplied by x; it's saying that the convolution of function f and function g forms a new function of x.

The main thing to note about convolutions at this point is that they are commutative: it doesn't matter which of our variables we subtracted from w. Here's the proof of that.

This is the convolution formula

$$\int_{-\infty}^{\infty} f(x)g(w - x)dx$$

Perform a u substitution. Let $u = w - x$. That means $x = w - u$. It also means that $\frac{du}{dx} = -1$, which means $-du = dx$.

$$= \int_{\infty}^{-\infty} f(w - u)g(u)(-du)$$

The limits of integration have changed. As x approaches infinity, u approaches negative infinity, and vice versa.

Multiply inside and outside the integration by –1, which doesn't change the value of the expression. Also reverse the order of the terms, which also doesn't change anything.

$$= -\int_{\infty}^{-\infty} g(u)f(w-u)\,du$$

Multiplying by –1 is the same as reversing the order of the integration. To see why this works, let's say that the value of the result when the top value of the integral is used is A, and the value of the result when the bottom value is used is B. The next step involves subtracting B from A: that is, A – B. If you multiply that by -1, you get B – A, which is exactly the same as reversing the limits of integration.

$$= \int_{-\infty}^{\infty} g(u)f(w-u)\,du$$

It doesn't matter what letter we use on the right; it disappears when the integral is calculated, anyway. So that means it doesn't matter what order we do the convolution in.

$$\int_{-\infty}^{\infty} f(x)g(w-x)\,dx = \int_{-\infty}^{\infty} g(u)f(w-u)\,du$$

And so we can see that convolution is commutative in the same way that multiplication and addition are.

$$(g * f) = (f * g)$$

We'll revisit convolutions when we come to Laplace transforms, then again later. It will make sense soon enough, if it doesn't yet.

Key points:

- Convolution involves taking two sets of terms or formulae, multiplying every possible combination, then adding together the results that meet a certain criteria.

- The formula for a convolution integral is

$$\int_{0}^{\infty} f(x)g(w-x)dx$$

- The formula $(f * g)[x]$ means a function of x, based on convolving two other functions (f and g).

- Convolution is commutative.

Laplace and Fourier transforms

> **Read this if** you want another way to solve challenging differential equations, or you would like to actually make sense out of moment generating functions when we come to them.

We have seen some situations in which it was helpful to introduce a completely different number plane to understand what was happening. This happened with the chain rule of differentiation, it happened when we looked at shifting real numbers to complex numbers, when we looked at logarithms and when we changed to polar coordinates. In all these situations we took what we were dealing with, changed its situation so that we could do more with it (or work more easily with it) then sometimes transformed it back.

In the same way, Laplace and Fourier transforms are ways to move functions to a different domain that behaves differently, to make the impossible possible and the difficult easier. The Fourier transform is a specific example of the Laplace transform.

The Laplace transform

Figure 158
Pierre-Simon Laplace
1749–1827

Pierre-Simon, Marquis de Laplace (1749–1827) (Figure 158) was a contemporary and fellow-teacher with Legendre. He was a French mathematician, astronomer and physicist. Laplace was one of the key figures in the development of statistics, and we will meet him again later. He shared Legendre's humility: as we shall see later, he gave credit to another mathematician, who now has one of the most famous names in statistics, for work that was probably largely his own. Laplace spent much of his time refining Isaac Newton's understanding of the mechanics of the solar system, to do away with Newton's idea that in order to be stable, celestial movements needed periods of divine intervention. He was the first to come up with the idea of gravitational collapse and black holes, and he came up with accurate equations to predict tides based on a range of complex interactions. Laplace achieved so much, in fact, that he has been called the "Isaac Newton of France". (You could be excused for calling Newton the Pierre-Simon Laplace of England, although Newton died 34 years before Laplace was born.)

Laplace was Napoleon Bonaparte's Minister of the Interior for all of six weeks, before he was sacked and replaced by Napoleon's brother. In his memoirs Napoleon said that the reason for sacking Laplace was that he saw subtleties everywhere and conceived only problems. In other words, he might have made a great Minister of the Interior if he'd been allowed to keep the job.

On the other hand, while the Laplace transform was named after him it was actually Euler who did most of the work to come up with this concept; Laplace played around with this idea, and started using it to solve challenging differential equations.

The fundamental idea is that you take a function to a different realm entirely, and change it so that it works quite differently. You do what you want to do with it, then move it back. You can think of it like taking something written on a piece of paper, putting it into a computer to do something entirely different with it, then printing the computer's output back onto paper; except that the Laplace transform does everything mathematically.

The Laplace transform involves multiplying a function of x by e^{-sx}, where s is a new variable introduced for the sake of the transform. You then take the definite integral of the result with respect to x, between zero and infinity. The notation, for some function of x, looks like this:

$$\mathcal{L}\{f(x)\} = \int_0^\infty e^{-sx} f(x)\, dx$$

$$= F(s)$$

The x disappears because it is a definite integral; after the integration, the x is replaced by infinity and by zero, which are the limits of integration. What remains is a function of s which, as an integral, is indicated by a capital F. This means $f(x)$ is the input function, and $F(s)$ is the output function.

You will note that there are some similarities with gamma function: both are definite integrals between 0 and infinity, both involve a newly-introduced variable as well as x, both involve raising e to a negative power, then multiplying the result by another function. Euler was 48 years older than

Laplace, so Euler's gamma function probably preceded the Laplace transform and inspired it.

What the Laplace transform is doing, though, as we said, is taking a function to a completely different domain, as well as making it a different function.

A Laplace transform can also be written using the Taylor series expansion of e^{-sx}:

$$\int_0^\infty e^{-sx} f(x)\, dx = \int_0^\infty \left[\left(\sum_{n=o}^\infty \frac{(-sx)^n}{n!}\right)\right] f(x)\, dx$$

In this expression, because of linearity, the function can be taken within the summation. In other words, the expression that is summed can be multiplied each time by $f(x)$. Likewise, also because of linearity, the integration can also be taken inside the summation. In other words, each term within the summation can be integrated separately if that is desirable.

Laplace transforms are reversable, and for each $F(s)$ there will only be one $f(x)$. That means we can apply a Laplace transform to a function of x, work with the resultant function of s, then work back to the new function of x. Working back to find the new function of x *is* called an **inverse Laplace transform,** and the notation is

$$f(x) = \mathcal{L}^{-1}\{F(s)\}$$

Unfortunately there isn't a single formula for inverse Laplace transforms. There are tables that you can look up, which have been calculated by doing Laplace transforms. You can use them to work backwards from a transform to a function[25]. (You may notice if you look up more about this that the original functions are normally written using t with the variable, rather than x. This is because Laplace transforms are used extensively when analysing electronic and other signals, and they are often used to convert functions in time (t) to functions in frequencies (s) and back again.) Another situation in which Laplace transforms become particularly useful is in differential equations. This is the sort of problem that does come up in applying mathematics to real world situations: you may know the rate of change of something, and how that is changing, what it is like under specific circumstances, but not what it actually looks like. For example, you might know how fast a pandemic is spreading, and you might also know the point where the rate of increase starts to change, but you might not have an equation for the spread of the pandemic itself.

If your interest is in the mathematics of physics and engineering, you will find some examples in an appendix on page 761.

There is one particular result that we will look at here rather than in the appendix, because we use it in statistics: the inverse Laplace transform of the product of two functions is the convolution of those functions. In other words:

Let

$$\mathcal{L}\{f(x)\} = F(s), and$$

$$\mathcal{L}\{g(x)\} = G(s)$$

then

$$\mathcal{L}\{f(x) * g(x)\} = F(s)G(s)$$

where the convolution is defined as

$$\int_0^x f(\tau)g(x - \tau)d\tau$$

Dealing with the limits of integration in proofs regarding convolution can sometimes be a bit convoluted, because the limits of integration can vary depending on the particular application that the convolution is used for. This proof contains one approach, which is appropriate for the limits of integration we are using here. Other proofs use different limits of integration, and have different approaches for dealing with them.

This is a Laplace transform applied to the standard formula for a convolution integral.

$$\int_0^\infty e^{-sx} \int_0^x f(\tau)g(x - \tau)d\tau \, dx$$

To solve this double integral, it helps if the limits of integration are the same. The set up for this is that the part of the formula shaded blue ($f(\tau)g(x - \tau)$) is a function of τ (tau). If we graph it, not worrying about the shape of the curve, we have τ as the horizontal axis, $f(\tau)$ as the vertical axis, and x as a value that τ is able to take.

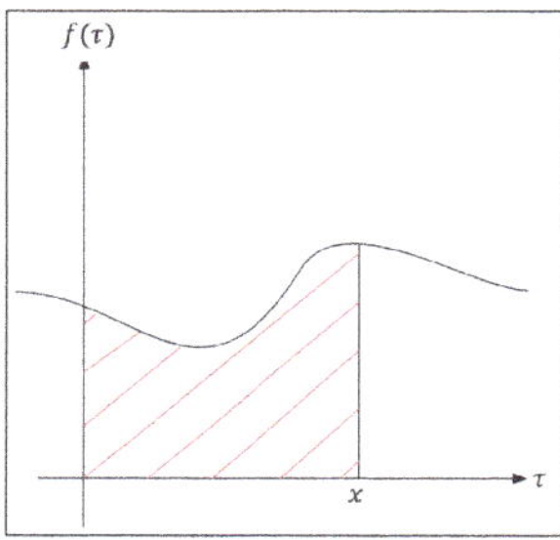

The integral will give the area under the curve, where τ can take values between 0 and x. We want to show the limits of integration for the inner integral (coloured orange) from zero to infinity. We can achieve this by using a sneaky but perfectly legitimate step known as a **unit step**

function. We define a new function $u(\tau, x) = \begin{cases} 1, & \tau \leq x \\ 0, & otherwise \end{cases}$ This means that if τ is less than or equal to x, the value of this function will be 1. If it isn't, the value of the function will be 0.

That solves our problem. We don't need to worry about calculating this unit step function within the integral; we already know that it is either 0, making the whole function equal to 0, or 1, in which case it disappears.

$$\int_0^\infty e^{-sx} \int_0^\infty f(\tau)g(x-\tau)u(\tau,x)\,d\tau\,dx$$

e^{-sx} is a constant with regard to τ, and so can be taken inside the integral. The order of integration doesn't matter with double integrals, so we can reverse the order.

$$\int_0^\infty \int_0^\infty e^{-sx}f(\tau)g(x-\tau)u(\tau,x)\,dx\,d\tau$$

$f(\tau)$ is a constant with regard to x, and so can be taken outside of the integral.

$$\int_0^\infty f(\tau) \int_0^\infty e^{-sx}g(x-\tau)u(\tau,x)\,dx\,d\tau$$

Now we'll do a change of variables. Let $v = x - \tau$. That means $\frac{\delta v}{\delta x} = 1$, which means $\delta v = \delta x$. It also means that $x = v + \tau$. We are working with a situation where $\tau \leq x$, so we can define v as always being zero or positive.

$$\int_0^\infty f(\tau) \int_0^\infty e^{-s(v+\tau)}g(v)u(\tau, v+\tau)\,dv\,d\tau$$

We defined the unit step function as $u(\tau, x) = \begin{cases} 1, & \tau \leq x \\ 0, & otherwise \end{cases}$ Now if $x = v + \tau$, and v is positive, then $\tau \leq x$. That means our unit step function will always equal 1, and we can ignore it.

$e^{-s(v+\tau)} = e^{-sv - s\tau} = e^{-sv}e^{-s\tau}$

$$= \int_0^\infty f(\tau) \int_0^\infty e^{-sv}e^{-s\tau}g(v)\,dv\,d\tau$$

$e^{-s\tau}$ is a constant with respect to v, and so can be taken outside of the inner integral.

$$= \int_0^\infty e^{-s\tau}f(\tau) \int_0^\infty e^{-sv}g(v)\,dv\,d\tau$$

All of the τ's are now under the one integral sign, and all of the v's are under the other, which means that we can make the first integral with respect to τ, and treat this as two integrals multiplied together

$$= \int_0^\infty e^{-s\tau} f(\tau)\,d\tau \int_0^\infty e^{-sv} g(v)\,dv$$

The particular variables don't matter, because all of them except for s disappear when the definite integral is calculated. What we have is the product of the Laplace transforms of the two functions.

$$= F(s)G(s)$$

And so the inverse Laplace transform of the product of two Laplace transforms is the convolution of those functions.

$$\mathcal{L}\{f(x) * g(x)\} = F(s)G(s)$$

$$\mathcal{L}^{-1}\{F(s)G(s)\} = f(x) * g(x)$$

This result is known as the convolution theorem of Laplace transforms, and we will be making use of it in Part 2.

The Fourier transform

Jean-Baptiste Joseph Fourier (1768-1830) was 19 years younger than Laplace. He was a physicist as well as a mathematician, and accompanied Napoleon Bonaparte's army to Egypt as a scientific advisor. Through his calculations on the dissipation of heat he calculated that the Earth should be a lot colder than it is, and postulated that the atmosphere may serve as an insulator. He is thus credited as the first person to discover the greenhouse effect, although, life being what it was at the time, he did not realise that human activity would ever be able to contribute to putting enough additional energy into the atmosphere to increase climate volatility. Fourier came up with a slight variation on the Laplace transform, and Fourier transforms are now probably used more than Laplace transforms.

Figure 159
Jean-Baptiste
Joseph Fourier
1768-1830

Fourier transforms are Laplace transforms where, instead of using the new variable s, the new variable is imaginary. Obviously any letter can be used to represent the variable, but it is usually indicated by the Greek letter omega (ϖ) multiplied by i ($\sqrt{-1}$).

So the formula is

$$\mathcal{F}\{f(x)\} = \int_{-\infty}^{\infty} e^{-i\varpi x} f(x)\,dx$$

$$= F(\varpi)$$

Multiplying the function by an imaginary number prior to integrating it has the effect of taking functions that move forward, and effectively rotating them around the origin. This is particularly useful in things like understanding signals made of multiple overlapping wave forms: Fourier transforms can be used to analyse them and separate them into their component parts. This type of calculation is used, for example, in graphic equalizers.

Because the Fourier transform uses an imaginary number, there are alternative ways of expressing it.

$$\mathcal{F}\{f(x)\} = \int_{-\infty}^{\infty} e^{-i\varpi x} f(x)\,dx$$

$$= \int_{-\infty}^{\infty} (\cos \varpi x - i \sin \varpi x)\, f(x)\,dx$$

Expressing it in this way helps us to see the relationship between graphs that extend longitudinally, and rotational graphs.

This can also be expressed as

$$\mathcal{F}\{f(x)\} = \int_{-\infty}^{\infty} \cos \varpi x f(x)dx - i \int_{-\infty}^{\infty} \sin \varpi x\, f(x)\,dx$$

$$= F(\varpi)$$

This approach may make the integration easier, and can also be used to get a more intuitive sense of the effect that the integration can have on a given function. Note that the i is a constant in integration with respect to x, which means that in the form above it can be taken outside of the integration.

If you are interested in thinking through the amazing places that Fourier transforms can take you, then it is worth diving into the information on the complex plane in an appendix on pages 745 and following.

Key points:

- $\mathcal{L}\{f(x)\} = \int_0^\infty e^{-sx} f(x)\, dx = F(s)$

- $\int_0^\infty e^{-sx} f(x)\, dx = \int_0^\infty \left[\left(\sum_{n=0}^\infty \frac{(-sx)^n}{n!}\right) f(x)\right] dx$

- $\mathcal{L}\{f(x) * g(x)\} = F(s)G(s)$

- $\mathcal{L}^{-1}\{F(s)G(s)\} = f(x) * g(x)$

- $\mathcal{F}\{f(x)\} = \int_{-\infty}^\infty e^{-i\varpi x} f(x)\, dx = F(\varpi)$

- $\mathcal{F}\{f(x)\} = \int_{-\infty}^\infty e^{-i\varpi x} f(x)\, dx$

$$= \int_{-\infty}^\infty (\cos \varpi x - i \sin \varpi x)\, f(x)\, dx$$

$$= \int_{-\infty}^\infty \cos \varpi x f(x)\, dx - i \int_{-\infty}^\infty \sin \varpi x\, f(x)\, dx$$

$$= F(\varpi)$$

Conclusion to Part 1: Fermat's proof of his last theorem?

Figure 160
Pierre de Fermat
1607-1665

Pierre de Fermat (1607-1665) was a friend of Blaise Pascal, who we met at the beginning of this book. They both had poor health and used to enjoy sending one another letters addressing probability problems associated with "games of chance"[26]. They tried to encourage one another to travel across France so that they could meet one another, but neither was well enough, and both died when they were relatively young. Laplace believed that Fermat was the true inventor of calculus, and that Newton refined it while Leibniz enriched the notation[27]. Fermat was, therefore, one of the forerunners of our modern understanding of both statistics and calculus. What he is best known for, however, is **Fermat's last theorem** , which made the Guinness Book of Records as the most difficult mathematical problem to solve, based on the number of unsuccessful attempts. He liked to scribble in the margins of *Arithmetica*, a textbook by Diophantus (from Alexandria, between about 200 and 300 AD). In the margins he came up with a number of theorems, and said he had proofs for them. Of this particular theorem he claimed to have an elegant and simple proof that was too large to write in the space available.

After Fermat's death his son republished *Arithmetica* with Fermat's observations and comments (Figure 161).

All of his other theorems have been solved long ago, but Fermat's last theorem defied a solution until Andrew Wiles finally proved it in 1994, using mathematics that had only just been developed, and that certainly wasn't available in Fermat's time. Wiles's solution definitely wasn't simple, and no one has been able to provide a proof of the theorem using mathematics that was even remotely available in Fermat's time.

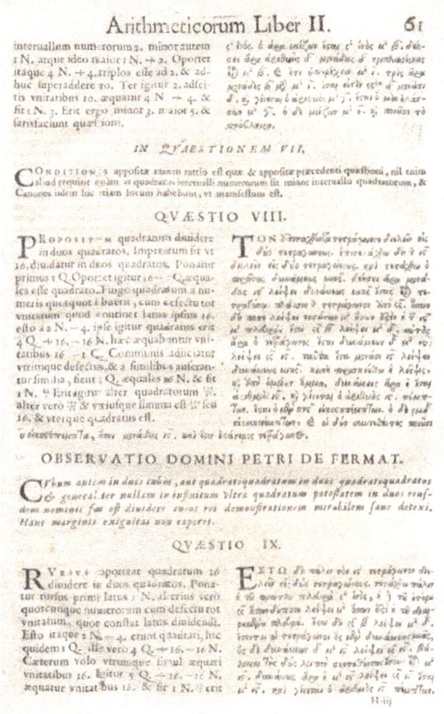
Arithmeticorum Liber II. 61

IN QVAESTIONEM VII.

CONDITIONIS appositæ eadem ratio est quæ & appositæ præcedenti quæstioni, nil enim aliud requirit quam vt quadratus interualli numerorum sit minor interuallo quadratorum, & Canones idem hic etiam locum habebunt, vt manifestum est.

QVÆSTIO VIII.

OBSERVATIO DOMINI PETRI DE FERMAT.

Cubum autem in duos cubos, aut quadratoquadratum in duos quadratoquadratos & generaliter nullam in infinitum vltra quadratum potestatem in duos eiusdem nominis fas est diuidere cuius rei demonstrationem mirabilem sane detexi. Hanc marginis exiguitas non caperet.

QVÆSTIO IX.

Figure 161

The theorem states that there are no positive integers for x, y, z and n that solve the equation:

$$x^n + y^n = z^n$$

where n is greater than 2.

Certainly no one has ever been able to find any. Every counting number forms part of a solution were $n = 2$; and many counting numbers are part of more than one solution. (If that surprises you see the YouTube How to Love Statistics Video about *The problem with counting Pythagorean triples*.)

This is the algebraic expression for Pythagoras' theorem. It can be solved, for example, by $x = 3$, $y = 4$ and $z = 5$:

$$3^2 + 4^2 = 5^2$$

There are no solutions at all for any higher numbers of n.

Here's a relatively simple, elegant proof. It uses differential calculus, which wasn't quite available in its current form in Fermat's time, but his thinking was clearly heading in that sort of direction. It is therefore quite possible that his "elegant and simple proof" was something like this.

Taking the case where n is equal to 3, start by assuming that a solution can be found.

This is our starting equation		$x^n + y^n = z^n$
If a solution can be found, and if $y > x$, then the equation can	Let n = 3	

be expressed in these terms.

If $x^n + y^n = z^n$ then let $y = x + a$ and let $z = x + b$. Let $x, y, z, a,$ and b be natural numbers (non-negative, non-zero integers).

$$x^3 + (x + a)^3 = (x + b)^3$$

Expand out the binomial terms in brackets.

$$x^3 + x^3 + 3x^2a + 3xa^2 + a^3 = x^3 + 3x^2b + 3xb^2 + b^3$$

Subtract x^3 from both sides.

$$x^3 + 3x^2a + 3xa^2 + a^3 = 3x^2b + 3xb^2 + b^3$$

Take the derivative of both sides with respect to x, using the addition rule and the power rule. We can take the derivative of both sides, because they are equal, so we can say that both sides equal $f(x)$. We therefore find $f'(x)$

$$3x^2 + 6xa + 3a^2 = 6xb + 3b^2$$

Take the derivative of both sides with respect to x, again.

$$6x + 6a = 6b$$

Divide both sides by 6.

$$x + a = b$$

In our original definition we said that $y = x + a$, which means $y = b$

$$y = b$$

Substitute $y = b$ back into the right hand side of the starting equation.

$$x^3 + y^3 = (x + y)^3$$

Expand the expression on the right.

$$x^3 + y^3 = x^3 + 3x^2y + 3xy^2 + y^3$$

Subtract $x^3 + y^3$ from both sides.

$$0 = 3x^2y + 3xy^2$$

This cannot be correct where x and y are counting numbers, and therefore there cannot be a solution with natural numbers for:

$$x^3 + y^3 = z^3$$

The same approach will work for any n greater than 3, and therefore we have just found a relatively simple, elegant proof of Fermat's last theorem!

Or not.

Why it doesn't work

The proof works the same way for any n greater than 3, but it also works for n equals 2. If you want the practice, you can try it for yourself. There are, however, as we saw earlier, infinite solutions where $n = 2$, such as

$$9 + 16 = 25$$

So, sadly, the "proof" can be used to prove something that isn't true. What is quite likely true, however, is that Fermat thought he had a proof (whether this one or something different) when, in fact, he did not.

To cement your understanding:

Can you see where the flaw is in the proof, before you read on?

The error in this "proof" goes to the sneaky use of the letter x. We originally defined it as a natural number, but we are used to seeing it used as a variable that can take any real value. Because it can only represent specific values, differentiation wasn't an appropriate technique to use: it didn't actually mean anything to differentiate this equation. In mathematical terms we say that the equation wasn't differentiable; but we did it anyway, and ended up with a proof that didn't work.

Two morals of the story

The first thing we can learn from this is that even the greats of mathematics can make mistakes. (While this may not have been the particular mistake Fermat made, it's likely that he made one.) Therefore don't take other people's word for it when it comes to statistics. Many of the people we will look at in part 2 made mistakes, at least at some point. There is no reason to believe that all of the mistakes have now been found; history suggests that they will keep happening. Be inquisitive, and make sure that you understand, and agree with, mathematical proofs for yourself.

There is, though, a second possibility that is more intriguing. Throughout the history of mathematics (and the history of invention generally) there have been great breakthroughs which appear obvious in hindsight, but which weren't obvious to anyone beforehand. The best-known example is the invention of the wheel. The Incas were ingenious enough to build Machu Picchu without it. They worshipped a sun-god, and their art had circles in it, so the circle was a familiar shape. In hindsight, did no one think to roll some rocks instead of dragging them? If they did roll the rocks, did they not make the jump to inventing the wheel? Clearly not. Other examples include:

Umbrellas	While parasols existed in ancient Egypt and China, using a waterproof version for rain was delayed until the 18th century in Europe. The technology was available, but the application wasn't obvious.
Paperclips	The modern paperclip wasn't patented until 1899, even though bent wire had existed for centuries. The simplicity of holding papers together took a surprising time to catch on.
Safety Pins	The safety pin was invented in 1849. The need for a mechanism of a clasp to shield the sharp end wasn't recognized until much later.
Bicycles	Despite the simplicity of the concept, bicycles didn't appear until the 19th century, thousands of years after the wheel was invented.
Thumbtacks	These weren't invented until 1900, even though people had been using nails and tacks for centuries. The idea of a flat surface to press with your thumb came surprisingly late.
Roller Suitcases	The invention of the wheel on luggage didn't occur until 1970, long after wheels and luggage both existed independently. Travelers carried their heavy suitcases instead of rolling them.

The ancients from Mediterranean regions used ropes with knots to measure distance. Did no one make the jump from that to a number line, and from there to assigning a number of 0 and using negative numbers? Again, they did not. It's only obvious looking back. Descartes' idea of crossing two number lines at right angles to form the Cartesian plane, which opened the way for all of the mathematics that came afterwards, was such a simple idea: why did no one think of it earlier? Because they didn't.

Fermat and Descartes were contemporaries, both living in France (although Fermat was born later 11 years later but died 25 years earlier than Descartes). He was unquestionably a genius with an extraordinarily original mind. So the intriguing possibility about Fermat's last theorem is this: perhaps he *did* have a simple proof as he claimed, using a simple, different approach to mathematics that he thought up but did not have the life-span or health to publicise, and no one has thought of it since, just because they haven't. Maybe it is just around the corner, waiting for you to discover or invent it.

Maybe there are unlimited possibilities like that.

Thinking of Pierre de Fermat is a good place to finish our exploration of the underlying mathematics behind statistics, because his correspondence with Pascal set the scene for what follows in Part 2.

Part 2:
Statistics & the Mathematics of Uncertainty

Introduction to Part 2

The predictability of randomness

There is a mystery at the heart of statistics: randomness can be predictable. This is a strange yet commonplace idea.

Figure 162

To appreciate the concept, consider this thought experiment. Imagine there is a grain silo with a valve at the bottom. You open it and let the grain spill out for 30 seconds. Individual grains will ricochet in random, chaotic ways. At the same time, the grain will predictably fall into a heap with a recognisable shape: roughly conical, higher in the centre but rounded, and scattered out around the edges. Every single time, the shape of the pile will be similar (Figure 162). Whether you are thinking of washing powder or sugar, you already know that granular material forms a similar shape. With the right measures and the right mathematics, it is possible to predict what proportion of grains will lie within a given distance of the centre.

Now imagine that one of the grains is purple. There is no way to be certain of where the purple grain will land. Because we can predict the shape of the pile, though, we can calculate what proportion of the grains will lie where. This lets us predict the probability that our purple grain will end up within a certain distance from the centre.

This underappreciated yet everyday phenomenon makes statistics workable. It can allow you to answer questions like: how probable is it that what you are seeing is important compared to a random fluctuation? What level of risk are you taking on? What are the chances that what you have found gives genuine insight into a broader picture rather than being an insignificant fluke?

Statistics is the powerful yet paradoxical application of mathematical precision to guesswork. It isn't so much about understanding the information you *do* have but about using the predictability of randomness to understand the information you *don't*. That makes it phenomenally useful in virtually every field of study, from **A**rtificial intelligence (one of the newer applications) to **Z**ymology (the study of fermentation) which was one of the earliest industrial applications of this approach.

Origins

Like a flower emerging from a spiky desert cactus, statistics had an unpromising start. It originally came together out of four streams of thought.

Strictly speaking **descriptive** statistics is about the information you *do* have, while **inferential** statistics is about the information you *don't* have. But modern computer-based graphic capabilities have made most descriptive statistics, apart from basic averages and quantiles (e.g. medians and deciles), largely redundant except in so far that they help to let us know how to apply inferential statistics to the data we have. It's easier and more helpful to get a computer to graph data than to try to come up with a meaningful statistical description of it. We won't therefore pay much attention in this book to the distinction between descriptive and inferential statistics.

The first grew out of the world of gambling and used mathematics to answer questions about the probability of uncertain outcomes by addressing the question: how many ways can something happen? The second came from an obscure (in every sense) unpublished paper found in the effects of a deceased Presbyterian clergyman, which was presented to the Royal Society by a friend and then promptly forgotten. The third stream involved using mathematics to understand the nature of errors in measurement, particularly attempts to find a correct value from a range of measures taken by imperfect people using basic technology. (This was important in the world of

The Royal Society (formerly the Royal Society of London for Improving Natural Knowledge) is the oldest scientific academy in the world. Its motto is *Nullius in verba*, which roughly translates as: 'don't take anyone's word for it'. This is a good motto for anyone studying statistics.

navigation by the stars. Even tiny errors in navigation could mean that ships were unable to know how far they were from dangerous shores. Precise astronomy was therefore quite literally a matter of life and death as well as being crucial for commercial and military success.) The fourth stream involved using physics (not statistics) to better understand and improve the moving parts of steam engines and other mechanical contraptions. We will follow the same paths, to understand how these ideas arose, so that we can make better use of them.

Towards the end of the 1800s, these unlikely streams were brilliantly brought together by people trying to understand more about Darwinian evolution. Some, in an idealistic but ultimately horrific attempt to improve the human race, sought to breed out imperfections and create 'better' humans. The horrors of Nazi death camps were rooted in the ideology of eugenics.

Fortunately, those who brought the streams together recognised that statistics had much wider applications. In the 20th and 21st centuries, it has become the basis of virtually all research that involves the collection of data, and without it, the world would be far less advanced.

First Stream: Gambling Beyond Intuition

---

*"It is remarkable that a science, which commenced with the considera-
tion of games of chance, should be elevated to the rank of the most important subjects of human knowledge."*

Pierre Simon, Marquis de Laplace[28]

It may be tempting to think that mathematical probability is just glorified common sense, but sometimes a closer examination reveals unexpected and counterintuitive results.

One example is the "three-door" (aka "Monty Hall") problem. In this scenario, a game show host shows you three identical doors. There is a car behind one and nothing behind the other two. If you pick the right door, you win the car. The host asks you to pick one door. Without telling you whether you are right or not, the host opens another door that does not have a car behind it. You then have a choice. You can stick with the choice you have made or change to the other door. Most people tend to intuitively think the probability will not change and so stick with the choice they have made. What happens, though, is that initially, you have a one-third chance of being right. When the host opens a door that you have not chosen and that does not have a car behind it, it has the effect of combining the two-thirds probability that it was behind one of the doors you did not select and assigning both of those probabilities to the door you did not pick. So you have a one-third chance of getting the car if you do not change your pick but a two-thirds chance if you do. Therefore, if you change your choice, you effectively double your chance of winning a car.

You can think of it like this. Imagine the host were to tell you that you can either stick with the choice you have made or open both of the other doors, and if the car is behind either of them, you get the car. In that case, the choice is easy: you have a one-third chance of getting the car if you stay with your existing choice or a two-thirds chance if you open the other two doors. That is the choice the host is really giving you. He is just saving you some effort by opening one of the other doors for you, and it is one that he happens to know does not have the car. But it does not matter at all who opens the door; the outcome will be the same. (If you are still struggling with this, you might want to leave it for now and return to it when we have explored the subjective nature of probability when we look at Thomas Bayes.)

A second counterintuitive example is Simpson's paradox. Suppose you are comparing the reviews of accountants to decide which one to go to. You look at two different review sites. You find that 90% of reviewers on site 1 give accountant A a good review, whereas only 80% of reviewers on site 1 give a good review to accountant B. You also find that 50% of reviewers on site 2 give accountant A a good review, whereas only 30% of reviewers on site 2 give accountant B a good review. The choice seems clear: choose accountant A. That is, until you review the actual data (Figure 163).

		Recommend accountant A		Recommend accountant B	
Review site 1	Positive reviews	36	90%	200	80%
	Total reviews	40		250	
Review site 2	Positive reviews	180	50%	50	30%
	Total reviews	360		150	
Total	Positive reviews	216	54%	250	62.5%
	Total reviews	400		400	

Figure 163

When you see the underlying numbers, you realise that out of 400 reviewers for each accountant, more preferred accountant B to accountant A. That is Simpson's paradox at work, and it is an example of why it can be a good idea to understand probability for yourself rather than assume that others are correct when they assure you that statistics back them up.

In a third example, in the 1990s and early 2000s, generous philanthropists gave about 2 billion US dollars to change larger schools into smaller ones because the research evidence showed that there was an unrepresentatively large proportion of smaller schools among the best-performing ones. A closer look at the data, however, shows that there is also an unrepresentatively large proportion of smaller schools among the worst-performing ones. Further analysis showed that school size really was not a factor in student performance. The actual issue was that smaller samples allow more extreme mean values, whereas larger samples decrease the chances of means at either extreme of the spectrum.[29] (We'll examine this in more detail.)

There is far more to probability than just intuitively following "common sense".

The history of trying to calculate probability goes back to the time people first started to gamble (in all probability). Whenever that was, it was certainly a long time ago.

"How many ways can something happen?" is the fundamental question of gambling and therefore of probability, and some of the writers on the subject from the 1600s and earlier (such as Blaise Pascal and Pierre de Fermat, in the correspondence referred to at the end of Part 1) dived enthusiastically into these challenging calculations. An issue they were particularly interested in was: if players agree to stop a game of chance before it is finished, how do you work out an equitable distribution of money based on their probabilities of winning?

Unfortunately, their work is hard for the modern reader to follow, partly because notation has changed[30], and partly because the "games of chance" themselves have changed.

What hasn't changed is the use of coins, playing cards and 6-sided dice, so that's where we'll start following their footsteps.

> The game of poker wasn't invented until the 1800s and contract bridge didn't achieve its current form until the early 1900s. The rules of games of the 1600s such as ambigu, brandeln, conquian, lanterloo and wit-&-reason are unknown to most of us today.

Probability basics

> **Read this if** you don't understand what fractions have to do with probability, you don't know how to calculate odds, you didn't know that $q = 1 - p$, you aren't familiar with using a tree chart to calculate probability, or you aren't confident about the difference between independent and conditional probabilities.

Describing probability

We are used to speaking about probabilities as percentages: "the forecast predicted a 40% chance of rain". If we think there is an even chance of something happening or not happening, we say it is "fifty-fifty". If we believe that there is no doubt about something we describe it as being "100% certain".

As we saw at the beginning of the book, a percentage is a way of writing a fraction, which is the same as a **proportion**. Probability can often be expressed as a proportion of the number of possible ways of achieving the outcome you are looking for, compared to the number of ways of achieving all of the possible outcomes. If there is no chance at all, that is 0. Complete certainty is 1. Anything else is a fraction between 0 and 1. If we are tossing an evenly-balanced coin, the chance of getting a head is

$$\frac{1}{2}$$

where the 1 in the numerator is the outcome we want (a head) and the two in the denominator represents the number of possible outcomes (one head and one tail).

The probability of something happening is often represented by the letter p, while the probability of it not happening is often given the letter q. Because probabilities are a fraction of 1, that means:

$$p + q = 1$$

$$q = 1 - p$$

A **different way to express probability** is in terms of **odds.** The odds of something happening is the probability that it will happen, divided by the probability that it won't happen. This is:

$$\frac{p}{q} = \frac{p}{1-p}$$

When speaking of odds, we normally express it as "odds of (p) to $(p-1)$". For example, the odds of getting a head with one flip of a coin are one-to-one.

Odds is a less intuitive concept than probability. It is used more in gambling (where it is mostly used to describe the amount that can be won rather than the probability of winning it) than in statistics. In statistics we usually stick to talking about probability. An exception is the a situation where an **odds ratio** is the best way to express something. (See page 384 and following.) If we were to look at the odds of throwing a 4 from a single role of a 6-sided dice, the probability, p, will be one sixth. The probability of not throwing a 4 will be five sixths. That makes the odds:

$$\frac{\frac{1}{6}}{\frac{5}{6}} = \frac{1}{5}$$

To cement your understanding:

What are the odds of winning a lottery buying 1 ticket if 5 million tickets are sold? (Hint: they aren't exactly one-to-five million.)

Tree charts

Suppose you want to work out the chances of flipping a coin three times, and getting a head, a tail then another head.

You can draw probabilities up as a tree chart, starting from the left and putting the probability/proportion of each outcome along the branches (Figure 164). For each possible outcome on the first throw, there are two possible outcomes on the

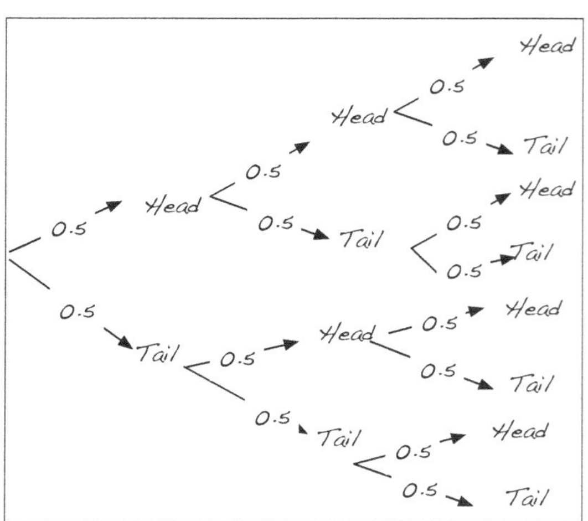

Figure 164

second throw. That means you have half a chance, or 0.5, of getting any particular outcome.

For each of these, there are two more outcomes, or one half a chance for each, on the third throw. That means that probability of getting any particular outcome after three throws is

$$\frac{1}{2} \cdot \frac{1}{2} \cdot \frac{1}{2} = \frac{1}{8}$$

$$= 12.5\%$$

If we don't care about the order of throws, however, there are three different ways to get two heads and one tail: head-head-tail, head-tail-head, or tail-head-head. That means to find the total probability we need to add the probabilities of each of these possible outcomes:

$$\frac{1}{8} + \frac{1}{8} + \frac{1}{8} = \frac{3}{8} \text{ (or 37.5\%)}$$

In other words, in the diagram at Figure 164, there are 8 possible end points on the right hand side. Each of these has one eighth of a chance of occurring. Three of these will give the result of two heads and one tail, if we don't care about the order; that is, if we are looking at the combinations rather than the permutations. (See page 155.)

As a short cut, to find probabilities of particular outcomes, multiply along the branches where both things are true, and add vertically where either one or the other is true. (If that isn't clear yet, it will become clearer when we look at the addition and multiplication rules).

To cement your understanding:

• Draw up the tree chart at Figure 164 for yourself, to confirm that it makes sense to you.

• What is the probability of throwing two heads in a row out of two throws? (Can you see why it is one quarter, or 25%?)

• What is the probability of throwing one head and two tails, if you don't care about the order? (Can you see why it is 37.5%?)

Diving into this a bit more deeply, people who study statistics can be divided into two groups: those who want to study statistics, and those who would prefer not to, but who have to study it in order to get qualified. There is also a second way they can be divided: those who enjoyed mathematics at school, and those who did not. There is even a third way they can be divided: those who are happy to accept what they are told at face value, and those who like to question so that they can achieve a deeper understanding.

Let's now assume hypothetically that we know the proportion of people who fit into each of these categories. We'll say that 90% of those who study statistics must do so to gain another qualification, while 10% actually wanted to study it. We'll also say, for the sake of the argument, that of those who study statistics, 40% enjoyed mathematics at school while 60% didn't. Finally, we'll assume that 70% of people studying statistics are happy to take things on trust, while 30% want to question and understand deeply. Figure 165 shows those percentages marked onto the branches of the tree chart.

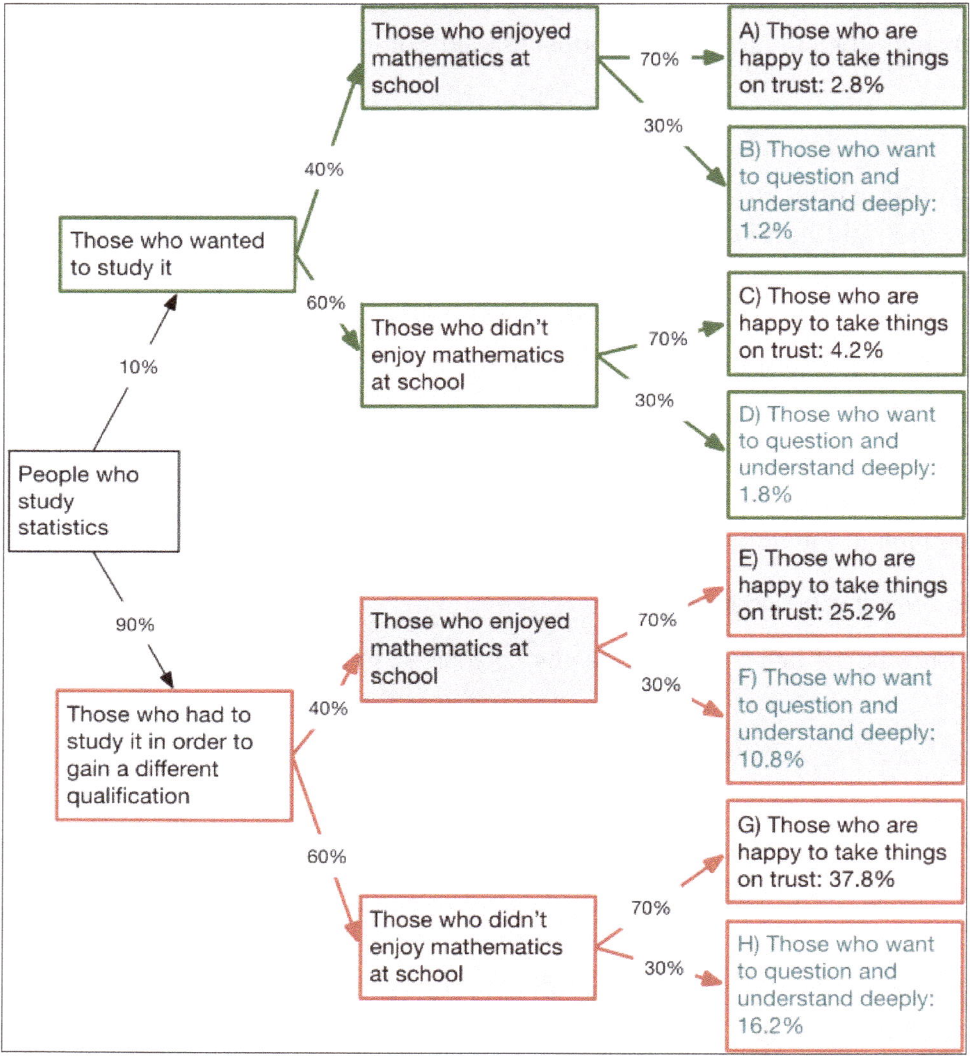

Figure 165

To find the probability that any given student will fit into one of these categories, we need to find the proportion of students in each one.

To find the proportion of students who fit into box A on the right of the diagram, we can follow along the top branches. The first branch tells us that 10% of students wanted to study it. *Of this* 10%, 40% enjoyed mathematics at school. To find what this represents as a proportion of all the students, we need to find the fraction *of* the fraction, which we can calculate by multiplying the two fractions together: 40% of 10%. This tells us that 4% of all students studying statistics (given our made-up assumptions) both wanted to study it and enjoyed mathematics at school. Of this 4%, 70% are happy to take things on trust. Multiplying these percentages, we can calculate that 2.8% of all the students fit into box A. (Remember, from when we first looked at fractions, that multiplying two fractions between 0 and 1 creates a smaller fraction.)

Undertaking the same calculation along each of these branches gives us the percentages written in the boxes on the right of Figure 165. If we add each of these probabilities together, they add up to 100%, or just 1 (same thing). So based on box G, 37.8% of statistics-students had to study it to gain a different qualification, didn't enjoy mathematics at school and are happy to take things on trust. That means, in turn, that there is a 37.8% chance that any randomly-selected statistics-student will fit into that particular category.

The order doesn't matter because order doesn't matter in multiplication or in addition. We could just as easily have started by dividing students initially into whether they prefer to trust or to question, then whether they wanted to study it or had to, and then whether they enjoyed mathematics at school. The proportions would have worked out the same.

To cement your understanding:

• Draw up the same tree chart, using the same percentages, but in a different order. Start with whether people are happy to take things on trust (70% are, 30% want to question), then add in whether they want to study it on its own (10%) or had to study it as part of another subject (90%), and finally add in whether they enjoyed mathematics at school (40% did, 60% didn't).

• Use a calculator to check that the percentages in the boxes at the end of each branch are the same as those in Figure 165.

• Add these percentages together, to confirm that it gets to 100%, or in other words, that you have included all of the possibilities.

We treated this exercise as though we were looking at **independent** probabilities. For example, we acted as though regardless of whether a student enjoyed mathematics at school, the chance that he or she wanted to study statistics at university would not change. An independent probability means that **knowing about one thing** tells you **nothing** about the probability of something else. That may not be the case, of course. If liking mathematics at school would change the probability of wanting to study statistics, we describe it as a **conditional probability**—the probability depends upon, or is contingent upon, some other condition. The tree chart is just as useful, except that for a conditional probability the order that you put things in might be important, because the probabilities on each branch depend on what has gone before.

To cement your understanding:

• Change the tree chart you have drawn, to show that, of those who wanted to study statistics, **95%** of them enjoyed mathematics at school, while **5%** didn't. Don't change the percentages of those who didn't want to study statistics. What you have done now is drawn up a conditional probability: the probability of whether people enjoyed mathematics at school varies, according to the subject they have chosen to study.

• Now recalculate the probabilities in the final boxes.

• Add the percentages down the page, to confirm that they still add up to **100%**.

• Can you see why, now that you have made the probabilities conditional, you can't just change the order of the tree chart?

Key points:

- Probabilities are expressed as a **fraction** (or percentage) where 1 is complete certainty and 0 is no chance.

- They can often be expressed as a **proportion** of the number of ways of achieving a specific outcome in the numerator, compared with the total number of possibilities in the denominator.

- The probability of something happening is often represented by the letter p, while the probability of it not happening is often given the letter q.

- $p + q = 1$

- $Odds = \frac{p}{q} = \frac{p}{1-p}$

- Probabilities can be drawn as a **tree chart**.

- An **independent probability** means that knowing the probability of one thing tells you nothing about the probability of something else.

- If this is not the case, it is described as a **conditional probability**.

- On a tree chart, the order doesn't matter if probabilities are independent, but it may if they are conditional.

The multiplication and addition rules

<table><tr><td>Read this if you aren't familiar with the multiplication and addition rules and when to use each, you don't know what these have to do with set theory, or you don't know how to predict the probability of getting a specific number of heads from a specific number of coin tosses.</td></tr></table>

The multiplication rule

The **multiplication rule** is an extension of the way we calculated these probabilities: multiplying along the branches. It says that that the probability of *both* A and B happening is the probability of A multiplied by the probability of B.

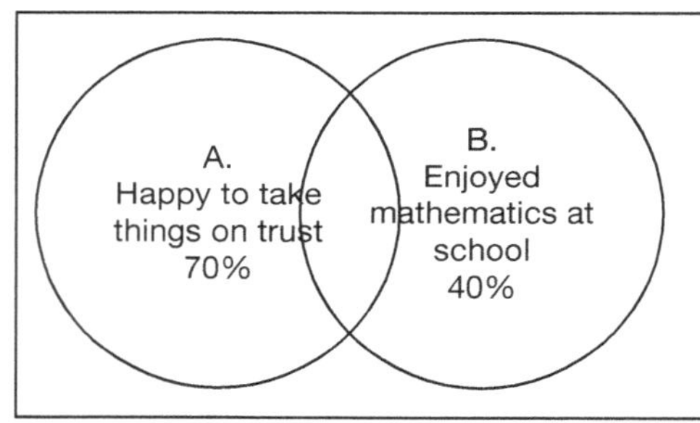

Figure 166

Based on the tree chart at Figure 165, some students were happy to take things on trust, and some enjoyed mathematics at school. We can show this as a Venn diagram (Figure 166). You can see that 70% of students were happy to take things on trust, while 40% enjoyed mathematics at school. The part we are interested in is the overlap: the chance that someone was happy to take things on trust *and* enjoyed mathematics. To find this, we multiply them together. We want the 70% of 40%, or 40% of 70% (which is the same thing, of course). This tells us that 28% of students are happy to take things on trust *and* enjoyed mathematics at school. (You can check this yourself from Figure 165 if you like: add boxes A and E.) Note that it only works out this neatly if the probabilities are **independent**. We will explore why that is later in the book.

You may remember from page 74 that we describe the area of overlap between these two groups as the **intersection** of the two sets, and we write the probability using this notation:

$$Pr(A \cap B)$$

Note that even though the probabilities are multiplied together, their product is smaller than either A or B. The chance of both these things happening is less than the chance that either one of them will happen.

The addition rule

The **addition rule** is an extension of the idea of adding vertically. If we want to find the probability that someone will *either* be happy to take things on trust *or* enjoyed mathematics at school, we could add these percentages, but that would give us 110% of students, which makes no sense. The problem is that we have double counted the 28% who fit into both categories, so we have to subtract them. That means that 82% of students either enjoyed mathematics at school, or are happy to take things on trust, or both. This is the total of both of the circles in the Venn diagram, less the overlap so we don't double count it.

In the Venn diagram what we are after is the total of the two circles. You may recall from page 74 that the total of both circles is the **union**. In this case we want the probability of the union, less one lot of the intersection. Using set notation, we write this as:

$$Pr(A \cup B) - Pr(A \cap B)$$

That's a bit to take in, so to sum it up:

For **both/and**, use the **multiplication** rule: find the **intersection of the** sets by multiplying the probabilities (if they are independent). This will be smaller than either of the probabilities that you are multiplying.

For **either/or**, use the **addition** rule: find the union of the sets less the intersection, by adding them then subtracting the product of their probabilities.

To cement your understanding:

• Looking back at Figure 165, if you survey a student at random, what are the chances that they wanted to study statistics and are happy to take things on trust, whether or not they enjoyed mathematics at school? Can you see two ways that you could work this out?

• Now if you survey two students at random, do you use the multiplication rule or the addition rule to work out whether they both fit into this category? And which rule do you use to work out if either of them fits into this category?

• Do the calculations above, to work out the percentages. (Did you get that there is a 14% chance that one of them does, but only a 0.49% that both of them do? If not, can you see where you went wrong?)

• If you want to find out the chances that either one of them does, but not both of them, how might you go about that? (Hint: There is no such thing as a "subtraction rule", but if there was…)

Fun fact:

In a class of 23 people, what is the probability that at least two people share a birthday? (Have a guess and note your answer before reading on.)

You can work this out using the probability we have covered. This is easier to work out by thinking of the probability that two people *don't* share a birthday.

If you think about the chance that you don't share a birthday with one other person, and you are the only person in the class, it is totally certain that you aren't sharing with another person. You could have your birthday on any of 365 days (ignoring leap years), and regardless, you won't share with another person (because there isn't anyone else).

$$\frac{365}{365} = 100\%$$

Now introduce a second person. For each of the 365 days when you could have a birthday, there are 364 days for their birthday when they won't share with you. That makes the probability, using the multiplication rule:

$$\frac{365}{365} \cdot \frac{364}{365} \approx 99.7\%$$

If you introduce a third person, there are 363 days for their birthday when they won't share with either of you. That makes the probability that none of you will share a birthday:

$$\frac{365}{365} \cdot \frac{364}{365} \cdot \frac{363}{365} \approx 99.2\%$$

Now if you keep adding people, on each occasion there will be one less birthday available for the person not to share with anyone else, so that by the time you get 23 people involved, it will be:

$$\frac{365}{365} \cdot \frac{364}{365} \cdot \frac{363}{365} \cdot \ldots \cdot \frac{344}{365} \cdot \frac{343}{365} \approx 49.3\%$$

If that is the probability of *not* having a shared birthday, and probabilities add to one, then the probability of having at least one shared birthday among the group is

$$100\% - 49.3\% = 50.7\%$$

which is slightly better than an even chance.

Beyond tree charts

Back to flipping coins, now suppose we want to work out the chances of flipping a coin eleven times and tossing nine heads. If this is starting to seem like simply adding irrelevant complexity and confusion, it's worth realising that this type of thinking underlies a great deal statistics. For example, this sort of question has similarities with working out the chances of whether people will or won't vote for a particular political candidate, if their choices are completely random. Once you know what the probabilities around random choices are, you have more meaningful information about what a survey might or might not be telling you. But we're getting ahead of ourselves.

To work out the chances of flipping a head exactly nine times from eleven throws, we could draw up a tree chart, but it would be enormous, and we'd probably get lost along the way. We could, if we really had to, draw up a table to count all the possible **combinations**, but that would just be hard work.

Fortunately, we've already seen a way to find the answer: it's an n choose r problem. How many ways can we pick 9 coins out of 11? As we've seen (page 145), the formula for this is:

$$\binom{n}{r} = \frac{n!}{r!\,(n-r)!}$$

In our case n is 11, and r is 9, so we get:

$$\binom{11}{9} = \frac{11!}{9!\,(11-9)!}$$

$$= \frac{11 \cdot 10 \cdot 9 \cdot 8 \cdot 7 \cdot 6 \cdot 5 \cdot 4 \cdot 3 \cdot 2 \cdot 1}{9 \cdot 8 \cdot 7 \cdot 6 \cdot 5 \cdot 4 \cdot 3 \cdot 2 \cdot 1 \cdot (2)!}$$

$$= \frac{110}{2}$$

$$= 55$$

Using this same approach, we can work out how many ways we can get any number of heads (and tails) from 11 throws:

Count of heads	11	10	9	8	7	6	5	4	3	2	1	0
Count of tails	0	1	2	3	4	5	6	7	8	9	10	11
Number of ways to arrive at that count	1	11	55	165	330	462	462	330	165	55	11	1

(Note that this is symmetrical: the same probabilities apply to 9 heads out of 11 throws as they do to 9 tails, as you would expect.) It's possible that the number of ways to arrive at each count will look a little familiar: $1, 11, 55, 165, 330, 426, 426, 330, 165, 55, 11, 1$. We've already seen this sequence a number of times, because it is a row of Pascal's triangle (see pages 78 and following).

As we've seen (pages 160 and following), this is the result we would have obtained if we had expanded out the binomial expression $(H + T)^{11}$ (and if you are lost at this point, now would be a good time to revisit that topic). You might also remember that the numbers of each row of Pascal's triangle add up to the relevant power of 2. Each time we tossed the coin we had 2 possible outcomes, and we did that 11 times, so to find the total number of possible outcomes we multiply 2 by itself 11 times. That is $2^{11} = 2,048$.

That means the probability of getting exactly 9 heads from 11 throws is:

$$\frac{55}{2,048} \approx 2.7\%$$

Now suppose that we want to know, not the probability of getting exactly 9 heads, but the probability of getting 9 or more heads. That means we need to use the addition rule to add the probabilities of getting 9, 10 or 11 heads. Using the same approach, it will be:

$$\frac{55}{2,048} + \frac{11}{2,048} + \frac{1}{2,048} = \frac{67}{2,048}$$

$$\approx 3.3\%$$

If we wanted to know the chance of throwing 8 or less heads, we could do it the hard way, calculating the chance of each outcome then adding them. The easy way, though, is to remember that probabilities add to 100%. We can see straight away that there is roughly a 96.7% chance of getting 8 or less throws.

To cement your understanding:

What are the chances of throwing three or more heads from five throws? (Hint: Use what you know from Pascal's triangle to work it out.) Did you get that there were

$\frac{26}{32} = 81.25\%$? If you didn't, can you see where you went wrong? If you can't, it's probably best to go back and read the material on permutations and combinations, and on the binomial theorem in Part 1. Then put that on the back burner while you read the next section and come back to it in the section after that when we look at the binomial theorem.

Key points:

- The **multiplication rule** says that that the probability of *both* A and B happening is the probability of A multiplied by the probability of B. (Because probability is a fraction, the multiplication rule creates a lower probability.)

- The multiplication rule represents the **intersection** of sets.

- The **addition rule** says that that the probability of *either* A or B happening is the probability of A plus by the probability of B, less the probability of both.

- The addition rule represents the **union** of sets (while avoiding double-counting the intersection).

- Finding out how many heads (or tails) you will get from a given number of throws is an n choose r problem. That means you can read it from Pascal's triangle.

Binomial probabilities with uneven chances

> **Read this if** you aren't confident about how or why the binomial theorem works with probabilities with uneven chances.

Getting back to gambling, now imagine that we were throwing a dice six times. In this case, we don't have an even chance of getting a particular number; we have a 1 in 6 chance with each throw of each die, and 5 in 6 chances of not getting it. The probabilities, mathematically are:

$$\frac{1}{6} + \frac{5}{6} = 1$$

Each time we throw the dice, based on the multiplication rule, the probabilities are multiplied. If we are throwing six times, the probabilities are:

$$\left(\frac{1}{6} + \frac{5}{6}\right)^6 = 1$$

To generalise this, let's let p be the fraction that represents the probability of success (in this case one sixth). We'll let q be the probability of failure (in this case five sixths), which will always also be equal to $1 - p$. Finally we'll let n be the number if **trials** (e.g. tosses of the dice or coin). That makes the formula, when we expand it out into a series using the binomial theorem (page 160 and following):

$$(p + q)^n = \sum_{r=0}^{n} \frac{n!}{r!\,(n-r)!} p^r q^{n-r}$$

where p is the probability of **success**, q is the probability of **failure** (and in probability $p + q$ will always add up to 1), n is the number of **trials** and r is the **number of successes**. Because p and q are already expressed as probabilities—they are fractions showing the number of ways something can happen to give success or failure over the total number of ways things can happen—then in the binomial expansion there is no need to divide the result by the number of possibilities; that has already been factored in; it's aways 100%.

> **To cement your understanding:**
>
> Pick a number for n (a small one like $1, 2$ or 3 if you don't want to spend much time on this; a larger one if you want more practice) and expand out $\left(\frac{1}{6} + \frac{5}{6}\right)^n$ to see the way it expresses the probability for any particular number of successes, r, and also to see that the result will always add to 1.

> If you are really observant, you might have noticed that as well as exchanging x and y for p and q compared to what we did on the binomial theorem in Part 1, the way the formula is written above (which is the way you will normally find it written in statistics text books) reverses the indices of the p and the q, so that the $n - r$ is the exponent for the q, instead of the p in the way it was expressed when we looked at the binomial theorem (which is the way it is normally expressed in mathematical text books). It works out the same way, except that when you write it out the terms come out in the opposite order. The order the terms come out in doesn't matter, in terms of reaching the answer. It is written this way to make it easier to calculate a Poisson distribution, which we will come to soon.

We can also write the formula omitting the q, because we know that the probability of failure (q) is $1 - p$, when p is the probability of success. So the formula then becomes:

$$1 = (p + 1 - p)^n$$

$$= \sum_{r=0}^{n} \frac{n!}{r!\,(n - r)!} p^r (1 - p)^{n-r}$$

Here is another way to think about why this works, which connects it to some of the other concepts we have covered.

When we looked at the binomial theorem (pages 160 and following), one way we thought about it was with a device similar to Figure 167. Recall that the numbers—which are also the numbers of Pascal's triangle—represent the number of paths a ball bearing could take, if dropped in the top, in order to reach that particular point. This is the same as the number of different ways you can flip a coin or roll a dice to get identical combinations.

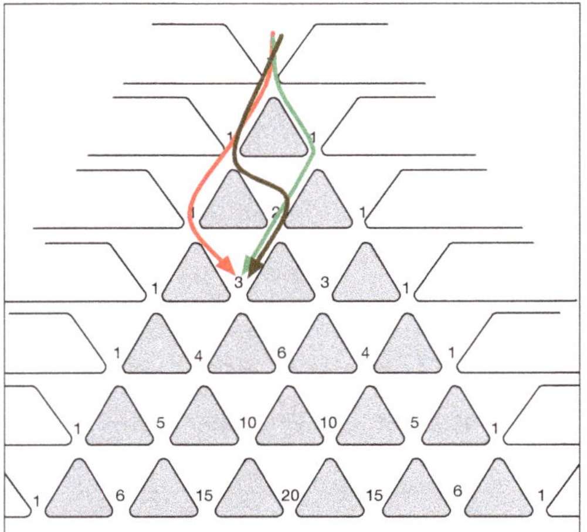

Figure 167

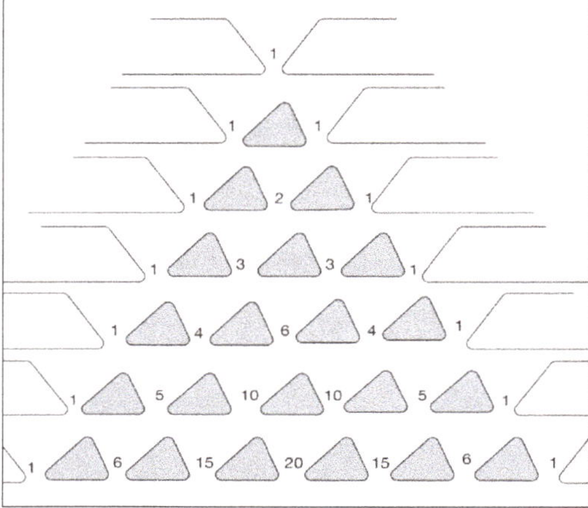

Figure 168

Now let's modify that idea a little bit, as per Figure 168. This time the triangles have been slanted to the side. If the point of the triangle were five sixths of the way across, then, on average, the ball bearing would fall to the left five out of six times, and to the right one out of six times. So now there are two different factors impacting on the probability that the ball bearing will come out at any particular space: the number of pathways that lead to that point (given to us by Pascal's triangle) and the cumulative effect of all of the choices of which way the ball will fall at each triangle: five sixths or one sixth. This cumulative effect is almost the same as a tree chart but drawn from the top down rather than the side. It isn't quite the same as the tree charts we have seen, because in this situation there are multiple pathways that lead to the same position, whereas previously each destination had its own pathway. The multiple tree charts are accounted for by the n choose r formula. We can still use the multiplication rule to work out the probabilities, though. We can multiply the probabilities of each path and multiply that total by the number of pathways that could lead to that point, divided by the total number of pathways.

Mathematically, we can express this as:

$$\frac{6!}{r!\,(6-r)!}\left(\frac{1}{6}\right)^{r}\left(\frac{5}{6}\right)^{6-r}$$

The first part of this formula (in red) is based on n choose r; the number of pathways. The second part (in blue) is based on the combined probabilities of getting to a particular position at the bottom. The positions are numbered from 0. We multiply them together, because for each combination of

$$\left(\frac{1}{6}\right)^{r}\left(\frac{5}{6}\right)^{6-r}$$

there are

$$\frac{6!}{r!\,(6-r)!}$$

ways to achieve it.

If you are working out the probability of throwing a 5, exactly three times, with ten dice-throws, the probability of success on each throw (p) will be one sixth and the probability of failure (q) in each throw will be five sixths. The number of throws (n) will be 10, and the value we are seeking (r) will be three. The overall equation will be:

$$\left(\frac{1}{6}+\frac{5}{6}\right)^{10} = \sum_{r=0}^{10} \frac{10!}{r!\,(10-r)!}\left(\frac{1}{6}\right)^{r}\left(\frac{5}{6}\right)^{10-r}$$

This equation sums up all the terms of the binomial expansion, but we are only interested in the term where $r = 3$, so we don't have to bother expanding the others out.

$$\frac{10!}{3!\,(10-3)!}\left(\frac{1}{6}\right)^{3}\left(\frac{5}{6}\right)^{7} \approx 0.15504 \ldots$$

So there is a slightly-better-than-15% chance of throwing a five no more and no less than three times with 10 dice throws.

Key points:

- When looking at repeating a process with two possible outcomes, where p is the probability of success with any single trial, q is the probability of failure, and n is the number of trials, the probability of any given number of successes (r) can be calculated using the binomial theorem:

$$(p+q)^n = \sum_{r=0}^{n} \frac{n!}{r!\,(n-r)!}p^r q^{n-r}$$

$$= \sum_{r=0}^{n} \frac{n!}{r!\,(n-r)!}p^r (1-p)^{n-r}$$

- This is the same as multiplying the number of possible ways of achieving a given outcome by the combined probabilities of success at each step, using the multiplication rule.

Probabilities with more than two options

> **Read this if** you don't know how a permutation with repetition differs from a permutation or a combination, you don't understand its formula or you don't understand about multinomial probabilities.

Permutations with repetition

You may remember (page 145) that the formula for the number of **permutations** (where the order matters), when selecting r items from n possibilities, is:

$$\frac{n!}{(n-r)!}$$

and that the number of **combinations** (where the order doesn't matter) is:

$$\frac{n!}{r!\,(n-r)!} = \binom{n}{r}$$

Now let's think about the number of permutations of the letters of the words "failure" and "success". Calculating the permutations of the letters of the word "failure" is straightforward. There are 7 letters, and so there are 7 possibilities for the first letter, and for each of these options there are 6 remaining letters, and so on, so the number of permutations is:

$$7 \bullet 6 \bullet 5 \bullet 4 \bullet 3 \bullet 2 \bullet 1 = 7!$$

$$= 5{,}040$$

(In this case, because we are selecting all the letters, $r = n$, which means that the denominator of the n choose r formula will be 0! which is equal to 1. That means the formula is just $n!$)

When we think about doing the same for the word "success", however, it becomes trickier. We are considering permutations, and so the order matters, but the letter "s" occurs three times, and the letter "c" occurs twice, and so the order for these letters is irrelevant. This is called a **permutation with repetition,** rather than a combination. In the same way that we needed to reduce the number of permutations to find the number of combinations, so we need to reduce our 7! by the number of possible permutations of 3 letters (3!) for the "s" and the number of combinations of 2 letters (2!) for the "c". (We could also show the number of combinations of letters that appear only once, but that would be 1, so it's superfluous.) That gives us:

$$\frac{7!}{3!\,2!} = \frac{7 \bullet 6 \bullet 5 \bullet 4 \bullet 3 \bullet 2 \bullet 1}{(3 \bullet 2 \bullet 1)(2 \bullet 1)1}$$

$$= 420$$

So there are 420 permutations of the letters of the word "success", compared to 5,040 permutations of "failure". If it isn't completely clear how we arrived at this, then imagine that we first wrote down all 420 permutations of the letters in "success", then decided to add a subscript to each "s" and "c", as in $s_1 s_2 s_3 c_1 c_2$. To find all of the permutations now we would have to multiply the 420 by the 3! permutations of "$s_1 s_2 s_3$" and multiply the result by the 2! permutations of "$c_1 c_2$". That would take us back to 5,040 permutations.

To turn this into a more general mathematical expression, we can designate the number of letters as n, and so

$$n = 7$$

There are, however, only 4 types, or categories of letters (s, u, c, e). We will call this number of categories k, and so

$$k = 4$$

Finally we will designate the number of individual letters in each category of letters as:

Letter	Number of occurrences	Designation of number of occurrences
s	3	x_1
u	1	x_2

Letter	Number of occurrences	Designation of number of occurrences
c	2	x_3
e	1	x_4

That means we can write this algebraically as:

$$\frac{n!}{x_1!\, x_2! \ldots x_k!}$$

This is called the formula for **multinomial coefficients.** (If a number only appears once, then its factorial is 1 and it makes no difference to the calculation.)

If we only had two options, this formula would become the formula for binomial coefficients $\binom{n}{r}$. In that instance x_1 would be r, and x_2 would be $(n - r)$.

(The formula for multinomial coefficients is sometimes written as:

$$\binom{n}{x_1, x_2 \ldots x_k}$$

However, this is really not much easier to write than the formula we used above and it is less obvious what it means, so we will avoid it.)

To cement your understanding:

Calculate the number of permutations of the word "look" in two ways: writing each of them out and counting them, then using the formula for multinomial coefficients.

Multinomial probabilities

Where there are more than two possible outcomes, instead of talking about **binomial** probabilities we call them **multinomial probabilities.**

Getting back to our preferences of students in regard to studying statistics (page 340), the tree chart we drew gave us the following probabilities:

Preferences	% of students
Wanted to study statistics and enjoyed mathematics at school.	4%
Wanted to study statistics, didn't enjoy mathematics at school.	6%

Preferences	% of students
Had to study statistics in order to gain a different qualification and enjoyed mathematics at school.	36%
Had to study statistics in order to gain a different qualification, and didn't enjoy mathematics at school.	54%
	100%

Now, for example, suppose that for some (admittedly unlikely) reason we wanted to know the probability that a class of 100 students is made up of exactly the numbers in the last column of the table below:

Preferences	
Wanted to study statistics and enjoyed mathematics at school.	5
Wanted to study statistics, didn't enjoy mathematics at school.	5
Had to study statistics in order to gain a different qualification and enjoyed mathematics at school.	40
Had to study statistics in order to gain a different qualification, and didn't enjoy mathematics at school.	50
	100

That may seem daunting but based on what we have learned so far, part of this problem is straightforward: we can use the multiplication rule to multiply all of these probabilities together, the number of times we are looking for them to occur. That gives us:

$$(4\%)^5 (6\%)^5 (36\%)^{40} (54\%)^{50}$$

(which, as you might expect, is a small number). It isn't quite that simple, though. That would tell us the chances of picking students in that order, but we don't care about the order, so we need to multiply this by the number of distinct sequences that we could use to select that particular outcome. This will be the formula for permutations with repetitions that we saw above.

So the probability is

$$\frac{100!}{5!\,5!\,40!\,50!}(4\%)^5 (6\%)^5 (36\%)^{40} (54\%)^{50} \approx 0.1548\%$$

(which is not much chance at all).

Expressing this in a general form, we can say that the probability is:

$$\frac{n!}{x_1!\, x_2! \dots . x_k!} {p_1}^{x_1} {p_2}^{x_2} \dots {p_k}^{x_k}$$

where n is the number events/objects/trials/whatevers, k is the number of categories (in this case the four different categories we are putting students into), x_1 etc represents the number of events/objects/trials/whatevers in each category (in this case students), and p_1 etc represents the probability of each of these (in isolation) occurring.

This is called the formula for **multinomial expansion.**

This reason it is called that, is if you expand out a multinomial expression, you can use this formula to calculate each of the terms. Even something as simple as $(A + B + C + D)^4$, which is not a large expression, expands out to:

$A^4 + B^4 + C^4 + D^4 + 4A^3B + 4A^3C + 4A^3D + 4B^3A + 4B^3C + 4B^3D + 4C^3A + 4C^3B + 4C^3D + 4D^3A + 4D^3B + 4D^3C + 6A^2B^2 + 6A^2C^2 + 6D^2A^2 + 6C^2B^2 + 6D^2B^2 + 6D^2C^2 + 12A^2BD + 12A^2CD + 12A^2BC + 12B^2AC + 12B^2AD + 12B^2CD + 1C^2BD + 12C^2AB + 12C^2AD + 12D^2AB + 12D^2AC + 12D^2BC + 6ABCD + 6ABCD + 6ABCD + 6ABCD$

To save a bit of time (actually, not that much) you can use the formula above to create each of the terms. The total of the exponents of each term will add up to whatever the exponent was on the original expression.

You can use this formula for binomial expansions and binomial probabilities, where k is equal to 2.

To cement your understanding:

Apply the formula for multinomial expansions to flipping a coin, to prove to yourself that the binomial probabilities are a specific case of multinomial ones or, to put it differently, multinomial probabilities are a generalisation of binomial ones.

Key points:

- Permutation with repetitions requires a different formula to either permutations or combination.

- The formula is

$$\frac{n!}{x_1!\,x_2!\,....\,x_k!}$$

  where n is the number of items, and x_1, $x_2 \ldots x_k$ are the number of items in each category.

- Where there are more than two possible outcomes, instead of talking about **binomial** probabilities we call them **multinomial probabilities.**

- The formula for multinomial probabilities is

$$\frac{n!}{x_1!\,x_2!\,....\,x_k!}\,p_1{}^{x_1}p_2{}^{x_2}\ldots p_k{}^{x_k}$$

  where $p_1, p_2 \ldots p_k$ are the probabilities of achieving each of the outcomes.

- The binomial theorem is a special case of the formula for multinomial probabilities.

Probability distributions and their terminology

Read this if you aren't familiar with any of the following terms or ideas: raw data, categorical variable, nominal variable, ordinal variable, success, random variable, frequency distribution, probability distribution, sample, sample space, histogram, population, parameter, discrete distribution, continuous distribution, symmetrical, skewed, unimodal, bimodal, multimodal, univariate, bivariate or multivariate.

"When I use a word', Humpty Dumpty said, in rather a scornful tone, 'it means just what I choose it to mean—neither more nor less."
Lewis Carroll 1832-1898

(As well as writing Alice in Wonderland and other children's stories, Lewis Carroll was a mathematician who wrote a book of probability puzzles[31].)

(If you are struggling with this, see the How to Love Statistics YouTube video on *Introduction to Probability Distributions*)

We are about to get into defining some terms and concepts around probability and statistics. If you have been reading for a while and you are tempted to just skim-read what follows, now would be a good time to take a break, and to return to this with a fresh mind. The material may seem like the sort of thing which you can be vague about, but if you don't get this clear you won't have the basic foundations for what follows.

Back to binomial probabilities and flipping coins, if we look at the number of possible outcomes from 4 throws, we can draw up the table at Figure 169. Column 1 lists the **raw data**. In this case it isn't expressed as a number, it is expressed as combinations of letters. We could call this a **categorical variable:** each category would be a different variable. Categorical variables can be **nominal** (only based on the name) or **ordinal** (based on an ordered list). Column 1 is ordinal: the categories are ordered, or sorted, in terms of the number of heads in each one.

Column 1 (raw data)	Column 2 (successes)	Column 3 (frequency distribution)	Column 4 (probability distribution)
Permutations	Number of heads in that combination	Number of permutations giving that number of heads	Probability of getting that number of successes
TTTT	0	1	6.25%
HTTT, THTT, TTHT, TTTH	1	4	25%
HHTT, THHT, TTHH, HTHT, HTTH, THHT	2	6	37.5%
THHH, HTHH, HHTH, HHHT	3	4	25%
HHHH	4	1	6.25%
Total:		**16**	**100%**

Figure 169

We could equally have shown it with pictures of coins, or by writing out the words in full. If the coins were stamped with a number on each side the raw data could have been expressed as a number. It will be either the outcome of a piece of research or, in this case, the outcome of working out what is possible.

Column 2 represents number **successes**. In probability a **success** is whatever you are looking at and you can define it in any way you like. We could have called a success a tail, in which case the numbers would have been the same except that the order would have been reversed: the column, counting down, would have the numbers 4, 3, 2, 1 and 0.

Column 2 can be represented by a **random variable**. Random variables are, by convention, represented by a capital: typically capital X, but any letter can be used. (We'll say more about random variables in a moment.)

Column 3 represents the **frequency distribution of the random variable:** it is the number of times that that particular number occurs. In this instance, that is represented by the number of permutations.

Column 4 represents the **probability distribution.** Altogether there were 16 possible permutations. Column 4 was obtained by dividing each of the frequencies in column 3 by 16, to show the percentage chance that any given throw of 4 coins will give that particular number of successes.

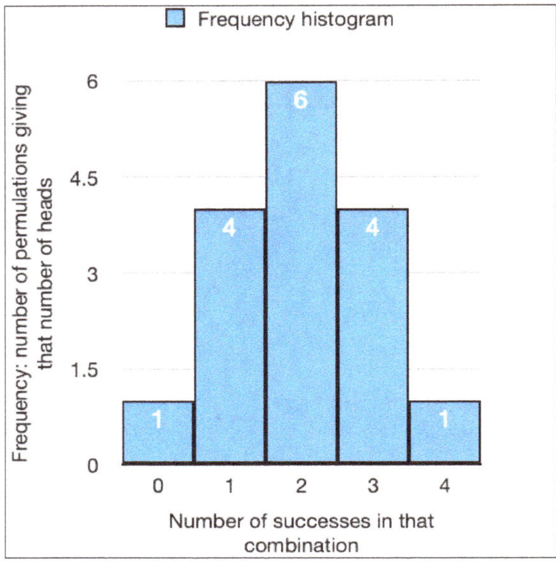

Figure 170

The **sample space** is the total possible outcomes: in this case, the 16 permutations of heads or tails possible in this experiment. (Sixteen, of course, is two to the fourth power, because there are two options and we are repeating the experiment four times. Recall that the sum of numbers along the rows of Pascal's triangle are powers of 2.) Note that the **sample space** isn't the number of possibilities, it's the actual possible outcomes.

Figure 170 shows a **frequency histogram** of the information in columns 2 and 3 of the table (so named by Karl Pearson, whom we will meet again. Technically, to be called a histogram the columns should touch one another. He first used the name based on graphing things that had happened at particular points of time: hence "historical...").

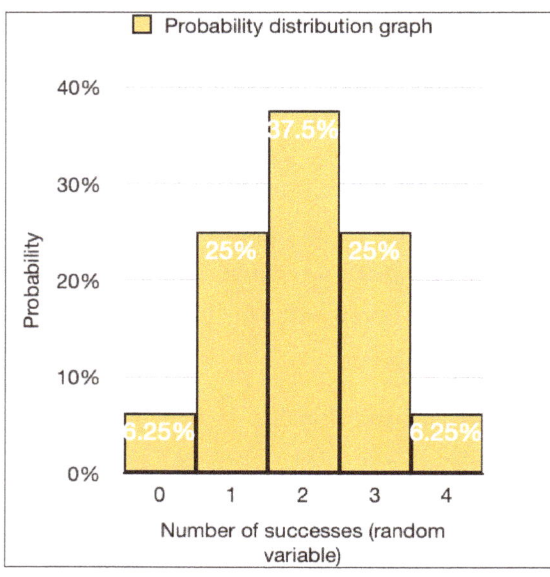

Figure 171

Figure 171 shows the **probability distribution graph** of the same information, drawn up from columns 2 and 4 of the table. It is almost exactly the same information as Figure 170, except that the scales are different, because in the probability distribution each of the frequencies has been divided by total number of possibilities. That means all the possibilities will add up to 1 (or 100%). The columns now show both the percentage of the total combinations that involve that particular outcome, and the probability of achieving that outcome, which amounts to the same thing. The numbers in a valid probability distribution graph will always add up to 1, or 100%. They must do so, because all possible options have to be shown, and there is a 100% chance of getting one of the possible outcomes.

A great deal of statistics is based on probability distributions of various sorts. Once you are confident that the data you have fits into a describable distribution, then you can understand the probability of a particular finding or outcome. As a general rule frequency tables and associated graphs deal with actual data, while probability distributions deal with what you would expect to see if all the factors affecting the outcome were **random** (we'll examine this concept shortly). This can be used to determine the **statistical significance** of data you are looking at: what are the chances that what you are seeing is random, compared to the chances that it is influenced by whatever you are investigating? We looked earlier at the way we can use this to evaluate the significance of experimental outcomes (when thinking about evaluating whether a coin was biased). It also has applications as diverse as predicting where an individual fits within a **population** on a particular **parameter**, assessing the true value from a range of slightly distorted findings, predicting probable future trends, and more.

A **parameter** is a measure that defines conditions, and in statistics it relates to populations. In statistics the word **population** doesn't necessarily refer to people, although that is where it began. It refers to the set of all members of any group you are interested in. It could be, for example, all the lightbulbs of a particular type ever manufactured, or all the atoms involved in a physics experiment. It is usually too hard to measure a whole population and if we could do so we might not need to apply statistics, because we would have certainty. Statistics is usually involved when a smaller **sample** of a population is examined. A sample is a subset of a population. The equivalent of a population parameter, within the context of a sample, is called a **statistic**.

Figure 171 shows a **discrete probability distribution**. The reason it is called "discrete" isn't because it knows how to keep its mouth shut; that would be "discreet". It's because the horizontal axis only allows for separate, or "discrete" numbers (usually integers, as in this case). The alternative, if it was a smooth curve, would be called a **continuous distribution.** This distinction between discrete and continuous distributions goes right back to two different ways of thinking about numbers, that we encountered at the beginning of the book. Numbers can refer to "how many" (which will generally lead to a discrete distribution), or can also refer to "how much?" (which will generally lead to a continuous distribution).

This particular graph is **symmetrical** because the values are the same to the right and left of the highest point, and also **unimodal** because there is only one peak. As you can imagine, things can get more complicated than that. Unimodal graphs can be **skewed** to the left or the right, and graphs can also have more than a single peak (**bimodal** or **multimodal**).

Artificial intelligence leverages probability distributions to determine how much attention to allocate to relationships between words in questions from users.

A distribution with a single variable is called **univariate**. Distributions can also be **bivariate** or **multivariate** if there are two or more variables involved.

Key points:

- Raw data can sometimes be **categorical**, meaning that each category is a different variable.

- Categorical variables can be **nominal** (only based on the name) or **ordinal** (based on an ordered list).

- In probability a **success** is whatever you are looking at, and can be defined in any way you like.

- Random variables are, by convention, represented by a capital: typically capital X, but any letter can be used.

- A **frequency distribution of a random variable** is the number of times that that particular number occurs.

- A **probability distribution is** similar, except that it shows the chance that any given outcome will occur.

- The **sample space** is the total possible outcomes.

- A **frequency histogram** graphs a frequency distribution in columns, with each of the columns touching.

- A **probability distribution graph** is similar to a frequency histogram, but scaled differently. It will always add to **1**.

- A **parameter** is a measure that defines conditions, and in statistics it relates to populations.

- In statistics the word **population** doesn't necessarily refer to people. It refers to the set of all of the members of any group you are interested in.

- A **sample** is a subset of a population. The equivalent of a population parameter, within the context of a sample, is called a **statistic**.

- Probability distributions can be discrete or continuous; symmetrical, or skewed to the right or left; unimodal, bimodal or multimodal: and univariate, bivariate or multivariate.

A deeper look at variables

> **Read this if** you are confused about the ideas of randomness, independence, *i.i.d*, or why you can't assume that all numbers are numerical variables.

It can be important to think in terms of variables rather than simply the subject of your research, because a single subject can be associated with many different variables. For example, if you are researching a group of people, you will usually collect several kinds of information about the same individuals.

You might want to record their place of birth. This is a **categorical variable**, because place of birth is a category, and it is **nominal**, because those categories have no natural ordering. You might also collect **ordinal variables**, which involve an order, such as ranking people by test performance (1st, 2nd, etc). At some point you will almost certainly collect data that either is a number or can be turned into numbers.

Variables that take numerical values are called **random variables**. This term can be misleading, because such variables are usually governed by probability and are not arbitrary. The word "random" reflects the fact that you do not know their values with certainty before conducting the investigation. As discussed earlier, the distinction between discrete and continuous distributions corresponds to discrete and continuous random variables. For example, the number of whole years a person has been alive is discrete, whereas the amount of time they have been alive, measured without restriction on units, is continuous.

If you were measuring people's different heights, the random variable (call it X) would be the abstract quantity associated with height. The individual heights of subjects may be represented by lower case x, often with subscripts: $x_1, x_2, \dots x_i, \dots, x_n$. Expressed in this way, x_i is usually used to indicate any particular example of a value, while x_n represents the final observation in the sequence.

X might represent the number associated with whatever you are looking at. In the example of our table showing the coin tosses, it would be the numbers that can be allocated to successes in column 2: the values of 0, 1, 2, 3 or 4. In this situation, column 3 would represent the frequency distribution of the random variable. Column 2 would be represented on the horizontal axis of a graph, and column 3 would represent the vertical axis.

You may hear that a random variable is a "function". In this context, but rather a rule that assigns numbers to experimental outcomes—for example, counting the number of heads and recording that count.

If, instead of recording the number of successes, you recorded the square of that number (0, 1, 4, 9, 16), this would still define a random variable. Similarly, if you record people's weights and heights and calculate body-mass index (weight divided by the square of height), BMI is also a random variable. Statisticians express this by saying that a function of a random variable is itself a random variable.

When you see X in a statistical formula, you can usually treat it as any other x in any type of mathematical formula, with the added understanding that X has an associate probability distribution.

For simplicity, you can almost always think of the random variable as whatever you are recording on the horizontal axis of a frequency distribution or a probability distribution, and the frequency or probability of that random variable will be recorded on the vertical axis.

Random variables can be constrained to particular ranges or types of numbers, such as integers: ($X \in \mathbb{Z}$). The numbers which can be included within the constraints are called the **support** of the variable.

Unpacking "randomness"

The idea of **randomness** is, unfortunately, not well-defined in the statistical literature. There are, in fact, contradictory definitions in some textbooks. One article from 2005 lists 5 completely different definitions from textbooks, all of which are incompatible[32]. Getting back to flipping a coin, there are factors that control whether it will land on its head or its tail: the initial position, the force and direction of the flip, the currents of the air, the temperature and humidity, the rotation of the Earth and probably more, yet the event is considered random. The way we will use the idea of "randomness" in this book (following the great Pierre Simon Laplace[33]) is that the multiple drivers of what happens are outside of our capacity to easily evaluate them and so the event is subject to probability. It does not, however, need to be subject to an even probability. If the coin was so weighted that 99.9% of the time it landed on tails, it can still be described as "random", because there is no absolute certainty about what will happen on the next throw.

Many (but not all) random variables are sometimes described as being **i.i.d.**, or **independent and identically distributed**. This is really a way of saying that they aren't functions of one another (they are independent, and so what happens in one occurrence of the variable doesn't tell you anything about what happens in the next one), and they are all described by the same probability function (they aren't parts of separate graphs).

Unpacking "independence"

When we are looking at the *theoretical* mathematics of probability distributions, it's an easy assumption that variables are likely to be independent unless we have specific reasons to think otherwise, and we'll see a lot of it in the pages that follow.

In the *real* world, once we start collecting data, however, this is rarely the case. For example, if two randomly-selected people who don't know one another are given the same psychological intervention on the same day, it would be easy to assume that they can be treated as independent apart from the intervention itself. To be truly independent, what you know about one subject should tell you nothing about the next subject. If, however, one of the subjects comes in running late and complaining about unexpected traffic and is clearly stressed about it, there is at least some chance that the next patient will be affected in a similar way. There could potentially be numerous **confounding** factors or **lurking variables** affecting both of them that mean they are not completely independent at all, such as: the weather on the day, seasonal allergens, or a news story that both were exposed to that created widespread levels of stress in that particular population. In real life few subjects are truly independent of others, and care needs to be taken in the use of statistics that assume independence. The mathematical results can vary dramatically when variables aren't independent. (We'll see more of this later.)

The reason that looking at independent variables is so important (and we will spend quite a bit of time looking at their behaviour) when they aren't all that common in real life, is mainly that the starting point in research is often to assume that variables are independent. For example, a researcher may hypothesise that spraying a particular type of fertiliser has nothing to do with the size of oranges. In other words, the starting assumption is that the size of oranges and the use of fertiliser are independent variables. If it turns out that assumptions of independence don't hold up—size is connected to fertiliser—then it is statistically likely that fertiliser has had an effect.

Unpacking "numerical" variables

We'll also return to this theme later, but just because something is represented as a number doesn't mean that it should be used as a numerical variable.

Temperature is one area where something recorded as a number needs care when used as a numerical variable. Forty degrees celsius is not twice as hot as twenty degrees celsius; the numbering scale is based on the convention that 0 degrees is the freezing point of water, and one hundred degrees is its boiling point, but that is an arbitrary convention. Years are based on numbers, but the year 2000 AD isn't twice as much as the year 1000 AD, and the year 1500 AD wasn't the "average" year of the last millennium.

You might think that no serious researcher would make a mistake like this, but there are areas where similar errors are made all the time, and the results published in peer-reviewed academic journals and taken seriously. If a researcher asks subjects to give their subjective experience a number— say, to rate their pain on a ten-point scale—they are using a number as a representation of something non-numerical. Just because a person rates their pain as six out of ten one day, and four out of ten a week later, you can't necessarily say that their pain is now two-thirds of what it was. If two people are reporting using the same scale, you can't necessarily assume that the numbers they report mean the same thing to them. A related area is questionnaires. Just because "agree" is marked with a 1, and "strongly agree" is marked with a 2, it doesn't mean that two people agreeing means the same thing as 1 person strongly agreeing. If "neither agree nor disagree" is marked with a 0, "disagree" is marked with a –1 and "strongly disagree" with –2, you can't say that if you have four people strongly agreeing and two people strongly disagreeing, that is the same as six people who neither agree nor disagree. (In these sorts of situation it is often, but not always, more appropriate to rank answers in order, and to ascribe an ordinal variable to them.)

At the same time, it can be perfectly legitimate to use numbers to describe phenomena which at first don't seem to be numeric. Allocating a 1 for a head and a 0 for a tail is perfectly legitimate; it won't give you any different results than just counting the number of heads or tails (as we will see in the section that follows.) It *can* be appropriate to allocate numbers to temperature, years, subjective experience or questionnaires; you just need to think about what you are doing.

Categorical variables (based on allocating categories rather than numbers) still need to be counted, and statistics can still help to understand the level of uncertainty around them; it isn't necessary to artificially turn them into numbers in order to gain the benefits of statistical analysis.

As is usually the case with statistics, once you understand what you are doing and why you are doing it, it can be incredibly useful. However, mindlessly applying approaches because others have used them, or because they seem to vaguely make sense, may end up misleading more than clarifying.

Key points:

- When the outcome you are looking at is a number or converted to a number, it is called a **random variable**. In this context "random" means that you don't know the value until you look into it.

- Random variables in statistics are usually identified by a capital letter, whereas specific incidence of the outcome are referred to with lower-case letters.

- You can think of random variables in statistics in the same way you can think of any other variables we looked at in part 1, except that they are subject to probability.

- Random variables are normally graphed on the horizontal axis, while their frequency or probability is graphed on the vertical axis.

- Random variables can be constrained by being within particular ranges, or by being particular types of numbers. The numbers which can be included within the constraints are called the **support** of the variable.

- The way we will use the idea of "randomness" in this book is that the multiple drivers of what happens are outside of our capacity to easily evaluate them and so the event is subject to probability.

- Many (but not all) random variables are sometimes described as being **i.i.d.**, or **independent and identically distributed**. That means what happens in one occurrence of the variable doesn't tell you anything about what happens in the next one, but they are all subject to the same probabilities.

- Independence is easy to assume in theoretical probability distributions, but more challenging in the collection of real-world data.

- Confounding factors or lurking variables are factors not accounted for in the results that alter the results.

- Just because something is represented as a number doesn't mean that it should be used as a numerical variable.

Describing probability distributions and their features

> **Read this if** you aren't sure what *Bin(n, p)* means, what *X~binomial* refers to, what $Pr(X = r) = PMF$ means, what the difference between PMF and PDF is, what a CMF is, or what $Pr(X \leq r_i) = \sum_{r=0}^{r_i} \frac{n!}{r!(n-r)!} p^r q^{n-r}$ refers to.

Different types of distribution can be written with their own short cuts. The binomial is written as *Bin(n,p)*, or sometimes *Binom(n,p)*, which means "a binomial distribution with two **parameters**: the power of the expansion (n) and the probability of success of each trial (p)". There is no need to make q (the probability of failure) a parameter because you can calculate that from p ($q = 1 - p$). Likewise there is no need to give r as a parameter, because each distribution will include a range of values for r. In this case that is what we are showing on the horizontal axis. So the expression *Bin(n,p)* contains all the information you need. If you input values for n and p you can map out the entire distribution.

This leads to an important feature of probability distributions. It is one of the things that makes statistics doable: a distribution is independent of the content. It doesn't make any difference whether you are looking at tossing a coin, rolling a dice, or working out the chances of a new drug being successful, once you know that it can be described as *Bin(n, p))* and you have values for the two parameters, you know exactly what it will look like. Of course, that assumes a lot. To be described as *Bin(n, p)* the variable needs to be i.i.d.: independent and identically distributed.

Which leads us to another feature of probability distributions in real-world applications: the distribution usually *approximates* reality rather than describing it exactly, and a large part of the task is evaluating the quality of the approximation.

There are different but related mathematical functions associated with probability distribution graphs. The first, for a **discrete distribution**, is the **probability mass function**, or **PMF**. This terminology suggests that whoever came up with it was thinking in terms of a graph with an actual solid mass (Figure 172). In a general form the PMF may be written as:

$$p(X) = Pr(X = x_i)$$

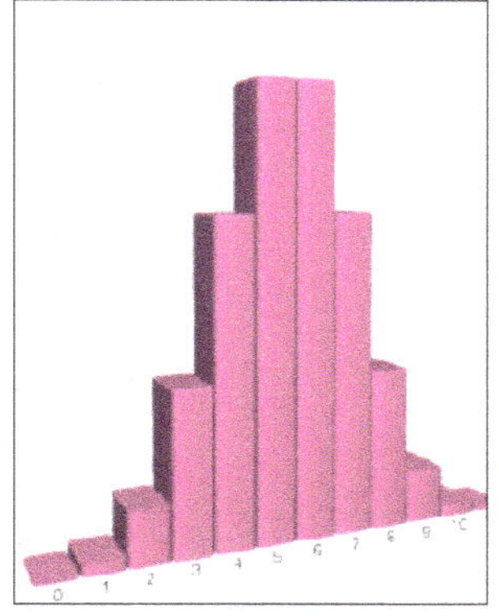

This is a mathematical way of writing that the function $p(X)$ indicates the probability that a random variable will equal any specific value. Note that the PMF does not give the value of the random variable (the horizontal value): it gives the value of the probability of having that variable (the vertical axis). In the case of the binomial distribution, the x_i—or the number of successes—is normally allocated the letter r.

Figure 172

So the PMF for a binomial distribution is the formula for any given term of a binomial expansion:

$$PMF = Pr(X = r)$$

$$= \frac{n!}{r!\,(n-r)!} p^r q^{n-r}$$

$$= \binom{n}{r} p^r q^{n-r}$$

It gives the probability of getting any particular combination from r successes out of n trials with p the probability of success with each trial, and q the probability of failure.

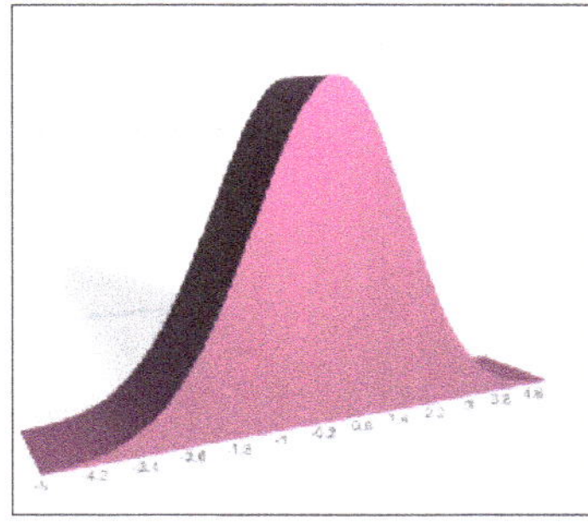

When a function such as this refers to a **continuous** distribution, for reasons lost in time it is called a **probability density function, or PDF** (not to be confused with post-script document format), rather than a PMF. Again, this terminology suggests the idea that the graph is a physical object in 3 dimensions (Figure 173). The reason behind this change of name is to do with the fact that in a continuous distribution, where there are in effect an infinite number of values that

Figure 173

can be placed on the horizontal axis, the probability of something happening at an exact number is negligible, to the point of approaching 0. So it can be said that it has no mass. Rather, probabilities in continuous distributions are always determined by calculating the area under the curve between two values.

Another formula is the **cumulative distribution function**, or **CDF** (Figure 174). It gives you the probability of getting outcomes between a range of values. By convention it starts from the lowest value on the horizontal axis, and stops at the value you are looking at. The CDF for the binomial distribution for values of r from 0 to some number r_i is the formula for a whole binomial expansion, up to that particular value of r_i:

$$Pr(X \le r_i) = \sum_{r=0}^{r_i} \frac{n!}{r!\,(n-r)!} p^r q^{n-r}$$

$$= \sum_{r=0}^{r_i} \binom{n}{r} p^r q^{n-r}$$

It is the probability that the random variable will be less than or equal to a specific value. Once again, note that this is not giving the value of the random variable: it is the probability that the variable will be less than or equal to a given number. In a discrete probability distribution like the binomial

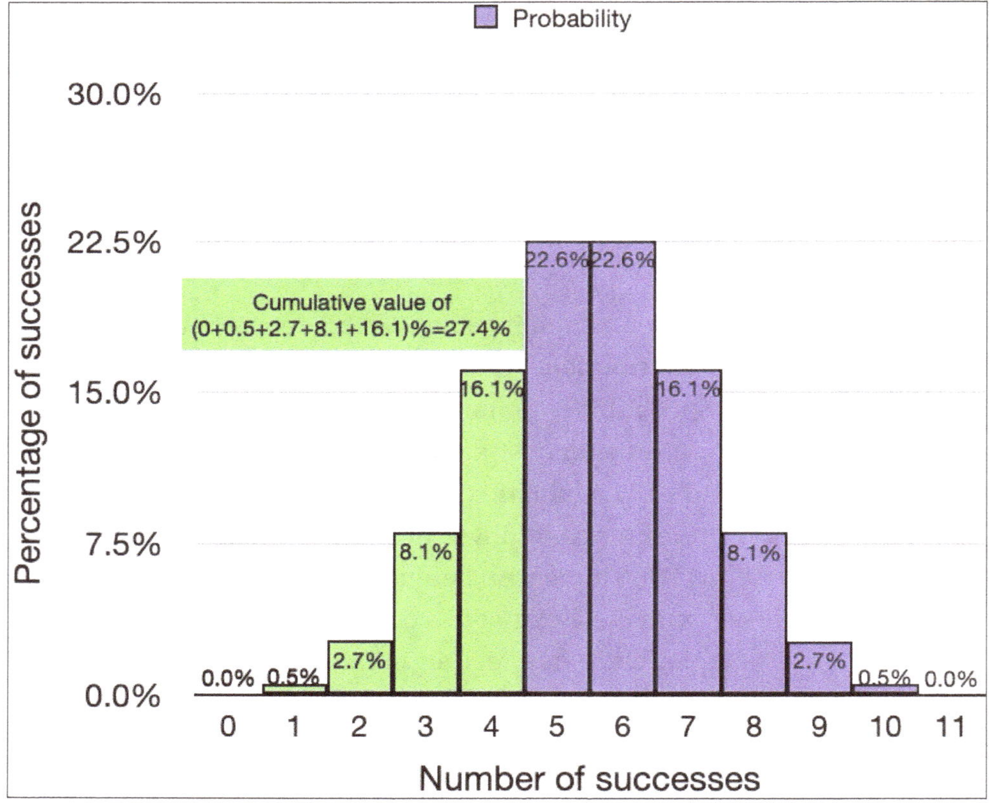

Figure 174

expansion, the CDF is the sum of the probabilities up to and including the value in question (although with particular distributions there may be other ways of calculating the CDF).

So in Figure 174 the CDF tells us that, for a binomal expansion with n of 11 and a probability of 50%, there is a 27.4% chance of having 4 or less successes.

Because probability distributions always add up to 1, it is possible to use this formula to obtain the probability of any particular range of values. To find the probability of obtaining more than r_i successes, subtract the CDF from 1. To find the probability of obtaining the total number of successes between two values, calculate the CDF for the range in question. To be valid, by definition a CDF covering all the support of X must equal 1.

In the case of continuous distributions, the summation function (Σ) is (usually) replaced by an integration, to find the area under the curve.

Key points:

- The expression **X~binomial** means that a random variable X has a binomial probability distribution.

- A binomial distribution is written as *Binom(n,p)*, which means "a binomial distribution with two **parameters**: the power of the expansion (n) and the probability of success of each trial (p)".

- A **probability mass function (PMF)** is used to describe a discrete distribution, while a **probability density function (PDF)** describes a continuous distribution.

- A PMF is written mathematically as $PMF = Pr(X = r)$

- The PMF for the binomial distribution is $Pr(X = r) = \binom{n}{r} p^r q^{n-r}$

- A **cumulative distribution function**, or **CDF**, gives you the probability of getting outcomes between a range of values. By convention it starts from the lowest value on the horizontal axis, and stops at the value you are looking at.

- The CDF for the binomial distribution for values of r from 0 to some number r_i is the formula for a whole binomial expansion, up to that particular value of r_i is

$$Pr(X \le r_i) = \sum_{r=0}^{r_i} \frac{n!}{r!\,(n-r)!} p^r q^{n-r}$$

Some discrete distributions

> **Read this if** you don't know how a binomial distribution works in practice, you aren't familiar with the Bernoulli, geometric, negative binomial, or hypergeometric distribution, or you aren't sure which side the peak is shifted to on a right-skewed graph.

Figure 175
Abraham de
Moivre
1667–1754

Abraham de Moivre's (1667–1754) **story was a sadly-familiar one**: a hugely-talented refugee who struggled to have his abilities recognised in his new country. He was another French mathematician[34], but as a Protestant (a Huguenot[35]) he escaped to England to avoid religious persecution. In England he changed his name from Moivre to de Moivre. He wanted a position teaching mathematics at a university but never obtained one, so made his money as an enlightenment-era version of a tutor and a free-lance financial planner. He worked from coffee shops for private clients, providing tutoring in mathematics and helping them make money by advising on problems like which bets made good sense and how much should be charged for annuities.

De Moivre was a friend of Isaac Newton (1642-1726). He said that he cut out pages of Newton's *Principia* and read them when he was waiting between clients. He was passionate about probability and wrote a 348-page book on it called *The Doctrine of Chances*[36]. While Fermat and Pascal wrote to one another about specific problems, de Moivre was more ambitious and covered many scenarios. One of his driving motivations was to teach gamblers that there was no such thing as luck (he saw random chance as completely different to luck); while another was simply to share the delights of probability with his readers. It was he (possibly following from others before him) who defined probability as a fraction of one:

> *"Wherefore, if we constitute a fraction whereof the numerator be the number of chances whereby an event may happen, and the denominator the number of all the chances whereby it may either happen or fail, that fraction will be a proper designation of the probability of happening"*[37].

He also defined **expected value** (although again, he may not have been the first):

> *"In all cases, the expectation of obtaining any sum is estimated by multiplying the value of the sum expected by the fraction which represents the probability of obtaining it."* [38]

(In modern gambling, the house always wins over the long term by ensuring that gamblers pay more than the expected value when they bet.)

De Moivre and others working on these problems at about the same time didn't explicitly name the distributions that follow. They presented different, specific problems within "games of chance". The following distributions are a more modern way of describing the sort of problems they were looking at.

Binomial distribution (again)

When tossing an evenly-balanced coin, the outcomes of heads or tails both have a 50% probability. For a different example, suppose we think about free throws in basketball, but with different players having differing levels of skill. (de Moivre wrote about card games with which we are no-longer familiar.) In a binomial you do n trials, with a probability of success each time of p. Because the probability of success each time is the same, the expected value is going to be np. (We'll delve into the idea of expected value later.) In other words, suppose you throw a basketball at a hoop, and your success rate is 60%. That is your probability of success. So to work out, on average, how many times you will get the ball through the hoop when you throw it 15 times, just multiply 15 by 60%. You will expect to be successful 9 times, on average.

Figure 176 graphs this probability distribution. Note that because your probability of success with each throw is greater than 50%, the whole graph has been shifted to the right along the x axis, which shifts the peak of the distribution to 9. In basketball terms, de Moivre would have said that if you throw the ball 15 times, and get paid a pound for each successful throw, then the expected value will be 9 pounds. If you pay 8 pounds for the opportunity then that represents good value, because it costs you less than the expected value, or expectation. If you pay 10 pounds, even though there is a maximum chance that you will win 15 pounds, the wager still represents poor value, and over time you will lose.

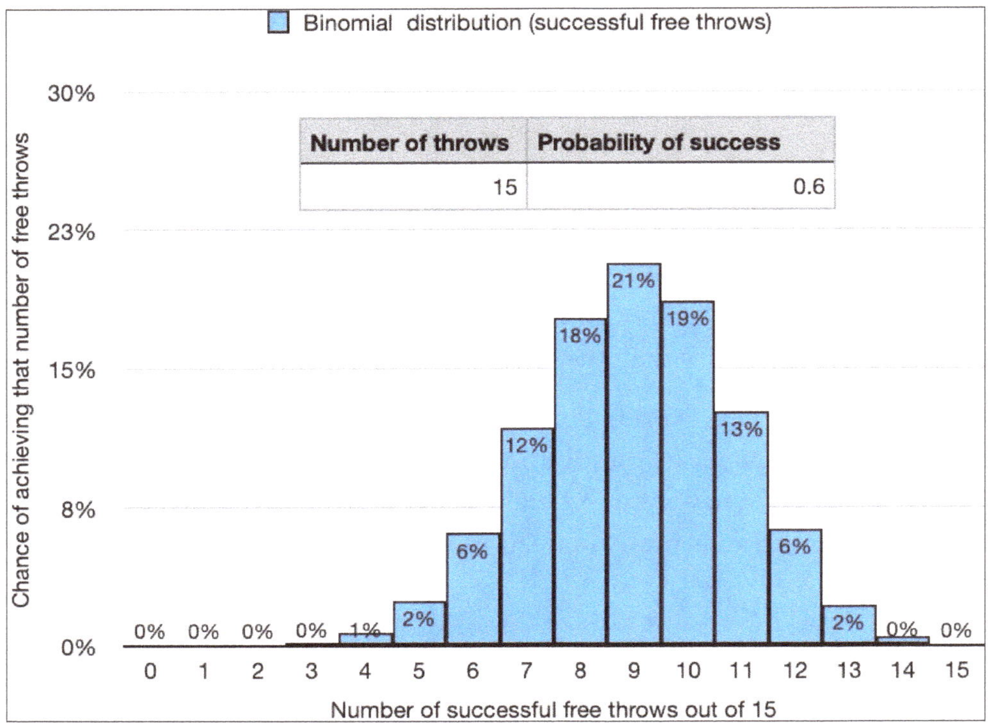

Figure 176

Figure 177 shows the probability of success over 15 throws for a player with a success rate of 10%. The distribution is still binomial: for each throw, there will either be a success or a failure, with no other options. The graph looks different to the earlier graphs, because the decreased probability shifts the graph to the left. The graph appears to be skewed, because it is meaningless to have less than 0 successful throws, so the support of the random variable X is the integers from 0 to 15: the possible number of successes. Note that the probabilities from 6 successful shots upwards are not really 0%; that is a rounding error, because the probabilities are very low. There is still a chance that someone with a 10% success rate will have 15 successful throws out of 15, but the probability is 0.0000000000001%, which would be too difficult to show on the graph.

(Of course, to be neat binomial distributions we would have to assume that the player always has the same probability of success. That may be true for a professional player who practices constantly, but most players are likely to have different factors affecting their success rate on different days, so that the probability on any given day won't be a known constant.)

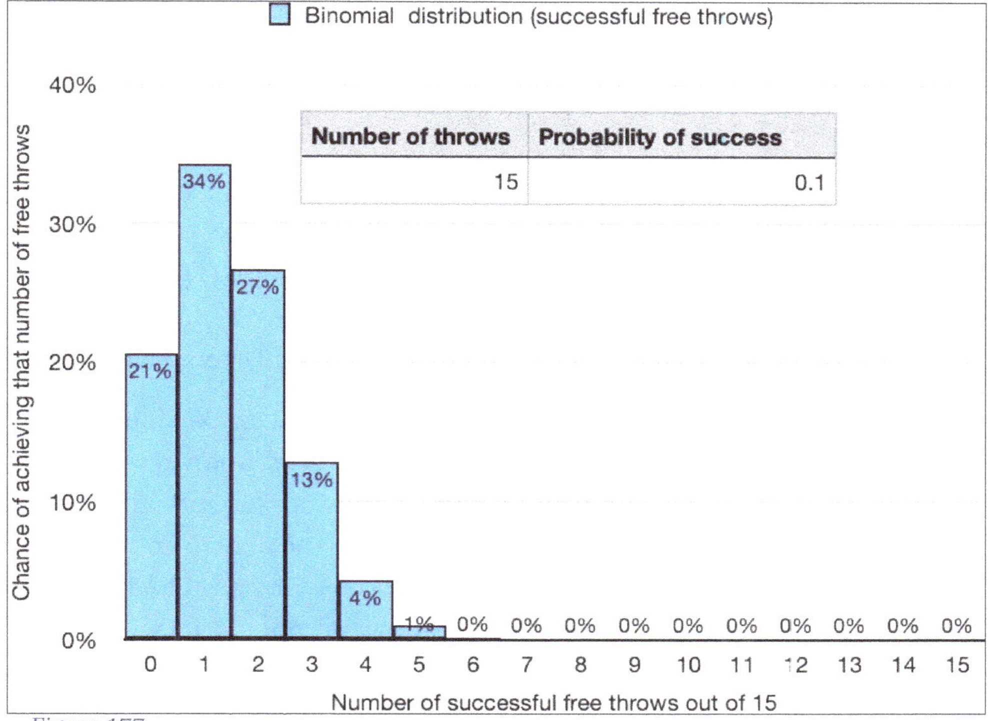

Number of throws	Probability of success
15	0.1

Figure 177

Other sorts of situations in which binomial distributions can be used are:

- A factory knows that a certain percentage of their products will be defective when they come off the production line. Applying that percentage as a probability, they can work out the distribution of probability of having a certain number of defects on any given production run and allocate quality-control inspectors accordingly.

- Insurance companies know that a certain percentage of storms will produce property-damaging hail. They can use a binomial distribution to calculate the distribution of probabilities that a certain number of storms will generate claims, and factor that into their pricing.

- A retail store knows the percentage of goods sold that are returned. Using this percentage as a probability, they can determine the distribution of probabilities that they will have a certain number of returns for a certain number of sales.

It is common to think about things in terms of averages: average free-throws in basketball, average house prices, average sales performance. We will explore averages and some of their complexities later, but hopefully you are getting a sense that when you start to look at things in terms of probability distributions a richer picture emerges.

Bernoulli distribution

Figure 178
Jacob Bernoulli
1665–1704

There were eight Swiss mathematicians and/or physicists named Bernoulli. The one this distribution is named after was Jacques/Jacob/James Bernoulli, (same person) (1665-1704)[39]. He did a lot of work on probability, particularly in a book called *Ars Conjectandi* (*The Art of Conjecturing*) published posthumously. A number of the things he wrote about started with this very basic idea.

The Bernoulli distribution is such a simple concept that it can be hard to think of it as a distribution at all, and yet many other probability distributions (including the binomial) are based on it. It is a straightforward yes-or-no choice and assigned a probability. So flipping a coin is a Bernoulli trial with a probability of one half, whereas rolling a dice in order to get a 3 is a Bernoulli trial with a probability of one sixth. Rolling a dice to see what number comes up, however, doesn't not constitute a Bernoulli trial, because there are more than 2 possible outcomes. Some of the other distributions we will encounter, as well as the binomial distribution, can be described as "a series of Bernoulli trials", but at the same time, a Bernoulli trial can be described, for example, as a binomial distribution with an n (number of trials) of one.

A Bernoulli distribution can be written as $X\sim\mathrm{Bern}(p)$, or in other words, a random variable (X) which has a Bernoulli distribution, with a probability of p. Unlike the binomial, the Bernoulli only has one parameter, its probability.

> When fields are defined for databases, they may be specified in different ways such as a "text" field or a "numeric" field. There are other options, one of which is "boolean": this refers to a field that can only have one of two choices. It is named after the Bernoulli distribution.

Geometric distribution

(The name **geometric distribution** comes from the fact that this distribution can be formed by a geometric progression.) The sort of problem that de Moivre, Bernoulli and fellow probability enthusiasts were interested in was: what is the probability of throwing the number two on the fourth throw of dice, and not earlier? This type of distribution has modern applications as diverse as the probability of publishing a successful book within a given number of books published (given the percentage of books that are successful, and assuming the other variables involved balance one another out to make the process random), the probability of making a sale after a certain number of calls, or the probability of finding a contaminated sample after a certain number of tests.

The binomial expansion won't quite give us this, although we are still looking at a series of Bernoulli trials with an i.i.d. random variable. The binomial will tell us the probability of throwing a 2 once out of 4 throws, but it won't tell us anything about the order of the throws. Thinking back to tree charts and the multiplication rule, however, this is actually easier to calculate than a binomial expansion.

To throw a two on the fourth throw and not earlier, we need to throw anything but a two on the first three throws. The probability of achieving this is

$$\frac{5}{6}$$

each time, and so we multiply those probabilities together three times. In other words, we have

$$\left(\frac{5}{6}\right)^3$$

chances. Then on the fourth throw we have a

$$\frac{1}{6}$$

chance of throwing a 2, so the solution to the problem becomes

$$\left(\frac{5}{6}\right)^3 \frac{1}{6} = 0.0964....$$

There is about a 10% chance of throwing a two on the fourth throw of a die, and not on the first three. We can generalise this. If the probability of something happening any time you try is designated as p, and p is a fraction of 1 (which represents certainty), then the probability of that thing not happening is $1 - p$. If we want to know if that will happen on the rth attempt, the formula for the probability mass function (PMF) is:

$$PMF = (1 - p)^{r-1}p$$

In other words, calculate the probability of the event not happening, multiply it by itself one less than the number of times you are interested in, then multiply that by the probability of success on any one occasion.

This gives us the probability mass function.

Looking at a different example, Figure 179 graphs the values that this approach gives for the probability (vertical axis) of having a success after a given number of attempts (horizontal axis). In this instance the probability of the event happening on any given attempt in this particular instance is estimated to be 0.7.

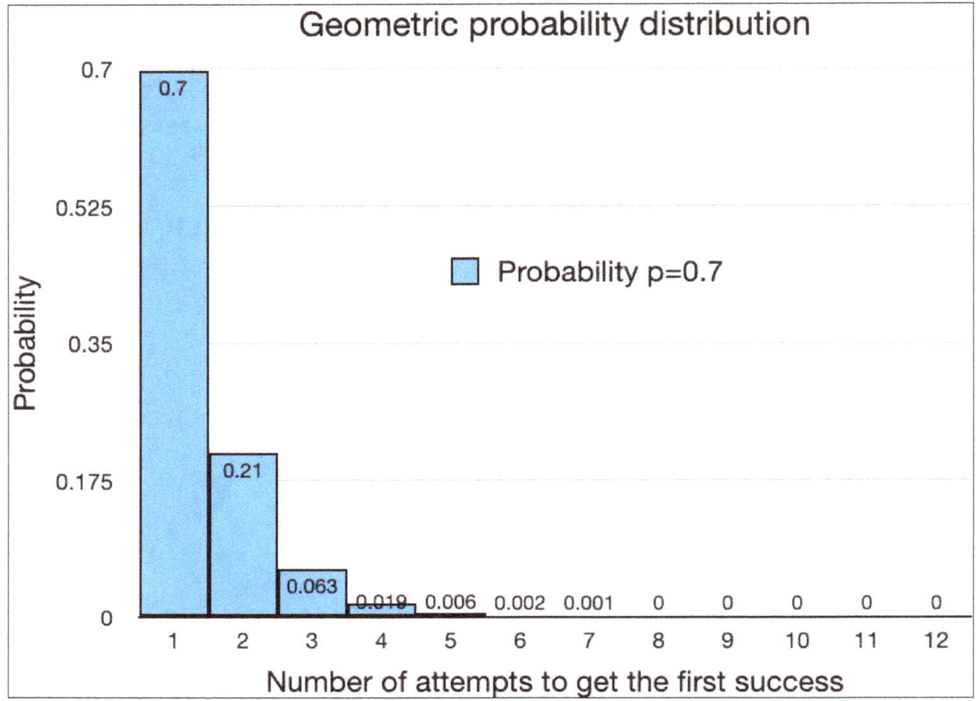

Figure 179

The graph shows 0.7 as the probability of success on the first attempt, 0.21 as the probability on the second attempt but not the first, and so on. As expected, the total probability for all possible outcomes is 1. In this case the probability distribution is supported for the natural numbers from 1 to infinity: $x \in \mathbb{N}$. It has to start at 1, because it makes no sense to have less than 1 attempt, and it goes to infinity because, in theory, you could go forever without having any success, although the chance of that happening would be vanishingly small. Like the binomial distribution, the geometric distribution is discrete (the graph is stepped). Unlike the binomial it is not symmetrical, it is right skewed. (We say right skewed, because the tail extends to the right. That means the peak is shifted to the left.)

The cumulative distribution function (CDF) for a geometric probability distribution could be obtained by calculating the formula for each of the values of r, from 1 to infinity, then adding them together in a series, but in this case that is overkill; there's an easier way. If we want to know the probability of success within the first r trials, then based on the multiplication rule we can first work out the probability that does *not* happen in the first three trials. That is the probability of *failure* $(1 - p)$ multiplied by itself r times. That is the same as saying that it takes more than r trials to achieve the first success. Then we can subtract that from 1, to give the probability it *does* happen in the first r trials. This gives us:

$$CDF = Pr(X \leq r)$$

$$= 1 - (1 - p)^r$$

Figure 180 is a line graph showing a geometric probability distribution with three different probabilities of success for each trial. As you can see, the higher the probability, the steeper the graph. If there is a higher probability, there is a higher chance of achieving success in a smaller number of trials. Regardless of the steepness of the graph, however, the sum of all possible values will always equal one.

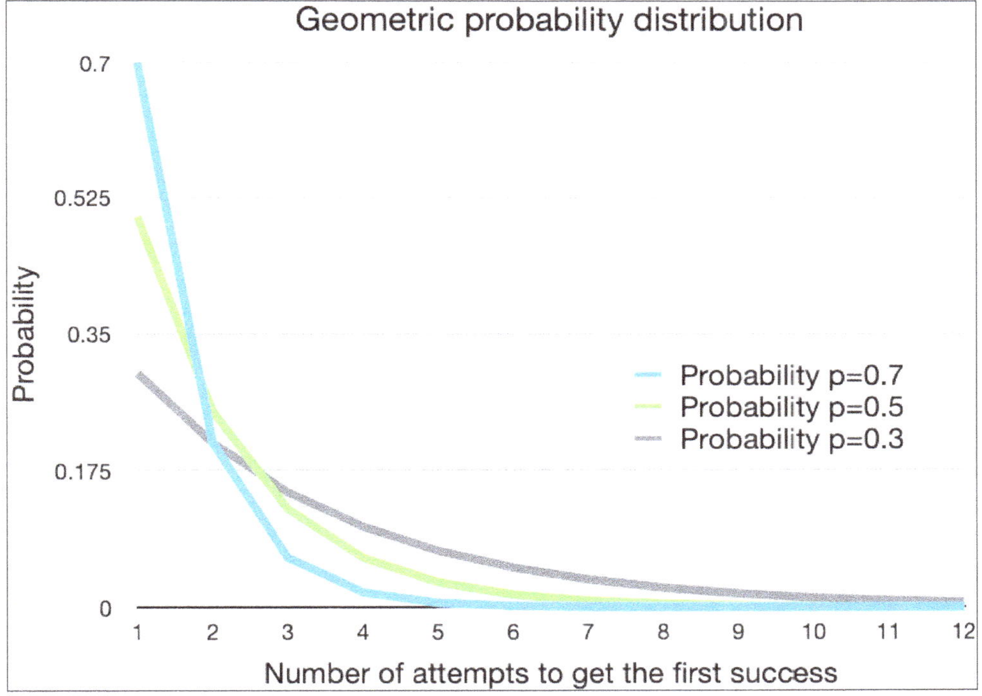

Figure 180

There is a mind-bending idea within the geometric and some other distributions: the idea that you can calculate a proportion of something infinite. This works because while there can *theoretically* be an infinite range of values on the horizontal axis, in practice the probability of nearly all of them is so small that their sum approaches zero, and so the sum of all of the possible values is equal to one. It's still a weird idea, though.

The symbol for the geometric distribution is $G(p)$.

One thing to note with the geometric distribution is that the way we calculated it was looking at the trial that will give the first success. Sometimes it is defined as the number of failures prior to the first success, in which case the formulae for the PMF and CDF have to be modified. If you are reading a paper that cites the use of a geometric distribution without saying which formula was used, you really won't know whether to believe the results unless you have the raw data to compute it for yourself. If you are the one using the statistic, and the software documentation doesn't tell you which basis was used for the formula, it is important to check it for yourself; otherwise the results it gives could be erroneous.

Negative binomial distribution

The negative binomial generalises the geometric distribution. Instead of just asking about probability that the *first* success will occur after a given number of trials, it looks at the probability of getting *any* particular number of successes after a given number of trials. So where the binomial distribution looks at the probabilities of different numbers of *successes* in a fixed number of Bernoulli *trials*, the negative binomial looks at the probabilities of different numbers of *trials* needed to get a fixed number of *successes*.

It can answer questions like: if salespeople have a fixed percentage of successful sales calls, what is the probability distribution of the number of calls they will have to make, to meet their target? That could, for example, tell a sales manager when they should intervene to find a way to get a salesperson to lift their performance, and when they should recognise that a particular poor performance was just an inevitable random fluctuation.

We can think about it in this way. In order to understand the probability that the rth success will occur on the xth trial, we just need to know two things. The first is the probability distribution of the one less than the rth success. The order of earlier successes doesn't matter here, so we can calculate the probabilities using the binomial distribution. The second thing we need is the probability that it will occur on the trial in question; and that is simply the probability of success on a single trial. Once we have those two results, we just have to multiply them together.

As a reminder, the PMF for the binomial (with the version that replaces q with $1 - p$) is

$$\frac{n!}{r!\,(n-r)!}p^r(1-p)^{n-r}$$

In this case we are, first, looking for the probability that we have $r - 1$ successes in $x - 1$ trials, so in the formula we need to replace n with $x - 1$, and r with $r - 1$.

That gives us:

$$\frac{(x-1)!}{(r-1)!\left((x-1)-(r-1)\right)!}p^{r-1}(1-p)^{(x-1)-(r-1)}$$

We can simplify that down a bit, because $(x-1)-(r-1) = x - r$.

$$\frac{(x-1)!}{(r-1)!\,(x-r)!}p^{r-1}(1-p)^{x-r}$$

That gets us to the point where we have $x - 1$ successes in $r - 1$ trials. To get an expression for the probability that we get x successes in r trials, we need to multiply that expression by the probability of success in a single trial (the final one), which is just p.

p^{r-1} multiplied by p is p^r.

$$\boxed{\frac{(x-1)!}{(r-1)!\,(x-r)!}p^r(1-p)^{x-r}}$$

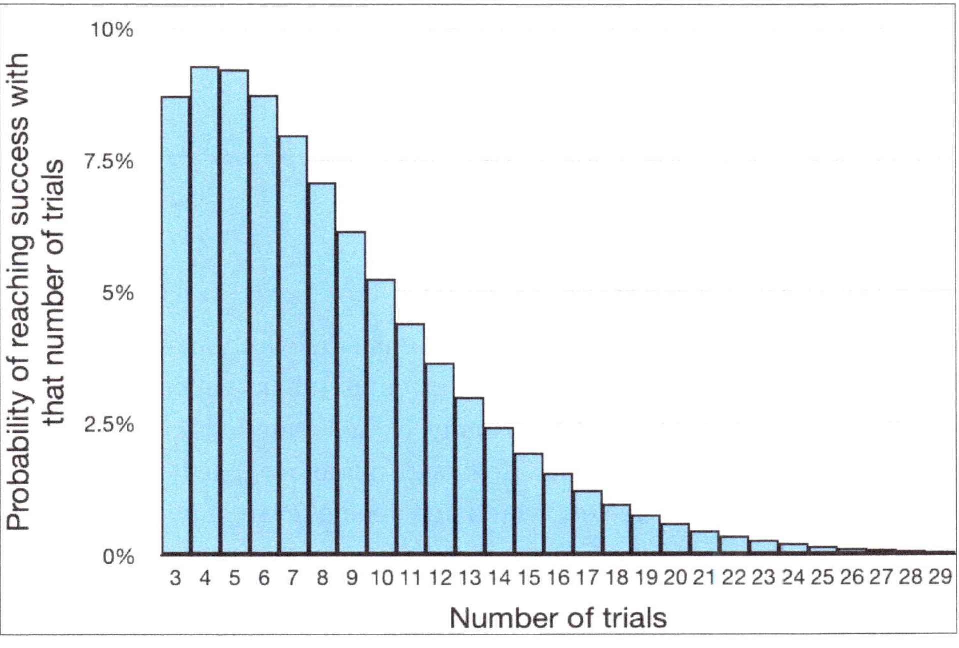

Figure 181

This is normally written with the shortcut of the n binomial r, or n choose r.

$$\binom{x-1}{r-1}p^r(1-p)^{x-r}$$

(It is called a "negative binomial" because, with the right algebraic manipulation—which we aren't going to do here—the coefficient can be made to look like a generalised binomial distribution with a minus one factored out.)

As you would expect, if you replace r with 1, then you get the formula for the geometric distribution: the probability of having 1 success on the xth trial.

$$\frac{(x-1)!}{(0)!\,(x-1)!}p^1(p-1)^{x-1} = (1-p)^{x-1}p$$

So the negative binomial distribution is a generalisation of the geometric, and the geometric is a special case of the negative binomial.

As with the geometric distribution, the graph of the negative binomial distribution is right skewed. The lowest value it can take is the number of successes you are looking for, because it is meaningless to think about achieving a success without a trial, and so you can't have less trials than successes.

For this reason the CDF of the negative binomial is obtained by summing the values from x to infinity:

$$CDF = \sum_{x=r}^{\infty}\binom{x-1}{r-1}p^r(p-1)^{x-r}$$

The symbol for this distribution is $NB(r,p)$.

Hypergeometric distribution

The hypergeometric is similar to the binomial distribution, in that it gives the probability distribution of getting a certain number of successes from a certain number of trials, or more usually, from a sample of a certain size, which amounts to the same thing. (It gets its name because it is associated with a **hypergeometric series**, which is a special case of a geometric series.)

The difference is that unlike the binomial distribution, the hypergeometric is **without replacement**. That means that once you have made a choice, that item is no longer available for choosing, so the trials are not independent. This has applications ranging from picking students from a class to join a study group, through dealing a hand of cards with 3 spades in it, to selecting samples of survey participants, to understanding the deterioration of the surfaces of electrodes. Most of us know a lot more about the first of these than the last, so we'll go with a study group as an example. The calculation for the PMF works like this:

$$\frac{(No.\,of\,combinations\,that\,will\,lead\,to\,success)(No.\,of\,other\,combinations)}{Total\,no.\,of\,samples\,that\,could\,possibly\,be\,chosen}$$

The parameters we start with for this are:

$$n \;=\; \text{the number in the sample.}$$

$$N \;=\; \text{the number in the source from which the sample is taken.}$$

$$R \;=\; \text{the number of options that will be considered ``successes''.}$$

$$N - R \;=\; \text{the number of options that will be considered ``failures''.}$$

$$r \;=\; \text{the number of successes in your sample.}$$

(In statistics N is often taken to be the number of people in a whole population, while n is taken to be the number of people in a sample.)

So in terms of our example, let's assume that you are a university lecturer, and you have asked another lecturer to recommend four students from their class, to help with your research.

There are 20 students in the class, you want 4 of them, there are 12 conscientious students you would be happy to have, and 8 lazy students you would rather not. As a bare minimum, you want 2 of the students you are allocated to be conscientious.

So that makes the parameters:

$$n \;=\; 4 \quad \text{the number in the sample, that is the students who will be working with you.}$$

$$N \;=\; 20 \quad \text{the number in the source from which the sample is taken; the size of the whole class.}$$

$$R \;=\; 12 \quad \text{the number of options that will be considered ``successes''; the number of conscientious students.}$$

$$N - R = 8$$ the number of options that will be considered "failures"; the number of lazy students.

$$r = 2$$ the number of successes in your sample that you will take as a bare minimum; the minimum number of conscientious students you would find acceptable.

To find the probability that you will be allocated a group that will meet your minimum requirements, the *denominator* will be the total number of different groups of four (i.e. samples) that could possibly be chosen. That's going to be our *n* choose *r* formula again:

$$n \text{ choose } r = \binom{N}{n}$$

$$= \binom{20}{4}$$

$$= \frac{20!}{4!\,(20-4)!}$$

$$= \frac{20 \bullet 19 \bullet 18 \bullet 17}{4 \bullet 3 \bullet 2}$$

$$= 4{,}845$$

So in total there would be 4,845 different combinations of students you could be allocated.

The *numerator* will be, first, the number of combinations that would give you 2 students you are happy with, out of the twelve available. We use the same formula again:

$$n \text{ choose } r = \binom{R}{r}$$

$$= \binom{12}{2}$$

$$= \frac{12!}{2!\,(12-2)!}$$

$$= \frac{12 \bullet 11}{2}$$

$$= 66$$

To get exactly 2 students you are happy with, though, in this particular scenario you will need to have 2 other students you would rather not have. We use the same formula:

$$n \text{ choose } r = \binom{N-R}{n-r}$$

$$= \binom{8}{2}$$

$$= \frac{8!}{2!\,(8-2)!}$$

$$= \frac{8 \bullet 7}{2}$$

$$= 28$$

We have to multiply the probability that we get two students we would be happy to have, with the probability that we get two students we would rather not, to cover all of the eventualities.

Putting all of that together, our PMF is:

$$Pr(X = r) = \frac{\binom{R}{r}\binom{N-R}{n-r}}{\binom{N}{n}}$$

Specifically in our case, the probability that you end up with *exactly* 2 students you are happy to work with comes to:

$$Pr(X = 2) = \frac{\binom{12}{2}\binom{8}{2}}{\binom{20}{4}}$$

$$= 0.381 \ldots$$

which is less than 50%. Fortunately that is only one of the outcomes you will be happy with. You will also be happy with having three, or even four conscientious students.

To find the probability that you will have an outcome you are happy with, we need the cumulative distribution function (CDF):

$$CDF = \sum_{r=Max(0,n-(N-R))}^{Min(R,n)} \frac{\binom{R}{r}\binom{N-R}{n-r}}{\binom{N}{n}}$$

The values we are summing to and from here are complicated. What this is saying is that the smallest number r (the number of successes) is the number we are sampling minus the number of failures in the population, unless that number is less than 0. In our case there are eight potential failures, but we are only sampling four, and four minus eight is less than 0. Therefore the minimum is zero. But if our sample was eight, and there were only four possible failures, our minimum number of successes wouldn't be zero, it would be eight minus four, or just four; we couldn't have less success than that.

The largest number of successes we can have is R or n, whichever is smaller, because r can't be larger than the number of successes available in the population, and it also can't be larger than the sample size.

The CDF and the probability distribution graph show that your chance of getting at least as many students in your study group as you are happy to work with (2, 3 or 4) is about 85%. Much better!

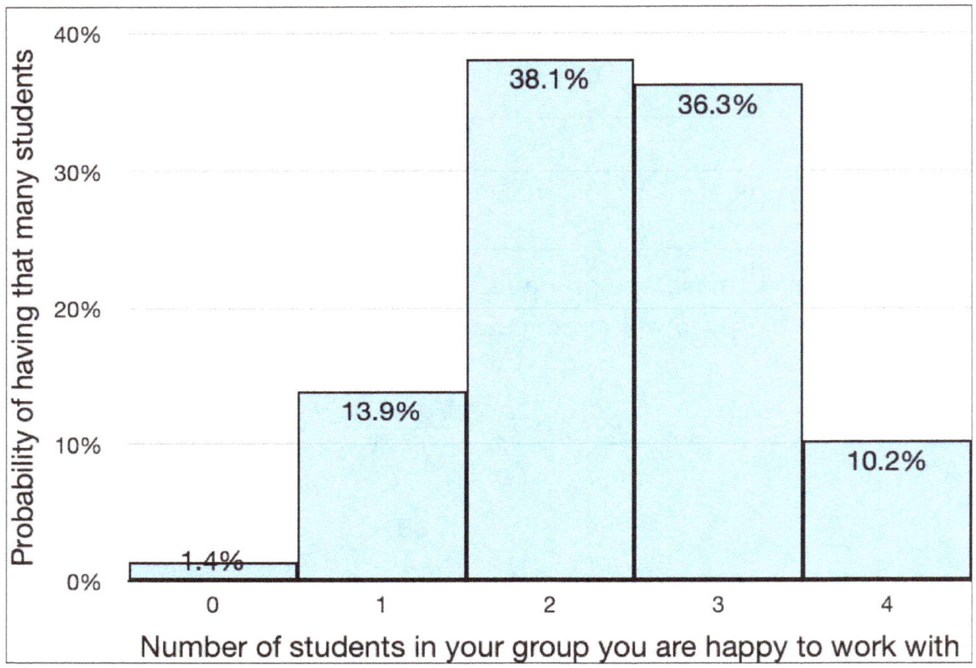

Figure 182 Note that percentages are rounded, meaning this does not quite add to 100%.

Categorical and Multinomial distributions

If you have been wondering about the connection between the sorts of probability distributions we have just been looking at, and the examples we saw earlier—of rolling a pair of dice, dealing from a deck of cards, and the eight different categories we put people who study statistics into—the categorical and multinomial distributions are where it all comes together. You may think that once we start putting subjects into categories then a distribution of probabilities is no longer relevant, but it is once you understand the multinomial distribution.

In the introduction, we said that the number of ways that something can happen is often two: the way you want them to (which is called a success) and the way you don't (called a failure). Of course, it often isn't that simple, and there are more than two options. Tossing a coin is a binomial choice; rolling a dice is a multinomial one, with even chances of success for each choice.

A categorical distribution is an extension of a Bernoulli. It's a single choice, where, unlike the Bernoulli, there are more than two categories. So, if you like, a Bernoulli is a special case of a categorical distribution (where there are only two options to choose from).

Deciding which shirt to wear today is a categorical choice, regardless of how many shirts you own. So is rolling a dice; each number is a different category.

Different categories can have different probabilities assigned to them. Dealing a single card to find whether you will get the ace of spades, the queen of diamonds or something else is still a categorical distribution, but the queen of diamonds and the ace of spades each have 1 chance in 52, while getting anything else has a probability of 50 in 52.

The multinomial distribution is to the categorical what the binomial is to the Bernoulli; it's a series of repeated trials, where the probability is the same each time. The probabilities are calculated using the multinomial expansion that we saw earlier (page 356). It isn't usually graphed, because there isn't just the one random variable that can be used on the horizontal axis. Each of the different possible situations under consideration is actually a separate random variable, so we need to express them as an ordered list of numbers, or a **vector**. (There is a lot to vectors, which are important in linear algebra, but at this stage the only thing that matters is that a vector is like a set of numbers, where the order matters.)

So the PMF for the multinomial is the formula we saw earlier when we were looking at the multinomial coefficient:

$$Pr(X_1 = x_1, X_2 = x_2, \dots X_k = x_k) = \frac{n!}{x_1! \, x_2! \, \dots \, x_k!} p_1^{x_1} p_2^{x_2} \dots p_k^{x_k}$$

where the random variable X_i represents the number of occurrences that lead to outcome i, k is the number of catergories, n is the total number of occurrences (i.e. $n = x_1 + x_2 \dots x_k$) and p_i is the probability of each occurrence.

The CDF is not usually written, because of the prohibitively large number of ways that all of the possible numbers of variables can occur.

The shorthand for this distribution is:

$$Mult\left(n, \vec{p}\right)$$

(where p is a list of probabilities that add to 1. The arrow indicates that it is a vector, in the sense of an ordered list of numbers.)

You may recall that when we were looking at the multinomial coefficient (page 356), we had eight categories, which each had their own probability of occurring. We then looked at the probability that each category would have a particular number of students in it. The probability was, as you would expect, incredibly small. What we didn't have was a way of looking at how likely it was that the numbers we looked at would come close to the number in each category that we were expecting. Understanding the multinomial distribution doesn't quite get us to that point, unfortunately. We'll see how to calculate that when we come to the chi square distribution.

Key points:

- Changing the probability of a binomial distribution shifts the graph horizontally.

- A **Bernoulli** distribution is a simple binary choice, and can be written as $X \sim Bern(p)$, or in other words, a random variable (X) which has a Bernoulli distribution, with a probability of p.

- A **geometric** distribution gives the probability of a first success after a given number of trials. The probability mass function (PMF) is

$$(1-p)^{r-1}p$$

and the cumulative distribution function (CDF) is

$$1-(1-p)^r \, .$$

- The **negative binomial** distribution generalises the geometric distribution. Instead of just asking about the probability that the first success will occur after a given number of trials, it looks at the probability of getting any particular number of successes after a given number of trials. The PMF is

$$\binom{x-1}{r-1}p^r(1-p)^{x-r}$$

and the CDF is

$$\sum_{x=r}^{\infty}\binom{x-1}{r-1}p^r(p-1)^{x-r} \, .$$

- The **hypergeometric** distribution is similar to the binomial distribution, in that it gives the probability distribution of getting a certain number of successes from a certain number of trials, or more usually, from a sample of a certain size, without replacement. The probability density function (PDF) is

$$\frac{\binom{R}{r}\binom{N-R}{n-r}}{\binom{N}{n}}$$

and the CDF is

$$\sum_{r=Max(0,n-(N-R))}^{Min(R,n)}\frac{\binom{R}{r}\binom{N-R}{n-r}}{\binom{N}{n}} \, .$$

- A **categorical** distribution is a single choice, where there are more than two categories.

- A **multinomial** distribution is a series of repeated trials where there are more than two categories, where the probability is the same each time. The shorthand for this distribution is: $\text{Mult}\left(n,\vec{p}\right)$. The PMF is

$$Pr(X_1 = x_1, X_2 = x_2, \ldots X_k = x_k) = \frac{n!}{x_1!\,x_2!\ldots.x_k!}p_1^{x_1}p_2^{x_2}\ldots p_k^{x_k}$$

Practical examples comparing these distributions

The list of subjects below is not exhaustive; what follows are just some examples of the way the distributions we have seen so far can be applied in different disciplines.

Field of Study	Binomial	Geometric	Hyper-geometric	Negative Binomial	Categorical / Multinomial
Astronomy / Space Science	Number of stars with planets in a fixed observational set	Number of stars examined before discovering the first one with a planet	Number of stars with rare features from a sample	Number of non-planet stars before finding a target number with planets	Counting celestial bodies by type (e.g., red dwarf, neutron star, gas giant)
Biology / Medicine	Number of patients responding to a treatment out of a fixed number	Number of failed drug trials before first success	Selecting individuals with a genetic trait from a population without replacement	Number of failed treatments before achieving a fixed number of successes	Count of patients by disease categories (e.g. cancer type) based on symptoms or biomarkers
Computer Science	Number of successful packet transmissions over a network	Number of attempts before a successful data packet transfer	Sampling infected files in a batch of downloads	Number of software bug reproductions before hitting a target count	Classifying emails into categories (spam, promotions, social, etc.)
Criminology / Forensics	Number of fingerprints matching a suspect out of a fixed number	Number of suspects investigated before finding the first match	Selecting evidence samples with a trait from a crime scene sample	Number of false leads before finding a fixed number of correct matches	Categorizing crimes by type or evidence by source
Ecology / Environmental Science	Number of tagged animals recaptured in a sample	Number of trap checks before catching the first animal	Drawing a fixed number of fish from a pond with some tagged (no replacement)	Number of untagged animals caught before finding a target number of tagged ones	Categorizing animal species observed in a region during a survey

Field of Study	Binomial	Geometric	Hyper-geometric	Negative Binomial	Categorical / Multinomial
Education / Psychology	Number of students passing a test out of a fixed number	Number of test attempts before first pass	Selecting students with a specific trait without replacement	Number of wrong answers before a fixed number of correct ones	Classifying students by learning style, performance level, or answer types
Epidemiology / Public Health	Number of individuals infected in a sample of fixed size	Number of people contacted before first infection occurs	Testing individuals for disease from a population without replacement	Number of uninfected individuals before finding a fixed number of infected ones	Classifying cases by variant of virus or severity of symptoms
Finance / Actuarial Science	Number of profitable investments in a fixed portfolio	Number of trades before the first profit	Choosing winning assets from a finite market sample	Number of losses before a fixed number of profitable trades	Classifying investments into risk classes or sectors
Linguistics	Number of correct syntactic judgments out of a set of test sentences	Number of words heard before encountering a new lexical item	Drawing rare constructions from a corpus without replacement	Number of non-target constructions before hitting a fixed target count	Classifying utterances into syntactic structures or parts of speech
Marketing / Consumer Research	Number of consumers who click an ad out of a fixed campaign	Number of consumers shown ads before the first click	Drawing loyal customers from a sample without replacement	Number of ignored ads before reaching a fixed number of click-throughs	Categorizing customers into behaviour profiles or product preferences
Operations Research / Logistics	Number of successful deliveries out of fixed attempts	Number of delivery attempts before first successful delivery	Drawing specific configurations of parts from stock without replacement	Number of unsuccessful dispatches before meeting a success target	Categorizing shipments by destination, type, or urgency
Political Science	Number of voters in a precinct voting for a candidate out of fixed number	Number of doors knocked before first supporter is found	Drawing a sample with a fixed number of voters favouring a certain party	Number of canvass attempts before finding a fixed number of supporters	Classifying voters into ideological or party affiliation categories
Quality Control	Number of defective items in a fixed batch tested	Number of items tested before finding the first defect	Choosing defective parts from a batch without replacement	Number of non-defective items before finding a fixed number of defects	Sorting manufactured items into defect types (e.g., size, colour, structural flaw)

Field of Study	Binomial	Geometric	Hyper-geometric	Negative Binomial	Categorical / Multinomial
Sales / Business	Number of successful sales calls out of a fixed number	Number of calls until the first successful sale	Drawing winning products in a promotion from a limited sample	Number of failed transactions before a fixed number of successful sales	Customer segmentation into categories (e.g., price-sensitive, brand-loyal, impulsive)
Sociology	Number of survey respondents adhering to a specific belief	Number of households visited before first supporter of a particular idea	Choosing a sub-category in a focus group drawn without replacement	Number of rejections before finding a fixed number of willing respondents	Categorizing attitudes into attitudinal groups
Sports Analytics	Number of successful penalty shots out of a fixed number	Number of failed attempts before first goal scored	Drawing a lineup with a set number of star players without replacement	Number of games without a win before a fixed number of wins occurs	Categorizing plays (e.g., pass, run, foul) or outcomes (win, draw, loss)

Relationships between probability distributions

Read this if you are confused about how the distributions we've looked at relate to one another, you don't know what a contingency table/ cross-tabulation/ crosstab is, you don't know the difference between a marginal and a joint distribution or how to calculate one from the other, you don't know how to define the independence between two marginal distributions, you aren't clear about conditional probability or how to symbolically describe it, you aren't clear about the law of total probability, you don't understand about relative risks or odds ratios, or you don't know what a case-controlled study has to do with either.

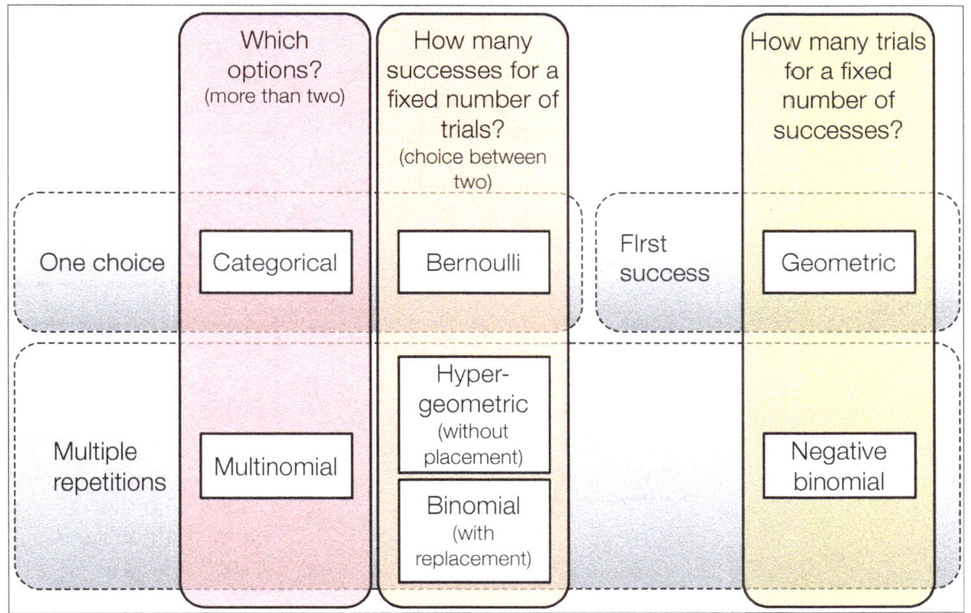

Figure 183

Figure 183 shows how the distributions we have seen so far relate to one another. We will build on this model as we progress, adding more distributions.

Once we start talking about working with two or more probability distributions at the same time, it's important to be clear about what we are looking at. In the following sections we are looking at the relationship between probability distributions, without changing the variables themselves. But what does that mean? (Later we will look at changing the variables themselves, by transforming them and/or adding them together.)

Marginal and joint distributions

(If you are struggling with this, see the How to Love Statistics YouTube video on *Combining probability distributions part 1*.)

When we looked at the multinomial distribution, we looked at having more than two categories in the one distribution. It is also often necessary to think about more than one whole distribution. We may, for example, want to measure two or more different variables from the same group of subjects in an experiment, such as energy consumed in food and blood pressure. When we first looked at tree charts, we looked at students studying statistics. To simplify things, if we just looked at two of the variables we considered then—the reasons for studying statistics and the enjoyment of mathematics at school—we can show these in the tree chart at Figure 184.

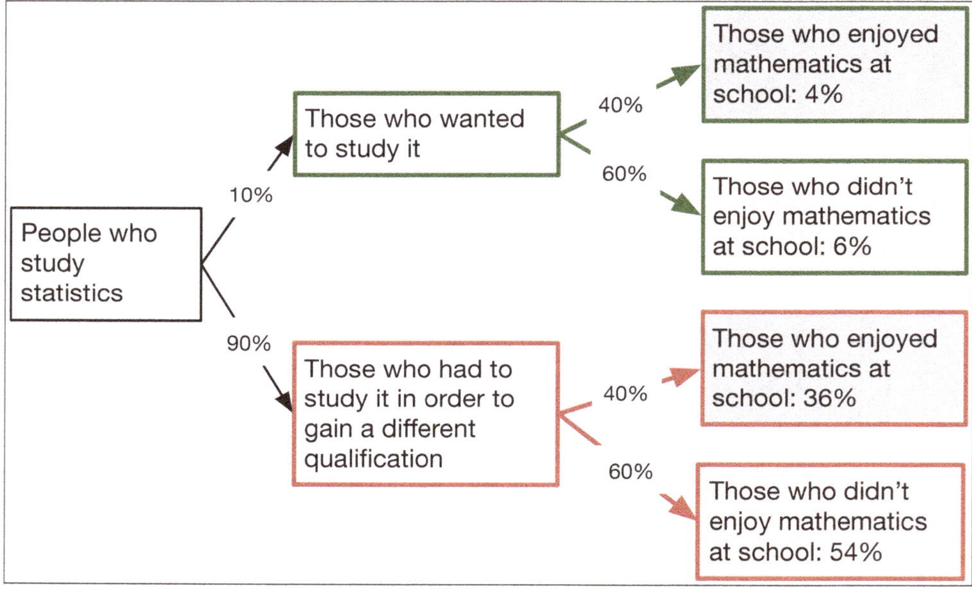

Figure 184

As we saw when we first looked at a tree chart like this (page 340), the percentages in the boxes on the right are obtained by multiplying along the branches to that point. So the 4% in the box on the top right (those who enjoyed mathematics at school and wanted to study it) is calculated by finding the 10% of students who wanted to study statistics, and multiplying it by the 40% of students who enjoyed mathematics; that is, 40% of the 10%.

We can think of each of these variables as having their own probability distribution. These are much simpler distributions than the others we have looked at so far (there are only two alternatives) but they are still legitimate probability distributions: each shows the probability of something happening (the probability that a random student will fit into one of two categories) and the percentages in each distribution add to 100%, meaning all of the options have been covered. (Yes, it's an oversimplification. In the real world there would be many shades of grey and more complex alternatives. It's shown in this simplistic way to illustrate some concepts.)

If we want to think about the relationship between these variables, we have to think about how the individual probability distributions for each of these variables interact.

		Distribution of reasons for studying statistics (*Y*)		
		Those who wanted to study statistics	Those who had to study it in order to gain a qualification	Marginal distribution of enjoyment of mathematics at school
Distribution of enjoyment of mathematics at school (*X*)	Those who enjoyed mathematics at school			40%
	Those who didn't enjoy mathematics at school			60%
Marginal distribution of reasons for studying statistics		10%	90%	100%

Figure 185

We said (hypothetically) that 10% of those who study statistics actually want to study it, while 90% must do so in order to gain another qualification. We also said that of those who study statistics, 40% enjoyed mathematics at school while 60% didn't. As well as displaying this information as a tree chart, we can also show it as a **contingency table** (sometimes called a **cross-tabulation** or **crosstab**), as in Figure 185. The 40% and 60% on the right *margin* is called, unsurprisingly, the **marginal probability distribution**: in this case, the marginal distribution of people who enjoyed mathematics at school. The 10% and 90% at the bottom show the marginal probability distribution of people who wanted to study statistics. Both these marginal distributions individually add up to 100%, as probability distributions must. We already knew these numbers, however. What we want to know about is what happens when we combine the distributions.

The yellow-shaded area within the heavy lines is known as the **joint probability distribution**. To calculate the joint distribution in this instance, we can multiply the numbers of each of the marginal distributions. So, for example, we can multiply the 40% on the right margin by the 10% on the bottom margin, to give us a 4% probability that someone wanted to study statistics, and enjoyed mathematics at school in the top left yellow square of Figure 186. (Remember that probabilities are fractions between 0 and 1, and multiplying these fractions creates a smaller fraction.) This is consistent with the multiplication rule, which we use when thinking about the probability of *both* things being true, and we represent as $P(A \cap B)$, or the probability of the intersection of the two sets. This is what we did in our tree chart at Figure 184, multiplying along the branches.

		Distribution of reasons for studying statistics (Y)		
		Those who wanted to study statistics	Those who had to study it in order to gain a qualification	Marginal distribution of enjoyment of mathematics at school
Distribution of enjoyment of mathematics at school (X)	Those who enjoyed mathematics at school	4%	36%	40%
	Those who didn't enjoy mathematics at school	6%	54%	60%
Marginal distribution of reasons for studying statistics		10%	90%	100%

Figure 186

You can see (in Figure 186) that the probabilities of the **joint distribution** (within the bold lines) also all add up to 100% (4% + 36% + 6% + 54%). The joint distribution is a legitimate probability distribution in its own right.

We'll call a random variable representing the distribution of enjoyment of mathematics at school X, and a random variable representing the reasons for studying statistics Y. A joint distribution probability mass function (PMF) of X and Y can be written as $Pr(X, Y)$ which means "the probability that both X and Y are true". Alternatively it can be written as $Pr(X = x, Y = y)$ which means "the probability that the random variable X will be equal to some specific value of x and that the random variable Y will be equal to some specific value of y".

$$f(X, Y) = P(X = x, Y = y)$$

(The lower case x and y are sometimes given with a subscript such as i to make it clearer that a specific instance is indicated.) So the 4% in the top left yellow box in Figure 186 shows the probability that any randomly selected statistics student will have enjoyed mathematics and school, and have wanted to study statistics.

The cumulative distribution function (CDF) for a joint distribution can be written as $Pr(X \leq x, Y \leq y)$ which means "the probability that the random variable X will be less than or equal to some specific value of x and that the random variable Y will be less than or equal to some specific value of y".

$$F(X, Y) = P(X \leq x, Y \leq y)$$

That isn't particularly relevant when we are only considering two options for each variable, but it becomes relevant with more complex distributions.

If we had started by knowing the numbers in this joint distribution, we could have calculated the marginal distribution by adding the rows and columns. And so, for instance, we could have calculated that 10% of people wanted to study statistics (using the left-hand yellow column of Figure 186) by adding the 4% of people who wanted to study statistics and enjoyed mathematics at school, to the 6% of people who wanted to study statistics and didn't enjoy mathematics at school. (It's quite neat the way this works out.) This is consistent with the addition rule, which we use when we want the probability that *either* situation is true, and we represent it as $P(A \cup B)$, or the probability of the union of the two sets.

It may seem counter-intuitive until you think about it, but you can see from Figure 186 that if we want to calculate the numbers in a marginal distribution and we start with the joint distribution, we need to add up the *other* distribution. That is, to find the marginal distribution of reasons for studying statistics, we keep the categories we already have for that distribution (the columns of our table) and add the totals for those who did and did not enjoy mathematics at school. In other words, the 10% of people who wanted to study statistics is made up of the 4% who wanted to study statistics and enjoyed mathematics, to the 6% who wanted to study statistics and didn't enjoy mathematics. We add the percentages of enjoyment of mathematics (the X values), in order to find the marginal Y value: in this case, 10%.

Dependent and independent probabilities

(If you are struggling with this, see the How to Love Statistics YouTube video on *Combining probability distributions part 1*.)

The binomial distribution works on an assumption of independence: what happens in each trial won't affect the probability of what happens in the next trial. The hypergeometric, on the other hand, is not based on an assumption of independence. Because there is no replacement—once you have made a selection, what you chose is not available for the next selection—then the probability changes each time. In those situations we are talking about the relationships *within* distributions. Here we are talking about the relationships *between* different distributions.

Our contingency table (Figure 186) worked out neatly because the proportions of the numbers in each row are the same, and the proportions of the numbers in each column are the same. In other words, the two probabilities didn't affect one another. Looking at the proportions of the vertical columns:

$$\frac{4\%}{6\%} = \frac{36\%}{54\%}$$

$$= \frac{2}{3}$$

The probabilities in each column match, because whatever the column heading is, the probability does not change, so the distributions are **independent**.

Likewise in the horizontal rows:

$$\frac{4\%}{36\%} = \frac{6\%}{54\%}$$

$$= \frac{1}{9}$$

Again, the probabilities in each row match.

If you think about it though, if people choose to study statistics because they want to pursue it as a career it is more likely that they also enjoyed mathematics at school. In Figure 187 let's say that, of those who had to study statistics in order to gain a different qualification, the percentages stay the same: 40% enjoyed mathematics at school, and 60% didn't. But of

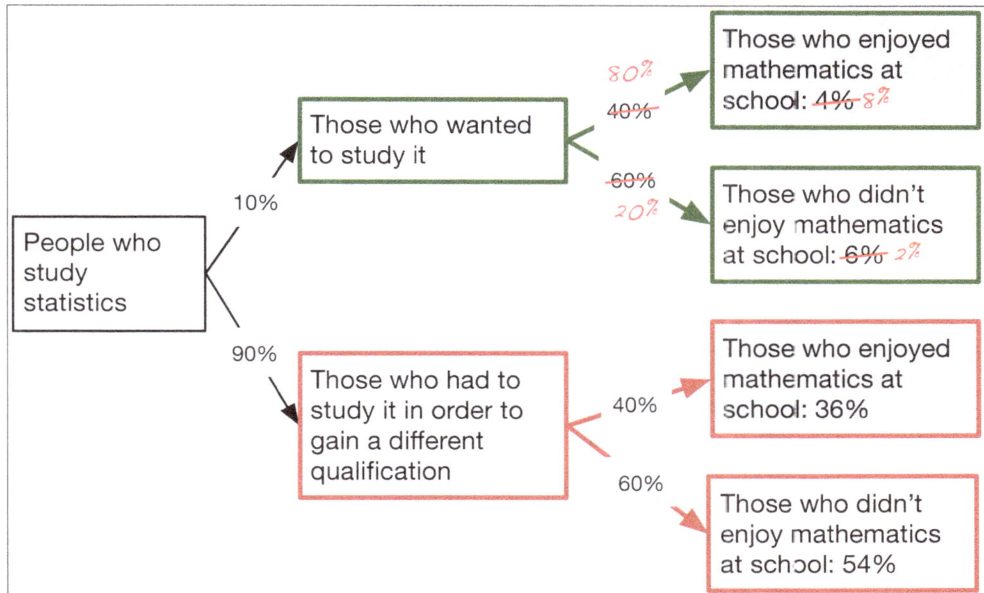

Figure 187

those who wanted to study statistics, let's say that 80% enjoyed mathematics and only 20% didn't. When we multiply that through, we find that 8% of all students wanted to study statistics and enjoyed mathematics at school (the top right box in Figure 187), and 2% of all students didn't enjoy mathematics but did want to study statistics.

If the probability that a given student enjoyed mathematics at school is to any extent **dependent** on whether they wanted to study statistics (or vice versa), then rather than say that the probabilities are independent, we say, not surprisingly, that they are dependent, or **contingent**.

When we make the same changes in our contingency table (Figure 188), we can see that we still have three legitimate probability distributions: the two marginal ones and the one joint one each add up to 100%. We can also see that we can find the marginal distribution by adding the relevant options from the opposite distribution. To find the marginal distribution of those who enjoyed mathematics at school, we can add those who enjoyed it and wanted to study statistics (8%) to those who enjoyed it and had to study it to gain another qualification (36%) giving a total of 44%.

What has changed is that the neat way that we could multiply the relevant numbers in the marginal distribution in order to calculate the joint distribution no longer applies. If you multiply 10% by 44% you don't get 8%. (You get 4.4%, which is very different.)

This is actually the statistical definition of independence: if, and only if, probabilities are independent, the probability of all of them happening is the same as multiplying their individual probabilities. So in Figure 188:

		Distribution of reasons for studying statistics		
		Those who wanted to study statistics	Those who had to study it in order to gain a qualification	Marginal distribution of enjoyment of mathematics at school
Distribution of enjoyment of mathematics at school	Those who enjoyed mathematics at school	~~4%~~ 8%	36%	~~40%~~ 44%
	Those who didn't enjoy mathematics at school	~~6%~~ 2%	54%	~~60%~~ 56%
Marginal distribution of reasons for studying statistics		10%	90%	100%

$$Pr(A \cap B) = Pr(A)Pr(B)$$
if and only if $Pr(A)$ *and* $Pr(B)$ *are independent.*

We will revisit this when we look at correlation and covariance, to see what dependence between distributions does to this formula.

More complex combined distributions

This was a particularly simple example of combining two or more probability distributions, but the same principles apply when we are considering distributions that are calculated by formulae, and when there are more than two distributions involved. Provided that the probabilities are independent, we can find the joint probability mass function/probability density function (PMF/PDF) by multiplying the two marginal PMF/PDF functions, and we can find the joint cumulative distribution function (CDF) by multiplying the two marginal CDF functions.

Figure 188

We can describe the joint PMF/PDF (with a discrete random variable) is:

$$f(X,Y) = P(X = x, Y = y)$$

In other words, we are considering the probability that X is equal to some particular value and that Y is equal to some particular value, at the same time.

Assume that X and Y are two random variables with their own probability distributions. The joint CDF is written as:

$$F(X,Y) = P(X \leq x, Y \leq y)$$

In other words, the joint cumulative distribution of two random variables (and there is no need for us to be limited to two) is equal to the probability that X is less than some specific value of x, together with the probability that Y is less than some specific value of y.

Provided that X and Y are independent (knowing something about one of them tells you nothing about the other one), then, using the multiplication rule (we are trying to find the probability that both things are true) the joint PMF/PDF is equal to the product of the marginal PMFs/PDFs:

$$Pr(X = x, Y = y) = Pr(X = x)Pr(Y = y)$$

Likewise the joint CDF is equal to the product of the marginal CDFs:

$$Pr(X \leq x, Y \leq y) = Pr(X \leq x)Pr(Y \leq y)$$

All of that means, among other things, that the formulae for joint probability distributions are likely to have two or more random variables in them. If there are two variables, and we also graph the value of the function on a third axis, that means the graph will be three-dimensional.

Getting to a joint CDF from joint PDFs in the continuous case is similar to finding the CDF from a single PDF: you integrate. The difference is that with joint PDFs, you need to undertake a double integration. Similarly, to find a joint PDF from a joint CDF, you find two partial derivatives.

In Figure 186 we could find the marginal distribution by adding all the possibilities that contributed to that result. Where probabilities are formulae we will be dealing with more than one random variable in the formula. In the case of cumulative functions, to find the marginal distribution of a function with a single variable, we take the limit as the other variable(s) approach infinity. That is, we integrate. In the case of PMFs, we sum up all possible values of the other variable(s), and with PDFs we integrate with respect to the other variable(s). If we are trying to find the marginal distribution of X, we integrate with respect to Y. That means, of course, that the Y values disappear after the integration, and we are left with a function of X.

We will see examples of this as we progress.

Ratios of joint and marginal distributions

Contingency tables are often used for reporting results in studies. Consider a simple 2 by 2 contingency table for a hypothetical medical study, showing a joint distribution of exposed and diseased, and the marginal distributions of exposure and of disease (Figure 189):

	Disease Present	Disease Absent	**Total**
Exposed	a	b	$\boldsymbol{a + b}$
Unexposed	c	d	$\boldsymbol{c + d}$
Total	$\boldsymbol{a + c}$	$\boldsymbol{b + d}$	$\boldsymbol{a + b + c + d}$

Figure 189

Here, 'a' represents the number of exposed individuals with the disease, 'b' the number of exposed individuals without the disease, 'c' the number of unexposed individuals with the disease, and 'd' the number of unexposed individuals without the disease.

The **relative risk** is a measure of the risk of a certain event occurring in one group compared to the risk of the same event occurring in another group. Using Figure 189 it is calculated as:

$$Relative\ risk = \frac{Risk\ in\ exposed\ group}{Risk\ in\ unexposed\ group}$$

$$= \frac{\frac{a}{(a+b)}}{\frac{c}{(c+d)}}$$

To cement your understanding:

Let $a = 3$, $b = 100$, $c = 2$ and $d = 150$. Complete a contingency table modelled on the one above. Calculate the relative risk of getting the disease, based on exposure.

On page 337 we talked about calculating odds. The **odds ratio** is, as you might expect, the ratio between the odds. So the formula, again using Figure 189, is:

$$Odds\ ratio = \frac{\frac{a}{b}}{\frac{c}{d}}$$

With an odds ratio, it actually doesn't matter whether you look at the ratio of the odds by horizontal rows or by vertical columns; the ratio ends up being the same.

Start with the odds ratio by rows, and multiply the numerator and denominator by b.

$$\frac{\frac{a}{b}}{\frac{c}{d}} = \frac{\frac{a}{b}}{\frac{c}{d}} \cdot \frac{\frac{b}{1}}{\frac{b}{1}}$$

$$= \frac{\frac{ab}{b}}{\frac{cb}{d}}$$

Divide the numerator and denominator by c.

$$= \frac{a}{\frac{cb}{d}}$$

$$= \frac{\frac{a}{c}}{\frac{b}{d}}$$

While this symmetry is a convenient fact that makes it easier to use odds ratios because you don't have to remember what to put into the rows and what to put into the columns, in general it is preferable to use relative risks than odds ratios, because, just as probabilities are easier to understand intuitively than odds, so relative risks are easier to understand intuitively than odds ratios. There are, however, specific situations where odds ratios are preferred.

A **case-controlled study** compares subjects who have an attribute of interest (cases)—usually a medical condition—with subjects who do not have the attribute (controls). It looks back retrospectively to compare how frequently the exposure to a risk factor is present in each group, to determine the relationship between the risk factor and the disease. In this situation there often isn't enough information available to work out the risk across a large population, in order to find the risks in an unexposed group so that you can calculate a relative risk. There is, however, enough information to calculate an odds ratio.

Odds ratios are also easier to use with certain statistical models.

To cement your understanding:

Using the same table for the previous exercise ($a = 3, b = 100, c = 2$ and $d = 150$) calculate the odds ratio.

Key points:

- A **contingency table**, also known as a **cross-tabulation** or **crosstab** allows us to compare two distributions.

- A **joint distribution** is obtained by **multiplying marginal distributions**. It is a probability distribution in its own right, adding to 100%. If the distributions are continuous, this involves a double integration.

- A joint distribution PMF or PDF of X and Y can be written as $Pr(X, Y)$ which means "the probability that both X and Y are true".

- A cumulative distribution function for a joint distribution can be written as $Pr(X \leq x, Y \leq y)$.

- A **marginal distribution** is obtained by adding the relevant values in a joint distribution. These relevant values belong to the other distribution.

- When probabilities are independent, the probability of all of them happening is the same as multiplying their individual probabilities. So:

$$Pr(A \cap B) = P(A)P(B) \text{ if and only if } P(A) \text{ and } P(B) \text{ are independent.}$$

- **Relative risk** is a measure of the risk of a certain event occurring in one group compared to the risk of the same event occurring in another group. It is calculated as:

$$Relative\ risk = \frac{Risk\ in\ exposed\ group}{Risk\ in\ unexposed\ group}$$

- The **odds ratio** is the ratio between the odds. Relative risk is usually a better measure than an odds ratio, but it can't always be used.

- A **case controlled study** is a situation in which an odds ratio can be calculated, but a relative risk often can't.

Second Stream:
an Obscure Paper by
an Obscure Clergyman

Change can be challenging, whether you're trying to change the world or the world is trying to change you. Very occasionally, though, an idea comes along which is just so right and whose timing is so perfect that the idea itself creates massive change. The change may not happen quickly, but the idea has so much power that, once it is articulated, the change is inevitable.

People somewhere were trying to move a block of stone, and then someone—we'll never know who—said: "Let's invent a wheel to make things easier!" That simple idea changed the world.

People had been building fantastic structures for thousands of years, doing the necessary mathematics by using geometrical shapes. For centuries, technological progress stagnated. From the Romans to the Middle Ages, little changed. Then René Descartes (1596–1690) came along and said: "Why don't we make a second number line and put it at right angles to the first, so that we can describe points and lines on a plane with algebra?" That simple rotation of a line was a key turning point in changing the world from what existed in the Middle Ages to the world we live in now.

Ideas like that don't come along all that often, but what follows is the story of no less than four such ideas, all from a single, hard-to-read, almost-forgotten paper—published after its author was dead.

Probability beyond gambling

Read this if you want to know what Thomas Bayes actually contributed to our thinking about probability and statistics, you aren't clear about where Bayes's rule comes from or how to apply it, you can't see how probability could be subjective, or you aren't clear about how, if you test positive for a common disease, when the test is 99% sensitive and 95% specific, you might only have a 29% chance of having the disease.

Figure 190

Figure 190 is not Thomas Bayes: it's Richard Price (1723–1791). If efficiency is about maximum useful achievement with minimal effort, then Price should win first prize for the most efficient contributor in the history of statistics (or, for his other accomplishments, possibly in the history of everything). Price was a radical Welsh Nonconformist minister, a member of the Royal Society, and a man whose superpowers included writing provocative pamphlets (one of which may have been instrumental in triggering the American War of Independence[40]), recognising the talents of others, being a good friend to intelligent and impactful people, and connecting the right people together.

His friends included George Washington, Thomas Jefferson, John Adams (second President of the United States), Thomas Paine (who wrote the two most influential pamphlets at the start of the American War of Independence), Comte de Mirabeau (a leader in the early stages of the French Revolution), the Marquis de Condorcet (one of the key thinkers of the Enlightenment), Prime Minister William Pitt the Elder, empiricist philosopher David Hume, Benjamin Franklin (a Founding Father of the United States and inventor of bifocals, among many other achievements), and Adam Smith (known as the Father of Economics). He was also a passionate early supporter of the idea of women's equality. Which is all (hopefully) very interesting, but none of that is why he was important to the world of statistics.

The family of another Protestant clergyman, Royal Society member and friend, Thomas Bayes (1701–1761), asked Price to look through Bayes's unpublished papers after his death and see if there was anything worthwhile in them.

Figure 191

Figure 191 probably isn't a portrait of Thomas Bayes either. It is normally used as his portrait but an analysis of the style of clothing suggests that a Presbyterian minister of the time probably wouldn't have been dressed in that way[41]. Bayes was not well-known in his time and probably never sat for a portrait. He was a keen amateur mathematician who published two religious tracts and two mathematical papers for the Royal Society, none of which is memorable[42].

If Price would have won the prize for the most efficient contributor in the history of statistics, Bayes would have come a close second. The only reason anyone has heard of him is that he also once wrote a document called *An Essay Towards Solving a Problem in the Doctrine of Chances*[43]. He never bothered to publish it. He saw it as an addition to de Moivre's work, addressing a particular question that de Moivre raised but didn't answer: if you have observed something happening in the past at a certain rate but don't know anything else about it, what is the probability that its future rate will lie between two other numbers?

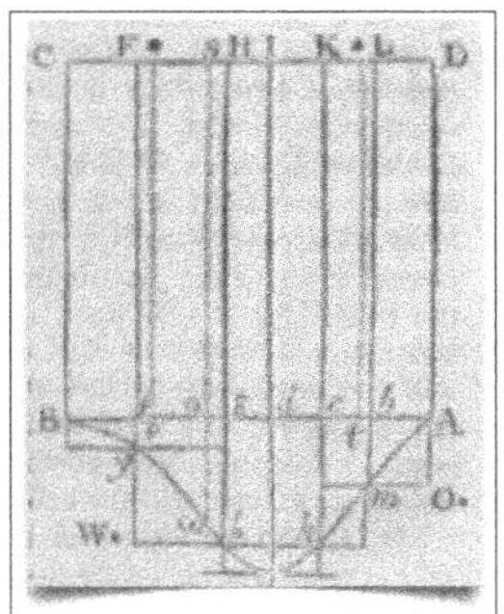

Figure 192

At first, that may not sound interesting, and Bayes himself apparently wasn't excited enough by it to push it forward. Price found it when he was going through Bayes's papers, saw something in it, made some introductory and concluding remarks about it, and introduced it to the Royal Society, where it was once again ignored for about 40 years. At first glance, it is just an extension of the sort of work de Moivre had been doing: giving an example of a problem in probability, writing some clear definitions about it (he outlined the addition and multiplication rules, but only to restate in one place work that others had done), writing some complex calculations about them, and not quite getting to a modern understanding[44]. The paper is difficult to follow (Figure 192), and there are still differing views about what Bayes was trying to say[45]. It's likely that the only reason Price understood the paper was that he and Bayes had talked through the ideas at an earlier stage.

Bayes has been wrongly credited with coming up with Bayes's theorem, but Bayes's theorem came after his time.

The idea that Price and Bayes had talked about Bayes's ideas is suggested by a comment from Price at the end of the essay: *"Mr Bayes, therefore, by an investigation that it would be too tedious to give here, has deduced from this rule another..."*

What makes Bayes's essay so influential, and Price so important in discovering it, is that it touched on and inspired others to think about no fewer than four new paradigms—or frameworks for conceiving things—that each eventually moved the field of probability forward. As is usually the case with new paradigms, they took a while to have an impact.

1. The first paradigm was picked up by Price in his introduction: Bayes was probably the first person to argue from a specific instance of probability to a more generalised idea. He showed that thinking about probability has much wider implications than just its contribution to gambling and pricing annuities.

    > *"Every judicious person will be sensible that the problem now mentioned is by no means merely a curious speculation in the doctrine of chances, but necessary to be solved in order to a sure (sic) foundation* **for all our reasonings concerning past facts, and what is likely to be hereafter."** (Emphasis added.)[46]

2. Bayes introduced the idea that that you don't need to know the absolute possibility of events, just their **relative probability**. De Moivre had touched on relative probabilities, but it was Bayes who explicitly pointed out that probability wasn't necessarily absolute, but could be **conditional**, relative to other factors. This idea was picked up by Pierre-Simon Laplace and led to Bayes's theorem.

3. Related to that, Bayes proposed that probability can change as more information becomes known. It is subjective, based on what is known at the time. This idea was also picked up by Laplace and ultimately led to Bayesian statistics. The first three points are largely based on the first section of Bayes's essay.

4. He also introduced the idea that probability can be used to understand an ongoing stable process, where past events can be used to predict what is likely to happen in the future. This idea was also picked up by Laplace, who, among other things, used it as the basis for calculations about the probability of the sun rising tomorrow. (He decided it probably would, but he couldn't be completely certain. And he put numbers

to his uncertainty.) It was then picked up by Siméon Denis Poisson, a pupil of Laplace. This is largely based on the second section of Bayes's essay, but Poisson's mathematics are more accessible to modern readers and based on a different approach from that which Bayes attempted. Poisson's work ultimately opened the way for the use of statistics in process improvement, workforce planning, physics, radiological oncology and other fields.

Pierre-Simon Laplace and statistics

As well as giving us the Laplace transform that we looked at in Part 1, Pierre-Simon, Marquis de Laplace was one of the most important thinkers in the history of statistics. Laplace was heavily influenced by Bayes's paper, although even he said of Bayes's work:

> *"...he has arrived at this in a refined and very ingenious manner, although a little perplexing."*[47]

Sometimes something we hear can be more thought-provoking if it isn't completely clear. We can spend more time trying to wrestle the meaning from it and may develop meanings of our own. This might be what Laplace went through when he read Bayes's paper. In 1812 he wrote a book of nearly 500 pages called *Analytic Probability Theory* (*Théorie Analytique des Probabilités*). In this book, while he did use gambling examples, he delved deeply into the application of probability: to demographics, to the calculation of probable errors in measurement, to research and to the probability of future events more generally.

> *"Strictly speaking it may even be said that nearly all our knowledge is problematical; and in the small number of things which we are able to know with certainty, even in the mathematical sciences themselves, the principal means for ascertaining truth induction and analogy are based on probabilities; so that the entire system of human knowledge is connected with the theory set forth in this essay"*[48].

In other words Laplace was the first person to have a sense of statistics as a wide-ranging discipline in the modern sense, although he didn't come up with that word.

The word "statistics" was originally used in English in 1791, in a large piece of analysis on the demography of Scotland. It has its origins in the Latin word status: fixed, appointed, to set up, resolve or judge. This word in turn gave rise to the idea of one who judges, which gave us statute and state.

Conditional probability

(If you are struggling with this, see the How to Love Statistics YouTube video on *Combining probability distributions part 2*.)

With a **conditional probability** the range of options is narrowed down before we think about how probable something is.

		Distribution of reasons for studying statistics (Y)		
		Those who wanted to study statistics	Those who had to study it in order to gain a qualification	Marginal distribution of enjoyment of mathematics at school
Distribution of enjoyment of mathematics at school (X)	Those who enjoyed mathematics at school	4%	36%	40%
	Those who didn't enjoy mathematics at school	6%	54%	60%
Marginal distribution of reasons for studying statistics		10%	90%	100%

Figure 193

Getting back to our original table (Figure 193) the **absolute probability** that a student will have enjoyed mathematics at school and will want to study statistics is 4% (top left yellow box). But **conditional probability** is different. This is the probability of one thing, *given that* another thing is true. The probability that someone will want to study statistics, *given that* they enjoyed mathematics at school is 10% (Figure 194).

	Distribution of reasons for studying statistics (Y)		
	Those who wanted to study statistics	Those who had to study it in order to gain a qualification	Marginal distribution of enjoyment of mathematics at school
Those who enjoyed mathematics at school	~~4%~~ 10%	~~36%~~ 90%	~~40%~~ 100%

Figure 194

In other words, if we are only considering those students who enjoyed mathematics at school and no other ones, as per Figure 194, then the total of all possibilities isn't 40% any more, it's 100%. We had to multiply 40% by 2.5 to get to 100%, and so the other probabilities in this row also get multiplied by 2.5.

That is, the 4% of all students is also 10% of those who enjoyed mathematics at school. We describe this as "the probability of A, given B" and write it as:

$$Pr(A|B) = 10\%$$

where A is wanting to study statistics, and B is enjoying mathematics at school.

Now let's think about the condition probability of $Pr(E|A)$; the probability of having enjoyed mathematics at school, given that a person is studying statistics. This is represented in Figure 195.

		Those who wanted to study statistics
Distribution of enjoyment of mathematics at school (X)	Those who enjoyed mathematics at school	~~4%~~ 40%
	Those who didn't enjoy mathematics at school	~~6%~~ 60%
		~~10%~~ 100%

Figure 195

The 10% of students who wanted to study statistics is now 100% of the students we are considering. 40% of these people enjoyed mathematics at school, while 60% of them didn't. So:

$$Pr(B|A) = 40\%$$

That means, in this instance, that the probability of B given A is four times the probability of A given B. As a general rule

$$Pr(A|B) \neq Pr(B|A)$$

It's easy to get confused about this, both when learning statistics and in real-world situations. It can have profound implications when we are thinking about how probable something is. We'll explore more of what this means, with some practical examples, when we come to look at Bayes's theorem.

We said a little earlier that we can define independence by saying that the probability of two things happening (the intersection of their individual probabilities) is the same as the product of their probabilities.

Another way we can express it is to say that:

$$Pr(A|B) = Pr(A) \, and$$

$$Pr(B|A) = Pr(B)$$

In other words, A and B don't affect one another in terms of probabilities. The probability of A, given B, is exactly the same as the probability of A, because whether or not B is true doesn't affect how probable A is.

To cement your understanding:

• Look back at Figure 193, where we assumed that the two probability distributions were independent, to confirm that $Pr(A|B) = Pr(A)$ in every case.

• Then look at Figure 188, where (more realistically) we assumed that the probabilities were dependent, to confirm that $Pr(A|B) \neq Pr(A)$.

• Confirm that, in either case, $Pr(A|B) \neq Pr(B|A)$.

Bayes Theorem

When we looked at the contingency table of student preferences, we noted that contingent probabilities were different from absolute probabilities. We saw that $Pr(A|B) \neq Pr(B|A)$. That is, the probability of A, given B is not usually the same thing as the probability of B, given A. Laplace (possibly over-generously) credited Bayes with this theorem about it:

$$Pr(A|B) = \frac{Pr(B|A)Pr(A)}{Pr(B)}$$

The meaning of these terms is:

$Pr(A|B) =$ the probability of A, given that B is true.

$Pr(B|A) =$ the probability of B, given that A is true.

$Pr(A) =$ the probability of A, regardless.

$Pr(B) =$ the probability of B, regardless.

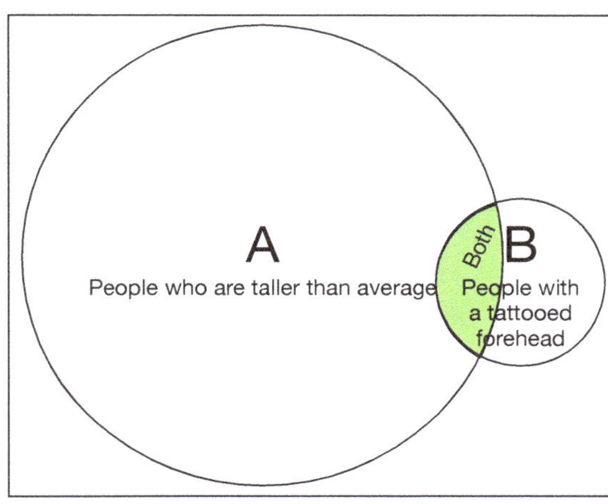

Figure 196

To explain it a different way, Figure 196 shows a Venn diagram with A (people who are taller than average) and B (people who have a tattoo on their forehead). There is an overlap, called Both (people who are taller than average and have a tattoo on their forehead). Both is represented by the green area in Figure 196. The important thing to remember is that we are talking about probabilities (which in this instance is the same as proportions), not numbers. So the number of people in Both doesn't change, but the *proportion of Both to A* is smaller than the *proportion of Both to B*. You may remember from when we were looking at set notation and Venn diagrams (page 74) that the probability of both represents the **intersection** of A and B, and we can write it like this:

$$Pr(Both) = Pr(A \cap B)$$

If you are meeting someone at the airport you have never seen, but whom you already know has a tattoo on her forehead, the probability that she is also taller than average is can be written as:

$$Pr(A|B) = \frac{Pr(A \cap B)}{Pr(B)}$$

That is, the probability that she is taller than average *given that* she has a tattoo on her forehead is the same as: the probability that she is both taller than average and has a tattoo on her forehead, over the probability that she has a tattoo on her forehead. Straightforward enough; it's just two ways of saying the same thing. Both sides of the equation are talking about the ratio of the green area to the area of the circle marked B in Figure 196. If you have heard that the person you are meeting has a tattoo on her forehead, you won't be particularly surprised to learn that she is also a little taller than average; that situation is not particularly improbable.

Likewise, if you are told that the person you are meeting is taller than average, the chance that she also has tattoo on her forehead can be written as:

$$Pr(B|A) = \frac{Pr\,(A \cap B)}{Pr(A)}$$

If all that you have heard about her is that she is a little taller than average, you are likely to be surprised to find that she has a tattoo on her forehead and that no one thought to mention it so that you could recognise her more easily. That is a much more unusual, and therefore improbable, feature. In other words, if the information you have initially is that she has a tattoo on her forehead, then the probability that she is both will be *higher* than if the information you have initially is that she is taller than average. The number of taller-than-average people with a tattoo on their forehead doesn't change, but from your perspective the probability of meeting such a person changes markedly, based on the information you had before the meeting.

A provocative idea that comes out of an example like this is the concept that what a person currently understands affects the probabilities they are dealing with. (This idea was particularly important to Laplace and recurs through his writings.) Bayesian probability is an approach to probability that factors in the subjective expectations of people who are determining probabilities. So if you rate your favourite sporting team as having a 60% chance of winning, and the game is over but you don't know what happened, then a Bayesian approach would be to say that for you, the probability is still 60%, whereas for someone who knows the score it will either be 1 (certainty of a win) or 0 (certainty of a loss or a draw). So probability is subjective. Someone who didn't accept the Bayesian idea of probability (sometimes called a frequentist) would be more likely to say that the game has been played, so probability doesn't come in to it. The outcome is certain, you just don't have the information yet to know what it is.

On the other hand—and this is another crucial idea underlying Bayesian statistics—you can't generally know whether you have all the relevant information. Bayes's rule forms the basis of an approach to statistics that involves making the best possible estimate of what you know now, doing some research, then using this to modify what you currently believe based on the new information. In this way knowledge is seen as being something that evolves and is modified, rather than being fixed.

Bayes did touch lightly on this point in his treatise, but it is questionable whether he would recognise current Bayesian thinking.

Back to "Bayes' Theorem", we said that

$$Pr(A|B) = \frac{Pr\ (A \cap B)}{Pr(B)}$$

Multiplying both sides of this equation by $Pr(B)$ gives us:

$$Pr(A|B)Pr(B) = Pr(A \cap B)$$

We also said that

$$Pr(B|A) = \frac{Pr\ (A \cap B)}{Pr(A)}$$

Multiplying both sides of this equation by $Pr(A)$ *gives us*

$$Pr(B|A)P(A) = Pr(A \cap B)$$

We can therefore say that

$$Pr(A|B)Pr(B) = Pr(B|A)Pr(A)$$

Dividing both sides by $Pr(B)$ gives us

$$Pr(A|B) = \frac{Pr(B|A)Pr(A)}{Pr(B)}$$

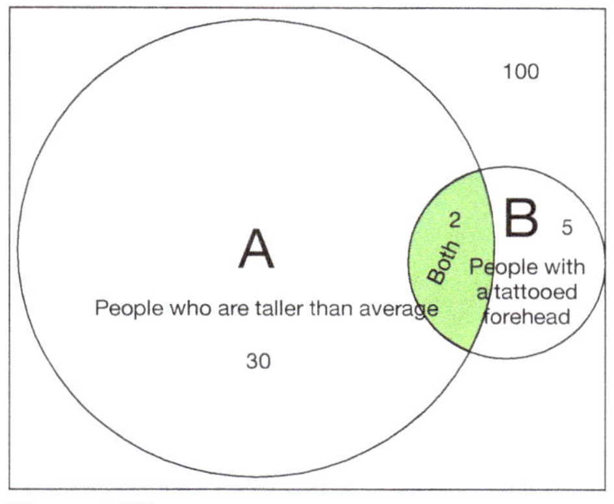

Figure 197

That is, it gives us Bayes's theorem. To get a clearer idea of that, let's go back to our Venn diagram but arbitrarily put some numbers in it (Figure 197). Assume our total population is 100 people. Let A (the number of people who are taller than average) = 30; B (the number of people with a tattoo on their forehead) = 5 and the number of people who fit into both categories = 2.

That means $Pr(A|B) = \frac{2}{5}$, $Pr(B|A) = \frac{2}{30}$, $Pr(A) = \frac{30}{100}$ and $Pr(B) = \frac{5}{100}$

Plugging those numbers into our formula for Bayes's theorem:

$$\frac{2}{5} = \frac{\frac{2}{30} \cdot \frac{30}{100}}{\frac{5}{100}}$$

$$= \frac{\frac{2}{30} \cdot \frac{30}{100}}{\frac{5}{100}}$$

$$= \frac{2}{5}$$

Which is clearly correct. It wouldn't matter what positive real numbers we put into our Venn diagram; it would still be correct; it's just mathematics.

To understand what that means in practice, imagine that you visit your doctor for a check-up. She orders some blood tests. When she has the results she tells you that you have tested positive for a disease. The doctor tells you that the test is a very accurate one: it is 99% **sensitive**, meaning that you if you have the disease there is a 99% chance that you will test positive. The test is also 95% **specific**, meaning that there is only a 5% chance that you will test positive if you don't have the disease. You feel very unlucky to receive this diagnosis, so you ask her how common the condition is. She tells you that it is actually very common for people in your age group; about 1 person in 50 has it; that's 2%.

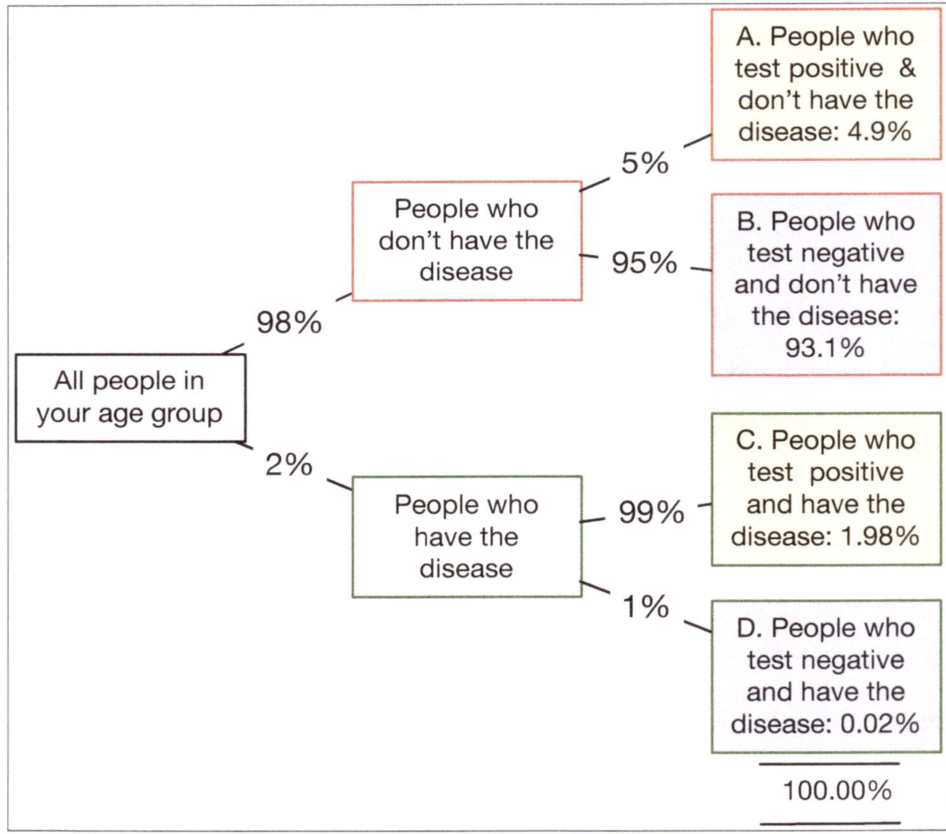

Figure 198

She wants you to start treatment. The treatment carries risks of side-effects and will also be quite inconvenient. You feel reluctant, so she suggests you go away and think about it and let her know what you decide to do.

None of that sounds good. The numbers are a bit confusing, but it sounds as though, having tested positive, you have a 95% chance of having the disease. Simply because you want the practice, though, you decide to draw up a tree chart (Figure 198).

This is a conditional probability: different percentages apply depending on whether or not you have the disease, which means that the order you draw the tree chart in matters. So you draw up the chart and multiply along the branches. Things suddenly look a whole lot better. Out of all the people in

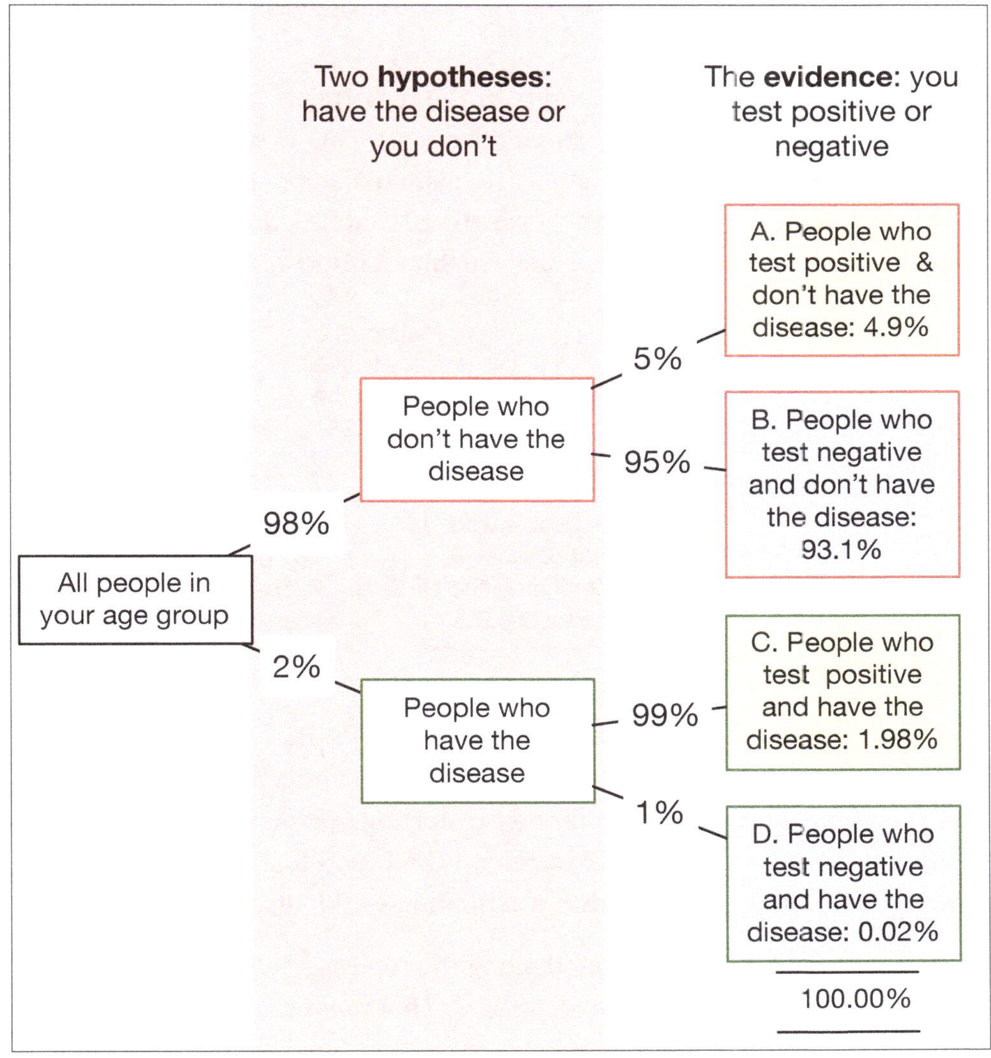

Figure 199

your age group, 4.9% will test positive even though they don't have the disease, whereas only 1.98% will test positive and have the disease.

So far so good, but now you want to know the probability that you don't really have the disease, having tested positive. You decide to call the choice between whether or not you are healthy two *hypotheses*, and then you call the results of the test the *evidence* that you have the disease (Figure 199).

We'll call the hypothesis that you don't have the disease H. So the probability that you don't have the disease is $Pr(H)$. The other hypothesis is that you do have the disease. We'll call that $\neg H$. (In probability, the $\neg$ symbol indicates that you are looking at the alternative to whatever follows it. So in this case, it is the alternative hypothesis.)

We'll call the evidence E, and so the probability that the evidence—the test—is positive will be $Pr(E)$.

You are interested in the probability of the hypothesis that you don't have the disease given the evidence (a positive test). This is written as $Pr(H|E)$. To find that, you need the probability that you tested positive but don't have the disease (box A in Figure 200), over the probability that you tested positive *whether or not* you have the disease (box A plus box C in Figure 200).

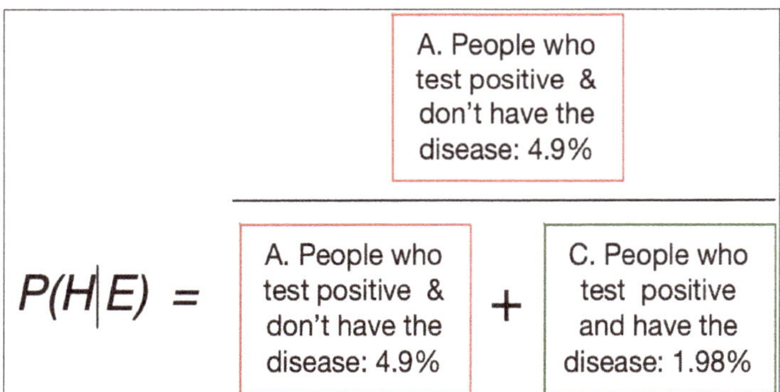

Figure 200

That gives you a probability of approximately 71% that you *don't* have the disease. This is counterintuitive given that the test only has a 5% false-positive rate, but it is correct. So rather than starting the risky and inconvenient treatment you ask your doctor to re-run the blood test. She runs it twice more. Sure enough, it shows that you don't have the disease.

To get to that point, we multiplied along the top row (98% • 5%) to get box A, and multiplied 2% by 99% to get box C. That means:

$$Pr(H|E) = \frac{5\% \bullet 98\%}{(5\% \bullet 98\%) + (2\% \bullet 99\%)}$$

$$= \frac{4.9\%}{4.9\% + 1.98\%}$$

$$\approx 71\%$$

In more general terms, the 98% is $Pr(H)$: the probability that you don't have the disease. The 5% is $Pr(E|H)$: the probability that you test positive (the evidence) given the hypothesis: that you don't have the disease. The 2% is the alternative hypothesis: that you do have the disease. And the 99% is the evidence given the alternative hypothesis; that you do have the disease. That can be written as:

$$Pr(H|E) = \frac{Pr(E|H)Pr(H)}{Pr(E|H)Pr(H) + Pr(E|\neg H)Pr(\neg H)}$$

Now box A plus box C is just the total evidence, whether or not you have the disease. It is the probability of testing positive, regardless. So it is just equal to $Pr(E)$, or the probability of the evidence.

That means we can write our formula as:

$$Pr(H|E) = \frac{P(E|H)P(H)}{Pr(E)}$$

You will note that this is just a restatement of Bayes's rule, replacing A and B with H and E.

Initially it looked as though you only had one chance in 20, or 5% chance of *not* having the disease. Once we added more information, however, the probability changed. Your probability of not having the disease shot up from 5% to about 71%, because we factored in more information. If *less* than 1 in 50 people had the disease, or if the test was *less* accurate, your probability of *not* having the disease would be even higher.

If that isn't what you would have intuitively thought, you aren't alone. A study in 2007 looked at how well doctors were advising patients about their risks of having breast cancer if they received a positive mammogram. They were provided with these three pieces of hypothetical information:

1) the probability that a woman in a certain region has breast cancer is 1%.

2) The test is 90% *sensitive*, meaning that if a woman has breast cancer the probability that she tests positive is 90%.

3) The test is 91% *specific*, meaning that if she does not have cancer the false-positive rate is 9%.

Based on this, if a woman has a positive mammogram, which is the best answer?

A. The probability that she has breast cancer is about 81%.

B. Out of 10 women with a positive mammogram, about 9 have breast cancer.

C. Out of 10 women with a positive mammogram, about 1 has breast cancer.

D. The probability that she has breast cancer is about 1%.

Only 21% of respondents indicated the correct answer. This is worse than you would have expected if they had answered at random[49]. Looking at a collection of numbers like this can get a bit overwhelming. Trying to remember whether to apply the addition rule or the multiplication rule or neither can also be quite challenging.

If you draw up a tree diagram and multiply along the branches, though, you can map it out quite easily. Note that this is an example of conditional probability: whether they fit into one category affects whether they fit into a different one.

If you do this (Figure 201) you find that (0.9% + 8.91%) = 9.81% will test positive, but only 0.9% of those will have cancer. In other words, for every 100 women presenting for testing, about 1 will have cancer but about 10 will test positive. There will be about a one in ten chance of having cancer with a positive mammogram. (Please note that these figures were hypothetical for the study and not based on the true percentages.) The correct answer was C.

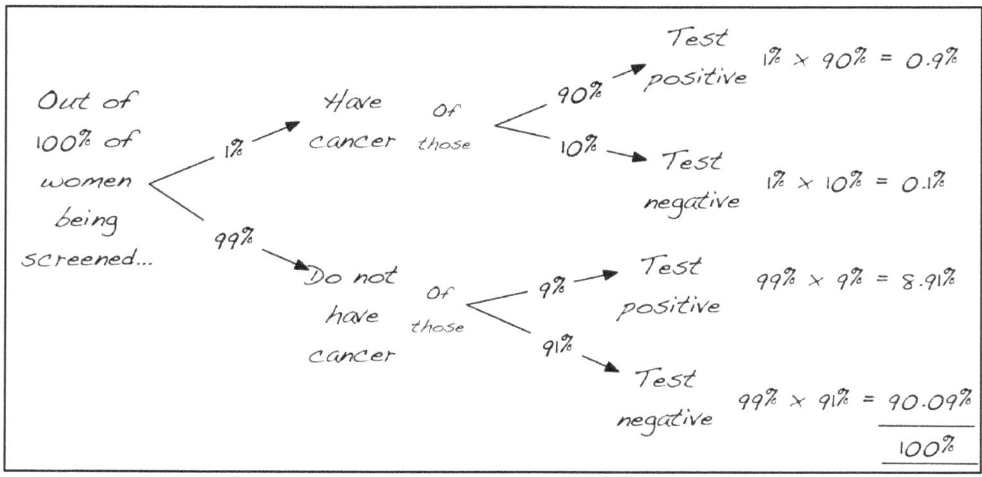

Figure 201

Here is a different view of the same issue, which might make it easier to understand. Suppose you attend a conference on statistics and psychology, and you learn that everyone there is either a statistician or a psychologist. You start talking to someone in the break, you find them really easy to talk to. Before you know it you are talking about something which has been stressing you. They seem incredibly interested and empathetic, and talking the problem through with them is very helpful.

You don't want to make a premature judgement about a group of people, though, so you arbitrarily estimate that there might be a 60% chance that they are a psychologist (and a good one), and there is a 40% chance that they are an extremely empathetic statistician. Before you can ask, you are called back into the next session where the compere announces that of the 1,000 people present, 900 are statisticians and 100 are psychologists. You realise that if there is a 40% chance that you were talking to an empathetic statistician, that means there are 360 people to choose from, whereas if you are talking to an empathetic psychologist there are only 60 to choose from. The ratio of empathetic psychologists to empathetic statisticians at the conference is 60:360, which is equal to 1:6. There was 6 times as much chance that you were talking to a statistician. (And of course, if you bothered to do that calculation it is almost certain that you are a statistician yourself).

To be fair to Thomas Bayes, this underlying idea of proportions changing and narrowing down your belief in the truth was at least suggested in the first part of his paper.

The law of total probability

Suppose we have a situation where there are four, and only four, possibilities, with no overlap between them, which we will call $B1$, $B2$, $B3$ and $B4$. They could be anything, but to make it more concrete we can think about the four possibilities relating to cancer diagnosis that we just looked at: they have cancer and test positive, they have cancer and test negative, they don't have cancer and test positive, or they don't have cancer and test negative.

Now let's consider another variable that could potentially overly any of these possibilities. We'll call it A. Again, it could be anything, but to be more concrete we can think about the possibility that the person smoked prior to the age of 21. We can represent these alternatives using Figure 202. You can see from this that, because the probabilities starting with B cover all of the

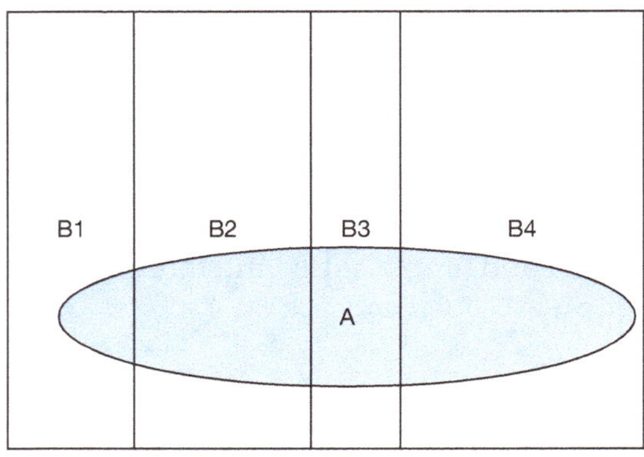

Figure 202

options and there is no overlap between them, the value of *A* is the same as the sum of all of the intersections between A and the other probabilities. In other words:

$$Pr(A) = Pr(A \cap B1) + Pr(A \cap B2)$$
$$+ Pr(A \cap B3)$$
$$+ Pr(A \cap B3)$$

Of course, it doesn't have to be four other possibilities that A intersects with; it could be any number. So if the number of intersections is *n,* we can write this as:

$$Pr(A) = Pr(A \cap B1) + Pr(A \cap B2) + \cdots + Pr(A \cap Bn)$$

This is known as the **law of total probability**.

We are talking about adding probabilities, but we could equally have described the total of $Pr(A)$ as the union of each of the areas that represents an intersection:

$$Pr(A) = Pr(A \cap B1) \cup Pr(A \cap B2) \cup \dots \cup Pr(A \cap Bn)$$

When we looked at Bayes's theorem and the Venn diagram of the probability of being above average in height and having a tattoo on the forehead (page 417) we saw that there is a third way to describe the intersection, using conditional probability:

$$Pr(A \cap B) = Pr(A|B)Pr(B)$$

That means we can also express the law of total probability, without using set notation, as:

$$Pr(A) = Pr(A|B1)Pr(B1) + Pr(A|B2)Pr(B2) + \cdots + Pr(A|Bn)Pr(Bn)$$

Because this last expression of the law involves multiplying and adding numbers, it tends to be the more useful version.

To understand what this can mean in practice, imagine that you are marketing manager who uses direct marketing.

You know that, in general, when you send messages to existing customers you get a 10% response rate. You have decided to target a new demographic. You are unsure of whether to use your existing messaging for the new group of potential customers, or to try something different.

The first thing you want is a benchmark, to see how well a sample of people in the new demographic respond to the existing messaging.

So you run a direct marketing campaign with 70% of the recipients who are existing customers, and 30% who are potential new customers. The overall response rate is 7%. What you want to know, without having to count the individual responses is, what was the percentage of responses from potential new customers?

Putting that in terms of the law of total probability, we have:

$$\Pr(response) = \Pr(response|existing)\,\Pr(existing) \\ + \Pr(response|potential)\,\Pr(potential)$$

In terms of the information above, that translates to:

$$7\% = (10\%)\,(70\%) + Pr(response|potential)(30\%)$$

When you do the algebra to work that out, you find that

$$Pr(response|potential) = 0$$

You probably had no responses at all from people who weren't existing customers. That makes the decision easy: you need to try something different.

Key points:

- Thomas Bayes's paper introduced four new ideas: probability was more important than just gambling, you don't need to know the absolute possibility of events, just their relative probability, probability can change as more information is known, and probability can be used to understand an ongoing, stable process.

- This means that probability can be subjective, even though it is defined mathematically.

- $Pr(A\,|\,B)$ describes a **conditional probability**: the probability of A, given B.

- Bayes's theorem states that

$$Pr(A|B) = \frac{Pr(B|A)Pr(A)}{Pr(B)}$$

- Understanding and using Bayes's theorem can dramatically change the subjective probability of events for us, as more information is known.

- The law of total probability states that if you have a set of events that cover every possibility and where there is no overlap (call them $B_1, B_2 \dots B_n$) then the probability of something else occurring (call it A) can be expressed as the sum of the probabilities of A occurring, given each B_i, multiplied by the probability each of B_i itself:

$$Pr(A) = Pr(A|B_1)\,Pr(B_1) + Pr(A|B_2)\,Pr(B_2) + \cdots + Pr(A|B_n)\,Pr(B_n)$$

$$= Pr(A \cap B_1) + Pr(A \cap B_2) + \cdots + Pr(A \cap B_n)$$

$$= Pr(A \cap B_1) \cup Pr(A \cap B_2) \cup \dots \cup Pr(A \cap B_n)$$

Bayesian Statistics

Read this if you aren't familiar with the statistical use of the terms prior, hyperparameters, prior elicitation, likelihood or posterior. Also read it if you aren't clear about the difference between Bayesian and frequentist statistics, and the relative advantages of each approach.

In the examples of meeting someone at the airport, the medical test and the empathetic conference attendee, we were looking at probabilities as simple *ratios*. Laplace (and Bayesian statisticians coming after him) took things a step further. He thought of Bayes theorem as a way to express the relationships between probability *distributions*. Note that all statisticians, whether or not they adopt a Bayesian approach, accept Bayes's rule: it's just algebra. It's after this point that people take differing approaches.

Under this approach $Pr(H)$ (which we dealt with in the previous section) is called **a priori understanding**, or just the **prior**: what you think you know now, usually expressed as a probability distribution. Prior distributions could be anywhere on a continuum from no-idea-at-all to fairly certain, but they are sometimes given one of three categories: **diffuse** (or **non-informative**), **weakly informative** or **informative.** These judgements may themselves be based on the researcher's personal judgement. They may sometimes resemble one of the distributions we have looked at, such as the reverse binomial or the hypergeometric. The parameters of the prior distribution (such as the probability and the number of trials in a binomial distribution) are the **hyperparameters.**

Constructing a prior distribution is done by a process called **prior elicitation**. This may involve interviewing experts, literature reviews, analysis of earlier findings (including **meta-analysis**, which involves numerically combining the findings from a number of studies), reviewing pre-existing data or anything else, really.

You might, for example, believe that the probability of throwing a certain number of heads out of a certain number of coin tosses can be expressed by a binomial distribution. You might also believe that the coin you are tossing gives an equal probability of giving you a head or a tail.

$Pr(E|H)$ from the previous section is called the **likelihood:** how likely it is that the new evidence is true, given your prior hypothesis. Likelihood calculations have a different meaning in statistics to what they do in everyday conversation: they aren't the same as probability. We saw earlier that to create a binomial distribution you only need to know three things: the fact that it is a binomial distribution, and two **parameters**: the probability of success with each trial (p), and the number of trials (n). A probability calculation takes the parameters, and uses them to find the probability of different events. A likelihood calculation takes events—the data—and the generic formula for a distribution, and uses them to find the parameters that will create the distribution that will most-closely fit the data. (Likelihood functions are used in non-Bayesian as well as Bayesian statistics.)

In Bayesian statistics each value for a parameter would lead to a different result, and so if you were looking at a range of discrete values for the likelihood, there would be a separate instance of Bayes formula for each of them. These separate instances are, however, normally expressed in a single formula (an integral or a sum).

You may see the range of potential parameters in the likelihood referred to as θ, and the evidence referred to as y or just as "data". (And yes, that can be confusing when to this point θ has usually referred to an angle and y has usually represented the value of a function. Best to just go with it.)

The **maximum likelihood** is the highest point of a distribution of possible likelihoods. Where the formula forms a curve (we will see some probability functions that form smooth curves shortly), we can find this point by calculating the point where the tangent to the curve is flat. In other words, we can differentiate the formula then evaluate it where the differential equals zero, as we did when we calculated the minimum value of a quadratic equation (page 240).

$Pr(E)$ from the previous section is the probability of the **evidence** you obtain by undertaking research. You might throw a coin 1,000 times and find out what the actual numbers came to. The likelihood ($Pr(E|H)$ is a conditional probability, and the evidence is not conditional. Because it is not conditioned on the hypothesis, it is a marginal probability. That means it can be obtained by integrating with respect to, or summing over all values of, the range of potential parameters. (Refer back to page 404 if this idea seems unfamiliar.)

In practice, $Pr(E)$ is normally just treated as a normalising constant to make sure that the posterior distribution has an area of 1. It is often ignored in Bayesian statistics, because different instances of Bayes formula, for different hypothetical parameters, are involved. When these are expressed as fractions, the denominators of each don't change and can be cancelled out.

$$\frac{P(H|E)_1}{P(H|E)_2} = \frac{\frac{P(E|H)_1 P(H)}{\cancel{Pr(E)}}}{\frac{P(E|H)_2 P(H)}{\cancel{Pr(E)}}}$$

The relationship between the prior distribution, the likelihood function and the posterior distribution can expressed as a proportion, in which the prior multiplied by the likelihood function is proportional to the posterior distribution:

$$\boxed{Prior \bullet Likelihood \propto Posterior}$$

$Pr(H|E)$ is the **a posteriori,** or just the **posterior:** a new probability distribution, based on the new information you have obtained.

So the theorem can be expressed:

$$\boxed{Posterior = \frac{Prior \bullet Likelihood}{Evidence}}$$

Example of a Bayesian analysis

Imaging you want to know the prevalence of a particular toxin in a population of birds. At the start of your analysis you have no idea whether it is present or not, so you have an uninformative prior: in this case, you decide it could be anything at all. The chances are completely even. You begin the process of trapping birds, and on the first day manage to test 4 of them. Tests can't detect the toxin in any of them.

Without using a Bayesian approach you would probably say that you now know that there isn't 100% contamination, but that you haven't tested enough birds to know what the true prevalence is. Using a Bayesian approach, however, you have at least enough understanding to update your ideas about the probability that a particular probability will apply. (This idea of a probability of a probability can be difficult to get your mind around; we'll explore it more shortly when we come to look at the beta distribution.)

To stay with a model we are familiar with, we'll treat this as a binomial situation: each bird either does or does not have the toxin.

We'll consider 11 possibilities. These represent the probability that, in the future, if we test any bird at random, it will be contaminated. If we knew the probability we would call it p as we have been doing, but the probability is itself an unknown, so we normally refer to it as theta (θ).

$$\theta \in \{0.0, 0.1, 0.2, 0.3, 0.4, 0.5. 0.6, 0.7, 0.8, 0.9, 1.0\}$$

We will use each of these potential probabilities in a table, based on the binomial theorem.

Once again, the formula for the binomial theorem for a single value of r is

$$\frac{n!}{r!\,(n-r)!} p^r (1-p)^{n-r}$$

We have tested 4 birds, so $n = 4$. We have 0 positive results, so $r = 0$. We will replace p with θ, because it is the probability that is the variable, and it could take on a range of values. That makes our formula for the likelihood

$$Pr(E|\theta) = Pr(0\ cases\ from\ 4\ birds)$$

$$= \frac{4!}{0!\,(4-0)!} \theta^0 (1-\theta)^{4-0}$$

$$= (1-\theta)^4$$

Value of θ	Likelihood $P(E \mid \theta) = (1-\theta)^4$
0.0	1.0000
0.1	0.6561
0.2	0.4096
0.3	0.2401
0.4	0.1296
0.5	0.0625
0.6	0.0256
0.7	0.0081
0.8	0.0016
0.9	0.0001
1.0	0.0000

Figure 203

Or, thinking of this a different way, the probability of contamination of a single bird is (the unknown) θ. Therefore, the probability of a single bird *not* being contaminated is $1 - \theta$. This has occurred 4 times, and the multiplication rule applies, so we multiply this probability by itself 4 times, or raise it to the 4th power: $(1-\theta)^4$.

Figure 203 gives a table of likelihoods based on the evidence. You can see from it that a likelihood is not a probability; the likelihoods don't add up to one. You can also see that the maximum likelihood it found is $\theta = 0$, meaning that, if there are no birds in our sample of 4 that are contaminated, the most likely scenario is a probability of 0; no birds are contaminated. There is no likelihood at all that there is a probability of 1, which would mean all the birds are contaminated. That couldn't be the case if we have found 4 that are not. We aren't finished yet, though.

We started with a prior assumption that there was an equal chance any probability could apply to our chances that a given bird we selected would be contaminated. That gives us a **uniform distribution**. We are considering 11 possible values, they are all equal and they must all add to one, so each of them is approximately 0.0909 …

To get the numerator we multiply by that number. To get the denominator we sum up all of the possible likelihoods (giving 0.126665). This gives us the distribution of posterior values and, as you can see in Figure 204, this is a probability distribution, and it does add up to 1.

Value of θ	Likelihood $P(E \mid \theta)$ $= (1 - \theta)^4$	Prior	Product of prior and likelihood (numerator)	Denominator (sum of all products)	Posterior (numerator over denominator)
0.0	1.0000	0.0909	0.0909	0.2303	0.3947
0.1	0.6561	0.0909	0.0596	0.2303	0.2590
0.2	0.4096	0.0909	0.0372	0.2303	0.1617
0.3	0.2401	0.0909	0.0218	0.2303	0.0948
0.4	0.1296	0.0909	0.0118	0.2303	0.0512
0.5	0.0625	0.0909	0.0057	0.2303	0.0247
0.6	0.0256	0.0909	0.0023	0.2303	0.0101
0.7	0.0081	0.0909	0.0007	0.2303	0.0032
0.8	0.0016	0.0909	0.0001	0.2303	0.0006
0.9	0.0001	0.0909	0.0000	0.2303	0.0000
1.0	0.0000	0.0909	0.0000	0.2303	0.0000
Sum			**0.2303**		**1**

Figure 204

So now, even with a sample of just four birds, we have been able develop a new probability distribution. We've moved from having no idea what the probability will be that any given bird will be contaminated, to a 39% chance that there is no probability at all ($\theta = 0$). We now think there is only a 2% chance that half of any birds sampled will be contaminated ($\theta = 0.5$).

The more we sample, the less influential the prior will be over the posterior. So even though we started with a completely open mind, if we kept testing for 2 more days, collecting just 12 specimens altogether (we calculate this by changing the likelihood calculation to $(1 - \theta)^{12}$), if none of them were positive for toxins we would be able to say that there was a 73% chance that no bird we sampled would be contaminated, a 21% chance that one in ten would be, a 5% chance that 2 in 10 would be, a 1% chance that 3 in 10 would be, and that there would be virtually no chance that any more than that would be.

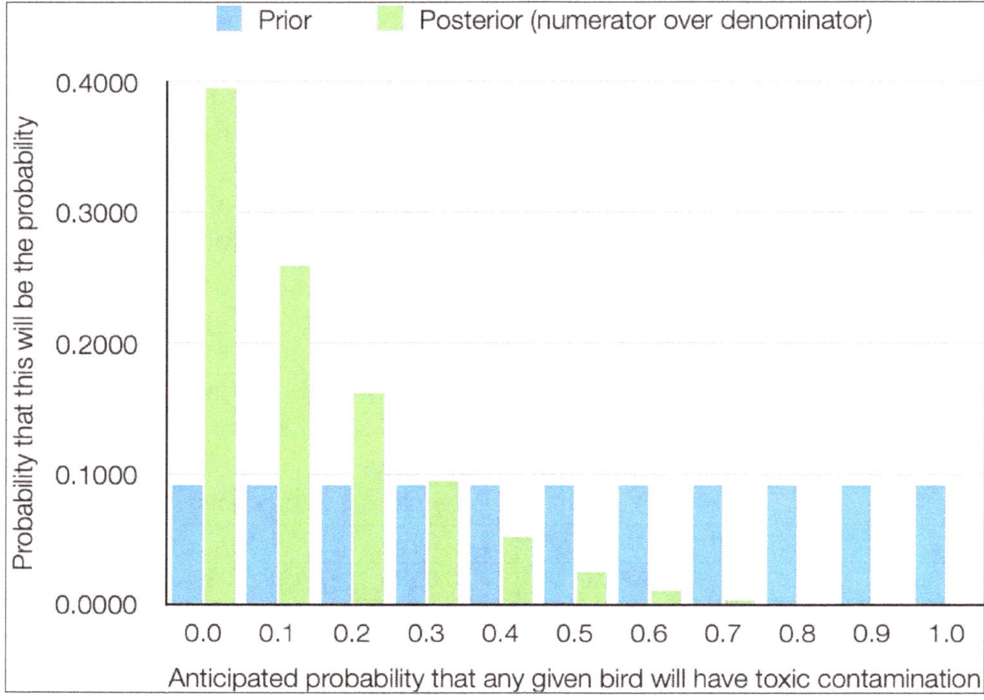

Figure 205

Figure 205 graphs our prior (uniform) and posterior distributions, showing how much the probabilities change based on sampling 4 birds that aren't contaminated.

We used a binomial distribution to develop this table, and other distributions are more commonly used with Bayesian statistics. We will see some of these shortly and will look at how different distributions can be used together when we come to consider **conjugate priors.**

> **To cement your understanding**
>
> - Copy the table at Figure 204 into a spreadsheet, replacing the calculated values with formulae as described above.
>
> - Then play with different options (different sample sizes; different values for the prior) to see what is the effect of these changes on the posterior distribution.
>
> - Use the binomial theorem to come up with different data for your likelihood calculations, based on finding one or more birds that test positive for toxicity.

- Copy the data (not the formulae) from the posterior column of your spreadsheet and paste it into the prior column. This would be your new prior distribution for the second day of your research if you chose to update it.

Of course a lot could change when you are out in the field doing research, and this could affect the amount of research you had to do. The following day you might trap four more birds and find that 2 were contaminated. That would change your likelihood column. You might also realise that the assumption of independence that underlies the selection of a binomial distribution could be flawed. (If you find one contaminated bird that might, in fact, increase your chances of finding a second one, because they might both be close to a contamination source.) In that case you might have to subdivide your search area and develop different models for each area.

This is actually an area where the use of a Bayesian approach shines. It allows you to modify your approach on the run as more information comes to hand, without invalidating the research you have undertaken to that point. There are other situations where it works well, and others where it may not be the best approach.

Bayesian vs Frequentist statistics

There can be disagreement between Bayesians themselves, as well as between Bayesians and non-Bayesians, about how to put this into practice. One 2-page article was titled *46,656 Varieties of Bayesians,* based on the combinations available from 11 different issues on which the author believed that Bayesians disagreed with one another[50].

Some of the difficulties lie with the complexities of the calculations. Only some forms of prior distribution are usable with likelihood calculations to get other forms of distribution. This leads to the widespread but suspect practice of revisiting and reworking the prior distribution based on the findings of the evidence in order to come up with something that would lend itself to being computed. For this reason, while Laplace's approach was probably the earliest form of statistics as we would recognise it, Bayesian statistics went out of fashion. For a long time most undergraduate statistics courses didn't teach it. Bayesian statistics is now growing in popularity again. This is largely because modern computers can crunch through complex calculations. That makes it easier to let prior elicitation come up with what it comes up with and posterior distribution come up with whatever the evidence suggests, without having to go back and doctor the prior to make it fit so that calculations can work.

Another common framework, known as **frequentist statistics**, evaluates hypotheses by examining how likely observed data are under specific assumptions. It begins with formulating something called a **null hypothesis** (H_0). This assumes there is no effect or no difference—essentially, that what you're studying has no real impact. Alongside this, you define an **alternative hypothesis** (H_1), which represents the possibility that there is an effect or difference.

The frequentist approach does not aim to prove a hypothesis true or false. Instead, it uses the observed data to assess the plausibility of the null hypothesis. By calculating a **p-value**, researchers estimate the probability of observing data as extreme as (or more extreme than) what was collected, assuming the null hypothesis is true.

> The use of the negative language of rejection to indicate a positive result and vice versa within frequentist statistics can be confusing when you first encounter it. The idea, at least, is that it is easy to get carried away by your enthusiasm to prove something. In order to counter that and be truly scientific, it is best to try to prove that what you are looking at doesn't actually do anything. If you *fail* in your attempt to prove that it does *nothing*, then you have a more trustworthy result. A Bayesian approach deals with the issue of bias differently: it starts by acknowledging it as part of the prior and then gets to the point of saying that all that has changed is a subjective perception of probability. Or at least, once again, that's the theory.

- If this probability is very low (typically below a predefined threshold like 0.05), the data are considered inconsistent with the null hypothesis, and researchers may reject H_0. This suggests, though does not prove, that the alternative hypothesis might be more plausible.

- If the probability is not sufficiently low, researchers fail to reject H_0, meaning there isn't strong enough evidence to favour the alternative.

Frequentist research sees results as part of an ongoing inquiry; rejecting the null hypothesis usually leads to further experimentation and hypothesis refinement, rather than definitive conclusions.

There is a lot of talk about a divide among statisticians between Bayesians and frequentists. These two approaches certainly each have a different philosophical underpinning and practical outworking, and it is possible for them to arrive at dramatically different conclusions; as we saw in the example of the medical test. The word "frequentist" was brought to prominence in 1977 in an article by Jerzy Neyman (1894-1981)[51]. Neyman developed the **confidence interval** (page 635) which is a component of most modern frequentist statistical analysis. In this article, Neyman describes a 3-step research process which actually reads a lot like a description of

Bayesian statistics. In other words, while there are some statisticians who would place themselves firmly in one camp or the other, it is probably more accurate to think of Bayesian and frequentist statistics as different perspectives and approaches than different camps; to recognise that each approach has its advantages; and to be able to select the most appropriate approach for whatever the uncertainty is that you are trying to understand.

Advantages of a frequentist approach	Advantages of a Bayesian approach
Doesn't assume that prior hypotheses are necessarily valuable. Allows for the possibility of completely rejecting old ideas on the basis of new scientific breakthroughs.	Recognises that new information doesn't emerge in a vacuum; it builds on what has gone before.
Simpler to understand and implement. This makes it more applicable to the sort of specific, narrow questions that are usually addressed in research journal articles. Much to the chagrin of really committed Bayesians, most scientific research written in journals is, in fact, based on more of a frequentist than a Bayesian perspective. The calculations behind Bayesian statistics can become quite challenging; especially for researchers who know how to use statistics software but don't really understand the mathematical principles behind statistics.	Attempts to engage with the real-world complexity of gaining new understanding.
It's easier to sell the idea that "this is what we learned through rigorous research, but more research is needed" than to try to sell the idea that "this is what we personally think is probable based on our (stated) assumptions and the data we have gathered so far, but the next time we look at this we will probably have a different set of underlying assumptions, so who knows?"	While it may not be easy to *sell* the idea that research findings are based on prior assumptions that are likely to change over time (or at least it is challenging to sell that to unsophisticated consumers of research) it is, in fact, usually more accurate to acknowledge this is the case, regardless of whether the research is overtly Bayesian. Given that that is the case, it will often be preferable to explicitly acknowledge that reality, and to build it into the research.

Advantages of a frequentist approach	Advantages of a Bayesian approach
Seeks to find objective truth, without being overly swayed by subjective perceptions. This is particularly relevant where you don't have information and *know* you don't have it: you have a completely open mind. In this instance it seems unnecessarily purist to simply invent a "non-informative" prior distribution in order to be able to use a Bayesian approach.	Recognises that there is no such thing as purely objective science. Seeks to bring prior beliefs into the open as part of the research process, where others can identify and critique them.
Defining probability in terms of what will occur if something is repeated with great frequency is more credible than defining it in terms of what the researcher subjectively believes to be true. There is an inherent danger in creating a complex mathematical overlay that is ultimately based on subjective opinion, because the degree of subjectivity may be obscured by the complexity of the analysis.	
It may be easier to get funding by promising to provide an answer than by promising to explore an issue.	It's usually more honest to promise to explore an issue than promising to deliver a definitive answer.
	Recognises that moving towards the truth is an ongoing journey of continuous learning rather than a destination.

Statisticians seem to be moving in the direction of starting with the research question rather than starting with the statistical approach.

In general, it's probably best to use a frequentist approach when:

Research characteristic	Reason	Example
Longitudinal research experiments when nothing is supposed to change.	Frequentist methods shine when the problem involves repeated sampling to ensure that nothing is going off track.	Manufacturing quality control, where the same process is repeated many times.
Limited prior information.	Frequentist methods avoid subjective priors, making them appealing when no strong prior knowledge exists.	Clinical trials for a new drug, where you cannot assume prior probabilities for outcomes.
Simple, well-defined hypotheses.	Works well with well-established null and alternative formulations.	Determining whether a coin is fair using a fixed threshold for the p-value.
Standardized testing.	Frequentist approaches are often used in regulatory or standardized environments where methodologies are pre-defined.	Approval processes often rely on frequentist analyses.
You have a specific hypothesis you want to test.	A frequent approach is set up for specificity and clarity. You know what the question is, and you want the clearest answer that statistics can provide. You can force a Bayesian approach to test a hypothesis, but it isn't a natural fit for the approach.	Finding out whether a change to a process materially affects the quality of the result.
You want to know a specific probability.	This approach will tell you whether a specific outcome is likely to have been achieved by random chance.	Determining whether a new product is effective.

In general it is probably best to use a Bayesian approach when:

Research characteristic	Reason	Example
Longitudinal research when change is anticipated.	Bayesian methods are ideal for problems requiring updates as new data arrives.	A repeated survey looking into the same issues, as time passes.
Incorporating prior information.	When prior knowledge, expert opinions, or historical data are available and relevant, and you want to be able to make use of what has already been learned.	Fine-tuning an approach that is already successful, testing incremental changes to assess the result.
Complex models or hierarchical structures.	Bayesian frameworks handle complex dependencies and hierarchies more naturally than frequentist methods.	Predicting outcomes in multi-level data, like student performance across schools.
Situations where subjective belief is important to the research.	Bayesian statistics are based on the idea that probability is about subjective beliefs.	What do people expect the economy to do?
Field research where there is a high degree of uncertainty about what you will encounter in the research process itself.	Bayesian methods let you modify your modelling on the fly, rather than having to commit ahead of time to a sampling approach.	The sort of research that we looked at, going out and measuring toxicity levels in birds in the wild.
You *don't* have a specific hypothesis to test; you have an open mind, and want to update your understanding of what is probable, based on more data.	This is what Bayesian statistics is designed for. You can force a frequentist approach to provide exploratory data, but that isn't its forte.	Finding out what a change to a manufacturing process will do to change the pattern of probable manufacturing errors.

Research characteristic	Reason	Example
You want to update a probability distribution with more helpful parameters.	The output of a Bayesian approach is a probability distribution, not a probability.	You know there will be benefits and harms n a new treatment, and you want a probability distribution of each, so that risks versus benefits can be carefully assessed.

One of the areas where a Bayesian approach is coming into its own is machine learning: an aspect of artificial intelligence that enables systems to automatically learn patterns from data and improve their performance on a task. Equations that reflect this sort of approach are used to allow computers to continuously update their understanding based on the acquisition of additional data. This is being used for everything from teaching computers to recognise cancerous skin lesions, to identifying traffic patterns on your GPS.

Figure 206
*Voltaire
(François-Marie
Arouet)*
1694–1778

In terms of human learning, perhaps Voltaire (1694–1778) said it best:

> *"Uncertainty is an uncomfortable position. But certainty is an absurd one."*

(Voltaire was not a statistician but was a contemporary of Bayes. It is unlikely they would have met. As a Presbyterian Bayes would probably not have agreed with Voltaire, who was an agnostic. But Voltaire's approach is a logical extension of modern Bayesian thinking; and, in fact, of statistics in general.)

Key points:

- Under a **frequentist** statistical approach, the **null hypothesis** is the hypothesis that the outcome from an intervention will be achieved by random chance.

- In research papers the null hypothesis may be indicated as H_0 and the alternate hypothesis is H_A or possibly as H_1.

- If we accept that an intervention creates a statistically significant outcome, that means we reject the null hypothesis.

- **Bayesian** statistics uses the basic thinking behind Bayes's theorem to start with a **prior distribution** rather than a null hypothesis. It then looks at evidence, and how that alters the prior distribution, to reach a **posterior distribution.**

- Bayesian statistics treats probability as **subjective** and based on the information available to the observer.

- It is probably most helpful to think of frequentist and Bayesian approaches as different emphases that are appropriate to different research requirements, rather than as irreconcilable camps.

More distributions

Read this if you don't know the difference between the beta, Poisson, exponential or gamma distributions or when to use them, you didn't realise you could graph a CDF, you assumed there couldn't be two different things called gamma distributions that will give different results on your computer, you don't know why Bayesians would love beta distributions, you're confused about the difference between beta and gamma distributions and functions, or you want to expand your views about what statistics could be useful for.

Beta distribution

One probability distribution that makes Bayesian statistics more workable is the remarkable **beta distribution**.

The beta distribution is sometimes described as a way of representing the **probability of probabilities**. Thinking back to our example of the basketball player's free-throw percentage (when we were looking at the binomial distribution on page 377), imagine that you were the coach running a selection trial. Different candidates will line up to take a range of shots, but the number of successful shots they make might not represent their ongoing average percentage. Their long-range average percentage will represent the probability that they will be successful at a free throw, and, as we saw, it can be represented as a probability distribution. How can you tell if one candidate is likely to be better than another over the long-run? How many shots should you ask each candidate to make, before you make your selection?

Each candidate's set of throws will have a binomial distribution (because there are only two options, success or a failure) where n is the number of throws, r is the number of times the ball goes through the hoop and p is the long-run probability that they will score from a shot. As we saw, the probability distribution of a candidate's shots will be given by the binomial probability density function (PDF):

$$\binom{n}{r} p^r (1-p)^{n-r}$$

When we looked at the binomial distribution, we assumed that the probability of success, p, was a given, and that the variables were n and r. Now as we did when we looked at Bayesian statistics, instead of using the number of successes as the random variable, we use the probability of success as the random variable. In other words, if three candidates score 6 out of 10 from free throws, then our r (number of successes) and our n (number of trials) are known (6 and 10 respectively). What we don't know is the p—the ongoing probability of success—for each player. This means we are calculating a **likelihood**; finding the probability as the variable, as discussed earlier.

Obviously there are an unlimited number of potential probabilities of success for any player, between 0 (absolutely no chance at all) and 1 (complete, 100% certainty of a successful score). The probability doesn't have to be a whole number. For the sake of clarifying the problem though, let's start by assuming that the (unknown) probabilities of success (scoring averages) for each of our players are the only possibilities. If we call these three unknown probabilities p_1, p_2, and p_3, then the probability distribution for the first candidate's scoring average can be described by the formula:

$$\frac{Distribution\ of\ 1st\ candidate's\ probability\ of\ success}{All\ possible\ distributions\ of\ probabilities\ of\ success}$$

That translates to:

$$\frac{\binom{n}{r} p_1{}^r (1-p_1)^{n-r}}{\binom{n}{r} p_1{}^r (1-p_1)^{n-r} + \binom{n}{r} p_2{}^r (1-p_2)^{n-r} + \binom{n}{r} p_3{}^r (1-p_3)^{n-r}}$$

We can divide the numerator and denominator by $\binom{n}{r}$ and it will disappear.

$$\frac{p_1{}^r (1-p_1)^{n-r}}{p_1{}^r (1-p_1)^{n-r} + p_2{}^r (1-p_2)^{n-r} + p_3{}^r (1-p_3)^{n-r}}$$

But of course, there aren't only three possible distributions; there is an infinite number. Where p is a probability it will be between 0 and 1

We can represent the sum of all possible values (the denominator) as an integration with respect to p, for values of p between 0 and 1. That means our formula will be:

$$\boxed{\frac{p^r (1-p)^{n-r}}{\int_0^1 p^r (1-p)^{n-r} \, dp}}$$

This denominator will be a constant, because the p will disappear when the definite integral is calculated. You may recall that this integral bears a resemblance to the beta function (page 310), whose formula is:

$$\int_0^1 t^{x-1}(1-t)^{y-1}dt$$

In fact, the formula

$$\int_0^1 p^r(1-p)^{n-r}dp = \beta(r+1, n-r+1)$$

The easiest way to write a formula for the PDF of the beta distribution is just calculate it based on :

$$\boxed{PDF\ Beta(a,b) = \frac{p^{a-1}(1-p)^{b-1}}{\beta(a,b)}}$$

where $a-1$ is the (known) number of successes, $b-1$ is the (known) number of trials minus the (known) number of successes, p is the random variable: the (unknown) probability of success, and $\beta(a,b)$ is the beta *function* (not to be confused with the beta *distribution*) of a and b, where a and b are any real numbers > 0.

You may recall that a beta *function* (page 310) can be expressed in terms of gamma functions:

$$\beta(a,b) = \frac{\Gamma(a)\Gamma(b)}{\Gamma(a+b)}$$

This means we can express the beta *distribution* as

$$\boxed{PDFBeta(a,b) = \frac{\Gamma(a+b)}{\Gamma(a)\Gamma(b)}x^{a-1}(1-x)^{b-1}}$$

(multiplying the numerator and denominator by $\Gamma(a+b)$). (We are now replacing p with x, to be more general.)

You may also recall that, for positive integers, the gamma function is equivalent to the factorial function:

$$\Gamma(n) = (n-1)!\ where\ n \in \mathbb{Z} > 0$$

This means that, where a and b are positive integers,

$$\boxed{PDFBeta(a,b) = \frac{(a+b-1)!}{(a-1)!\,(b-1)!}x^{a-1}(1-x)^{b-1}}$$

where $a - 1$ is the (known) number of successes (which we have been calling r), $b - 1$ is the (known) number of trials minus the (known) number of successes (which we have been calling $n - r$), x is the random variable (which we have been calling p: the unknown probability of success), and $\beta(a, b)$ is the beta *function*.

Back to the case of our candidates for the basketball team, if someone scores 6 from 10 throws, then

$$a = 6 + 1(i.e.\,7) \text{ and } b = 10 - 6 + 1(i.e.\,5)$$

The probability distribution graph will look like Figure 207. It resembles a bell-shaped curve centred around 0.6, and that isn't a co-incidence. (We shall see later why the probabilities of so many phenomena form a bell-shaped curve of this particular shape.)

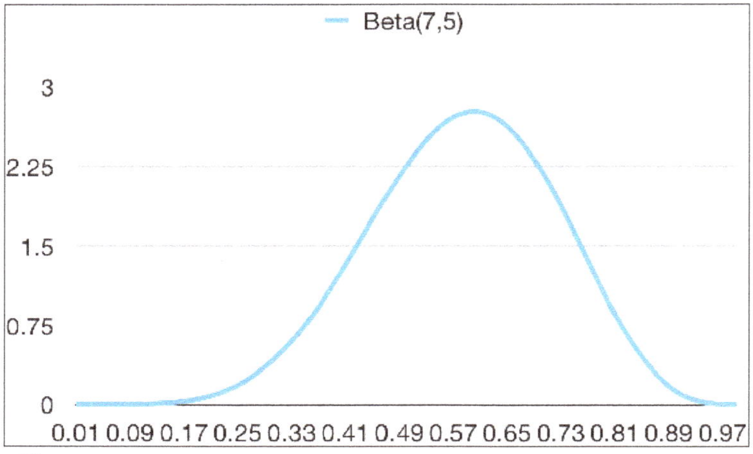

Figure 207

Now suppose that, instead of asking each player to shoot 10 times, the coach asked each player to shoot 100 hoops, and a number of them managed to score 60 times. That gives the same percentage of success, but this time the beta distribution will be

$$a = 60 + 1(i.e.\,61) \text{ and } b = 100 - 60 + 1(i.e.\,41)$$

Figure 208 compares beta(7,5) with beta(61,41). (The numbers for the blue line, beta(7,5) haven't changed; it looks lower than it does in Figure 207 because it is scaled differently). You can see that the distribution still shows a bell-shaped curve centred around 0.6. The area under the curve—the total probability—is still equal to 1. The difference is that the distribution with the larger numbers (or, in statistical terms, the larger sample size) has a narrower base. In other words, as you would expect, the long-run free-throw percentage of candidates in our trial is much more likely to be close to 60% when candidates have demonstrated the same result over a larger number of throws. Using a beta distribution allows you to calculate just how likely it is. That means a coach could use this analysis to work out, if there were a lot of candidates to assess, how many throws to ask each of them to make, before being confident of understanding their ongoing probable success rate.

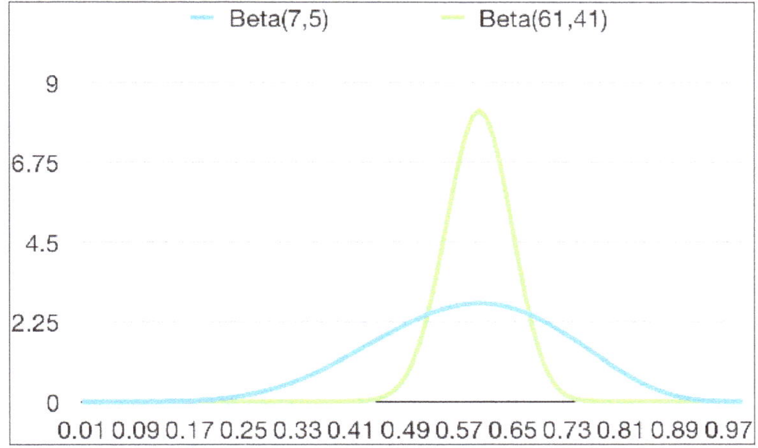

Figure 208

This is the first continuous distribution we have looked at. As we saw earlier (page 372), for continuous distributions we talk about the probability *density* function, or PDF, rather than the probability *mass* function. We also saw that, while it is customary (and normally easier to envision) to work with the PDF of continuous variables, the probability of something happening at an exact number is negligible, to the point of approaching zero. So it can be said that it has no mass. Rather, probabilities in continuous distributions are always determined by calculating the area under the curve between two values: in practice, we always use the cumulative distribution function (CDF).

The thing with the beta distribution that makes it particularly attractive for use in Bayesian statistical analysis, however, is that, depending on the values of *a* and *b,* it can mimic other distributions, not just a bell-shaped curve. (For this reason it is sometimes called a family of distributions, rather than a distribution.) This means it can be used to represent both the prior and the posterior distribution and, because it has the same basic structure, the calculations are manageable.

In Figure 209, for example, you can see that $Beta(1,1)$ represents a uniform distribution; all probabilities are equally likely (blue line). This is often used in situations where researchers have no pre-conceived idea of what the present situation is and still want to use a Bayesian analysis. *Beta*(0.5, 0.5) is U-shaped (orange line). This is called a **Jeffries prior**, and in some situations may be more appropriate than a uniform distribution for a completely unknown prior. *Beta*(2,1) and *beta*(1,2) are sloping lines (green and grey), while *beta*(7,1.7) (red line) is heavily skewed to the left (the tail is to the left, so the peak is to the right).

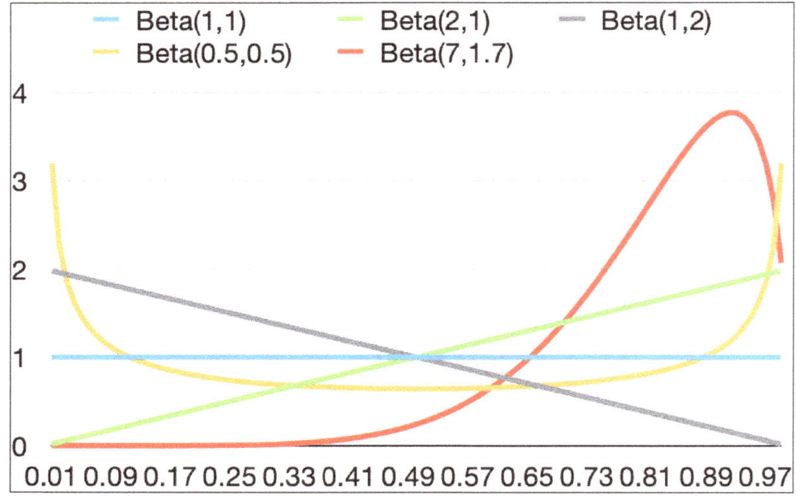

Figure 209

Now in graphs of the PDF like this, the value of the vertical axis isn't terribly helpful. Using $\beta(a, b)$ in the denominator acts as a **normalising constant**; in other words, it ensures that the cumulative value will add up to exactly 1. But that doesn't help in terms of the vertical axis of the PDF. The beta distribution is supposed to give a probability of probabilities, which should mean that the height of the curve could not be greater than 1, but it usually is. For this reason, while it is useful to graph the PDF to gain an understanding of the shape of the distribution, when the beta distribution is used in practice it is almost always done using the CDF. This is so much the case that some spreadsheets don't have a function for the beta PDF, only the CDF.

To calculate the CDF, we need to integrate the PDF. The PDF has a beta function in the numerator, and that is already an integral, but it is a definite integral which means we can treat it as a constant (if a and b are constant). We can therefore take it outside of the integration. That gives us:

$$\frac{1}{\beta(a, b)} \int_0^1 x^{a-1}(1-x)^{b-1} dx$$

The problem here is that $\int_0^1 x^{a-1}(1-x)^{b-1}$ is the function $\beta(a,b)$. So what we have is

$$\frac{\beta(a,b)}{\beta(a,b)}$$

That does, at least, prove that the integral is equal to 1, but it doesn't help us in practical terms. The way around this is to use a variable for the top limit of the integration, which can be anywhere between 0 and 1:

$$CDF = \frac{1}{B(a,b)} \int_0^{x=z} x^{a-1}(1-x)^{b-1} dx \, for 0 \leq z \leq 1$$

We call $\int_0^{x=z} x^{a-1}(1-x)^{b-1} dx$ the **incomplete beta function**, and it is written as $\beta(z; a, b)$. The CDF can therefore also be written as:

$$CDF = \frac{\beta(z; a, b)}{\beta(a,b)} \, for 0 \leq z \leq 1$$

To this point we have seen graphs of probability mass functions/probability density functions (PMFs/PDF), where the CDF represents the area under the curve. But it is equally possible to graph CDFs. Figure 210 graphs the CDFs for each of the beta distributions that we viewed above. You will see that the curves always start at zero and end at one, on both the horizontal and vertical axes, as we would expect from something graphing the probability of probabilities. The curves always increase within that range, because they are showing the accumulated probabilities. It isn't as

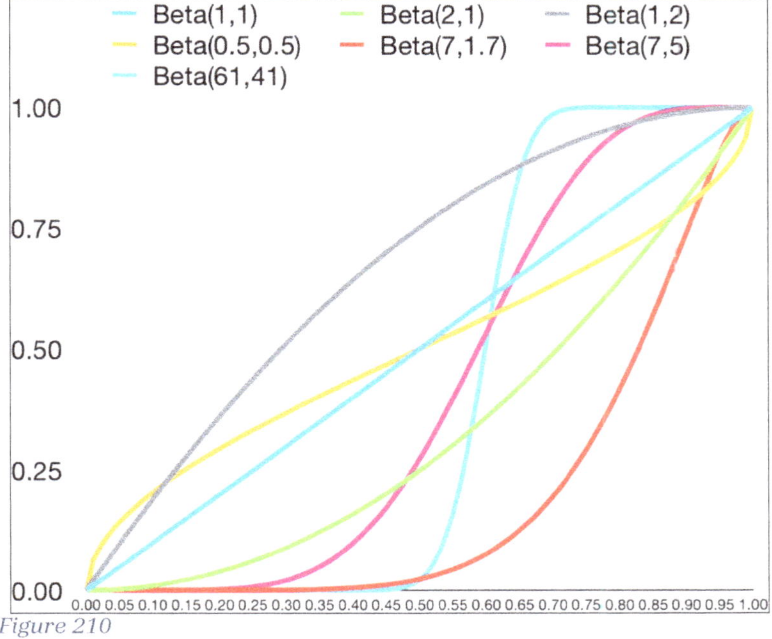

Figure 210

immediately obvious what the shapes of the distributions are in this sort of graph as it is with a graph of the PDF, but if you compare Figure 210 with Figure 208 and with Figure 209, you will be able to understand how they relate.

It isn't always necessary to use a beta distribution in Bayesian analysis, but it is the most commonly used approach.

The Poisson distribution

We've seen that three of the four new paradigms to come out of Bayes's essay—an expanded view of the potential role of probability, the idea that that you don't need to know the absolute possibility of events, just their relative probability, and the idea that probability can change as more information becomes known—were picked up by Laplace and ultimately led to Bayesian statistics and a modern understanding of statistics generally. The other key idea—that binomial thinking can be used to predict future possibilities within a stable but random process—was also picked up by Laplace and used in an analysis of the probability that the sun will rise tomorrow, among other things.

Figure 211
Siméon Denis Poisson
1781-1840

It was a student of his, Poisson, who was able to take this idea into a new, productive and practical direction.

Siméon Denis Poisson (1781-1840) was a French professor of mathematics who was mentored by Laplace (and others). He spent much of his time and energies on the application of mathematics to physics, making significant contributions in the understanding of electricity, magnetism, analytical mechanics, thermodynamics and understanding of fluid flows. Almost as an aside in one of his papers, and in relation to these sorts of physical problems, he developed what has now become known as the **Poisson probability distribution**.

The power of the Poisson distribution is that it enables us to understand the probability of randomly fluctuating but stable processes (which was the idea that Thomas Bayes was reaching for). This is used to understand the probability of a particular number of events occurring within a given period (for example, calls to a call centre) and also the probability of events within a physical area (for example radiation exposure to adjacent areas in cancer treatment). It can therefore help us to guess whether a fluctuation we are seeing is really random (Figure 212).

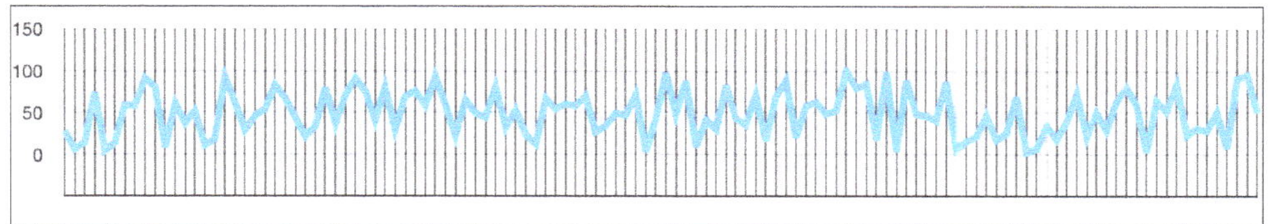

Figure 212

Poisson realised that having a stable process like this will let you work out the average number of events that will occur within a given time period or physical area, but he also knew that the actual events would be spread around this average in a random way. Ideally he would have been able to express the probabilities of a given number of events in a probability-density graph like the ones we have seen for binomial distributions; something like Figure 213.

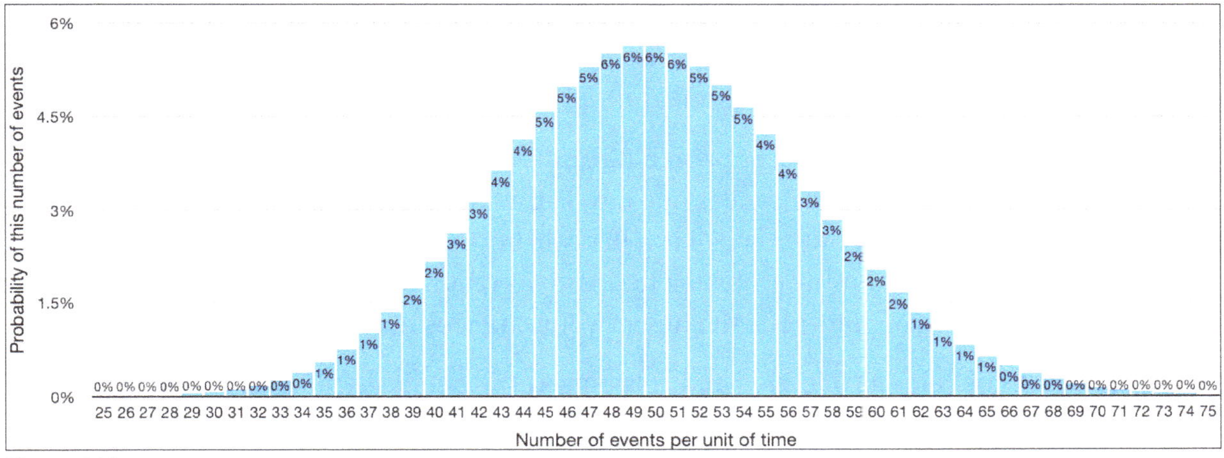

Figure 213

A probability distribution like this wouldn't just tell you that you only had a 6% chance of landing exactly on the average of 50 events (in this particular instance). You could use a cumulative distribution function (CDF) to see that you only had a 4% chance of having more than 65 events in this period. (That won't be obvious from the figures on the graph, because they are rounded to the nearest whole number to save space.) If we are talking about an ongoing process, that isn't extreme. It's about 1 in 25, so you could expect it to happen on average once in every 25 time periods. But the chance of it happening for two periods in a row is 1 in 400, while the chance of it happening 3 times in a row is 1 in 8,000 (by the multiplication rule). Depending on the length of the period, that might or might not make it important. If the time period is days, that would make it a once in 22-year event if the spread was truly random.

The more you think about this issue, the more applications you are likely to imagine for a statistical tool that could achieve this. As just three examples:

1) You make a change to a website, and over the next hour the number of customers who spend more than 10 seconds on a page goes down. Should you change it back again? How long should you keep waiting to see if there is a trend, and at what point do you have enough information to know that you have a problem? The sooner you make the right choice, the more successful your web site is likely to be.

2) You are running a café and you realise that the number of coffees you sell each day seems to fluctuate a lot, and so it is hard to get the staffing levels right. Are the variations in coffee sales truly random, or are they improbably large? If the swings are too wide to be random, there are factors that you have yet to understand. That would tell you that it may be worthwhile looking for the unseen drivers of the fluctuations, so that you could staff your cafe more appropriately.

3) You graph the number of people diagnosed with a disease each week, and there is a higher number one day. Is that higher number likely to be the start of a new outbreak, or is it likely to be just a random fluctuation? You could wait for the graph to change visibly, but if it is a new outbreak, the sooner you intervene the higher your chances of saving lives.

Poisson considered this idea and recognised the similarity of a graph like this with a binomial distribution. He thought about the problem in this way. Suppose we have an average number of events per period. We'll call that average λ (the Greek letter lambda). Now suppose we divide this time interval up into sub-intervals (n in number). For example, suppose that instead of looking at whole hours he looked at 15-minute subintervals, $n = 4$ (Figure 214).

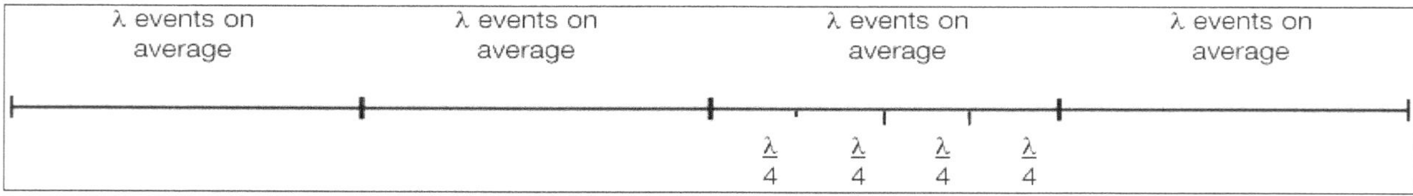

Figure 214

If we break our hour down into 15 minute periods, we say that there is only one quarter of the probability that it will happen in the subintervals as there is of it happening in the whole interval; our probability for a sub-intervals is

$$\frac{\lambda}{4}$$

or more generally if there are n subintervals, it is

$$\frac{\lambda}{n}$$

In each of these subintervals we have a chance of events happening or not happening. Now assume, for the sake of simplicity, that you can at most have one event in a sub-interval. That means whether you have something in an interval is a Bernoulli choice; yes or no, you do or you don't.

Just as with our example of tossing coins, there is only one way to have events in all four subintervals. There are four ways to have events in only three subintervals (with no event in any one of the four subintervals):

event, event, event, no event
event, event, no event, event
event, no event, event, event
no event, event, event, event

There are six ways to get events in two of the four subintervals:

event, event, no event, no event
event, no event, event, no event
no event, event, event, no event
event, no event, no event, event
no event, event, no event, event
no event, no event, event, event

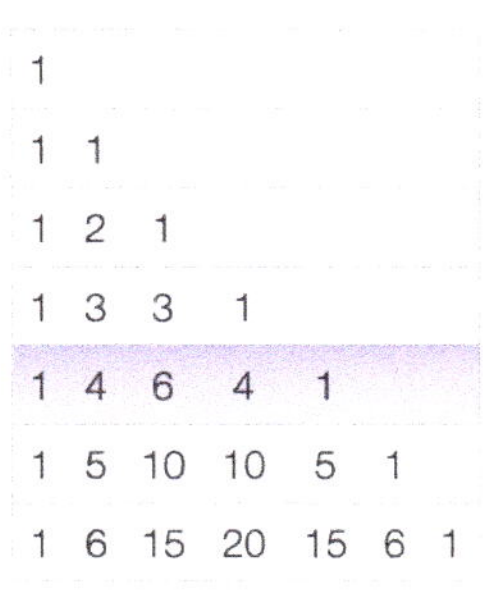

Figure 215

and so on. In other words, we can think about the problem as a binomial expansion, and we could have read it from Pascal's triangle (Figure 215).

There is still a problem, though, as you might have realised. What if you have more than one event in a subinterval (for example more than 1 customer in a 15-minute period)? The solution is to break the subinterval down into a shorter amount of time, and to put a huge number of them into a whole interval. We can do this by making n into a really large number. In the case of an hour, we could break it down into 60 minutes, or 3,600 seconds. Or fractions of seconds. Or, in fact, we should just keep going, so that the number of subintervals approaches infinity. Eventually we would get to the point where, whatever you were measuring, you could only possibly have either one or none in each subinterval: a binomial issue.

From then on, this just becomes a problem we can address through a binomial expansion. As a reminder, the formula is:

$$(p + q)^n = \sum_{r=0}^{n} \frac{n!}{r!\,(n - r)!} p^r q^{n-r}$$

where p is the probability of success (or having an event), q is the probability of failure (or not having an event), n is the number of trials (or sub-intervals) and r is the number of successes (or events).

<table>
<tr><td>

Note that the Poisson distribution is not the same as the binomial when n approaches infinity. We mentioned earlier when we were looking at de Moivre, and we will explore later in more detail, the limit of the binomial as n approaches infinity is the normal distribution. What makes the Poisson different is that we also have an n which is approaching infinity within a term of the expression being expanded to the power of n. That is

$$\left(\frac{\lambda}{n} + 1 - \frac{\lambda}{n}\right)^n$$

rather than

$$(p + q)^n$$

</td><td>

(As mentioned on page 350 and following, Poisson had to slightly modify this formula to make what happens next work. He raised p to the power of r instead of $n - r$, and did the opposite to q. If we are expanding this out, that will change the order we write the terms in, but that doesn't affect the actual values involved, which is the important part.)

We replace p, the probability of success with

$$\frac{\lambda}{n}$$

and we replace q, the probability of failure with

$$1 - \frac{\lambda}{n}$$

</td></tr>
</table>

because all probabilities add up to 1 by definition. We won't need to sum up the terms for a cumulative distribution function, so the sigma notation can go. We know that they will all add to 1. r will represent each particular number of successes for which we want to find the probability, and the only values of r that will be relevant will be the ones relatively close to λ, the average value. Now we need to see what happens when n approaches infinity.

For each value of r we are interested in, the formula will be:

$$\left(\frac{\lambda}{n} + 1 - \frac{\lambda}{n}\right)^n = \lim_{n \to \infty} \frac{n!}{r!\,(n - r)!} \left(\frac{\lambda}{n}\right)^r \left(1 - \frac{\lambda}{n}\right)^{n-r}$$

<table>
<tr><td>

There will be r terms in this expansion. (We saw why this is when we looked at factorials

</td><td>

Now $\dfrac{n!}{(n-r)!} = n(n - 1)(n - 2) \ldots (n - r + 1)$

</td></tr>
</table>

> on page 145.) As n approaches infinity each of the numbers from 1 to r in this expression will become trivial by comparison, and the expression will just equal n multiplied by itself r times, or n^r.

The n^r in the numerator and denominator cancel one another out.

$$\lim_{n \to \infty} \frac{n^r \lambda^r}{r! \, n^r} \left(1 - \frac{\lambda}{n}\right)^{n-r}$$

This is the expression for each instance of r using the property of adding and subtracting indices.

$$\lim_{n \to \infty} \frac{\lambda^r}{r!} \left(1 - \frac{\lambda}{n}\right)^{n} \left(1 - \frac{\lambda}{n}\right)^{-r}$$

Now as we saw on page 268

$$\lim_{n \to \infty} \left(1 - \frac{\lambda}{n}\right)^{n} = e^{-\lambda}$$

while

$$\lim_{n \to \infty} \left(1 - \frac{\lambda}{n}\right)^{-r} = 1$$

(because as n approaches infinity, the fraction with the constant λ over n approaches zero, and 1 raised to any power still equals 1).

That makes the probability mass function for each term r of the Poisson distribution:

$$PMF = \frac{\lambda^r e^{-\lambda}}{r!}$$

(This is the formula used to produce the graph at Figure 213 on page 451, with λ set to 50 and the whole numbers from 25 through 75 used as r values.)

The CDF is therefore:

$$CDF = \sum_{r=0}^{\infty} \frac{\lambda^r e^{-\lambda}}{r!}$$

Even though only a handful of values for r will be relevant, we can add up all values between 0 and infinity, because all are theoretically possible.

The sum of possible values will always equal 1, as they should in a valid CDF. To prove this:

Because a series has the property of linearity (see page 133) and $e^{-\lambda}$ is a constant with respect to r, we can take $e^{-\lambda}$ outside of the summation. What we are left with inside the summation is the Taylor series expression of e^{λ}.

$$e^{-\lambda} \sum_{r=0}^{\infty} \frac{\lambda^r}{r!}$$

This involves multiplying e^{λ} by its reciprocal, which always equals 1. Or, if you prefer to think of it this way, adding the indices gives 0, and anything raised to the power of 0 equals 1.

$$e^{-\lambda} e^{\lambda} = 1$$

So we can confirm that it is a valid probability function.

Suddenly the constant e makes its appearance from nowhere. When you think about it, that formula is a wildly improbable piece of probability.

Statistics is inherently paradoxical because it involves the application of mathematic precision to guess work. The Poisson formula is paradoxical, even for statistics. It is a specific example of the binomial formula (as n approaches infinity) yet looks nothing like it. It is another of those probability distributions that finds a proportion of something infinite. It indicates that regardless of the type of process, provided it is stable as an overall process and the individual events are random, all processes of all types with a given average of events per time interval will follow the same pattern of event-probability. If they don't follow that pattern, then they aren't completely random. Even the concept of being random and yet having a stable average is paradoxical. We'll see a lot more of this last paradox as we move forward.

The exponential distribution

The **exponential distribution** is related to the geometric, and the distinction is the same as one we saw at the start of the book: between using numbers to count *how many* and using them to measure *how much*.

You may remember (page 380) that the geometric distribution asks how many trials until the first success, and that it's probability mass function (PMF) is:

$$PMF = (1 - p)^{r-1}p$$

where p is the probability of success and r is the number of trials.

By contrast the exponential asks how much of something (usually time) before the first success.

Figure 216 uses an exponential distribution to graph the probability of seeing a car within a certain amount of time on a quiet road. Other examples could include the length of telephone conversations and the amount of time until the next earthquake. (It's called an "exponential distribution" because it is based on the reciprocal of exponential growth).

We can derive the formula for the exponential starting with the geometric. The geometric's cumulative distribution function is, as we saw earlier,

$$Pr(X \le r) = 1 - (1 - p)^r$$

where p is the probability of success with each trial and r is the number of trials.

To change the geometric distribution to the exponential, we need to convert the idea of discrete trials to that of continuous time.

To do that, we start by cutting up each unit of time into n sub-intervals, as we did with the Poisson distribution (Figure 217).

As per the Poisson, if there are 4 subintervals in each unit of time and if we expect λ events in an interval of time, then in each sub-unit we expect

$$\frac{\lambda}{4}$$

events. If there are n subintervals then each will have, on average,

$$\frac{\lambda}{n}$$

events.

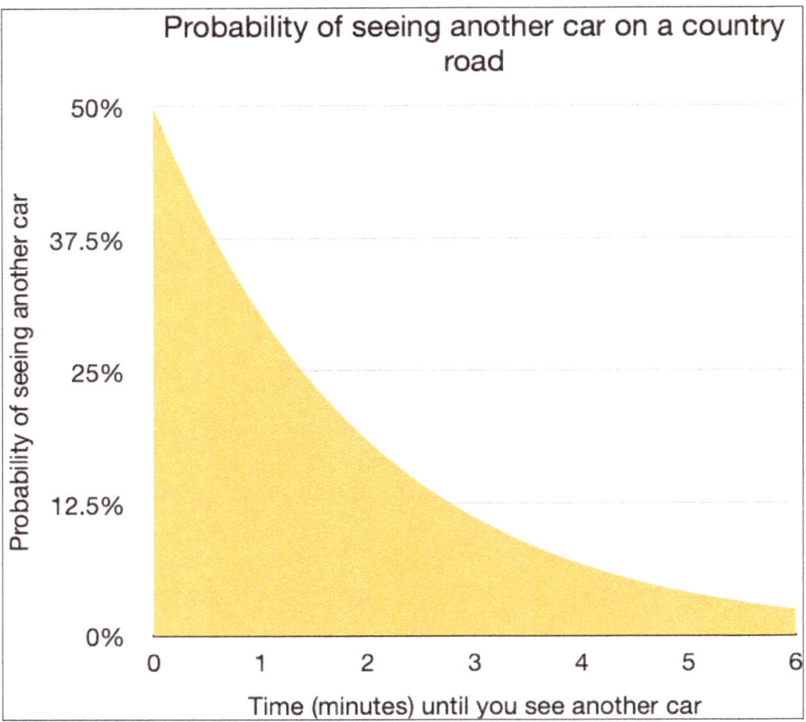

Figure 216

Thinking of this now as a geometric distribution, let's imagine that we have a metronome (a device with an audible click, where the timing can be set by the user) that ticks within each subinterval of time. Each tick of the metronome will be a trial. Then using the geometric distribution the probability that an event will occur within r ticks of the metronome will be

$$1 - \left(1 - \frac{\lambda}{n}\right)^r$$

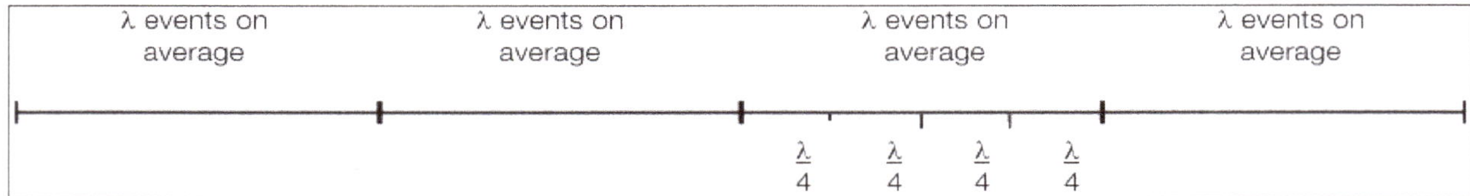

Figure 217

If r is the number of events before a success and we want to convert from events to time, then we can make $r = nx$, where n is still the subintervals that we have divided our time interval into, and x is an amount of time needed for n subintervals to equal r.

That gives

$$1 - \left(1 - \frac{\lambda}{n}\right)^{nx} = 1 - \left[\left(1 - \frac{\lambda}{n}\right)^{n}\right]^{x}$$

Now imagine that our metronome ticks faster and faster until it is going infinitely fast. That means n will approach infinity. As we saw when we first looked at Euler's number (page 268):

$$\lim_{n \to \infty} \left(1 - \frac{\lambda}{n}\right)^{n} = e^{-\lambda}$$

Which makes our equation

$$CDF = 1 - e^{-\lambda x}$$

Once again, Euler's number has appeared as if from nowhere.

As we mentioned earlier (page 373), instead of being called a probability mass function (PMF), the formula for the curve of a continuous distribution like the exponential is called a **probability density function**, or **PDF**.

To find the PDF from the CDF, we need to take the derivative of the CDF, just as we would take the integral of the PDF to find the CDF.

For $1 - e^{-\lambda x}$, using the chain rule

$$\boxed{PDF = \lambda e^{-\lambda x}}$$

Exponential probabilities start at zero (it can't be less than no time until an occurrence) and continue to infinity. In other words, it would theoretically be possible to wait forever to see another car. Possible, but not likely. Like the beta distribution, the exponential distribution is continuous: it draws a smooth curve. (It could take any amount of time for an event to occur.)

(It is also possible to derive the formula for the PDF and CDF of the exponential distribution by starting from the Poisson. The Poisson looks at events within a timeframe; the exponential looks at the time between the events.)

The gamma distribution

We've seen that the geometric distribution gives the probability of the first occurrence of something in a specific period of time, while the negative binomial gives the nth occurrence. We've also seen that the exponential distribution does something similar, except that time is measured continuously rather than in fixed periods. The gamma distribution does the same thing for the exponential that the negative binomial does for the geometric:

it looks at probabilities associated with how long before a specified number of occurrences, rather than just looking for the first occurrence.

It's remarkable that, given its importance in statistics, in its simplest form the CDF of the gamma distribution is just formed by placing one gamma function over another. Earlier (page 302) we saw that the gamma *function* is:

$$\Gamma(n) = \int_0^\infty x^{n-1} e^{-x} dx$$

We also saw that, when n is a positive integer, this function calculates factorials.

The CDF of the gamma *distribution* is

$$Pr(X \le x) = \frac{1}{\Gamma(n)} \int_0^\infty x^{n-1} \lambda^n e^{-\lambda x} dx$$

So when $\lambda = 1$, what we have is

$$\frac{\Gamma(n)}{\Gamma(n)}$$

As this is equal to 1, and it is always positive, we can see immediately that it is a valid CDF. In order to use it in practice an incomplete gamma function is used (which we also looked at on page 304), so that the bounds of integration represent the area of interest. This CDF is often written as:

$$P(X \le \alpha) = \frac{1}{\Gamma(n)} \int_0^\alpha \frac{(\lambda x)^n e^{-\lambda x}}{x} dx$$

(x^{n-1} is, of course, the same as $\dfrac{x^n}{x}$.) The λ^n is a constant with regard to x, and so can be taken outside of the integration.

The formula for the PDF is the same thing without the integration:

$$Pr(X = x) = \frac{1}{\Gamma(n)} \frac{(\lambda x)^n e^{-\lambda x}}{x}$$

for all $x > 0$.

Important warning about the gamma distribution: like the geometric, the gamma distribution is sometimes expressed in different ways. Instead of using n and λ, it is sometimes written with alpha and beta (α and β). In this instance α replaces n, but β replaces *the reciprocal* of λ. When that happens the PDF is written as:

$$Pr(X = x) = \frac{1}{\Gamma(\alpha)\beta^{\alpha}} \frac{x^{\alpha}e^{-\frac{x}{\beta}}}{x}$$

This can, of course, cause some confusion. For example, you may put gamma~(α, β) into a software package and not know the basis that is being used to calculate it: is it based on raising e to a multiple of λ, or to a multiple of the reciprocal of λ? These two approaches will give a different result, and it will probably not be specified by a software package, unfortunately. You will probably have to run some test data, before you know what you are looking at[52]. You can read more of this if you search online under "alternative parameterization of the gamma distribution".

The gamma distribution links a number of other distributions. This table shows the way that it relates to some of the distributions we have seen so far:

	Based on number of separate trials	Based on amount of time or space
Probable number of successes	Binomial	Poisson
Probable number/amount of events/time until the first success	Geometric	Exponential
Probable number/amount of events/time to the nth success	Negative binomial	Gamma

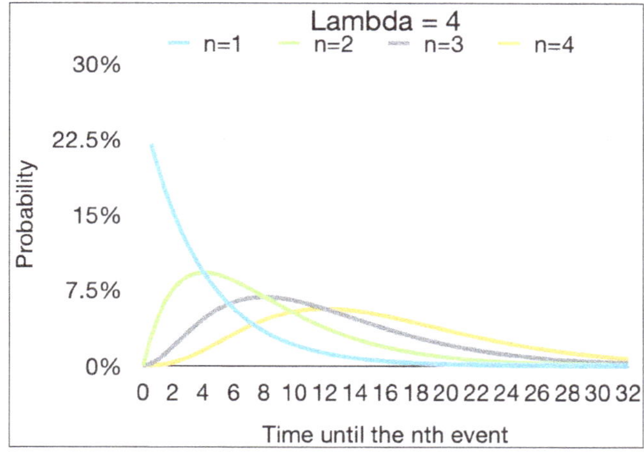

Figure 218

So the gamma distribution tells us how much continuous time will probably pass (or how much distance will be traversed) until the nth success. If the time you need to wait for a single success is given by the exponential distribution, then this is the same as summing up a random variable that has n different exponential distributions: an exponential distribution for each success until you reach n of them. The result when you do that is the gamma distribution. It will, in fact, be $Gamma(n, \lambda)$ where n is the number of successes we are waiting for, and λ is the average rate of success per period of time.

None of that proves, however, that this function represents the sum, n times, of events which are exponentially distributed. There are a couple of different ways to do this, but the easiest uses a **moment generating function (MGF)**, which we haven't come to yet, but we will (page 532). After we have covered that we will prove that the gamma distribution is what we get when we add n exponential distributions. Assuming for the present that it is true, when we come to graph the distribution for different values of n, we get Figure 218. In this case we are looking at λ set to 4: there are an average of 4 events per unit of time.

Notice that when $n = 1$, the (blue) line resembles an exponential distribution. That is exactly what we would expect, because the exponential distribution gives us the probabilities of time until the first event. If we use 1 as the value of n in the distribution, we get:

$$\frac{1}{\Gamma(n)}\frac{(\lambda x)^n e^{-\lambda x}}{x} = \frac{\cancel{1}}{\cancel{\Gamma(1)}}\frac{(\lambda \cancel{x})^1 e^{-\lambda x}}{\cancel{x}} = \lambda e^{-\lambda x}$$

which is the formula for the exponential distribution. As the number of events we are waiting for (say the number of cars we expect to see) increases, so the time when we expect to see that number of events shifts out. If we were waiting for 1 event, we might wait a certain amount of time. If we were waiting for 10 of them, we would have to wait longer.

To actually calculate probabilities using a gamma CDF, there is no generalised way to integrate the function. It used to be calculated by tables, but now you would just look it up on a computer. To understand how a computer would calculate it, we need to use the Taylor-series expansion of e. You might sometimes see this called a **power series.**

Start by expressing a gamma distribution in this form.

$$P(X \le \alpha) = \frac{\lambda^n}{\Gamma(n)}\int_0^\alpha x^{n-1}e^{-\lambda x}dx$$

Where n is an integer (as it usually will be) it can be written, as we've seen, as the factorial of one less than itself. When it is some number and a half, as we also saw it can be written as a function of $\sqrt{\pi}$.

$$= \frac{\lambda^n}{(n-1)!}\int_0^\alpha x^{n-1}e^{-\lambda x}dx$$

Use the Taylor infinite series formula for $e^{-\lambda x}$. This is usually expressed in terms of n, but that letter is already used in the formula, so we will use k as the integer that

$$= \frac{\lambda^n}{(n-1)!}\int_0^\alpha x^{n-1}\left(\sum_{k=0}^{\infty}\frac{(-\lambda x)^k}{k!}\right)dx$$

is increased by 1 each time the series is calculated.

x^{n-1} is a constant with regard to the summation, and so can be taken inside the summation.

$$= \frac{\lambda^n}{(n-1)!} \int_0^\alpha \left(\sum_{k=0}^\infty x^{n-1} \frac{(-\lambda x)^k}{k!} \right) dx$$

By linearity the integration can also be brought inside the summation. $\frac{-\lambda^k}{k!}$ is a constant with regard to the integration, and so can be kept outside of it (but inside the summation).

$$= \frac{\lambda^n}{(n-1)!} \sum_{k=0}^\infty \frac{(-\lambda)^k}{k!} \int_0^\alpha x^{n-1} x^k dx$$

Gather the exponents of x.

$$= \frac{\lambda^n}{(n-1)!} \sum_{k=0}^\infty \frac{(-\lambda)^k}{k!} \int_0^\alpha x^{k+n-1} dx$$

Use the reverse power rule (add one to the exponent and divide by the new exponent) to calculate the integral.

$$= \frac{\lambda^n}{(n-1)!} \sum_{k=0}^\infty \frac{(-\lambda)^k}{k!} \left(\frac{x^{k+n}}{k+n} \Big|_0^\alpha \right)$$

When $x = 0$, $\frac{x^{k+n}}{k+n}$ will also equal zero, so there is no need to subtract it.

$$\boxed{= \frac{\lambda^n}{(n-1)!} \sum_{k=0}^\infty \frac{(-\lambda)^k}{k!} \left(\frac{\alpha^{k+n}}{k+n} \right)}$$

While it would be painful to do it manually, this expression can be used to approximate the area under a gamma distribution for any λ, any positive-integer-valued n, between 0 and any desired value of α.

Key points:

- The **beta distribution** can be thought of as finding the probability of a probability.

- This makes it useful in **likelihood** calculations.

- It is important in Bayesian statistics, partly because it can mimic other distributions.

- The **Poisson distribution** considers the probability of a certain number of events within a given timeframe or area, given an ongoing, steady-state process that shifts randomly around a fixed amount.

- The **exponential distribution** looks at the probability of particular timeframe between events.

- The **gamma distribution** looks at the probability of a timeframe between a specified number of events.

- There are **two forms** of the gamma distribution (or actually two ways it can be parameterized) and it might not be apparent from software you are using, so you will probably need to test it.

- Calculating the CDF involves the use of a Taylor series, also known as a **power series.**

Distributions in practice

The major probability distributions are available as spreadsheet functions. Using a spreadsheet can sometimes be more helpful than using statistical software, at least in the early stages of a research undertaking. It allows you to play with and get a feel for the probabilities around your data in a more flexible (but less powerful) way than most statistics packages. Once you have understood what you are looking at, it can be more helpful to switch to using a statistics package.

You can make use of them to calculate probabilities in your work situation, or anywhere else, without knowing any more about probability than we have covered so far. These functions will also ask for your parameters. (The spreadsheet will guide you about the parameters to use). ("Cumulative" in spreadsheet formulae is asking whether you want the probability mass function/probability density function (PMF/PDF) or the cumulative distribution function (CDF) calculated: "No" for the former, "Yes" for the latter.)

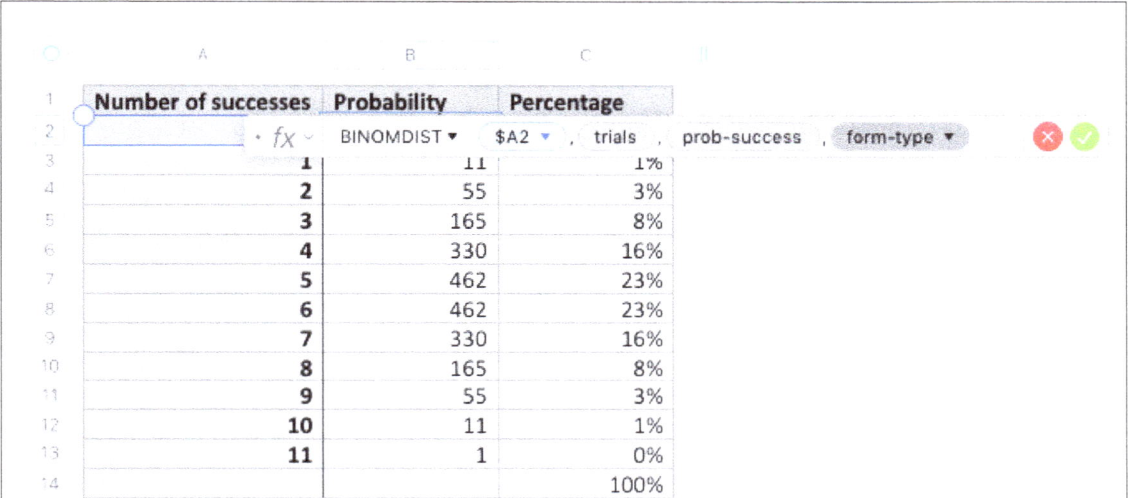

Figure 219

To create a distribution on a spreadsheet, you can put different numbers for one of your parameters into a column, then apply the formula to each of the cells in that column as per Figure 219. (This was done using the

Numbers spreadsheet, but Microsoft Excel works in a way that is almost identical.)

These are examples of how the distributions we have just covered can be used in some different disciplines. There are, of course, many more disciplines that make use of statistics.

Field of Study	Beta Distribution	Poisson Distribution	Exponential Distribution	Gamma Distribution
Astronomy / Space Science	Modelling belief about probability of finding a planet in a star system	Number of observed meteor showers per month	Time between supernovae or pulsar emissions	Time until multiple bursts or stellar events
Biology / Medicine	Modelling the probability of treatment success based on prior outcomes (Bayesian inference)	Modelling number of mutations per cell or per DNA strand	Time until next heart attack, given a constant risk rate	Time until multiple adverse reactions occur in a patient
Computer Science	Modelling belief about the probability of system failure	Number of requests per second to a web server	Time between incoming packets in a network	Total time to process a batch of user requests
Criminology / Forensics	Prior belief about a suspect's guilt probability	Number of crimes reported per week	Time between reported crimes	Time until a fixed number of crimes are reported
Ecology / Environmental Science	Prior belief about the proportion of a habitat occupied by a species	Number of rare species sightings per day in a park	Time between earthquakes or animal sightings	Time until multiple extreme weather events occur
Education / Psychology	Modelling belief about proportion of correct answers in a population	Number of errors per page in student essays	Time between learning milestones in skill acquisition	Time to reach a certain number of learning breakthroughs
Epidemiology / Public Health	Probability distribution for disease prevalence given sample data	Number of new cases of a disease reported per day	Time between reported infections or emergency calls	Time until a certain number of infections in an outbreak

Field of Study	Beta Distribution	Poisson Distribution	Exponential Distribution	Gamma Distribution
Finance / Actuarial Science	Belief about a company's default rate on loans or claims	Number of defaults or claims per time period	Time between claims or trades	Total time until multiple claims or risk events occur
Linguistics	Belief about proportion of utterances using a certain structure	Number of grammatical errors per essay	Time between occurrences of a target construction in speech	Time until a speaker produces several specific vocal sounds
Marketing / Consumer Research	Updating probability models about true click-through rate of an ad	Number of ad clicks or views per hour	Time between purchases by a user	Time until a user buys multiple products
Operations Research / Logistics	Fine-tuning delivery processes and updating models	Number of deliveries per hour	Time between truck arrivals at a depot	Time until a target number of deliveries is made
Political Science	Belief about a candidate's true support level based on poll data	Number of legislative proposals introduced per session	Time between policy changes or speeches	Time until several major political events happen
Quality Control	Modelling the probability a product meets specification after a new process is introduced	Number of defects on a length of material or per unit produced	Time between failures of machines on an assembly line	Time until a certain number of failures occurs
Sales / Business	Modelling customer conversion rate as a probability distribution	Number of customer arrivals at a service desk per hour	Time between customer arrivals or product purchases	Waiting time for a certain number of customer arrivals
Sociology	Modelling belief about proportion of population holding a particular opinion	Number of protests or social events per week in a region	Time between social unrest events	Time until multiple events of a given type (e.g., arrests) occur

Field of Study	Beta Distribution	Poisson Distribution	Exponential Distribution	Gamma Distribution
Sports Analytics	Prior belief about a player's success rate (e.g., shot accuracy)	Number of goals or fouls per match	Time between goals or substitutions	Time to observe a fixed number of goals scored

Recap of distributions so far

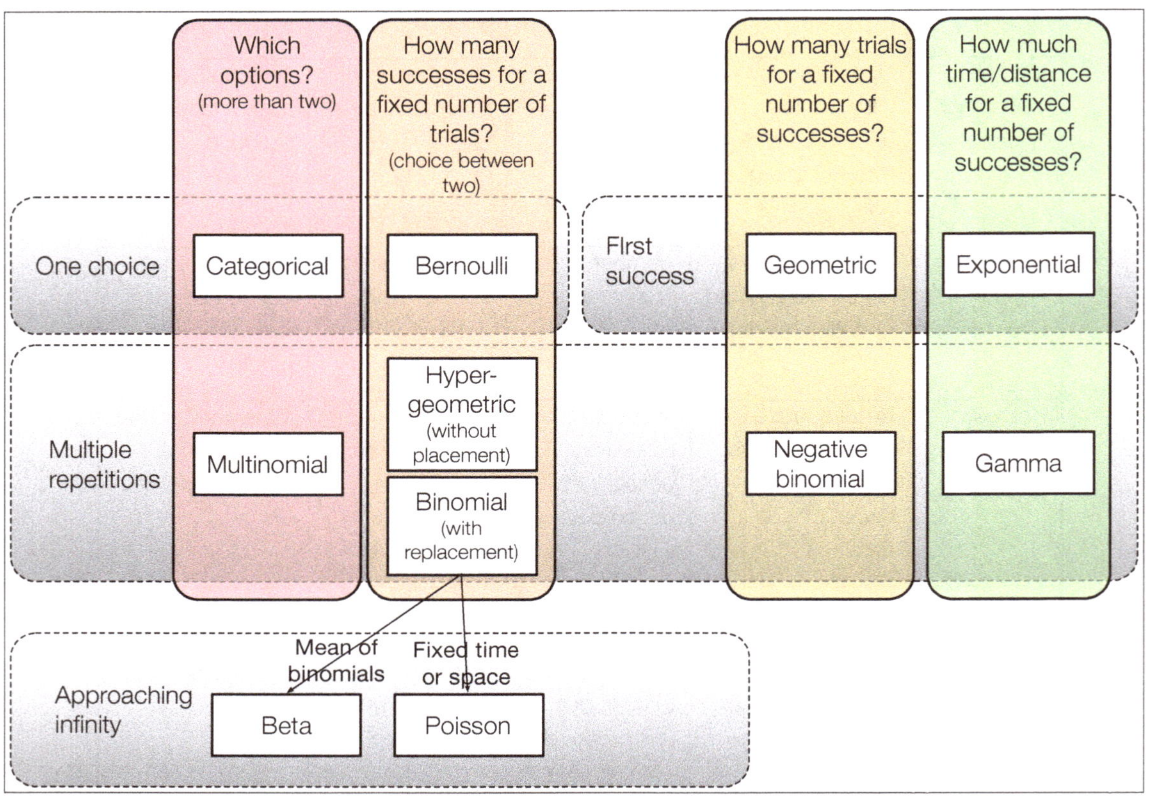

Figure 220

Distribution	Question answered	Can be thought of as	Spreadsheet function
Bernoulli: $Bern(p)$	Success or failure?	Binomial with n of 1. The basis for other yes/no distributions.	

Distribution	Question answered	Can be thought of as	Spreadsheet function
Binomial: $Bin(n, p)$	How many ways can you have success with a given number of trials, with replacement?	Hypergeometric with replacement.	=BINOMDIS
Geometric: $G(p)$	How many trials will it take to achieve one success?	The negative binomial where $r = 1$	(You might need to use the negative binomial, depending on your spreadsheet)
Negative binomial: $NB(r, p)$	How many trials will it take to achieve a given number of successes?	Generalisation of the geometric for any number of successes.	=NEGBINOMDIS
Hypergeometric: $H(N, a, n)$	How many ways can you have success with a given number of trials, without replacement?	Binomial without replacement.	=HYPGEOMDIST If your spreadsheet doesn't have this function, you can use the =COMBIN function to create each of the n choose r expressions, and use them to build the formula.
Multinomial: $Mult\left(n, \vec{p}\right)$ (where $\vec{p}$ is a list of probabilities that add to 1)	What is the probability of a particular group of different events/ trials/ objects which fit into different categories and which have their own probabilities?	Generalisation of the binomial, with more than 2 terms.	=MULTINOMIAL This command gives the multinomial coefficent. You input the different values of x_i. You then still need to work out the probabilities.

Distribution	Question answered	Can be thought of as	Spreadsheet function
	When k is 2 and n is 1, the multinomial distribution is the Bernoulli distribution. When k is 2 and n is bigger than 1, it is the binomial distribution. When k is bigger than 2 and n is 1, it is the categorical distribution.		
Beta: $Beta(a, b)$ *where* a 0	What is the probability of a particular probability?	Mean of binomials.	=BETADIST This gives only the cumulative values.
Poisson: $Pois(\lambda)$	How many successes will occur in a given period of time or area in space?	(Sort of) binomial where $n \to \infty$	=POISSON.DIST
Exponential: $Expo(\lambda)$	How long until the first success?	Continuous geometric, with time instead of trials. Poisson looking for zero successes.	=EXPON.DIST
Gamma: $Gamma(n, \lambda)$	How long until the nth success?	Generalisation of the exponential	=GAMMADIST

Conjugate priors in Bayesian statistics

Some prior distributions are easier to work with than others in Bayesian statistics. These are called **conjugate priors**. Of the distributions we have seen (and the normal that we haven't looked at yet) the following distributions work well together, mathematically. (There are some others, which we haven't covered.)

Conjugate Prior	Likelihood	Posterior
Beta	Bernoulli or Binomial	Beta
Beta	Geometric	Beta
Beta	Negative Binomial	Beta

Conjugate Prior	Likelihood	Posterior
Gamma	Poisson	Gamma
Gamma	Exponential	Gamma
Normal	Normal (known variance)	Normal

There can be a temptation to wait until some evidence is obtained, see what form it takes, then to determine the form of the prior so that it will be conjugate and the mathematics will be doable. The danger of this is that it may defeat the purpose of using a Bayesian approach. There are better ways to go about it, but conjugate priors are still sometimes used to give researchers a sense of where the experiment is going.

When conjugate priors aren't appropriate, different computer-based algorithms can be used to crunch through a lot of numbers, to find the most suitable posterior distributions.

These include the Markov chain Monte Carlo (MCMC). (We'll meet Markov later, but won't go into Markov chains, which are basically ways to generate a sequence of random variables where the future state depends only on the present state, not on the events that precede it). Monte Carlo methods refer to a broad class of algorithms that rely on random sampling to estimate numerical results. One of those classes of algorithm is the Metropolis-Hastings; another is Gibbs sampling.

There are other approaches, used in different situations.

Transforming random variables

> **Read this if** you don't understand the idea of transforming a random variable, you don't know how you would go about it, or you don't know what a monotone function is (or why it is important in transforming a random variable).

We've thought about the relationships between different distributions, but now we'll think about changing the variables themselves.

Suppose we have a random variable with a continuous probability distribution, and now we want to know what happens to that probability distribution if we apply some function to the variable. For example, if we were collecting data about a group of people, we might find out their heights and their weights. Each respondent's body mass index is their weight divided

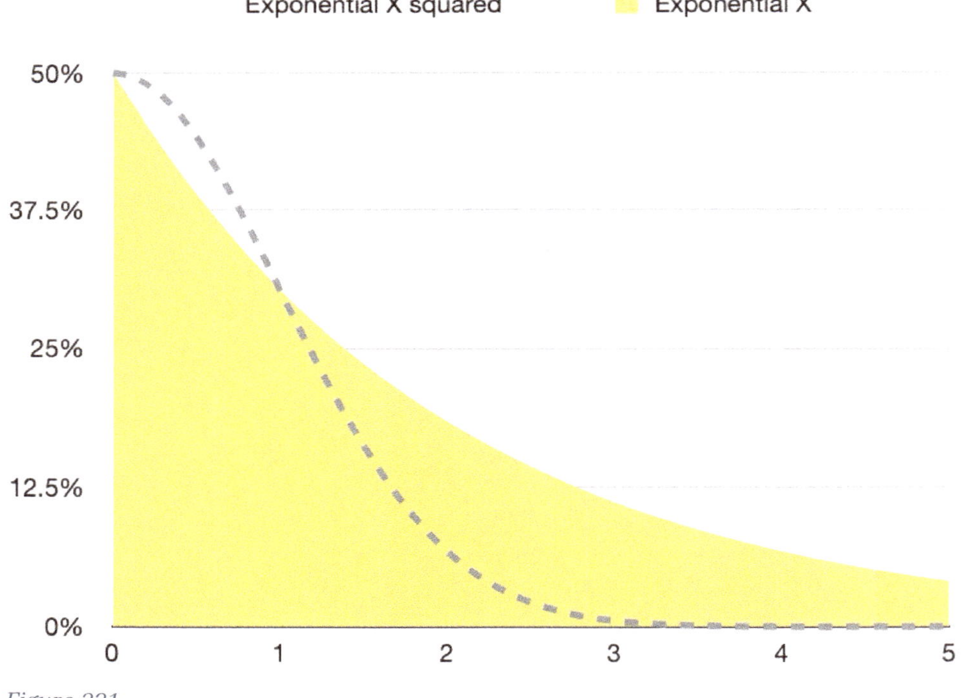

Figure 221

by the square of their heights. Before we do that calculation, we might want to understand the distribution of the square of people's heights.

The function of a random variable is also a random variable. Let's make a new random variable, V. This represents the square of X, which represents people's height.

$$V = X^2$$

Unfortunately, we can't just insert V in place of X; it gives us a different distribution altogether. Figure 221 shows an exponential distribution, overlayed with an exponential distribution of V, or X^2. As you see the whole shape of the curve is different.

Resolving this can often be straightforward. Think about the inverse function of the current variable, so that you can write the equation in terms of the new variable. In other words, if V is X^2, then X is $\sqrt{V}$. So put $\sqrt{V}$ into the equation in place of X and the problem might be solved. There are two situations where we need to do more: where the curve of the new function is horizontal or changes direction at any point, or if we are creating a probability density function (PDF) for a continuous function.

Monotone functions

Where the curve becomes flat or changes direction, the value of the new variable might represent more than one of the original variables. For example, $V = X^2$ draws a parabola. If $V = 4$, X might equal 2 or -2. In this situation it might be necessary to specify some limitations that the second variable has to fit within. When we are looking at people's heights the value has to be positive, so we don't have a problem.

When a function doesn't become flat or change direction (or at least not in the area we are interested in) we describe it as being **strictly increasing**, or **strictly decreasing**. Another word for this is **monotone.** In practical terms this will mostly be the case when transforming random variables, but it is something to be aware of.

Remember, it doesn't matter what the slope of the PDF is, and it doesn't have to be monotone. It's the function of X that gives us the new variable that has to be monotone. Where it isn't, we usually have to specify different transformations of the PDF for different ranges of values of the new variable.

The Jacobian for a transformed PDF

The second problem, where we are creating a new PDF, is a bit trickier. To think this through, let's start by remembering that to find a continuous probability, the PDF doesn't get us very far (even though it is usually used as a way to describe and draw a curve). The probability of something happening at *exactly* one specific point in a continuous curve is vanishingly small, so we always use the area under the curve between two values.

What we want to be able to say is that the cumulative distribution of probabilities of V will be the same as the cumulative distribution of probabilities of X at any equivalent point. In other words:

$$Pr(X \leq x) = Pr(V \leq v)$$

As we saw earlier, we can address this by putting the inverse of the new function of X into the original PDF. In the example we've been using,

$$V = X^2$$

so we can know that, using the inverse function,

$$X = V^{\frac{1}{2}} \text{ (the square root of } X\text{)}$$

The cumulative distribution function (CDF) of an exponential distribution is:

$$Pr(X \leq x) = 1 - e^{-\lambda x}$$

Writing the CDF in terms of V, we have

$$Pr(V \leq v) = 1 - e^{-\lambda v^{\frac{1}{2}}}$$

That's *almost* all we have to do for the CDF, whether the random variable is continuous or discrete. We will also have to write the limits of integration in terms of the inverse function of X, and to check whether anything about the calculation would make them change. (No change is needed for this particular situation; X and V will both have support from 0 to infinity. It's also all we have to do for a discrete probability mass function (PMF). (For the continuous PDF, however, we have an additional complexity.

We can find the PDF by taking the derivative of the CDF. In the case of an exponential distribution (without any transformation involved), when we do that the 1 at the beginning disappears, and we need to find the derivative of $-e^{-\lambda x}$ using the chain rule. The outer function is $e^{-\lambda x}$, and it is its own derivative. The inner function is $-\lambda x$. Its derivative is $-\lambda$. The two negatives cancel out, and we are left with

$$Pr(X = x) = \lambda e^{-\lambda x}$$

But for our *transformed* PDF, the chain rule will give us a different result. If we take the derivative of the CDF with respect to v using the chain rule, we get:

$$PDF = \lambda \frac{1}{2} v^{-\frac{1}{2}} e^{-\lambda v^{\frac{1}{2}}}$$

$$= \frac{\lambda e^{-\lambda v^{\frac{1}{2}}}}{2v^{\frac{1}{2}}}$$

We've had to multiply by

$$\frac{1}{2v^{\frac{1}{2}}}$$

which happens to be the derivative with respect to v of $X = V^{\frac{1}{2}}$

We can express that in more general terms like this. Let's assume we have a continuous random variable X, which is distributed according to the CDF $Pr(x)$, and has the PDF $pr(x)$. Let's also assume that we have another variable we'll call v, and that it is the value of a function of x:

$$v = f(x)$$

x is the inverse function of v. $\qquad\qquad x = f^{-1}(v)$

State the CDF in general terms. $\qquad\qquad CDF = Pr(x)$

Replace x with the inverse function of v. This is what we did when we replaced x with $\sqrt{v}$. This means that v is now the variable in the formula. $\qquad CDF = Pr\big(f^{-1}(v)\big)$

Take the derivative of both sides. On the right use the chain rule. $p\big(f^{-1}(v)\big)$ is the derivative of the outer function; the lowercase p indicates the PDF. $\qquad PDF = pr\big(f^{-1}(v)\big)\dfrac{d\big(f^{-1}(v)\big)}{dv}$

$x = f^{-1}(v)$ The new PDF is written in terms of v.

$$PDF = pr\left(f^{-1}(v)\right)\frac{dx}{dv}$$

To transform the PDF of a continuous random variable X using a different variable, which is its own function of X, we need to replace the x in the PDF with the inverse function of the new variable, then multiply by the reciprocal of the derivative of the function of X that we are going to use to transform the random variable. The

$$\frac{dx}{dv}$$

becomes the Jacobian of the transformation. (We looked briefly at Jacobians on page 289, in relation to transforming to polar coordinates with calculus.)

Another way to think of this is that we really need to make sure the rate of change of X is the same as the rate of change of V. In other words:

$$Pr(v)dv = Pr(x)dx$$

which means:

$$Pr(v) = Pr(x)\frac{dx}{dv}$$

The Jacobian standardises these rates of change. Remember, this is only relevant for continuous random variables.

This raises one final complexity to resolve: if the function we are using to transform the variable is strictly increasing, then the derivative will be positive, while if it is strictly decreasing, the derivative will be negative; and yet PDFs always have to be positive (because probabilities have to be positive). We solve this by putting the absolute value sign around the Jacobian. To see why this is legitimate, let's look at the simple case of a linear transformation (the function we are using to transform the random variable draws a straight line), with two different situations: one where the line has a positive slope, and the other where it has a negative slope.

Once again, we'll let the original CDF $= P(x)$ and the original PDF $= p(x)$. This time we'll let the function for the variable we want to use to transform the PDF be

$$v = mx + b$$

which means

$$x = \frac{v - b}{m}$$

$$\frac{\delta x}{\delta v} = \frac{1}{m}$$

Case 1: Let m be positive.

If m is positive the line will slope forwards. Based on the work we did above, we can say that the transformed PDF is

$$p(v) = \frac{1}{m} p\left(\frac{v - b}{m}\right)$$

Because m is positive, we can write this as

$$p(v) = \left|\frac{1}{m}\right| p\left(\frac{v - b}{m}\right)$$

Case 2: Let m be negative.

If m is negative the line will slope backwards.

$$P(v) = Pr(V \leq v)$$

Substitute $mX + b$ for V. $\qquad\qquad = Pr(mX + b \leq v)$

Subtract b from both sides of the inequality. (We looked at inequalities on page 28.) $\qquad = Pr(mX \leq v - b)$

At this point, when we divide both sides of the inequality by m, as m is negative we need to reverse the direction of the inequality. $\qquad = Pr\left(X \geq \frac{v - b}{m}\right)$

Now we work with the inverse of the CDF. Probabilities add to 1. X has to be either, on the one hand, greater than or equal to what is on the right of the inequality, or it has to be less than or equal to it. Together, this covers all of the possibilities, and so the total adds to 1.

$$Pr\left(X \geq \frac{v - b}{m}\right) + Pr\left(X \leq \frac{v - b}{m}\right) = 1$$

Subtract $Pr\left(X \le \frac{v-b}{m}\right)$ from both sides.

$$Pr\left(X \ge \frac{v-b}{m}\right) = 1 - Pr\left(X \le \frac{v-b}{m}\right)$$

$$P(v) = 1 - Pr\left(X \le \frac{v-b}{m}\right)$$

Take the derivative of both sides.

$$p(v) = -\frac{1}{m}p\left(\frac{v-b}{m}\right)$$

$\frac{1}{m}$ is negative. The two negatives cancel one another out.

$$p(v) = \left|\frac{1}{m}\right|p\left(\frac{v-b}{m}\right)$$

In both cases—whether the slope is backwards or forwards—the absolute value of the Jacobian will be correct.

So, if a continuous PDF is $p(x)$ and we are transforming the random variable with the function

$$V = f(X)$$

then the new PDF, written in terms of V, will be:

$$\boxed{\text{PDF} = pr\left(f^{-1}(v)\right)\left|\frac{dx}{dv}\right|}$$

As an example of this, let's look at the probability function

$$\text{PDF} = \frac{1}{\sqrt{2\pi}}e^{\frac{-x^2}{2}}$$

(This is called the **standard normal** PDF. We will encounter it a lot as we proceed, and we'll need the result that follows when we come to look at the chi square distribution. For now, just treat it as an example of transforming a random variable). We will replace the random variable X with $V = X^2$.

That makes the inverse function $x = \sqrt{v}$. Its derivative, which will be the Jacobian, is given by:

$$\left|\frac{dx}{dv}\right| = \left|\frac{1}{2}v^{-\frac{1}{2}}\right|$$

$$= \left|\frac{1}{2\sqrt{v}}\right|$$

As we saw earlier, the function

$$V = X^2$$

is not monotone. It is, however, symmetrical around 0, as is

$$e^{\frac{-x^2}{2}}$$

That means we can just worry about values where $x \geq 0$, then multiply the result by 2. And so, replacing x with the inverse of v (i.e. $\sqrt{v}$), our new PDF is

$$2\frac{1}{\sqrt{2\pi}}e^{\frac{-v}{2}}\left|\frac{1}{2\sqrt{v}}\right| = \frac{1}{\sqrt{2\pi v}}e^{\frac{-v}{2}}$$

Recall from when we looked at the gamma function on page 302 (not to be confused with the gamma distribution), that

$$\Gamma\left(\frac{1}{2}\right) = \sqrt{\pi}$$

That means we can write our formula for a PDF

$$\frac{1}{\Gamma\left(\frac{1}{2}\right)\sqrt{2v}}e^{\frac{-v}{2}}$$

That translates as the PDF of a distribution

$$V \sim gamma\left(\frac{1}{2},\frac{1}{2}\right)$$

Our new random variable, based on squaring a random variable that has a standard normal distribution, is gamma of one half, one half. Alternatively, if we are thinking of a gamma distribution using the other convention, where

$$\lambda = \frac{1}{\beta}$$

it will be

$$\sim gamma\left(\frac{1}{2},2\right)$$

This will become important when we look at the chi square distribution.

Key points:

- When you apply a function to a random variable you still have a random variable, but if you insert the new variable into the old probability formula, the probability distribution might not have the same shape.

- To overcome this, use the inverse of the new function, so that you have a function of the new variable which equals the original one. This can then be substituted into the original formula.

- For a CDF you also need to consider whether the levels of integration will change.

- To do this, you need to either make sure that the function which transforms the random variable is monotone (strictly increasing or strictly decreasing), or else specify an appropriate range of values.

- When applying this to a continuous PDF, start by using the process above in the CDF. Then differentiate to find the PDF, but multiply by a Jacobian. Where the function that transforms the variable is $y = f(x)$, the Jacobian is

$$\left|\frac{dx}{dy}\right|$$

- If we transform the standard normal PDF by squaring the random variable, the result is a gamma distribution.

Summing random variables

> **Read this if** you aren't clear what it actually means to sum random variables or, if you do know what it means, you don't know how you would go about it.

(If you are struggling with this, see the How to Love Statistics YouTube video on *Introduction to convolutions for adding random variables*.)

When we looked at transforming a random variable, we were starting with a single variable. When we looked at the gamma distribution, we described it as a sum of exponentials. We said that we would provide the proof once we had covered moment generating functions (and we will), but before we do, we need to think about what it actually means to add random variables together.

It involves adding two variables to create a third, then working out a new probability distribution that applies to the new variable. Imagine this scenario. You are running a small business and have two telephone sales representatives who work from home. As an experiment, you want to know if they will be more efficient working from the same space (more chance to learn from one another, better use of their individual strengths, you will have a better idea of how they are spending their time) or should they keep working individually (fewer interruptions, less commuting time)? The representatives themselves don't express a strong preference either way. You could wait for a decent amount of time for average sales performances to come in, but the figures fluctuate and you want to know as soon as possible which approach is going to be best. You want the advantages of using a probability distribution so that, for example, you can see how much risk there is in the assumptions you are making.

There are different ways you could approach this, but you decide to calculate a probability distribution based on the number of sales per day. You know their average number of sales per day; you expect the fluctuation around this to follow a Poisson distribution. A check back through their figures confirms that this is the case. You then graph the probability distribution of their sales (Figure 222).

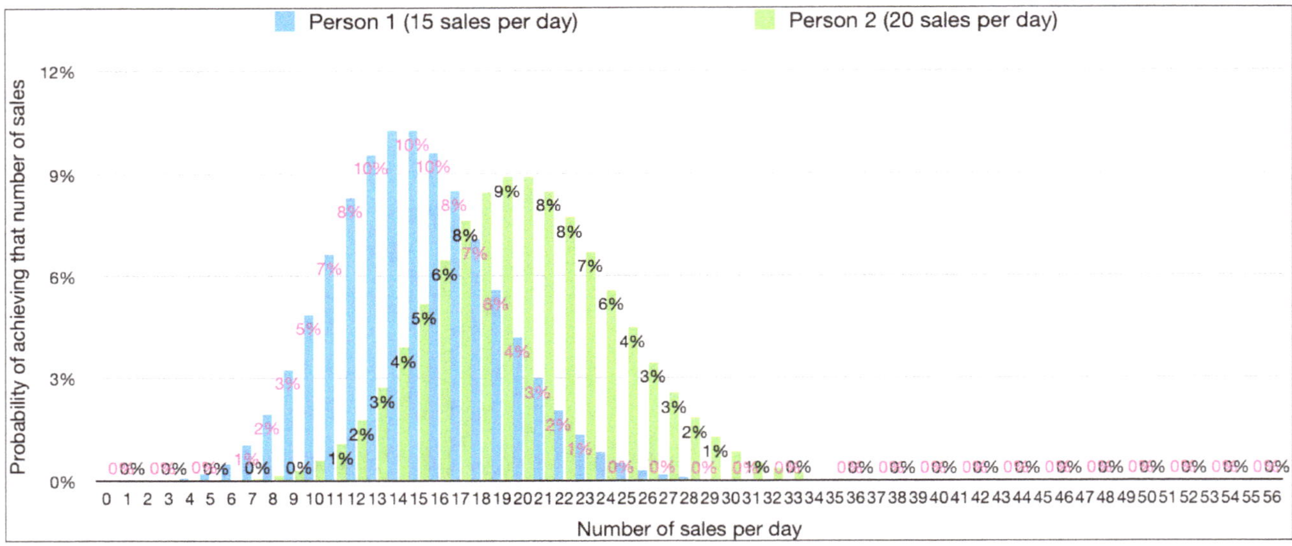

Figure 222

Your null hypothesis is that bringing the two people together will make no difference to their sales performance. So to start off you need a theoretical probability distribution that expresses their combined performance at present. The average number of sales per day is a different random variable, with a different Poisson distribution, for both of them. You need to add their performances together and to find a theoretical distribution for their combined sales total, so that you have something to compare to their actual results.

When they work together it will be more difficult to assess their individual performances, and you are less interested in that anyway. For example, the better sales performer might spend time advising the weaker one, and they might work jointly to make some sales, so looking at their individual performances won't really tell you what you want to know. You really want to know how their combined performance will compare to a theoretical situation where there is no change.

In other words, you need to be able to sum two independent random variables, then work out a probability distribution of the sum. You could add up the probabilities for each value, but that wouldn't give you a valid probability distribution; the values would add to 2, not 1 (because each currently adds up to 1).

You could average the two graphs, but that wouldn't give you the right answer either. You want the probability of their combined performances, not of their average performances. To make it easier to conceptually understand what is happening, we'll assume for the moment that the maximum number of sales that either can make in a day is two, and that we have a probability distribution showing 0, 1 or 2 sales for each of them (Figure 223). We will assign the variable X to Person 1's sales, and Y to Person 2's. Their probability distributions are:

Value of X (Person 1)	0	1	2	Value of Y (Person 2)	0	1	2
Probability of having that value	20%	50%	30%		55%	15%	30%

Figure 223

The support of each of the variables is from 0 to 2. Both are valid (if very simplistic) probability distributions, because the probabilities add up to 100%. Getting back to what we originally said about random variables, they are just like any other variable in mathematics, except that they follow a probability distribution. Now we are looking at a new variable, which we will call W, which will represent the sum of X and Y. What we need to find is the probability distribution of W.

$$W = X + Y$$

To work out the probability that W will take any particular value, we'll manually draw up the possibilities, to see what that looks like.

First, we want the probability that $W = 0$. That is, the probability that neither makes a sale on a day. From Figure 223 we can see that that this means *both X and Y* have to equal 0. Now this is where it gets a bit mind bending. We are *adding* the random variables, but we want to know the probability that *both* things are true: that means, according to the multiplication rule, we need to *multiply* the probabilities that X *and* $Y = 0$. Mathematically, we can express all of that as:

$$Pr(W = 0) = Pr(X + Y = 0)$$

$$= (20\%)(55\%)$$

$$= 11\%$$

(The colour coding will be relevant for Figure 224 when we get there.)

Now we want the probability that $W = 1$. There are two ways that this can happen: *either X equals 0 and Y equals 1, or X equals 1 and Y equals 0*. So thinking back to our multiplication and addition rules, for the *both–and* possibilities we need to multiply the probabilities, and for the *either–or* part of this, we need to add the probabilities. That means we have:

$$\Pr(W = 1) = \Pr([X = 0 \; and \; Y = 1] or [X = 1 \; and \; Y = 0])$$

$$= (20\%)(15\%) + (50\%)(55\%)$$

$$= 3\% + 27.5$$

$$= 30.5\%$$

Next we want the probability that $W = 2$. There are three ways that this can happen: *either X equals 2 and Y equals 0, or X equals 1 and Y equals 1, or X equals 0 and Y equals 2*. That gives us:

$$
\begin{aligned}
\Pr(W = 2) \quad &= \quad \Pr([X = 2 \; and \; Y = 0] or [X = 1 \; and \; Y = 1] or [X \\
&\qquad = 0 \; and \; Y = 2]) \\[6pt]
&= \quad (30\%)(55\%) + (50\%)(15\%) + (20\%)(30\%) \\[6pt]
&= \quad 16.5\% + 7.5\% + 6\% \\[6pt]
&= \quad 30\%
\end{aligned}
$$

To cement your understanding

Continue working through these probabilities, using the same approach, and noting that it is also possible for W to equal 3 or 4. Check that the answers you get are the same as those shown in Figure 224.

Value of Y	Value of X	0	1	2
	Probability	20%	50%	30%
0	55%	11%	27.5%	16.5%
1	15%	3%	7.5%	4.5%
2	30%	6%	15%	9%

Figure 224

What we end up doing is multiplying every possible combination between the values in the distribution of X and the distribution of Y, then adding each of the combinations that lead to the same probability of W. We can organise the information as per Figure 224.

> **To cement your understanding:**
>
> Look at Figure 224 and note that this is a contingency table that showed marginal and joint distributions, similar to one we saw when we first looked at marginal and joint distributions (page 398). Add up the rows, columns, and distributions, to get a sense of the way the distributions relate to one another.

If we add the probabilities along every diagonal (meaning the combinations that lead to the same values, when added), the eventual distribution is what we see in Figure 225:

Value of W	0	1	2	3	4
Probability of having that value	11%	30.5%	30%	19.5%	9%

Figure 225

What we are looking for is the probability of any given outcome when they are combined. That means first multiplying every possible combination, then adding those that all lead to the same result. That gives us the probability of any given result.

In other words, what we have performed is a **convolution**. (If this isn't fresh in your memory, refer back to page 101 and page 312.) We find the probability distribution of the sum of two random variables, by performing a convolution of their probability mass functions or probability density functions (PMFs or PDFs): we multiply every possible combination (because we are interested in either/or), and then add the results that correspond to each value (because we are interested in both/and).

In the case of a continuous random variable this is represented by an integral. If $Pr_1(x)$ is the PDF of X, $Pr_2(y)$ is the PDF of Y, $Pr_3(w)$ is the PDF of W, and $X + Y = W$, then we can say that:

$$Pr_3(w) = Pr_1(x) * Pr_2(y)$$

Recall that an asterisk is the operator that denotes convolution. To reduce the number of variables we are dealing with, because $X + Y = W$, we also know that $Y = W - X$ (subtracting X from both sides).

That means we can write our convolution integral as

$$Pr_3(w) = Pr_1(x) * Pr_2(w - x)$$
$$= \int Pr_1(x)Pr_2(w - x)dx$$

We'll look at a more mathematical proof that the PDF of the sum of two random variables is obtained by a convolution of their PDFs a bit later, when we look at moment generating functions, but hopefully this gives you an intuitive sense of why it works the way it does.

This will be a definite integral; the limits of integration will be the support of W. When we looked at convolution integrals we also looked at dummy variables. In this instance x (or y if we reverse the order because, as we also saw, convolution is commutative) takes the role of the dummy variable, which disappears after the definite integral is calculated. What we are left with is a function of W.

It's Important to note that the proof of this will assume that the variables are **independent**. We can't say that the PDF of the sum of two random variables is the convolution of their PDFs/PMFs if they aren't.

Returning to our scenario, we are looking at discrete distributions, not continuous ones. We can assume here that they are independent, because our null hypothesis says that their working together will make no difference to their sales performances, and we are using this as our starting benchmark. If their combined sales don't meet what we expected when we assumed they were independent, then we will reject our null hypothesis: something will have happened as a result of having them working in the same space.

To create a formula for the PMF of W, we need to create a convolution sum. Recall that the PMF of a Poisson distribution is:

$$Pr(X = x) = \frac{\lambda^x e^{-\lambda}}{x!}$$

Where λ is the average number of events per unit of time, and x is the number of events. In our case our events are sales, our unit of time is days, and we need to sum them.

This is the basic form of a convolution sum for discrete variables. The supports of both X and W are from 0 to ∞, so we could have made this a sum to infinity. However, it works out more neatly if we sum to W, which is technically at least as correct, because the support of the equation has to be the support of W (whatever that actually

$$Pr(W = w) = \sum_{x=0}^{w} f(x)g(w - x)$$

is), and it works out more neatly as we progress.

This is our sum, with the formulae for the two Poisson distributions added in, with their different values for λ. Convolution is commutative (page 312), so it wouldn't have mattered which λ we paired with x, and which with $w - x$. It's a really ugly-looking formula if you start plugging values in, but fortunately it works out neatly.

$$= \sum_{x=0}^{w} \frac{20^x e^{-20}}{x!} \frac{15^{(w-x)} e^{-15}}{(w-x)!}$$

Take the constants outside of the summation, and rearrange the terms differently. What we now have is really close to a binomial expansion of $(20 + 15)^w$ (fortunately). Not a binomial probability; just a binomial expansion.

$$= e^{-20} e^{-15} \sum_{x=0}^{w} \frac{1}{x!\,(w-x)!} 20^x 15^{(w-x)}$$

All we need to do is put a $w!$ in the numerator, and balance that by putting the same thing into the denominator outside the summation.

$$= \frac{e^{-35}}{w!} \sum_{x=0}^{w} \frac{w!}{x!\,(w-x)!} 20^x 15^{(w-x)}$$

Now it is a binomial expansion of $(20 + 15)^w$

$$= \frac{e^{-35}}{w!} (20 + 15)^w$$

What we end up with is a Poisson distribution with a λ of 35: in other words, the mean is the sum of the means of the individual sales performances.

$$Pr(W = w) = \frac{35^w e^{-35}}{w!}$$

So now you have a probability function for the combined performance of your sales people (Figure 226). You'll be able to track sales each day and quickly assess the probability that any changes you see to the sales totals are just random fluctuations, compared to being a change (for better or worse) brought about by the new working arrangements.

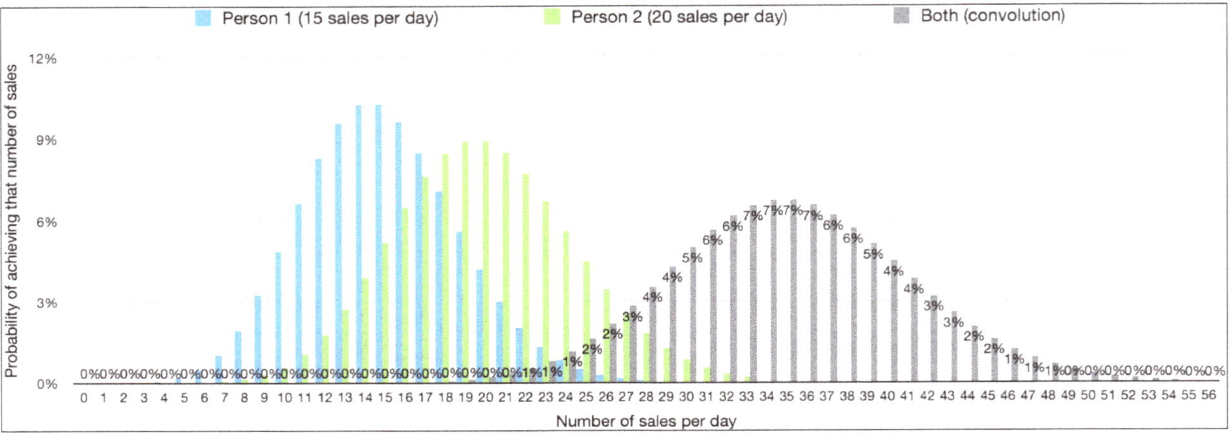

Figure 226

You could probably have guessed that it would be a Poisson distribution based on the sum of the averages of the individual sales performances, but it isn't always that obvious. There are other ways you could have looked at this problem, and convolution sums and integrals don't usually work out that neatly.

In fact, when adding identically distributed random variables the result is usually, but not always, the same type of distribution in the convolution as it is in the original distributions. The exceptions are as you might expect from what we have seen: the sum of Bernoulli's is binominal, the sum of geometrics is negative binomial, and the sum of exponentials is gamma. However, when the original distributions have the same basic formula but different parameters, things can be a little different. When you add two exponentials with different values for λ, you don't get a gamma, you get a hypoexponential, the PDF of which is

$$\frac{\lambda_1 \lambda_2}{\lambda_1 - \lambda_2} \left(e^{-x\lambda_2} - e^{-x\lambda_1} \right)$$

(We won't be examining the hypoexponential in this book.) It doesn't often happen that we need to add two random variables that have different types of distribution altogether, but if you ever do, you can still use convolution.

Key points:

- You sum random variables by convolution of their PDFs.

- This creates a joint distribution.

- For continuous distributions, the formula for that is

$$Pr_3(w) = Pr_1(x) * Pr_2(w - x)$$

$$= \int Pr_1(x)Pr_2(w - x)dx$$

- When adding identically distributed random variables the result is usually, but not always, the same type of distribution in the convolution as it is in the original distributions.

Third Stream: Errors in Measurement

When Lieutenant James Cook of His Majesty's Royal Navy set out from England in 1768 in a small sailing ship to travel around Cape Horn, with secret orders to cross the Pacific and explore the coastlines of New Zealand and the great Southern continent (later known as Australia), the British Admiralty put out a cover story to fool the French. Publicly, Cook was sent to observe the transit of Venus across the sun. To us that may sound like a feeble excuse for taking a small wooden sailing ship and her crew around Cape Horn and to the other side of the planet, but at that time it was perfectly reasonable. The more precisely ships could navigate by the stars, the safer they were and the more successful they could be in war and commerce. To allow sailors to navigate precisely, astronomers had to make their readings as accurate as possible, based on a range of different markers, with the handmade instruments and timepieces available at that time. Navigators often had to work from the deck of a moving ship, making the exercise more challenging. Astronomers and mathematicians therefore focussed energy on the quest to find the true reading from a group of scattered ones. Astronomical objects move and need to be observed from different locations, and so collecting as many readings as possible was not always feasible. Astronomers often had to work with what they had.

Furthermore, the idea of gaining as many readings as possible to get a more accurate idea of the true value was initially not universally accepted. Some people argued that different people had different skill levels and so involving too many people in the process could just lead to contaminating the data. They argued that a few readings taken by the best person would be a more accurate guide.

Leonhard Euler, one of the greatest mathematicians ever, failed in his attempt to address this problem adequately. In 1748, the Academy of Sciences in Paris offered a monetary prize for the best paper to explain certain apparent anomalies that had been observed in the orbits of Jupiter and Saturn. Euler wrote a 123-page paper, which was full of brilliant mathematics. He came up with an equation that had 13 separate terms in it, some of them quite involved, to explain the phenomenon in question. His paper was judged the best entrant and won the prize. Unfortunately, when he came to check his work with empirical observations that had been made over many years, he was unable to get his data to match.

The problem was that, thinking as a mathematician, Euler tried to work with a few variables at a time because he was aware that the maximum value of errors would tend to increase as more incorrect data was added, which is the way that mathematicians tend to think about such issues. The paradigm shift that came gradually and allowed statistics to flourish was a focus by Pierre-Simon Laplace on the *probable* value of combined observations rather than the *maximum* possible degree of error[53].

This gradual conceptual shift led to three important concepts of statistics: averages, least squares analysis and the error-distribution probability curve.

Averages

Read this if you can't remember the difference between a mean, median and mode; you aren't sure which of these would be best to use with a categorical variable; you glance through the mathematics and it isn't clear to you; you don't know why the sum of differences from the mean will always equal zero; or you don't know which average will be shifted where when a distribution is skewed.

When we looked at symmetrical, unimodal probability distributions earlier (page 360), the highest pointed showed the "average" of the distribution, but that is a word with a few different meanings.

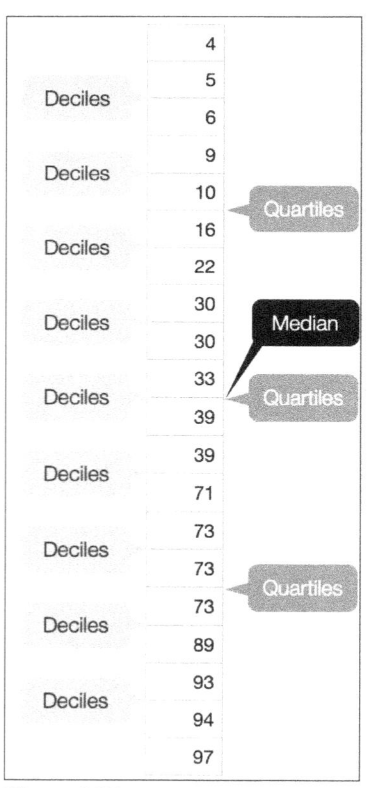

Figure 227

The **mode** is the most-common score (as in mode = fashion; most popular). In the column of numbers in Figure 227 the mode is 73, because it is the only number that occurs 3 times. The mode is usually most helpful for categorical data. For example, the mode would allow you to determine the roofing material of the "average" house. Other forms of average have no meaning in this instance. In a probability distribution, the mode shows the most probable outcome, or the highest point of the curve.

The **median** is the split point in the middle, once numbers are sorted. There will be the same number of data points on each side of the median. In Figure 227 the median lies between 33 and 39. (Figure 227 shows an even number of data points. If there is an odd number the median will be one of the data points. If there is an even number it will fall between two of them; in that case the **mean** of the two numbers closest to the middle is used; and we'll look at what a mean is soon.) One advantage of using a median is that it isn't affected by particularly large or small numbers. It tends to be used for housing prices and examination scores, and other areas where the priority is to understand where a particular datum point sits in relation to the other data points.

When we divided sorted data like this we call it using **quantiles**. The median is the quantile that is used when a set of sorted data is divided in half. Other commonly-used quantiles are **quartiles** (breaking the set of sorted data into quarters with an equal number of data points in each), **quintiles** (breaking the sorted data into fifths) and **deciles** (breaking the sorted data into tenths). When the median is used there can sometimes be additional ways to understand where one piece of information fits with the rest, such as the minimum, the maximum, the inter-quartile ranges.

Figure 228
Tycho Brahe
1546–1601

Tycho Brahe (1546-1601) was a Danish aristocrat who practiced alchemy and astrology, and who didn't own a telescope. He was also, though, an astronomer who was passionately interested in improving the accuracy of astronomical readings. While he didn't invent the concept of the arithmetic **mean**, he appears to have been the first person to make use of the idea of taking a number of readings, then find the mean of them in order to select the most accurate one.

The mean is obtained by adding up the different numbers and dividing by however-many of them there are. So the mean of the data points in Figure 227 is 45.3, obtained by adding up their values (giving 906) and dividing by the number of data points (that is, 20).

The mean can be written like this:

$$Mean = \frac{x_1 + x_2 + x_3 \ldots x_n}{n}$$

where $x_1, x_2, x_3 \ldots x_n$ are the individual data points, and n is how many data points there are.

Alternatively, it can be expressed as:

$$Mean = \frac{\sum_{i=1}^{n} x_i}{n}$$

$$= \sum_{i=1}^{n} \frac{x_i}{n}$$

$$= \frac{1}{n} \sum_{i=1}^{n} x_i$$

These are, of course, all ways of writing the same thing.

The mean is sometimes written with the letter $\bar{x}$ (pronounced x bar), particularly for the mean of a sample rather than a whole population, and so the distance of any point, x, in a set of data from the mean can be indicated by $x - \bar{x}$.

It is also sometimes signified with the Greek letter μ (mu), particularly (but not exclusively) when it represents the mean of a whole population.

The mean will lie in middle of a set of data, so some of the differences will be negative and some will be positive. If you add them all together, they will always equal 0. We can write this as

$$\sum_{i=1}^{n}(x_i - \bar{x}) = 0$$

We can see why this is so in this way:

By definition.
$$\bar{x} = \frac{x_1 + x_2 + \cdots + x_n}{n}$$

Multiply both sides by n.
$$\bar{x}n = x_1 + x_2 + \cdots + x_n$$

Subtract $\bar{x}n$ from both sides.
$$0 = (x_1 + x_2 + \cdots + x_n) - \bar{x}n$$

Because $-\bar{x}n$ is the same as n lots of $-\bar{x}$ and we know we have n data points, we can distribute $-\bar{x}$ among the data points.
$$0 = (x_1 - \bar{x}) + (x_2 - \bar{x}) + \cdots + (x_n - \bar{x})$$

And so the sum of differences from the mean always equals zero.

The graph of a frequency distribution often has a unimodal (single-humped), symmetrical shape, similar in shape to those of the binomial distribution. For shapes of this nature, it doesn't matter what definition of average is used: the mode will equal the median which will equal the mean. This is, however, not necessarily the case.

For example, Figure 229 shows the graph of a unimodal distribution which is skewed to the right (meaning that the **tail** is longer on the right; the peak shows on the graph as lying to the left). In this case the mean and the median will be shifted to the right, with the mean usually further to the right than the median. In a probability distribution of course, this indicates that the mode will be the most probable outcome.

There are two limitations in using an average of readings to find the most likely true value. The first is that, particularly with astronomy, it often wasn't a simple number that was needed. It was an equation that would describe the movement of a celestial body, and so the question became: how do you use a series of imperfect measures to find an equation for a curve that best describes them? The second is that it might have only been possible to take a very small number of readings: how do you make an educated guess at the correct number from that?

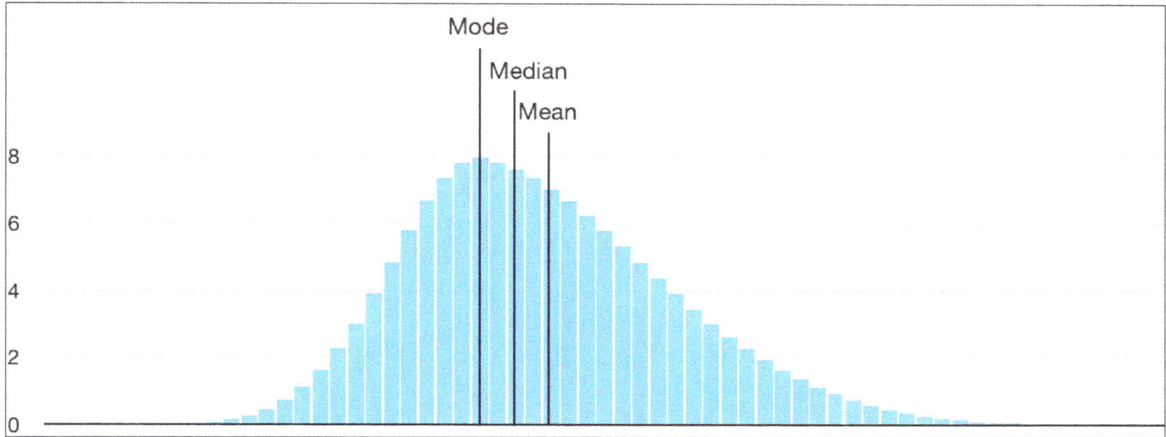

Figure 229

Key points:

- The **mode** is the most-common score (best with categorical data), the **median** is the middle score once numbers are sorted and the **mean** is obtained by adding up the different numbers and dividing by however-many of them there are.

- The median divides a distribution into two **quantiles**. Other commonly-used quantiles are **quartiles** (breaking the set of sorted data into quarters), **quintiles** (breaking the sorted data into fifths) and **deciles** (breaking the sorted data into tenths).

- The mean can be expressed mathematically by

$$Mean = \frac{x_1 + x_2 + x_3 \ldots x_n}{n}$$

$$= \frac{\sum_{i=1}^{n} x_i}{n}$$

$$= \sum_{i=1}^{n} \frac{x_i}{n}$$

$$= \frac{1}{n} \sum_{i=1}^{n} x_i$$

- The mean is sometimes written with the letter $\bar{x}$ (pronounced x bar), particularly for the mean of a sample rather than a whole population, and so the distance of any point, x in a set of data from the mean can be indicated by $x - \bar{x}$. It is also sometime signified with the Greek letter μ (mu), particularly (but not exclusively) when it represents the mean of a whole population.

- The sum of differences between each datapoint from the mean will always equal 0.

$$\sum_{i=1}^{n} (x_i - \bar{x}) = 0$$

- In a right-skewed distribution the mean and median will be shifted to the right, with the mean normally shifted further.

The least squares method

> **Read this if** you don't know what the least squares method is, how to calculate it and why it is used. Also read it if you don't know what a regression coefficient or a regression line are.

We met Carl Friedrich Gauss (1777–1855) in part 1, in a few different contexts. Gauss liked to develop final mathematical solutions without showing his working, because he saw that as inefficient. So when he claimed to have been the first to have developed these solutions, questions remain, because he wasn't the first to publish them.

In 1801 Italian astronomer Giuseppe Piazzi briefly observed an asteroid/dwarf planet which he named Ceres. Using the instruments of the time he was only able to make three inaccurate observations and was unable to derive a mathematical formula to account for its orbit. The asteroid went behind the sun, and Piazzi was unable to predict where it would be when it emerged. At the age of 24 Gauss solved the problem when others could not.

He said later that one of the techniques he used was the **least squares** method. (The first *written* explanation of this technique was by French professor of mathematics Adrien Marie Legendre (1752–1833), in 1805[54]. We met him in in relation to the Eulerian integrals.) The idea behind this method is that you develop an equation in a general form that you think will describe the shape of the line or curve you are looking for (in the case of an orbiting body, that would be an ellipse), develop a formula for the distance of each of your observations from that equation, then square each of the differences.

Squaring them achieves two things: first, it eliminates negative values (because otherwise they would probably just cancel out the positive ones) allowing values on either side of where you think the true value will lie. Second, it creates a curve that can be analysed with differential calculus to find the slope of the curve at any point. This can be used to find values for the equation which minimise the squares of the distances of each of your measures from it. In other words, you find an equation that best fits the data you have without necessarily matching it precisely.

From a mathematical standpoint, this was revolutionary thinking. Mathematics is all about *precision*; yet here (and in the rest of statistics) mathematics was being used to make the best-possible educated guess in a situation of *imprecision*.

Here is an example of how that can work in practice. In this instance, to keep it (relatively) simple, we are using the least squares method to find the equation of a straight line that best fits a set of points on a Cartesian plane (Figure 230). The main modern application of this is to find the relationship between two paired, numerical variables. Say, for example, we want to know the relationship between teenagers' ages and their height. We can record their ages on the *x* axis and their height on the *y* axis. Each student will therefore be represented by a point with an *x* and a *y* coordinate. Obviously there will be a lot of variation in height compared to age. What we want to know is what equation will give us a straight line that best fits the data. (The related, and usually more important question of how closely the available data then approximates to that line—the correlation of the measures—is something that came later in the history of statistics, and we will look at it later in this book.)

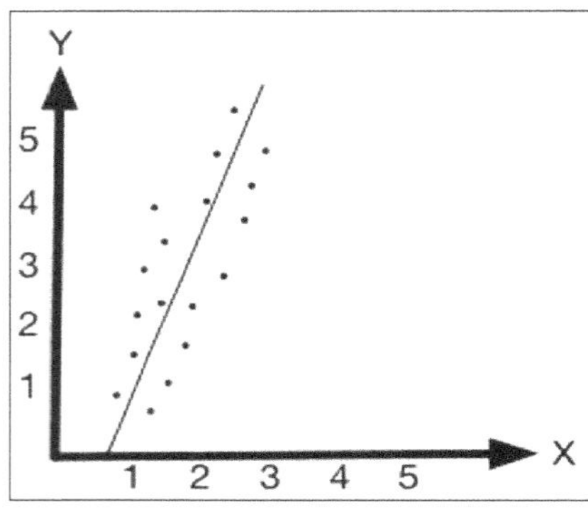

Figure 230

As a reminder about material we covered earlier (pages 194 and following), an equation on the *x-y* plane for a straight line will be in the form of

$$y = mx + b$$

where *m* and *b* are constants. This line of best fit is known as the **regression line** (Figure 231), for reasons we will come to in a later section. The slope (the vertical change in response to a horizontal change) will be given by *m*, because this constant determines how much *y* changes for a given change in *x*. If *m* is negative, the line will slope backwards (as it does in Figure 231). The elevation of the whole line is given by *b*: where $x = 0$, *y* will equal *b*. That will be the point where the line intersects the *y* axis, meaning that the whole line will be lifted vertically by an amount of *b* (or will drop by *b* if it is negative).

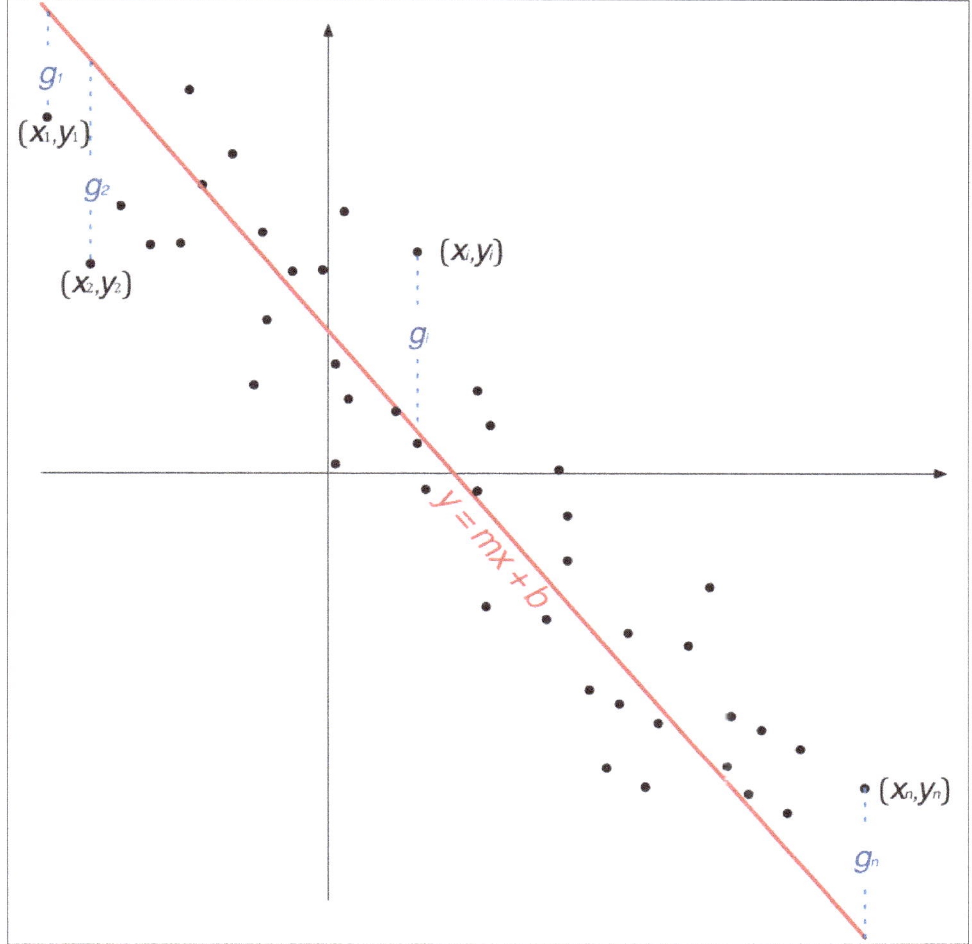

Figure 231

We'll call each of our actual, observed values (x_i, y_i) and we'll call $\bar{x}$ the mean of all of the x_i values, and $\bar{y}$ the mean of all of the y_i values. The difference between each of the observed values and the values on the regression line is called the **residual**, and the aim is find an equation for the line that minimises the absolute value of the residuals. We will designate each of these residuals as g for "gap". (We could have used "e" for "error", but we use e already. We could have used "r" for "residual", but r gets used for something connected but different, as we will see later. So for now we will stick with g).

We could work with either the y (vertical) or the x (horizontal) gaps, but the convention is to use the y gaps, so we'll go with that. You can see from Figure 231 that if we kept each of the x values as they are for our different data points, and only changed the y values by the amount of the gap, we could get each point onto the line. That means, among other things, that to find the line of best fit we only need to think of one set of gaps, not different sets for x and for y.

The line we are seeking will pass through the point $(\bar{x}, \bar{y})$. We said in the previous section that, in terms of finding the true value from a set of imperfect measurements, taking the mean will be a good place to start. The same thing applies when we are seeking an equation that gives the best fit. We can prove that our line will pass through the means of the x and y values like this:

Each y value will be m times its corresponding x value, plus the b from our equation, plus some residual amount that represents its distance (which may be positive or negative) from the line.

$$y_1 = mx_1 + b + g_1$$
$$y_2 = mx_2 + b + g_2$$
$$\cdots$$
$$y_i = mx_i + b + g_i$$
$$\cdots$$
$$y_n = mx_n + b + g_n$$

Add all of the equations representing our different x-y pairs together. Note that m is a constant, and so, by the property of linearity, can be taken outside the summation.

$$\sum_{i=1}^{n} y_i = m \sum_{i=1}^{n} x_i + \sum_{i=1}^{n} b + \sum_{i=1}^{n} g_i$$

b is a constant, so the sum of n lots of b is just nb.

$$= m \sum_{i=1}^{n} x_i + nb + \sum_{i=1}^{n} g_i$$

Divide both sides by n.

$$\frac{\sum_{i=1}^{n} y_i}{n} = m \frac{\sum_{i=1}^{n} x_i}{n} + \frac{nb}{n} + \frac{\sum_{i=1}^{n} g_i}{n}$$

$\dfrac{\sum_{i=1}^{n} y_i}{n} = \bar{y}$ and $\dfrac{\sum_{i=1}^{n} x_i}{n} = \bar{x}$ by the definition of "mean".

$$\bar{y} = m\bar{x} + b + \frac{\sum_{i=1}^{n} g_i}{n}$$

Now the expression

$$\frac{\sum_{i=1}^{n} g_i}{n}$$

gives us the mean distance from the **regression line** of each of our y values. As we are drawing a line that minimises these distances, we definitely want it to pass through any points where the distances cancel one another out and sum to zero: the total-distance-from-the-line of points above the line is exactly equal to the total-distance-from-the-line of points below the line. If we want our line to pass through here, that means our line $y = mx + b$ will pass through the point where

$$\bar{y} = m\bar{x} + b + (0)$$

Therefore the mean of the x values and the mean of the y values are on the regression line, at a point where the mean of the gaps is zero.

Now m and b are constants. They will remain the same regardless of the values of x_i and y_i. If we subtract $m\bar{x}$ from both sides we get:

$$\bar{y} - m\bar{x} = b$$

To give us an equation for our regression line, we need to find the values of m and b that will further minimise the sum of g^2. To (hopefully) further avoid confusion and to make it easier to distinguish the x and y values of our data points from the formula for the (as yet undetermined) line of best fit, we will follow convention and put hats over the variables that describe the line, as in $\hat{x}$ and $\hat{y}$. These are called, not surprisingly, x hat and y hat.

Each gap will be the difference between the observed value, y_i, and the hypothetical value of $\hat{y}$ on the line.

$$g_i = y_i - \hat{y}$$

The value of x_i will be the same for both y_i and for $\hat{y}$ (refer to Figure 231), so if $y = mx + b$ in general, then at the location of y_i we can say that $\hat{y} = mx_i + b$.

$$\hat{y} = mx_i + b$$

Based on the two equations above, this is the equation for each y gap.

$$g_i = y_i - (mx_i + b)$$
$$= y_i - mx_i - b$$

We said above that $\bar{y} - m\bar{x} = b$

$$g_i = y_i - mx_i - (\bar{y} - m\bar{x})$$
$$= y_i - mx_i - \bar{y} + m\bar{x}$$

Shuffle the terms, and factor out $-m$ from both terms that contain it. This equation applies to each of our data points.

$$g_i = (y_i - \bar{y}) - m(x_i - \bar{x})$$

At this point, square both sides of the equation, then sum these squares of the individual data points together.

$$\sum g_i{}^2 = \sum\{(y_i - \bar{y}) - m(x_i - \bar{x})\}^2$$

Expand out the binomial on the right-hand side of the equation.

$$= \sum\{(y_i - \bar{y})^2 - 2m(y_i - \bar{y})(x_i - \bar{x}) + m^2(x_i - \bar{x})^2\}$$

$$\sum g^2 = \{\sum(y_i - \bar{y})^2\} - 2m\{\sum(x_i - \bar{x})(y_i - \bar{y})\} + m^2\{\sum(x_i - \bar{x})^2\}$$

Use the linearity attribute of series to show this as three separate series, and to take the constants outside of the summations (and use colours to help keep track of things).

Now this is the clever bit.

Up until now we have been treating m as a constant, while x and y are variables. But we are seeking the value of m in this instance. This means we can treat each of these summed amounts as constants, and m as the variable. Let

$$\sum(y_i - \bar{y})^2 = C$$

$$-2\sum(x_i - \bar{x})(y_i - \bar{y}) = B$$

$$\sum(x_i - \bar{x})^2 = A$$

The equation now becomes

$$\sum g^2 = Am^2 + Bm + C$$

$$= f(m)$$

If we drew this up on a separate coordinate plane with m as the horizontal axis and $f(m)$ as the vertical it would form the shape of a parabola. Earlier (page 240) we used differential calculus to determine the minimum value of this parabola (not the minimum value of m, but the minimum value of $f(m)$ which will be the point where the slope of the tangent to the curve is flat: that is, it equals zero). We saw that it is found where

$$m = \frac{-B}{2A}$$

This is the value of m that will minimise $\sum g^2$, or the sum of the squares of the residuals.

Substituting back again gives us

$$m = \frac{--2\sum(x_i - \bar{x})(y_i - \bar{y})}{2\sum(x_i - \bar{x})^2}$$

$$\boxed{\begin{array}{c} Regression\ coefficient = m \\[6pt] = \dfrac{\sum(x_i - \bar{x})(y_i - \bar{y})}{\sum(x_i - \bar{x})^2} \end{array}}$$

So now we have an equation that will tell us what m should be to give us a regression line, or a line of best fit, through our collection of data points. This is called the **regression co-efficient** for a straight line. Using this value we can calculate the value of b, based on $\bar{y} - m\bar{x} = b$. These two values together will give us the formula for our regression line: $y = mx + b$. So the complete formula becomes:

$$y = x\frac{\sum(x_i - \bar{x})(y_i - \bar{y})}{\sum(x_i - \bar{x})^2} + \bar{y} - \frac{\sum(x_i - \bar{x})(y_i - \bar{y})}{\sum(x_i - \bar{x})^2}\bar{x}$$

$$\boxed{y = (x - \bar{x})\frac{\sum(x_i - \bar{x})(y_i - \bar{y})}{\sum(x_i - \bar{x})^2} + \bar{y}}$$

where x and y are the variables of the regression line, x_i and y_i are each pair of data points, and $\bar{x}$ and $\bar{y}$ are the means of the x_i and y_i values respectively. If you had to work it out without a calculator it would be time-consuming and tedious, but you could do it.

What you will really do, of course, is just plug the data into a computer to get the results of this formula, but now you know what the computer is doing.

There will virtually always be a line that offers the best fit for a set of data points, but it might not be a particularly close fit. Sometimes another equation will give a curve that offers a better fit for the data. The same approach—the least squares method—can be used to find formulae for approximating data to different types of curve.

Key points:

- The **least squares method** is used to develop an equation that provides the best fit between a set of data points.

- The basic idea is to use a generalised equation for the type of line or curve you think will be most appropriate, develop a formula for the sum of the squared distances of each of your points from the theoretical curve, then use calculus to develop an equation that will minimise the total of the squared distances.

- This line is called the **regression** line (or regression curve) and the difference between each point and the line is called the **residual**.

- The formula for a straight regression line is

$$y = (x - \bar{x}) \frac{\sum(x_i - \bar{x})(y_i - \bar{y})}{\sum(x_i - \bar{x})^2} + \bar{y}$$

- The part in green is the **regression coefficient**.

An error probability distribution curve and the quest for a breakthrough

> **Read this if** you don't understand how errors consistently behave, you don't know why they behave consistently, you can't imagine how you could express that mathematically, or you didn't know that there were two completely different ways of thinking about the problem that, independently, eventually led to the normal distribution/bell-shaped/Gaussian curve.

Least squares allows us to find an equation for a line or curve that best fits messy data. The next challenge was: how do you find the most likely true value when you have a very small number of readings? At first this sounds an impossible task. What makes it doable is the fact that random errors tend to cluster in a predictable pattern.

This thinking lead to the quest for a great breakthrough of measurement: finding an equation that would describe how errors in measurement would tend to consistently behave. If you had such an equation in general form, you could potentially apply the least squares method to even a very small set of data (say three readings), to find the specific equation that would give the most probable description of the pattern errors of measurement of whatever you were looking at. From there, you could potentially make a very realistic estimation of the true value.

This is another area where Laplace did the work and largely solved the problem, while someone else received the credit. What statisticians usually call a **normal distribution**, and psychologists tend to call a **bell-shaped curve**, mathematicians call a **Gaussian distribution**. There is, however, limited evidence that Gauss was actually involved[55]. Laplace largely solved it in 1774, three years before Gauss was born.

All else being equal (i.e. not caused by a consistent bias such as a clock that runs slowly), errors tend to be symmetrically distributed around a correct value. The frequency plot is usually bell-shaped, just as we saw with the Poisson distribution (which had not been developed at this point in history), and with the binomial distribution.

Binomial approach

Abraham de Moivre had already developed an equation that described a bell-shaped curve that seemed to meet the requirements. He used something similar to what was later called Stirling's approximation (which we won't be covering) to look at what happens to the binomial as n approaches infinity. He used hyperbolic logarithms, which are the same as logarithms to base e but were discovered before e was defined. He knew that the square root of 2π (which he called "the ratio of the circumference to the radius") was involved. Laplace was familiar with de Moivre's work and gave him credit for his thinking about it[56].

Gauss knew a lot about errors. He was once commissioned by King George IV of England, who also ruled Hanover, to survey the entire kingdom of Hanover. A story about him says he made the surveyors take 14 separate readings of each landmark, so that he could use the pattern of the errors to determine the true value. He was definitely more task-focused than people-focused. Yet another story says that when he was told his wife was dying he waited to finish the equation he was working on before he went to her bedside.

Laplace knew that one way to think about the distribution of errors was as a binomial problem. Think back to the device we have thought about previously in relation to the binomial (Figure 232). (Just when you thought it would never appear again in statistics, Pascal's triangle makes another appearance.)

Now let's do a thought experiment. Imagine that, instead of putting ball bearings in at the top, you pour in sand. The numbers drawn in Figure 232 show the number of different paths that a grain of sand can take to reach that point, and therefore the amount of sand that accumulates under each space. (We saw something similar, with a fuller explanation, when we were looking at the binomial theorem on pages 160 and following.) At each obstacle a falling grain of sand hits, there is a Bernoulli choice of whether it will fall left or right. The choices of whether the grains of sand go to the left or the right of the stops are likely to balance one another out; roughly half will go one way, and half the other. So the grains of stand still end up falling into a binomial distribution, just has ball bearings would.

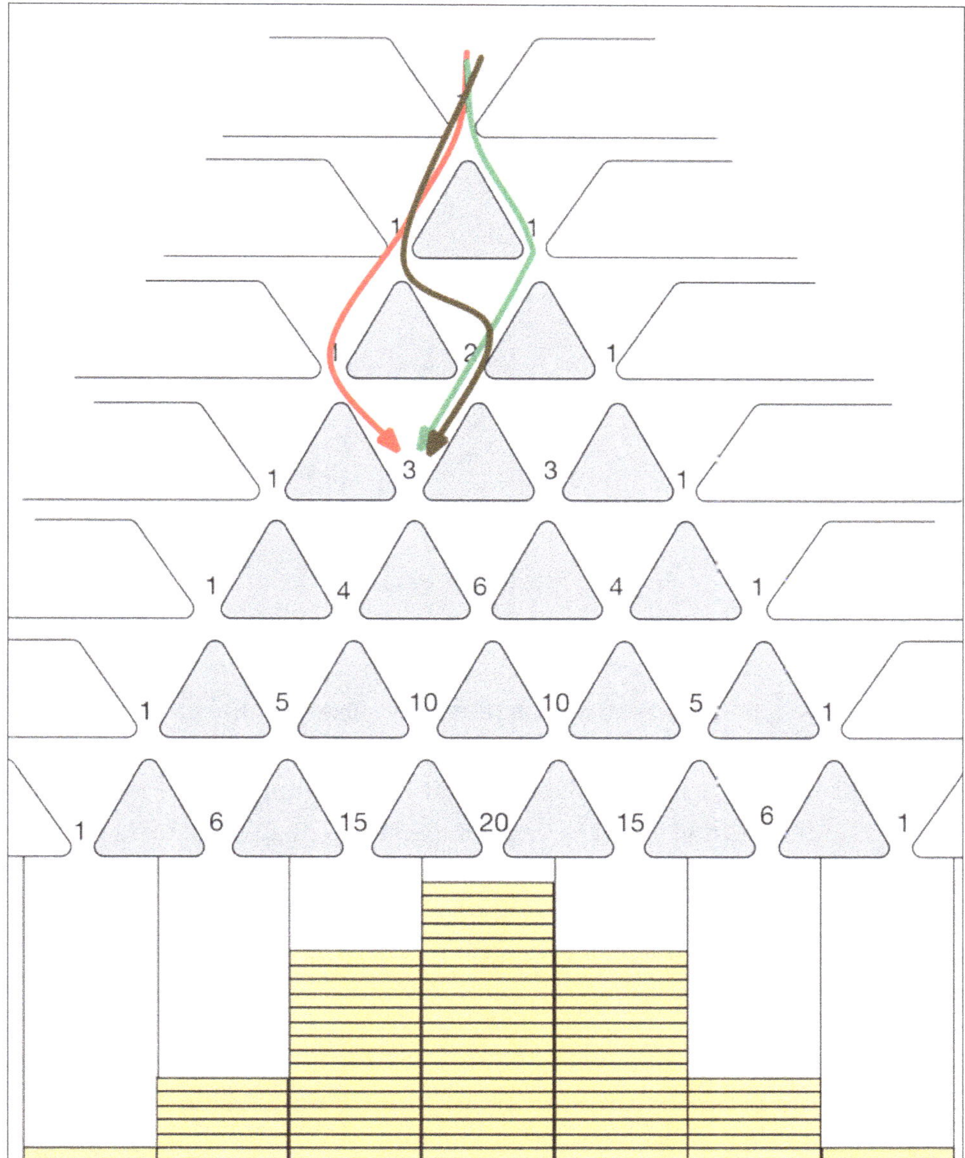

Figure 232

Now imagine we made the grey triangles much smaller, and put many more of them in. Imagine that instead of 6 rows, we had two hundred. Or two trillion. It won't take long before it starts to resemble a smooth, bell-shaped curve, except that it would still be stepped, not continuous. Now imagine that the stops keep getting smaller until they disappear, that the barriers between the columns at the bottom do as well, and that, when the grains fall, they are sticky enough to stay where they land and not roll away.

Imagine also that the sand was still kept between two close panes of glass; in other words, we are only looking at it in two dimensions.

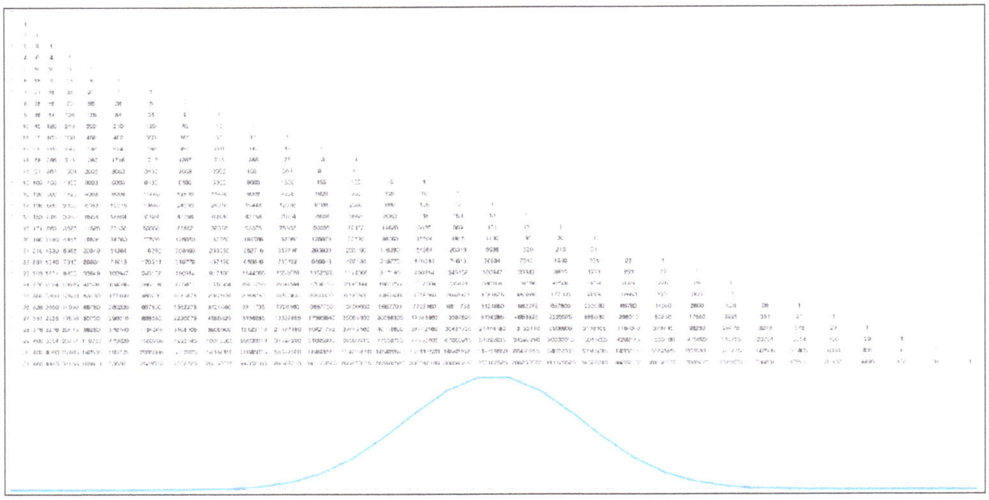

Figure 233

Figure 233 shows what happens when we just take Pascal's triangle to 31 rows. You can see that, even at this point, the graph it makes is starting to look smooth and bell-shaped. Laplace's thinking seems to have been that errors will work out in a similar way. At any instance there is a simple choice, about whether the error will be closer to or further from the true value. If you think about the effect of an accumulation of a myriad of such choices, you wind up with a smooth error-distribution curve. (We'll leave the mathematical proof that a binomial distribution eventually becomes a normal one until later.)

Laplace had an easier way to get to the same formula. He thought about a formula for an error-distribution curve in an entirely different way and reached the same mathematical conclusion.

Differential approach

Laplace thought about this problem in terms of the effect that the frequency of each reading, *together with* the distance of the reading from the true value, had on the *rate of change* of the curve.

The distance of errors in measurement from a true value is an example of a very common phenomenon: one where data points are both clustered and scattered, at the same time (Figure 234). Clustering and scattering are two different ideas that work together, and Laplace worked out how to express these two ideas with a deceptively simple mathematical equation, using calculus.

Figure 234

He realised that when data are randomly **clustered** the rate of change of the distribution curve was proportional to the negative distance of x from the true value. You can see from Figure 235 that when x is at zero, a line drawn as a tangent to the curve would have a slope of zero. That is, it would be flat. As x very slightly increases in a positive direction the tangent line would start to slope backwards; it would have a negative rate of change. As x decreases slightly from 0 it would start to have a positive slope.

Mathematically we can write this as

$$\frac{dy}{dx} \propto -x$$

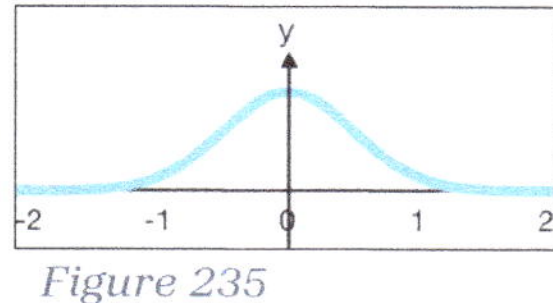

Figure 235

In terms of **scattering**, as x gets further from the true value, the frequency of errors (or the y value) is reduced. This means, once again, that the slope of the curve decreases as the y value decreases. In other words the rate of change of the curve is also proportional to the frequency of readings that measure that particular value: the closer one reading is to another, the more likely it is there will be more of them close by. Mathematically, we can describe this as saying that the rate of change of the curve is proportional to y.

$$\frac{dy}{dx} \propto y$$

The rate of change of the curve is negatively proportional to the closeness to the true value, *and* directly proportional to the frequency of other values. That is both a challenging idea to wrap your mind around, and yet a simple description for any data that are clustered and scattered in a random way, at the same time.

Putting those two tendencies together into an equation for the slope of the curve was simple.

$$\frac{dy}{dx} = -Axy$$

where A is some positive constant.

That is the essence of what has become one of the fundamental ideas in statistics. It's called the **normal distribution** curve, not in the sense of "what normally happens" (although it is, in fact, what normally happens) but in the sense of "distribution around the norm (or mean)." (It's also called a **Gaussian** distribution, but it's doubtful that Gauss had much to do with it.) Clustering and scattering in a random way doesn't just apply to errors, but actually, of course, turns out to be a common phenomenon in a very wide range of situations, which is a reason why this type of curve is so important in statistics (there are other reasons, which we will get to).

From this simple equation, finding the equation for the curve is an exercise in integral calculus using a separable differential equation, and we worked through this equation as an example on page 298.

We found that that the equation for the curve is given by:

$$y = Ce^{\left(\frac{-Ax^2}{2}\right)}$$

where A is some positive constant and C is another constant. (You may note here that there is a similarity with the equation for the Poisson distribution, which also has a not-dissimilar shape. Both include raising e to a negative power.)

This assumes that the correct value, or the mean of the observations, is zero. In fact it could be anything. We can account for this by writing the equation as

$$y = Ce^{\left(\frac{-A(x-\mu)^2}{2}\right)}$$

where A is some positive constant, C is another constant and μ is the mean of all the values. (μ can, of course, be written as $\bar{x}$.) That will shift the equation by an amount of μ along the x axis, but otherwise leave it unchanged. That way where x is equal to the mean, $x - \mu$ will equal zero.

Where C increases, the height of the peak of the curve will increase, and where A increases the sides of the curve will flatten out, but none of that makes it easy to make use of the curve until we understand what A and C are. That will be more challenging.

We do have two other pieces of information we can use. We aren't looking here for the distribution of errors in any single situation; we are looking for a theoretical model for a probability distribution of all possible random errors. There is no limit to how large an error could theoretically be, and so the edges of the curve can stretch in either direction to infinity. We also know that for a probability distribution, the total of all values must equal one.

We can therefore say that the area under this curve will equal 1, and we can use integral calculus to hopefully find values for C and A, with limits of integration between negative and positive infinity.

$$\int_{-\infty}^{\infty} Ce^{\left(\frac{-A(x-\mu)^2}{2}\right)} dx = 1$$

There isn't any straightforward way to solve that definite integral. In other words, there is no formula which, when differentiated will give this equation. Fortunately we don't actually need to complete the integration, because we already know the result (which is 1). What we need to do is manipulate the algebra of this equation, to find values for C and A, and there are ways to do that which you have already covered if you made it through the calculus section.

First take the constant C outside of the integration.

$$C\int_{-\infty}^{\infty} e^{\left(\frac{-A(x-\mu)^2}{2}\right)} dx = 1$$

Then use a substitution.

Let

$$w = \sqrt{\frac{A}{2}}(x - \mu)$$

Treat μ (mu) as a constant.

$$\frac{dw}{dx} = \sqrt{\frac{A}{2}}$$

Take the reciprocal of both sides.

$$\frac{dx}{dw} = \sqrt{\frac{2}{A}}$$

Multiply both sides by dw.

$$dx = \sqrt{\frac{2}{A}}\, dw$$

Square both sides of $w = \sqrt{\frac{A}{2}}(x - \mu)$

$$w^2 = \frac{A}{2}(x - \mu)^2$$

Substitute these values back into our integration

$$C\int_{-\infty}^{\infty} e^{-w^2} \sqrt{\frac{2}{A}}\, dw = 1$$

Take the constant $\sqrt{\dfrac{2}{A}}$ outside of the integration. We still aren't there in terms of having something we can integrate easily, though.

$$C\sqrt{\frac{2}{A}}\int_{-\infty}^{\infty} e^{-w^2}\,dw = 1$$

We have arrived here at the Gaussian integral, which we saw on page 305 has a solution of $\sqrt{\pi}$. The solution we have seen involved converting to polar coordinates and using a Jacobian, and that hadn't been thought of when Laplace was working on this. He came up with his own approach, which can be found in an appendix on page 758. He arrived at the same answer: $\sqrt{\pi}$

$$C\sqrt{\frac{2}{A}}\sqrt{\pi} = 1$$

Square both sides.

$$\frac{2C^2\pi}{A} = 1$$

Multiply both sides by A, divide both sides by 2π then take the square root of both sides. That means we have a way to express C in terms of A.

$$C = \frac{\sqrt{A}}{\sqrt{2\pi}}$$

So the formula for the error curve at this point became

$$y = \frac{\sqrt{A}}{\sqrt{2\pi}}e^{\left(\frac{-A(x-\mu)^2}{2}\right)}$$

This is another of those pieces of wonder-inducing mathematics. It's a general formula for the way that clustered data randomly scatters, and it turns out that it just happens to have three of the best-known irrational numbers in it:

$$e, \pi, \sqrt{2}$$

You may remember from the introduction that we spoke about the paradoxical predictability of randomness, and gave the illustration of grains being poured out. You can now see that the shape will be, seen from any side, roughly a bell-shaped curve fitting a normal distribution. Seen from above it will look like a circle, dense in the centre and thin at the periphery. These two shapes—the circle and the bell-shaped curve—are recurrent shapes of randomness, which may help to explain why e and π recur so often in statistical equations.

That doesn't quite get us to the modern formula for the normal distribution curve because we still don't know what A represents, and neither did anyone else until the end of the 1800s, but it gave them enough to work with.

Laplace was never able to work out what the constant, A, was equal to. In his words:

> *"The law of the possibility of the errors of observations introduces into the expressions of these probabilities a constant, whose value seems to require the knowledge of this law, which is almost always unknown. Happily this constant can be determined from the observations."*[57]

The formula that is used today includes the standard deviation (twice), and that's a concept that wasn't developed until about 60 years after Laplace died (and about 35 years after Gauss died, even though he somehow gets the credit for the whole thing). To get any further than this, though, we must overcome another obstacle: is this formula trustworthy?

Key points:

- There are at least two ways to understand how errors behave: as a binomial problem as n approaches infinity, and as a combination of clustering and scattering, and the impact those two ideas have on the slope of the distribution curve.

- The ideas of clustering and scatting are encapsulated in the simple formula
$$\frac{dy}{dx} = -Axy$$
where A is some positive constant.

- Treating this as a separable differential equation, then undertaking some complex integration (including the use of the Gaussian integral) this eventually leads to the formula for the curve
$$y = \frac{\sqrt{A}}{\sqrt{2\pi}} e^{\left(\frac{-A(x-\mu)^2}{2}\right)}$$

- That is as far as people were able to progress in the quest for an error-distribution curve, until more ideas about distribution curves (which we are yet to come to) were developed.

Empirical confirmation

Figure 236
Francis Galton
1822–1911

Getting a formula for the error curve to this point required both some heroic assumptions about the nature of the way the curve changes, and some fancy mathematical footwork. That, and the improbable fact that it contains e, π and $\sqrt{2}$ would conspire to cause us to naturally mistrust this formula as the foundation on which to build a statistical edifice. The situation might have remained that way if it wasn't for two scientists who were both inspired by Charles Darwin (1809–1882).

Francis Galton (1822–1911) was a first cousin of Darwin's. He has been described as "perhaps the last of the gentleman scientists"[58]. He had the financial resources to explore places and topics of his own interests, which ranged from visiting (then) remotest Africa, through collecting data from weather stations, through psychology, anthropology, sociology, education and fingerprints (on which he wrote a book). He was no great mathematician, but he did love to collect data, find the patterns within it, and see the way that patterns from one set of data could mirror patterns from another. For example, he wrote to European weather stations to collect data about barometric readings and wind patterns that he used to develop the understanding of high-pressure systems.

Galton was a firm believer in evolution and was particularly interested in eugenics (and in fact invented the term): the idea that we could help the process of natural selection along in humans and thus build a better human race. (And yes, sadly, the Nazis did look directly back to his work as part of their justification for death camps.)

Galton has been credited with introducing the ideas of regression and of correlation. He did recognise the importance of these issues, and wrote a eugenics-oriented paper called "Regression towards Mediocrity in Hereditary Stature", in which he said that physical characteristics in children are related to, but tend to be less extreme than those characteristics in their parents. He saw this as "regressing" towards the mean of the population, which is why the line of best fit we saw when we looked at least squares calculations is now called the **regression line**. Galton did little, however, to further the mathematical understanding of these subjects. In his own words:

> *"I can only say that there is a vast field of topics that fall under the laws of correlation, which lies quite open to a competent person who cares to investigate it."*

Likewise, he has been credited with development of the idea of variance. He did use the word to describe the fact that genetic traits were variable in their outworking in future generations, but he did not contribute to the mathematical understanding of the topic[59].

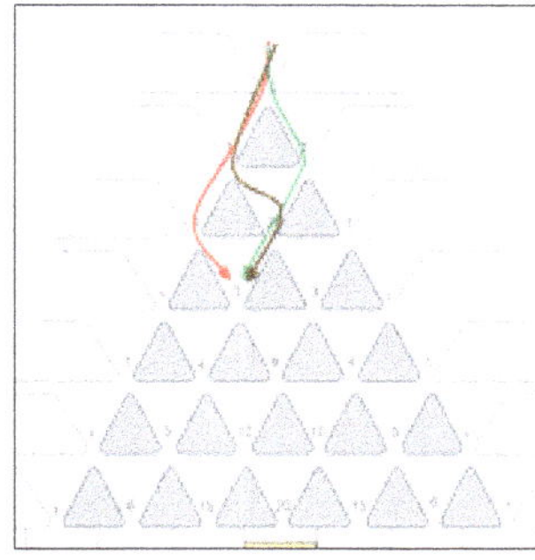
Figure 237

Galton became an enthusiastic champion of the normal distribution curve to describe natural phenomena in the data he was collecting in a whole range of seemingly disparate areas, and it was he who developed a version of the device we have seen for demonstrating the actions of the binomial theorem, which he called a **quincunx** (Figure 237).

The American John Dewey (1859-1952), now best known for his classification of subject matter used in libraries, reviewed Galton's work on normal distribution and saw the applicability of this sort of thinking. In 1889, reviewing one of Galton's books, he wrote:

> *"It is to be hoped that statisticians working in other fields, as the industrial and monetary, will acquaint themselves with Galton's developments of new methods, and see how far they can be applied in their own fields"*[60].

> The name "quincunx" means an arrangement of 5 objects in the pattern of the number 5 on dice. Galton's device must have looked like this, but placed at an angle, although there were more than 5 objects in it.

Marine biologist Raphael Weldon (1860-1906) was another scientist who developed empirical data about the normal distribution curve. Weldon was a true scientist, in that he wasn't convinced of either the truth or falsehood of Darwin's theory of evolution: he just wanted to gain the data and the insights to satisfy his curiosity one way or the other. He was particularly interested in finding whether it was possible to find data that could support Darwin's idea that animals can change through natural selection, not just in a way that emphasises characteristics that improve survival *within* a species, but he wanted to find whether there was evidence of, as Darwin predicted, ongoing gradual change *from* one species *to another*.

Figure 238
Raphael Weldon
1860-1906

He never did find the convincing evidence that this can happen, but on the way he collected vast amounts of obscure biological measures, such as taking 23 separate readings from different parts of the anatomy of 1,000 Mediterranean shore crabs. He recognised that many of these measures followed a normal distribution, and he wanted to know if there was evidence, from within his data, of two overlapping normal distribution curves, as one species separated from another. He realised that the issue was fundamentally one of statistics, and that he needed the help of a mathematician to assist him to analyse his data. In 1890 he was appointed to the Chair of Zoology at the University College, London. The man he enlisted to help him was a colleague at the college and a friend, as well as being a friend of Galton's: Karl Pearson.

Before we get to Pearson though, another stream of thought had to develop. It was based, of all things, on the physics (not the statistics) of the steam engine and other mechanical contraptions.

Fourth Stream: The Physics of Mechanical Contraptions

While the internal combustion engine didn't arrive until the 1860s, the first steam engine pumped water from a mine in Spain in 1698. It was inefficient and dangerous, and went through significant improvements before it could power railways and the industrial revolution.

Some of that improvement came from practical mechanical considerations, like the work of James Watt, while other aspects grew out of mathematics and physics, following on from the work on forces of rotation (initially applied to astronomy) by Isaac Newton, Leonhard Euler, Pierre-Simon Laplace and many others.

One of the areas that made a difference was the understanding of the torque needed to overcome rotational momentum around a central point, and around a point which was off-centre. None of which seems to have much to do with statistics; or it didn't until a Russian professor of mathematics/inventor of contraptions became intrigued by probability distributions.

Figure 239
Pafnuty Chebyshev
1821-1894

Pafnuty Chebyshev (1821-1894) was a Russian professor of mathematics at the University of St Petersburg. (To prevent things from becoming too simple, if you are looking him up, the spelling of his name can be transcribed from Russian characters in no less than 10 different ways: Chebyshev, Chebysheff, Chebychov, Chebyshov, Tchebychev, Tchebycheff, Tschebyschev, Tschebyschef, Tschebyscheff or Čebyčev.) He once objected to being described as a "splendid Russian mathematician", on the basis that he was a "world-wide mathematician", which was true. His first two papers appeared in French, and most summers he travelled to centres of mathematical learning in Europe and interacted with the leading mathematicians of the time. One of his early papers dealt with the work of French mathematician Poisson[61]. (While Euler was Swiss, he had also spent time working at the University of St Petersburg many years before, so Chebyshev was part of an impressive tradition.)

Chebyshev claimed that mathematics should always be focused on practical problems. One of his contributions was a mechanical calculator (Figure 240). It is possible, based on his comment about being a world-wide mathematician rather than a Russian one, that he may have been quite sensitive about having his contribution under-rated. If he was, he probably had the right to be. He is deservedly well-regarded within Russia, and is seen as the father of Russian mathematics. (You could argue that the reason that the USSR won the early part of the space race against the USA—being the first to get an object into space, a dog into space then a man into space—was that Russian mathematics was more advanced, and that the person most responsible for this was Chebyshev[62].)

Figure 240

Elsewhere in the world, while mathematicians from Western Europe are often well-known for smaller contributions to statistics than his, it was Chebyshev who did much of the deep mathematical thinking about probability distributions that paved the way for the birth of modern statistics, yet he isn't particularly well known outside of Russia[63]. In fact, he is probably still better recognised for his contributions to engineering than to statistics.

It isn't possible to know for certain how his mind worked or what conceptual models he used to develop his ideas, but it seems highly likely that Chebyshev worked, at least in part, by imagining probability distributions that were free-standing three-dimensional structures that could be balanced, twisted, and distorted; and that he thought about the mathematics that made this possible. In this (or some other) way, he made five outstanding contributions to thinking about probability:

1. his work on **expected value** (picking up on the work of De Moivre),

2. **moment generating functions (MGFs),**

3. **moments** of probability distributions,

4. the **Markov inequality** (somehow named for one of his students, yet Chebyshev wrote about it first), and

5. the **Chebyshev inequality** (which is possibly his smallest contribution, but is the one which his name is associated with).

Let's look at each of these in turn.

Expected value

Read this if you can't think of how you would find the "mean" of a skewed continuous probability distribution (or even what that would actually represent); or you don't know how to add, subtract or multiply expected values.

"Mean" of a probability distribution

When we looked at averages we glossed over some complexities. In particular, what does it mean to have the "mean" of a probability distribution? It is easy to see what that looks like in a discrete, symmetrical distribution like the binomial or the Poisson. But what does it mean in a skewed distribution like the geometric? How do you calculate it for a skewed continuous distribution like the exponential? It isn't as though we can just add up the values and divide by the total in this case. If Chebyshev conceptually envisaged a probability distribution as a three-dimensional object, he could have

Height (cm)	Number of players
174	2
175	1
177	4
178	3
181	2
184	1
Total	**13**

Figure 241

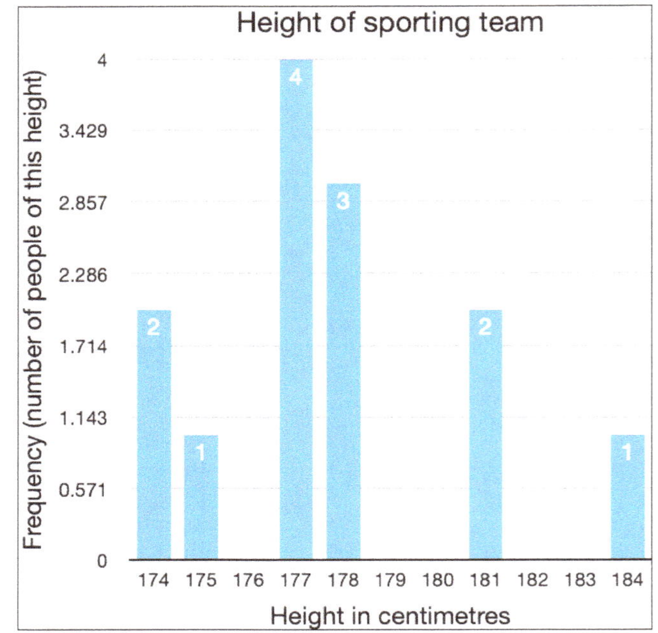

thought of the mean as the "balance point". But how do you find this mathematically?

To find the mean of a set of data that is expressed as a *frequency distribution*, you can multiply each of the values on the horizontal axis by the frequency with which they occur (the height on the vertical axis), then adding the results, before dividing by the total number of data points. This is simply another way to add each of the different data points together, then dividing by their total number.

So with Figure 241 we can work out the mean height of the members of a sporting team by adding each of their heights and dividing by the number of players. We can also calculate the same thing by multiplying each of the heights by the frequency with which it occurs and then adding the result, then dividing by the number of data points in this way:

$$\frac{(2 \bullet 174) + (1 \bullet 175) + (4 \bullet 177) + (3 \bullet 178) + (2 \bullet 181) + (1 \bullet 184)}{13} = \frac{2311}{13}$$

$$= 177.8cm$$

It's the same thing: we've still, ultimately, just added up all of the values and divided by the total number of values.

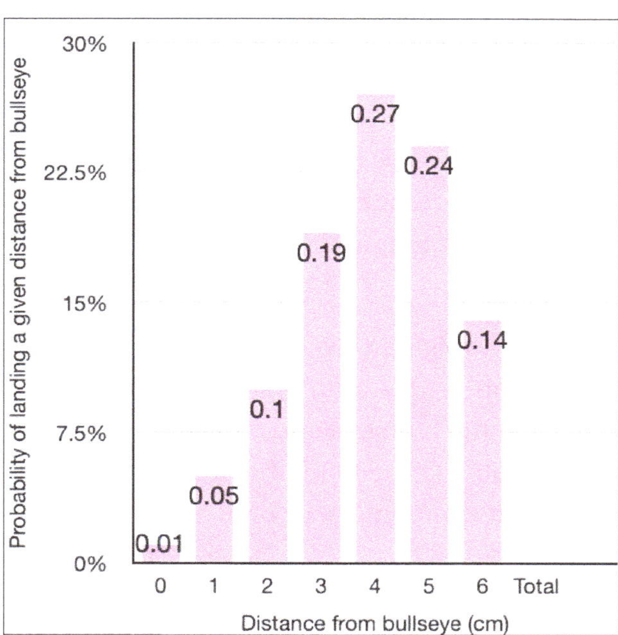

Figure 242

If, instead of having a specified number of players against the relevant heights, we had a *function* that described the frequency of people against the relevant heights, we could say that in a *discrete, frequency* distribution:

$$\bar{x} = \frac{\sum xf(x)}{n}$$

where x is each value, $f(x)$ is the function that tells us the frequency of readings at that value (the height of the graph), n is the total number of data points, and $\bar{x}$ is the mean.

If, instead of this being a *frequency* distribution it is a *discrete probability* distribution, as in Figure 242 (which looks at the probability of a mediocre darts player landing a certain number of centimetres from the bullseye), then you can calculate it the same way. Multiply the values on the horizontal axis by the probabilities on the vertical axis, and sum them together .

Chebyshev (and possibly others before him) called this the **expected value** or the expectation of the distribution, using the same terminology as de Moivre (but not quite the same meaning, as de Moivre was talking about money).

Distance from bulls eye (cm)	Probability	Total of distance multiplied by probability
0	0.01	0
1	0.05	0.05
2	0.1	0.2
3	0.19	0.57
4	0.27	1.08
5	0.24	1.2
6	0.14	0.84
Total	1.0	3.94cm

Figure 243

So the expected value—the "mean" place where this player will land a dart—is 3.94 cm from the bullseye (Figure 243). Expected value of a distribution of a random variable X has the notation $E(X)$. So we should use $E(X)$ when talking about a theoretical probability distribution, $\bar{x}$ for the mean of a sample of actual values, and μ as the mean of a population, and as we will see when we come to look at linearity and summing the expected values of random variables, the concepts, while similar, aren't completely interchangeable.

Note that when we are talking about the expected value of a probability distribution we aren't talking about the mean of the frequencies on the vertical axis (which would always lie between 0 and 1); we are talking about the "mean" of the data points on the horizontal axis, based on their expected frequencies.

Expected value is sometimes called a probability-weighted average. One way to think of it is as an average, but with each of the values given a weighting for its probability. (Note a distance from the bullseye of zero still influences the expected value, because it contributes to the total of 100%. If there were no scores on zero, the other probability percentages would be higher.)

With an expected value for a probability distribution there is *no need to divide by the n* as we did with a frequency distribution, because n is always 100%, and is *already factored in*. In other words, we could write:

$$(1\% \bullet 0) + (5\% \bullet 1) + (10\% \bullet 2) + (19\% \bullet 3) + \cdots etc$$

as

$$\left(\frac{1}{100} \bullet 0\right) + \left(\frac{5}{100} \bullet 1\right) + \left(\frac{10}{100} \bullet 2\right) + \left(\frac{19}{100} \bullet 3\right) + \cdots$$

which is the same as

$$\frac{1 \bullet 0}{100} + \frac{5 \bullet 1}{100} + \frac{10 \bullet 2}{100} + \frac{19 \bullet 3}{100} + \cdots,$$

which is the same as

$$\frac{1 \bullet 0 + 5 \bullet 1 + 10 \bullet 2 + 19 \bullet 3 +}{100} \cdots$$

The general form of the formula we used above for the **expected value of a discrete probability distribution** is therefore:

$$E(X) = \sum_{i=1}^{n} x_i \, p(X)$$

$$= \sum_{i=1}^{n} x_i \, Pr(X = x)$$

Where $p(X)$ is a way of writing the **probability mass function** of X. Once again, remember that the probability function will be a fraction between zero and one, so this function involves multiplying the sum of values by a fraction.

Let's, for example, think about the expected value of a binomial distribution. We saw when we were looking at it earlier (page 377) that the expected value is np.

We could also work out that the expected value of a binomial distribution is np using the formula for expected value of a discrete probability distribution, although the algebra isn't simple. But it's a good illustration of the concept, so we'll do it anyway!

Because r is traditionally used as the variable on the horizontal axis in a binomial, it takes the place of x in the formula. So we've just multiplied the formula for the binomial cumulative distribution function (CDF) by r.

$$E\big(Bin(n,p)\big) = \sum_{r=0}^{n} r \frac{n!}{r!\,(n-r)!} p^r (1-p)^{n-r}$$

At this point note that for the first term of this expansion (in red), where $r = 0$, the whole term will be multiplied by 0 and will therefore equal 0.

$$= \left(0\, \frac{n!}{0!\,(n-0)!}\, p^0 (1-p)^{n-0} \right) + \sum_{r=1}^{n} r\, \frac{n!}{r!\,(n-r)!}\, p^r (1-p)^{n-r}$$

In other words, we can ignore the term where $r = 0$, and start at $r = 1$.

$$= \sum_{r=1}^{n} r\, \frac{n!}{r!\,(n-r)!}\, p^r (1-p)^{n-r}$$

$\dfrac{r}{r!} = \dfrac{1}{(r-1)!}$ (divide numerator and denominator by r).

$$= \sum_{r=1}^{n} \frac{n!}{(r-1)!\,(n-r)!}\, p^r (1-p)^{n-r}$$

n and p are constants (because r is the variable for the summation). Take them outside of the summation, and balance this by multiplying by $\dfrac{1}{np}$ within the summation.

$$= np \sum_{r=1}^{n} \frac{1}{np}\, \frac{n!}{(r-1)!\,(n-r)!}\, p^r (1-p)^{n-r}$$

$\dfrac{n!}{n} = (n-1)!$ and $\dfrac{p^r}{p} = p^{r-1}$

$$= np \sum_{r=1}^{n} \frac{(n-1)!}{(r-1)!\,(n-r)!}\, p^{r-1} (1-p)^{n-r}$$

Substitute $(n-r)$ for $(n-1) - (r-1)$, which is the same thing. (It involves subtracting 1 then adding one)

$$= np \sum_{r=1}^{n} \frac{(n-1)!}{(r-1)!\,\big((n-1)-(r-1)\big)!}\, p^{r-1} (1-p)^{((n-1)-(r-1))}$$

To simplify what we are looking at, let

$$= np \sum_{r=1}^{n} \frac{a!}{b!\,(a-b)!}\, p^b (1-p)^{(a-b)}$$

$a = (n-1)$ and $b = (r-1)$

In terms of the limits of summation, $b = (r-1)$, which means that when $r = 1, b = 0$.

$$= np \sum_{b=0}^{n} \frac{a!}{b!\,(a-b)!}\, p^b (1-p)^{(a-b)}$$

$b = (r-1)$, and so $r = b+1$. When we come to the last iteration, when $r = n$, that means $b + 1 = n$. So therefore $a = \big((b+1)-1\big) = b$. We can therefore take the summation to a.

$$np \sum_{b=0}^{a} \frac{a!}{b!\,(a-b)!}\, p^b (1-p)^{(a-b)}$$

The expression in the summation is a binomial expansion with a probability of p, and so it is equal to 1. That leaves us with np as the expected value of a binomial probability distribution.

$$E\big(Bin(n,p)\big) = np$$

Expected value of continuous variables

Where probability is described by an equation that creates a smooth curve, summing discrete values will not work. Instead, Chebyshev multiplied x by the value of the function, and then integrated it rather than summing it. This is essentially the same as summing a number of infinitely narrow data points.

The general form of this formula for the expected value of a **continuous probability distribution** is:

$$E(X) = \int_{-\infty}^{\infty} xp(X)\,dx$$

For example, let's calculate the expected value of an exponential distribution:

$$p(X) = \lambda e^{-\lambda x}$$

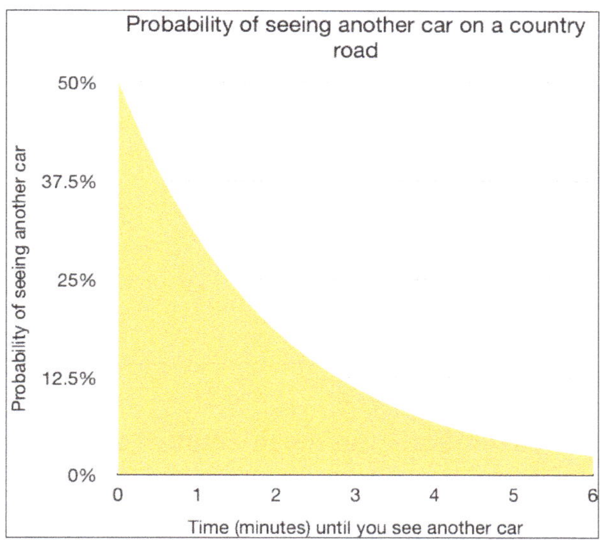

Figure 244

As we did when we first looked at this distribution, let's say that this represents the probability that you will see another car at a certain time, driving on a particular country road (Figure 244). On average there will be λ cars in whatever time you are looking at, but what is the average time you will wait before seeing another car?

The formula for the expected value will be

$$E\big(\lambda e^{-\lambda x}\big) = \int_{0}^{\infty} x\lambda e^{-\lambda x}\,dx$$

To calculate what that is:

Use integration by parts.

$$\int_0^\infty x\lambda e^{-\lambda x}\, dx$$

As a reminder, the formula for integration by parts is

$$\int u\frac{\delta v}{\delta x}\delta x = uv - \int v\frac{\delta u}{\delta x}\delta x$$

Use x as u and $\lambda e^{-\lambda x}$ as $\frac{\delta v}{\delta x}$. That makes $\frac{\delta u}{\delta x}=1$ and $v=-e^{-\lambda x}$. (If you differentiate this using the chain rule you get $\lambda e^{-\lambda x}$).

$$\int_0^\infty x\lambda e^{-\lambda x}dx = \lim_{x\to\infty}x(-e^{-\lambda x}) - 0(-e^{-\lambda 0}) - \int_0^\infty -e^{-\lambda x}1dx$$

The first expression on the right-hand side, $\lim_{x\to\infty}x(-e^{-\lambda x})$, causes a bit of a problem. It is the same as $\lim_{x\to\infty}\frac{-x}{e^{\lambda x}}$. It approaches infinity in the numerator and the denominator. However, the term on the bottom approaches infinity much faster than the term on top, so the whole expression approaches zero. We can confirm that using L'Hopital's rule (page 255). Using this, we see that taking derivatives of the top and the bottom will give us $\frac{1}{\infty}$, which approaches 0.

$$\int_0^\infty x\lambda e^{-\lambda x}dx=-\int_0^\infty -e^{-\lambda x}\, dx$$

Multiply the whole integration by the constant $-\frac{1}{\lambda}$ and multiply inside the integration by $-\lambda$ to keep everything balanced.

$$E\left(\lambda e^{-\lambda x}\right)=\tfrac{1}{\lambda}\int_0^\infty \lambda e^{-\lambda x}\, dx$$

Now the formula inside the integral is the volume of our density function, and we already know that its integral between 0 and infinity is equal to 1.

$$E\left(\lambda e^{-\lambda x}\right) = \frac{1}{\lambda}$$

That makes the expected value of this distribution—the mean amount of time in which you are likely have an event—equal to

$$\frac{1}{\lambda}$$

where λ is the mean number of events per whatever unit of time you are looking at. If λ is 2 cars per minute, then as you would expect on average it will take you 2 minutes to see a car (Figure 245). If λ is 5 cars per minute, then on average it will take you 12 seconds (one fifth of a minute) until you see the next car.

Linearity and the sum of expected values

Being either series or integrals, expected values have the property of **linearity**. It's easy to assume that that is enough to account for the fact that the expected value of the sum of random variables is equal to the sum of their individual expected values. They are both series, and if you add two series you get another series.

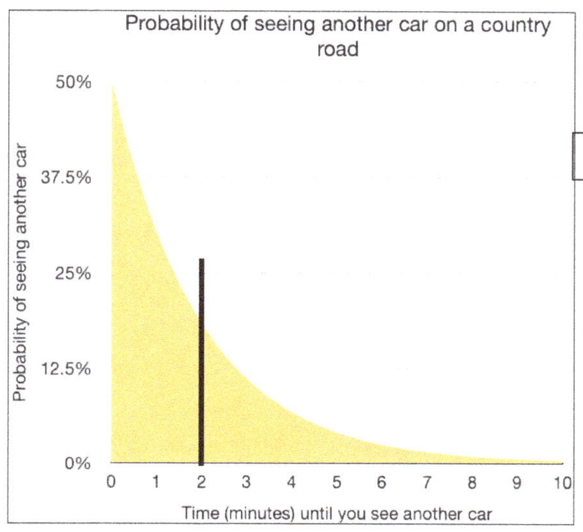

Figure 245

$$E(X + Y) = E(X) + E(Y)$$

It isn't quite that simple, though. This is a case where the expected value of a random variable (which is, remember, a variable subject to a probability distribution) isn't quite the same as the mean of a set of numbers. This formula doesn't necessarily apply to means in general. For example, the mean of the set of numbers $1, 2, 3, 4$ and 5 is 3. The mean of the set of numbers 100 and 200 is 150. The mean of 3 and 150 is 76.5. But if we take the mean of all of the numbers that produced these two means: $1, 2, 3, 4, 5, 100$ and 200 we get 45, which is obviously very different to 76.5. So the mean of two sets of numbers, when they are added together, isn't necessarily the same as the sum of their means.

This formula *does* apply when there are the same number of elements in both sets of numbers; that is, where the n in the calculation of the mean is the same. But it doesn't apply where you are looking at the means of sets of numbers that have a different number of elements.

Here is a proof that it *does* apply to expected values. This one assumes the variables we are looking at are continuous.

This is the formula for the expected value of the distribution of the sum of X and Y. It's a double integral, because there are two variables: it's a joint distribution.

$$E(X + Y) = \iint (x + y)P(X = x, Y = y)dxdy$$

Distribute the probability to the two variables. (A probability is a number; specifically, a fraction.)

$$= \iint xP(X = x, Y = y) + yP(X = x, Y = y)dxdy$$

Use linearity to separate the two integrations and exchange the order of one of them. Use brackets to clarify what is happening.

$$= \int \left(\int xP(X = x, Y = y)dy\right)dx + \int \left(\int yP(X = x, Y = y)dx\right)dy$$

$$= \int x\left(\int P(X = x, Y = y)dy\right)dx + \int y\left(\int P(X = x, Y = y)dx\right)dy$$

x can be treated as a constant in terms of the integration with respect to y and vice versa.

In a joint distribution with two variables, integrating the joint probability with respect to one of the variables gives the marginal distribution of the other variable: in this case $P(X = x)$ and $P(Y = y)$ respectively. (Refer to the section on adding distributions on page 404.)

$$= \int xP(X = x)dx + \int yP(Y = y)dy$$

$$\boxed{E(X + Y) = E(X) + E(Y)}$$

The approach would be similar if we were using discrete variables, except that we would use summation instead of integration. It can also be done, using a similar method, with one discrete and one continuous variable.

When we looked at adding random variables (page 481 and following) we thought about a situation where you were looking at the combined performance of two salespeople. After we had convolved their individual sales performances (both of which followed a Poisson distribution) we saw that we could, much more simply, have added the averages (which we now know are expected values) of their performances, and assumed that they would also give a Poisson distribution.

Linearity also means that

$$E(Cx) = CE(x)$$

In other words, you can take a constant outside of the expectation calculation.

If we were subtracting one of the variables and one of the expected values, the result would be the same. You can think of it as multiplying one by -1, then adding them. The -1, as a constant, can be factored out until the end.

$$E(X - Y) = E(X) - E(Y)$$

You may want to do this, for example, if you were conducting an ongoing research project and one section of the subjects had to be removed part of the way through.

Product of expected values of independent variables

Provided that two random variables are independent, the expected value of their product is the same as the product of their expected values:

$$E(XY) = E(X)E(Y)$$

This doesn't flow automatically from the property of linearity, although it is related to it. Here's the proof.

Recall that our basic formula for a probability density function (PDF) is $P(X = x) = p(X)$.

$$E(XY) = \sum_{x,y} x_i y_i P(X = x \text{ and } Y = y)$$

So this is saying that the expected value of the product of the variables can be calculated by changing the formula for expected value, basing it on the fact that now we are looking at the probability that two things are true. Because x and y are independent (what happens to one tells you nothing about the other) we can use the multiplication rule:

$$P(X = x \text{ and } Y = y) = P(X = x)P(Y = y)$$

So we are adding the sum for all values of x and y.

$$E(XY) = \sum_{x,y} x_i y_i P(X = x)P(Y = y)$$

This is the same as first summing the terms with x in them, and then including this sum each time we add up the terms with y in them.

$$E(XY) = \sum_{y} \sum_{x} x_i y_i P(X = x)P(yY = y)$$

Factor out the terms that depend on y.

$$E(xy) = \sum_y y_i P(Y = y) \sum_x x_i P(X = x)$$

$$E(XY) = E(X)E(Y)$$

So the expected value of the product of two *independently distributed* random variables is the same as the product of their individual expected values. Remember that only applies if x and y are independent.

Once again, this same reasoning applies if the variables are continuous, but we use double integrations instead of double series.

LOTUS

The wonderfully-names LOTUS (although, writing in French and Russian, Chebyshev didn't call it that) stands for the **Law of the Unconscious Statistician**. LOTUS says that:

$$E\big(g(X)\big) = \int_{-\infty}^{\infty} g(X)p(X)\,dx$$

In other words, the expected value of some function of X (where X is a random variable with a probability density function) can be obtained by using the formula for the expected value of X for that probability density function, but replacing the X with whatever function you are trying to find the expected value of. The same thing works with discrete random variables, except that the integration is replaced by summation.

It isn't completely clear how it came to get its name, but it seems to be connected to the fact that some people give it as a definition of expected value, rather than what it is: a consequence of the definition. It's probably the case that someone said it once, but the name was so enjoyable that it took root.

Total of distance multiplied by probability (cm)	Total of distance multiplied by probability (inches)
0	0
0.05	0.019685
0.2	0.07874
0.57	0.224409
1.08	0.425196
1.2	0.47244
0.84	0.330708
3.94	**1.551178**

Figure 246

To understand why it works, think back to our table of probabilities of different throws in darts (on page 522). Now imagine that instead of measuring in centimetres, we measured in inches (Figure 246). 1cm $\approx$ 0.3937 inches. The probabilities don't change. The numbers change, but they are following the same rules of probability, and so the expected value in inches is the same, except that the unit has changed to inches: $3.94 \bullet 0.3937 \approx 1.5511178$. What LOTUS is really saying is that if you apply a formula to x that won't change the probabilities, so it won't change the expected value, except by the amount that the function of x changes x.

Key points:

- Expected value is the equivalent of a mean of a probability distribution.

- When we are talking about the expected value of a probability distribution we aren't talking about the mean of the frequencies on the vertical axis (which would always lie between 0 and 1); we are talking about the "mean" of the data points on the horizontal axis, based on their expected frequencies.

- $E(X) = \sum_{i=1}^{n} x_i\, p(X) = \sum_{i=1}^{n} x_i\, Pr(X = x)$

- $E(X) = \int_{-\infty}^{\infty} xp(X)\, dx = \int_{-\infty}^{\infty} x\, Pr(X = x)$

- $E\big(Bin(n,p)\big) = np$

- $E(Exp(\lambda)) = \frac{1}{\lambda}$

- $E(X + Y) = E(X) + E(Y)$

- $E(Cx) = CE(x)$

- $E(X - Y) = E(X) - E(Y)$

- $E(XY) = E(X)E(Y)$ if, and only if, X and Y are independent.

- LOTUS says that $E\big(g(X)\big) = \int_{-\infty}^{\infty} g(X)p(X)dx$

Moment generating functions

> **Read this if** you have never studied moment generating functions, or if you have studied them but still don't get it, or if you sort of get it but can't see the point.

One of Chebyshev's interests was in the mechanics of forces of rotation, and of the torque needed to overcome momentum in rotational mechanical systems. (This is also called the "moment of inertia", in the sense of momentum, not a moment in time.) This was important for improving the efficiency of engines and other mechanical contraptions, and particularly where a linear movement needed to be converted to a rotatory one, and vice versa.

In this sort of situation Fourier transforms become important because they provide a way to take a function whose graph moves from left to right, and instead cause it to move in a circular motion around zero (page 321). We can imagine that one snowy winter's day in the mathematics department of the St. Petersburg University,

> **If you have studied moments of distributions** previously it might seem strange to look at **moment generating functions** before looking at what moments are, but this is almost certainly the way Chebyshev came to it, and it makes more sense this way.

when there wasn't much to do inside (no electricity) and it was too cold to go outside, Chebyshev was thinking about the formula for expected value while thinking about Laplace's work on probability, and noticed how much his expected value formula looked like the formula for a Laplace transform. (If that isn't meaning much to you, you might want to revise it on page 316.)

$$E(X) = \int_{-\infty}^{\infty} x p(X)\, dx \qquad \text{Formula for expected value of } X$$

$$\mathcal{L}\{p(X)\} = \int_{0}^{\infty} e^{-sX} p(X)\, dx \qquad \text{Formula for a Laplace transform of } X$$

$$\mathcal{F}\{p(X)\} = \int_0^\infty e^{-i\varpi X} p(X)\,dx \quad \text{Formula for Fourier transform of } X$$

In fact, using LOTUS, if you make e^{-sX} your function of X, you get

$$E(e^{-sX}) = \int_{-\infty}^\infty e^{-sx} p(x)\,dx$$

And if you make $e^{-i\varpi X}$ your function of X, you get

$$E(e^{-i\varpi X}) = \int_{-\infty}^\infty e^{-i\varpi x} p(x)\,dx$$

The only difference between these and the Laplace and Fourier transforms was that the lower bound of his integration was negative infinity rather than zero.

Possibly after some scribbling, he realised that he could replace $-s$ in the Laplace transform with a new dummy variable, t.

$$\int_{-\infty}^\infty e^{tx} p(X)\,dx = E(e^{tx})$$

You may remember from page 316 and following that it's quite common to use Taylor series with Laplace transforms, to solve them. That is probably what Chebyshev did next.

Use a Taylor-series expansion of the exponential power.

$$E(e^{tx}) = E\left[\frac{t^0 x^0}{0!} + \frac{t^1 x^1}{1!} + \frac{t^2 x^2}{2!} + \cdots \right]$$

$$= E\left[1 + tx + \frac{t^2 x^2}{2!} + \frac{t^3 x^3}{3!} + \frac{t^4 x^4}{4!} + \cdots \right]$$

Take the derivative with respect to t of what is inside the expectation on both sides, treating the powers of x as constants.

$$E\left(\frac{\delta(e^{tx})}{\delta t} \right) = E\left[x + x^2 \frac{t^1}{1!} + x^3 \frac{t^2}{2!} + \cdots \right]$$

Evaluate where $t = 0$. It is usual, of course, to evaluate a Taylor series at the point where the variable equals zero.

$$E\left(\frac{\delta(e^{tx})}{\delta t} \right) = E\left[x + x^2 \frac{0^1}{1!} + x^3 \frac{0^2}{2!} + \cdots \right]$$

All terms except the first disappear, because they involve multiplication by 0.

$$E\left(\frac{\delta(e^{tx})}{\delta t}\right) = E[x]$$

So he invented a new type of transform for probability density functions (PDFs), which he could have called the Chebyshev transform but didn't. He actually called it the **Moment Generating Function (MGF)** for reasons we will come to soon.

It was defined as:

$$\text{MGF} = \int_{-\infty}^{\infty} e^{tx}p(X)dx$$

By calculating this integral with respect to x (after which the x disappears because it is a definite integral), then taking the derivative of the result with respect to t, he had a new way to calculate the expected value of a distribution. If, as per Fourier, he kept the i in multiplying the t, he called it a characteristic function. These are sometimes used in place of moment generating functions (MGFs). (The main reason for sometimes using characteristic functions rather than MGFs is that MGFs don't exist for all distributions, whereas characteristic functions do.)

MGFs may not seem like much of an achievement. Instead of using the formula for expected value:

$$\int_{-\infty}^{\infty} xp(X)\,dx$$

he used

$$\int_{-\infty}^{\infty} e^{tx}p(X)dx$$

which looks harder to integrate (although it isn't always). He then had to take the derivative of the result with respect to t and evaluate it where $t = 0$, which is certainly at least a couple of extra steps. There are other benefits, however, of having an MGF once you have worked it out. Before we get to those, let's have a look at how it works in practice for finding an expected value, using the exponential distribution. (Note that from here on we mostly don't need to use Taylor series with MGFs or characteristic functions; they were there to demonstrate why MGFs work, and they are almost certainly how Chebyshev came to the idea, but the actual calculation usually works from the integration.)

Replace $p(X)$ with the formula for the probability density function (PDF) of the exponential distribution.

$$\int_0^\infty e^{tx} \lambda e^{-\lambda x}\, dx$$

Take λ, which is a constant, outside the integration, and combine the exponents of e, factoring out the x.

$$= \lambda \int_0^\infty e^{x(t-\lambda)}\, dx$$

Multiply inside the integration by $t - \lambda$ which is a constant with regard to x, and multiply outside the integration by the reciprocal of $t - \lambda$ so that the value of the expression doesn't change.

$$= \frac{\lambda}{t-\lambda}\int_0^\infty (t-\lambda)e^{x(t-\lambda)}\, dx$$

The integral of $(t-\lambda)e^{x(t-\lambda)}\, dx$ is $e^{x(t-\lambda)}$, because the derivative of $e^{x(t-\lambda)}$ is $(t-\lambda)e^{x(t-\lambda)}$ by the chain rule, treating $(t-\lambda)$ as a constant.

$$= \frac{\lambda}{t-\lambda}\left\{\left(e^{\infty(t-\lambda)}\right) - e^{0(t-\lambda)}\right\}$$

$$= \frac{\lambda}{t-\lambda}(-1), \text{ provided that } t < \lambda$$

λ will always be positive, because it represents the number of events that occur on average in a given time frame. If $t < \lambda$, then $(t-\lambda)<0$. This means $e^{\infty(t-\lambda)} \to 0$ because it is the same as raising e to negative infinity times a constant, which is the same as having e to the power of infinity in the denominator. If $t = \lambda$, then $(t-\lambda) = 0$, $e^{\infty(t-\lambda)} = 1$, which makes the integral equal 0. If $t > \lambda$ then $(t-\lambda)>0$ and the integral diverges to infinity. We can get away with the constraint that t is less than λ, because we will end up evaluating the derivative of this expression at the point where $t = 0$, so we can imagine t being as small as we like.

Multiply the numerator and denominator by –1. This has the effect of reversing the t and the λ in the denominator and leaving the numerator positive.

$$MGF = \frac{\lambda}{\lambda - t}$$
$$\text{where } t < \lambda$$

So now we have a formula for the MGF of an exponential distribution. Calculating it wasn't completely straightforward, but it wasn't hideous, either. Once we have it, calculating the mean of the distribution is straightforward.

Write the MGF like this.

$$\frac{\lambda}{\lambda - t} = \lambda(\lambda - t)^{-1}$$

Take the derivative with respect to t, using the chain rule. (The inner function is $w = \lambda - t$, and its derivative is –1. The outer function is λw^{-1}, and its derivative is $-\lambda w^{-2}$. The negatives cancel one another out.

$$f'(t) = \lambda(\lambda - t)^{-2}$$

Let $t = 0$ and divide the numerator and denominator by λ.

$$= \frac{\lambda}{(\lambda - t)^2}$$

This result is the expected value of an exponential distribution.

$$\boxed{E(X) = \frac{1}{\lambda}}$$

Note that this isn't just taking an integral then reversing it again by taking the derivative. The key thing is that the new variable t is added, the integral is taken with respect to x, then the derivative of that is taken with respect to t. In this way, once t is set to 0, you get the expected value of the original formula, rather than just getting the original formula.

The MGF is symbolised by

$$M_X(t)$$

For continuous distributions it uses this formula:

$$\boxed{M_X(t) = \int_{-\infty}^{\infty} e^{tx} p(X) dx}$$

where $p(x)$ is the PDF.

For discrete distributions it uses the formula:

$$\boxed{M_X(t) = \sum_{x=-\infty}^{\infty} e^{tx} p(X)}$$

where $p(X)$ is the probability mass function (PMF). This gives a third function to use to describe a probability distribution, in addition to the density/mass function and the cumulative distribution function.

It has many more uses than just finding the mean of a distribution. Each probability distribution will have a unique MGF, and so if you can prove something with an MGF, it is proven for the distribution. It has some features that we will see as we progress that can make it particularly easy to work with. Probably the biggest advantage, though, is what it showed Chebychev when he took further derivatives.

Before we get to that, the following table shows these functions for the distributions we have seen so far.

Distribution	PDF/PMF	MGF
Binomial: $Bin(n, p)$	$\dfrac{n!}{r!\,(n-r)!}p^r q^{n-r}$	$(q + pe^t)^n$
Geometric: $G(p)$	$q^{r-1}p$	$\dfrac{pe^t}{1 - qe^t}$ for $t < -\ln(1 - p)$
Negative binomial: $NB(r, p)$	$\dfrac{(x-1)!}{(r-1)!\,(x-r)!}p^r(1-p)^{x-r}$	$\left(\dfrac{q}{1 - pe^t}\right)^r$ for $t < -\ln p$
Hypergeometric: $H(N, a, n)$	$\dfrac{\binom{M}{x}\binom{N-M}{k-x}}{\binom{N}{k}}$	Does exist, but requires a special function called a hypergeometric function which we won't be covering.
Poisson: $Pois(\lambda)$	$\dfrac{\lambda^r e^{-\lambda}}{r!}$	$e^{\lambda(e^t - 1)}$
Exponential: $Expo(\lambda)$	$\lambda e^{-\lambda x}$	$\dfrac{\lambda}{\lambda - t}$ For $t < \lambda$
Beta: $Beta(a, b)$	$\dfrac{x^{a-1}(1 - x)^{b-1}}{\beta(a, b)}$	Exists but formula is too long to write here.
Gamma: $Gamma(n, \lambda)$	$\dfrac{1}{\Gamma(n)}\dfrac{(\lambda x)^n e^{-\lambda x}}{x}$ where n and $\lambda > 0$	$\left(\dfrac{\lambda}{\lambda - t}\right)^n$

Distribution	PDF/PMF	MGF
Laplace's error distribution:	$\dfrac{\sqrt{A}}{\sqrt{2\pi}}e^{\left(\frac{-A(x-\mu)^2}{2}\right)}$	We'll get to this too, once we know what the A represents.

MGFs form exponential curves, and when the value of $t = 0$, the value of the undifferentiated function equals 1 (Figure 247). (Characteristic function graphs form circular shapes on the complex plane.)

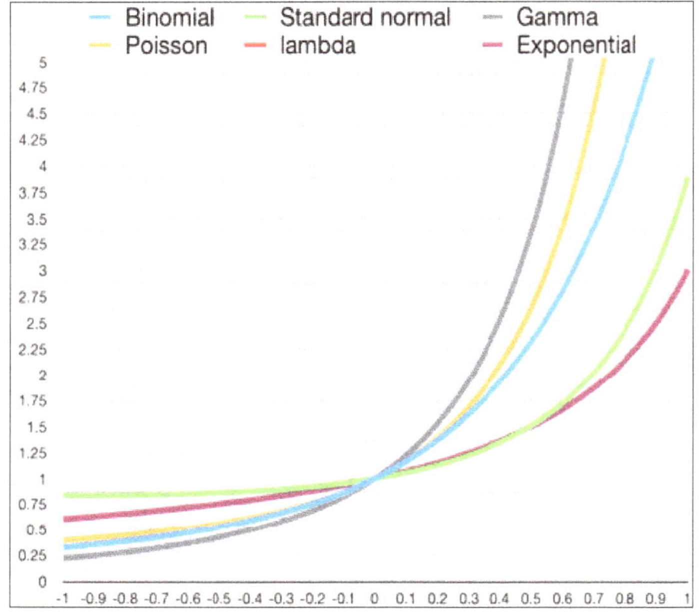

Figure 247

Back to the Taylor series expansion, we have seen that:

$$MGF = \int_{-\infty}^{\infty} e^{tx}p(x)\,dx$$

$$= E(e^{tx})$$

$$= E\left[\frac{t^0x^0}{0!} + \frac{t^1x^1}{1!} + \frac{t^2x^2}{2!} + \cdots\right]$$

By taking the first derivative, this gave Chebyshev the expected value of X. If you think about what Taylor series are good for, you will see that it was a short step from there to realise that the second derivative of the MGF, evaluated where $t = 0$, would give him the expected value of X^2 and, in fact, that the nth derivative evaluated at $t = 0$ would give him the expected value of X^n.

But what use was it to know the expected value of *different powers* of a random variable? We'll get to that, but before we do, we can use the MGF to prove a couple of results that we have put off until now.

Proof independent random variables can be added by convolution

We saw (page 481) an explanation of why convolution allows us to find the probability distribution of two independent random variables when they are added, but we held-off the mathematical proof until we covered moment generating functions (MGFs).

One of the ways we can express an MGF is:

$$\text{MGF} = E(e^{tx})$$

That means the MGF of the sum of two random variables, X and Y, will be:

$$E\left(e^{t(X+Y)}\right) = E(e^{tX+tY})$$

$$= E\left((e^{tX})(e^{tY})\right)$$

Based on what we saw earlier about the products of expectations of independent variables, this is equal to

$$E(e^{tX})E(e^{tY})$$

That means that the MGF of the sum of two independent random variables is equal to the product of their individual MGFs.

At this stage it is worth remembering that the MGF is similar to the Laplace transform, except that they have different levels of integration and the sign is reversed. In this instance that makes no difference. The inverse transform of the MGFs will be equal to the convolution of the original probability density functions. (See page 316 and following for proof that the inverse Laplace transform of the product of two Laplace transforms is the convolution of those functions.) That means the probability function of two independent random variables, when they are added together, can be obtained by performing a convolution of their functions.

Proof that the gamma distribution is the sum of exponentials

We looked at the gamma distribution earlier but put off the proof that it really does represent what it is supposed to (a sum of exponentially distributed random variables) until now. (The gamma distribution was probably first understood by Laplace, not Chebyshev, and it conceptually belongs with the Poisson and exponential distributions, although it is also closely related to the beta, the normal and the chi-square distributions. The reason for leaving the proof until now is that the easiest proof uses the moments of distributions, which hadn't been discovered/invented when Laplace was around, although he may have done something quite similar using his own Laplace transform).

We saw that the MGF of the exponential distribution is

$$\frac{\lambda}{\lambda - t} \text{ for } t < \lambda$$

So if we are adding n of these random variables, that will equate to

$$\left(\frac{\lambda}{\lambda - t}\right)^n$$

(based on the fact that to add random variables we can multiply their MGFs, as we saw above on page 539).

Plug the formula for the gamma distribution, $\frac{1}{\Gamma(n)} \frac{(\lambda x)^n e^{-\lambda x}}{x}$, into the formula for the MGF, $\int_{-\infty}^{\infty} e^{tx} p(X)\, dx$. Note that, because it is a gamma distribution it only has support for $x = 0 \to \infty$, so those are the limits of the integration.

$$\int_0^{\infty} e^{tx} \frac{\lambda^n x^{n-1} e^{-\lambda x}}{\Gamma(n)}\, dx$$

Take the constants with respect to x outside of the integration.

$$= \frac{\lambda^n}{\Gamma(n)} \int_0^{\infty} e^{tx} x^{n-1} e^{-\lambda x}\, dx$$

Gather the exponents of e, and write $t - \lambda$ as $-(\lambda - t)$ (same thing). For this to converge, $(\lambda - t) > 0$. That will mean for positive values of x (which are the only ones we are considering in this integration) $e^{-(\lambda - t)x} = \frac{1}{e^{(\lambda-t)x}}$ which will approach 0 as $x \to \infty$. λ is always positive, based on what it represents. It isn't a problem insisting that t be very small, because the derivatives of MGFs are evaluated at $t = 0$.

$$= \frac{\lambda^n}{\Gamma(n)} \int_0^{\infty} x^{n-1} e^{-(\lambda-t)x}\, dx$$

Now we will do a substitution. Let

$$u = (\lambda - t)x$$

$$\frac{\delta u}{\delta x} = \lambda - t$$

$$\frac{1}{\lambda - t}\delta u = \delta x$$

$$and \quad \frac{u}{(\lambda - t)} = x$$

Substitute those values.

$$\frac{\lambda^n}{\Gamma(n)}\int_0^\infty \left(\frac{u}{(\lambda - t)}\right)^{n-1} e^{-u}\frac{1}{\lambda - t}\delta u$$

$\left(\frac{1}{(\lambda-t)}\right)^{n-1}\frac{1}{\lambda-t} = \frac{1}{(\lambda-t)^n}$ which is a constant with respect to u, and can therefore be taken outside the integration.

$$= \frac{\lambda^n}{(\lambda - t)^n\Gamma(n)}\int_0^\infty u^{n-1}e^{-u}\delta u$$

$\int_0^\infty u^{n-1}e^{-u}\delta u = \Gamma(n)$, by the definition of the gamma function. The gamma functions cancel one another out.

$$= \frac{\lambda^n}{(\lambda - t)^n\cancel{\Gamma(n)}}\cancel{\Gamma(n)}$$

$$\boxed{= \left(\frac{\lambda}{\lambda - t}\right)^n}$$

So we have proved that the MGF of the gamma distribution is the same as the function representing the sum of n exponentially distributed independent and identically distributed (i.i.d.) random variables. If the MGFs are the same, so are the probability density or mass functions, and the cumulative distribution functions.

If we were using the alternative form of the gamma distribution (known as the **alternative parameterization**) we would have replaced λ with $\frac{1}{\theta}$. (The Greek letter theta, θ is, by convention, used for this.)

Replace λ with $\frac{1}{\theta}$

$$MGF = \left(\frac{\lambda}{\lambda - t}\right)^n = \left(\frac{\frac{1}{\theta}}{\frac{1}{\theta} - t}\right)^n$$

Multiply the numerator and denominator by θ.

$$= \left(\frac{1}{1 - \theta t}\right)^n$$

There is no need to show the numerator raised to the power of n, because it will still just equal 1.

$$\text{MGF} = \frac{1}{(1 - \theta t)^n}$$

$$= (1 - \theta t)^{-n}$$

This is the form of the MGF that is commonly used in statistics, and it is the one we will use going forward.

Key points:

- A **moment generating function (MGF)** is a slight modification of a Laplace transform, while a **characteristic function** is a slight modification of a Fourier transform.

- $\text{MGF} = \int_{-\infty}^{\infty} e^{tx} p(X) dx = M_X(t)$

- By performing the integral with respect to x, then differentiating the result with respect to t and evaluating the result where $t = 0$, you can find the **expected value** of a probability distribution. This is proved using a Taylor series.

- Like other transforms, each probability distribution will have a unique MGF, and so if you can prove something with an MGF, it is proven for the distribution.

- We've seen a proof that independent random variables can be added by convolution, and that the gamma distribution is the sum of exponential distributions.

Moments of probability distributions

Read this if you don't know what each of the moments of a distribution refer to, you don't know how to calculate them using the moment generating function (MGF), or you don't know much about the different ways the idea of variance can be expressed or manipulated.

It is easy to forget that ideas like the variance (which we're about to come to) and the standard deviation (which we aren't far away from) came relatively late in the development of statistics, even though they are now often taught as some of the earliest concepts in statistics courses. But the work we have seen so far occurred without anyone thinking about them, or seeing the need to. It was only after Chebyshev started thinking about MGFs, and the way that a Taylor series could be used to find expected values of higher powers of X, that they came to prominence.

Using a moment generating function to find the first moment

Chebyshev must have seen parallels in this work with the work he was doing regarding the physics of moving machinery parts (and he probably used Fourier transforms for that). He seems to have thought of *angular momentum*, as though the distributions were rotating on an axis. He therefore used the same sort of language he would use for describing the forces of moving parts. In physics a moment of force is the same thing as the torque: it is the distance from the pivot point times the force applied (so that the further away the force is applied to the pivot point, the greater the leverage).

Chebyshev called the expected value, or the mean, the **first moment** of a distribution. It is sometimes symbolised by m_1'. As mentioned earlier (page 520), you can envisage this as the point of balance of a distribution (Figure 248).

Figure 248

For what follows we'll use the moment generating function (MGF) of the binomial distribution.

Apply the formula for an MGF to the binomial. The support is from 0 to n, rather than from minus infinity to plus infinity.

$$\sum_{x=-\infty}^{\infty} e^{tx} p(x) = \sum_{x=0}^{n} e^{tx} \binom{n}{x} p^x q^{n-x}$$

Both p and e^t are raised to the power of x, so they can be combined.

$$= \sum_{x=0}^{n} \binom{n}{x} (pe^t)^x q^{n-x}$$

$\sum_{x=0}^{n} \binom{n}{x} (pe^t)^x q^{n-x}$ is the binomial expansion of $(pe^t + q)^n$

$$\boxed{\text{MGF of binomial} = (pe^t + q)^n}$$

To find the first moment—the mean or expected value—we take the first derivative of this MGF, then evaluate it where $t = 0$.

Take the first derivative of the MGF with respect to t, using the chain rule.

$$f(t) = (pe^t + q)^n$$
$$f'(t) = pe^t n(pe^t + q)^{n-1}$$

Evaluate where $t = 0$

$$= pe^0 n(pe^0 + q)^{n-1}$$

$q = 1 - p$ which means that $(p + q)^{n-1} = 1$

$$= pn(p + q)^{n-1}$$

$$f'(0) = np$$

$$\boxed{E(X) = np}$$

This is, of course, the same result as we calculated earlier (page 525) using a different method. Calculating, then using the MGF, in this instance, was a bit easier, although that isn't always the case.

The second moment: variance

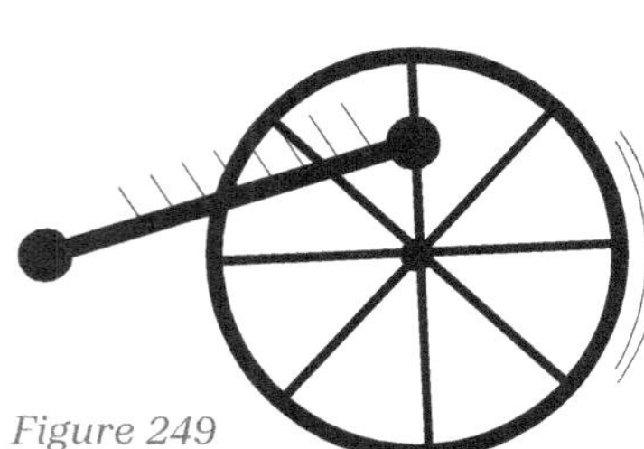

Figure 249

In mechanical engineering the "second moment of area"/ "area moment of inertia"/ "second area moment" is based on how widely spread the object being rotated is from its central point of rotation (Figure 249). The more widespread it is, the greater the inertia is. Second moments of area are calculated based on the mass multiplied by the square of the radius. If the axis of rotation is shifted, the second moment will change, because the radius of the furthest point of the object to the pivot will change.

So when Chebyshev realised that the new transform he had invented made it easy to calculate the expected value of X^2 he called it the **second moment of the distribution** (m'_2). It can be calculated by

$$E(X^2) = \int_{-\infty}^{\infty} x^2 p(X) \, dx \ OR \ \sum_{-\infty}^{\infty} x^2 p(X)$$

depending on whether the distribution is continuous or discrete.

If we subtract a constant from x before undertaking the calculation, it moves the whole distribution horizontally along the x axis, but doesn't change its shape. The first moment tells us where the centre of the distribution is. The second moment, as in physics, tells us how widely the data are spread from one another.

When a constant is subtracted, the general formula for the nth moment of a distribution is written as

$$E((x - A)^n) = \int_{-\infty}^{\infty} (x - A)^n p(X) \, dx \ OR \ \sum_{-\infty}^{\infty} (x - A)^n p(X)$$

Where $A = 0$ that doesn't tell you much that is useful about the second moment, because it isn't about the centre of the distribution. What is much more relevant with the higher moments is where they are in relation to the middle of the distribution, so they are usually calculated where A is the expected value. This is called a **central moment** of a distribution.

If we step away from probability distributions for a moment, and just think of frequency distributions, then we *could* express moments using the shortcut of the formula for the distribution:

$$\frac{\sum x f(x)}{n}$$

But to understand what is happening it can be helpful to just think about the individual data points.

So

$$m'_1 = \sum \frac{(x_i - E(X))}{n}$$

$$= 0$$

In other words, where A is set to the mean of the distribution, then the first moment sums up all of the distances from the mean and divides by the total number of data points, and we are left with 0. That's why for the first moment A is set to 0, and the formula just becomes the formula for the mean:

$$\sum \frac{(x_i)}{n}$$

The **second central moment**, however, can be expressed as

$$m'_2 = \sum \frac{\left(x_i - E(X)\right)^2}{n}$$

This is now giving us a measure of the average of the square of the distance of each data point from the mean.

This, let it be said, is not a great indication of how widespread a set of data is, because the sum of the squares of distances is going to exaggerate how far the data points are from one another. It is, however, very useful in a number of calculations involving probability distributions. It became known later as the **variance.**

It can also be expressed as:

$$m'_2 = E\left[\left(x - E(X)\right)^2\right]$$

In other words the expected value, or the mean, of squared distances from the mean.

Alternative form of the variance

We can use the MGF of the binomial distribution to find the second moment like this:

We calculated above that this is the first derivative of the binomial MGF:

$$f'(t) = npe^t(pe^t + q)^{n-1}$$

Calculate the second derivative, using the chain rule and the multiplication rule

$$f''(t) = n(n-1)(pe^t + q)^{n-2}(pe^t)^2 + n(pe^t + q)^{n-1}pe^t$$

Evaluate where $t = 0$ (which means $e^t = 1$)

$$f''(t) = n(n-1)(pe^t + q)^{n-2}(pe^t)^2 + n(pe^t + q)^{n-1}pe^t$$

$$= n(n-1)(p+q)^{n-2}p^2 + n(p+q)^{n-1}p$$

Remember that $p + q = 1$

$$= n(n-1)(p+q)^{n-2}p^2 + n(p+q)^{n-1}p$$

$$\boxed{E(X^2) = n(n-1)p^2 + np}$$

That gives us the moment around zero, but we actually want the central moment around the mean or expected value, i.e. around the first moment: $E[(x - \mu)^2]$

This can be expanded out as follows:

$$E\left[(x - E(x))^2\right]$$

Expand out the binomial term.

$$= E\left[x^2 - 2xE(x) + (E(x))^2\right]$$

Use linearity to change the expression to the addition of three expected values.

$$= E[x^2] + E[-2xE(x)] + E\left[(E(x))^2\right]$$

$E(x)$ is a constant, so it can be taken outside of the expectation in the second and third terms. It's squared in the third term, so it needs to be taken out squared. The expected value of 1^2 is 1.

$$= E[x^2] - 2E(x)E(x) + (E(x))^2 E 1^2$$

In the third term, remember that while x is a random variable, $E(x)$ is a number; it's a constant. The expected value of a constant is the constant; that is all it can be. The expected value of $(E(x))^2$ is $(E(x))^2$.

$$= E(x^2) - 2(E(x))^2 + (E(x))^2$$

This is a useful result in its own right, that we will use further in future.

$$\boxed{E[(x - \mu)^2] = E(x^2) - (E(x))^2}$$

In other words, the variance of X is equal to the mean of the square of X minus the mean of X, squared. In the first term we square it first, then find the expected value. In the second we find the expected value first, then square it.

So in terms of the MGF, we are looking for:

The expected value, or a first moment, of the binomial is, as we saw earlier, np. We are subtracting the square of this.

$$E(x^2) - \left(E(x)\right)^2 = n(n-1)p^2 + np - (np)^2$$

$$= (np)^2 - n(p)^2 + np - (np)^2$$

$$= \cancel{(np)^2} - n(p)^2 + np - \cancel{(np)^2}$$

Factor out np. Recall that, for the binomial, $1 - p = q$ where q is the probability of failure.

$$= np(1 - p)$$

So we now have a formula for the variance of a binomial, arrived at using the MGF.

$$\boxed{E[(x - \mu)^2] = npq}$$

You may note that, through the round-about route of mechanical engineering and physics, the variance brings us back to an idea that was crucial to our third stream: using the sum of squared distances from the mean. And so our disparate streams start to combine. The variance is, in fact, so important that it came to pave the way for the formula for the normal distribution to be completed, which opened up the world of statistics for widespread use.

Table of first and second moments

Distribution	Mean $E(x)$	Variance $E\left(X - E(X)\right)^2$
Binomial	np	npq
Geometric	$\dfrac{1}{p}$	$\dfrac{1-p}{p^2}$
Negative binomial	$\dfrac{pr}{1-p}$	$\dfrac{pr}{1-p^2}$
Hypergeometric	$\dfrac{KM}{N}$	$\dfrac{KM}{N}\left[\dfrac{(N-K)(N-M)}{N(N-1)}\right]$

Distribution	Mean $E(x)$	Variance $E(X - E(X))^2$
Poisson	λ	λ *(The mean and variance of the Poisson are the same)*
Exponential	$\dfrac{1}{\lambda}$	$\dfrac{1}{\lambda^2}$
Beta	$\dfrac{a}{a + b}$	$\dfrac{ab}{(a + b)^2(a + b + 1)}$
Gamma	$\dfrac{a}{b}$	$\dfrac{a}{b^2}$

Working with the variance

When working with the distributions of more than one random variable, sometimes all we need is the relationship between their expected values and their variances, which are often much simpler to work with mathematically than probability functions. We looked earlier (page 527) at some of the aspects of manipulating expected values: some of the implications of linearity, summing means and multiplying them.

Unlike expected value, the squaring means that we *cannot* assume that variance has the property of linearity. Let's look, for example, at what happens when we multiply the variance by a constant.

We saw above that variance can be written in this way:

$$Variance = E(x^2) - \big(E(x)\big)^2$$

Now replace x with cx, where c is some constant.

$$Variance = E((cx)^2) - \big(E(cx)\big)^2$$

Expectation has the property of linearity, so we can treat the constants like this.

$$= E(c^2x^2) - \big(cE(x)\big)^2$$

$$= c^2E(x^2) - c^2\big(E(x)\big)^2$$

$$\boxed{Variance(cx) = E((cx)^2) - \big(E(cx)\big)^2}$$

$$= c^2 \left\{ E(x^2) - \big(E(x)\big)^2 \right\}$$

When we multiply the variance by a constant, the constant gets squared when we take it outside the variance calculation. One other thing we need to think about with variance is that, as a series, provided that two variables are independent, the sum of their variances is the same as the variance of their sums.

$$variance(X + V) = variance(X) + variance(V)$$

You can see why this is when you think about variance as a sum (or an integration), but it is important to pay attention to the condition that the variables need to be independent. Earlier we said that, while it is common to have independence in theoretical distributions and situations, in real-life applications independence is much rarer. Let's look at an example where they aren't independent.

In this situation the variables are completely dependent. If we know what one is, we know exactly what the other is, whereas to be independent, knowing something about one would tell us nothing about the other.

$$variance(3x + 2x)$$

Based on our analysis above: square the constant when it is taken outside the integration.

$$variance(3x + 2x) = variance(5x)$$
$$= 25\,variance(x)$$

If we assume, in this instance, that the sum of the variances will be the same as the variance of the sums, then take the squared constants outside the variances, we get a completely different answer.

$$\neq variance(3x) + variance(2x)$$
$$= 9\,variance(x) + 4\,variance(x)$$
$$= 13\,variance(x)$$

So trying to use the rule that the sum of variances equals the variance of the sums, in the case above, gave an answer that wasn't just slightly off; it was almost half of the true answer.

The fact that the distances from the mean are squared means that there is no such thing as a negative variance. Variance is an absolute value.

We proved earlier:

$$Var(-Y) = E([-Y]^2) + (E[-Y])^2$$

Factor out the constant -1 from with the second expected value (by linearity).

$$= E([-Y]^2) + (-1E[Y])^2$$

When the negatives are squared, they will become positive.

$$= E([Y]^2) + (E[Y])^2$$

$$= Var(Y)$$

$$Var(-Y) = Var(Y)$$

This means that **variance of the difference between two random variables** is a bit unexpected (and we will need this result when we come to look at one of the most-commonly used statistical tests, the two-sample t test).

$$Var(X - Y) = Var(X + (-Y))$$

We proved above: the variance of the sum of random variables is the sum of their variances.

$$= Var(X) + Var(-Y)$$

As we saw immediately above, $Var(-Y) = Var(Y)$

$$Var(X - Y) = Var(X) + Var(Y)$$

So the variance of the difference between two random variables is the sum (not the difference) of their variances. This makes sense when you think that, if you are thinking about the difference between random variables (and remember, the difference between a random variable in statistics and any other variable in mathematics is that a random variable has its own probability distribution), then how widely spread the probabilities of each of them is will factor into how widely spread the differences between each value will be.

The third moment: skewness

The **third central moment** of a distribution is called the **skewness.** In a uniform distribution the skewness should equal zero. If a unimodal distribution is right skewed (the tail of the distribution extends to the right) we say it has a **positive skew,** whereas if the tail extends to the left we say it has a **negative skew.**

To understand what is happening here it is helpful once again to think about individual data points: in this case taking the mean of their cubed differences from the mean. The formula will be:

$$m'_3 = \sum \frac{(x_i - \mu)^3}{n}$$

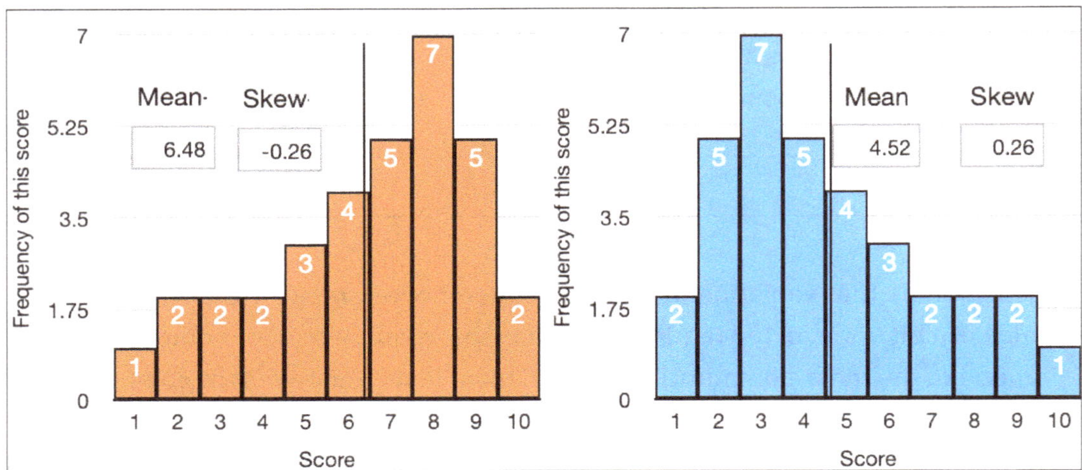

Figure 250

Figure 250 shows two graphs of two frequency distributions. The orange one on the left is skewed to the left, (has a negative skew), the blue one on the right is skewed to the right (has a positive skew). Figure 251 graphs the cubed distances from the mean of the individual data points of each of

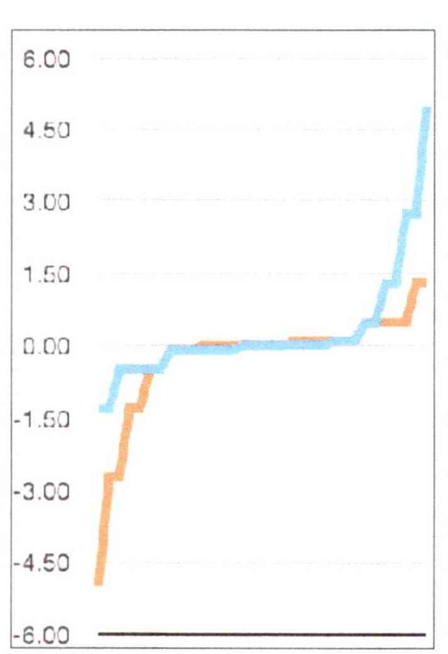

Figure 251

these graphs; the orange line shows the negatively-skewed graph and the blue line shows the positively-skewed one. This graph has roughly the same shape in each case. Positive numbers, when raised to the third power become larger positive numbers, while negative numbers raised to the third power become larger negative numbers. But in the negatively-skewed data-set (the orange line) the negative side to the left of the graph is steeper, and the mean of these cubed differences from the mean has shifted to the left. In the positively-skewed data set the mean of the cubed distances from the mean has shifted to the right. When calculating the skewness of a distribution the convention isn't just to report a number. In order to create some consistent scaling, so that one distribution can be compared to another, the skewness is divided by the variance, which, in order to give consistent units in the numerator and denominator is raised to the power of 3 divided by 2. So the formula for skewness becomes:

$$Skewness = \frac{m'_3}{(m'_2)^{\frac{3}{2}}}$$

The fourth moment: kurtosis

The **fourth central moment** of a distribution is called the **kurtosis.** This is a measure of how thick the tails of a distribution are. Two distributions could have the same mean and the same variance, but one could still have more of its values clustered around the centre, while another could have more of the data on the periphery. Kurtosis picks this up. By raising distances from the mean to the 4th power and taking the average, it exaggerates those values that are further from the mean, relative to those that are not. (Raising something to the 4th power draws something similar to a parabola with exaggerated steepness. Figure 252 compares the graphs of X^2 in blue, with that of X^4 in green.) In order to provide a meaningful comparison of one distribution versus another this needs to be standardised, and this is done in two different ways. First, it is divided by the square of the variance. Second, the number 3 is subtracted from it. This is the kurtosis of a normal distribution. Distributions with a kurtosis of $>$ 0 have thicker tails than the normal distribution while distributions with a kurtosis of $<$ 0 have thinner tails than the normal distribution. For this reason kurtosis is sometimes referred to as **excess kurtosis.** It is usually designated with the lower case of the Greek letter gamma (γ). So the formula for excess kurtosis is:

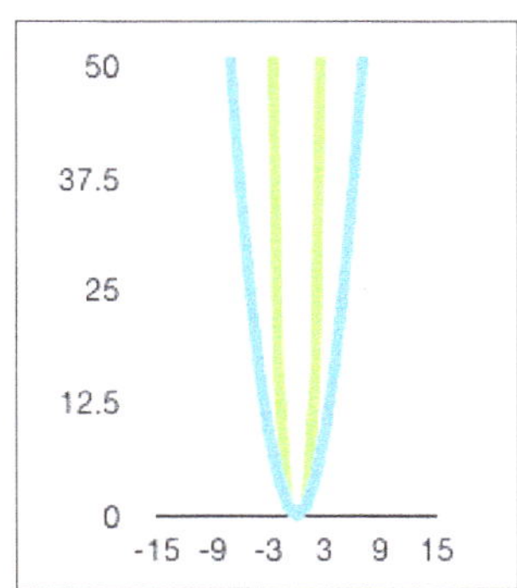

Figure 252

$$\gamma = \frac{m'_4}{(m'_2)^2} - 3$$

Distributions with thicker tales than a normal distribution are sometimes described as **leptokurtic.** Distributions with similar tales to the normal are **mesokurtic,** and distributions with thinner tales are sometimes called **platykurtic.**

Higher moments, and matching data to probability distributions

One of the uses of the idea of moments is to determine how closely some data obtained from a piece of research matches a theoretical probability distribution. We can calculate the different moments from the data we have collected. Then we can compare it with the mean, variance, skewness, kurtosis and higher moments from a known probability distribution. This can help to answer one of the fundamental questions of statistics: is what we are seeing attributable to random probability, or is there some other factor at work, such as the hypothesis being tested in an experiment?

The first four moments of the gamma distribution, using the MGF

The moments of a gamma distribution will be particularly useful when we come to look at the chi square distribution, and this is a convenient place to look at them.

Start with the MGF. We'll use this format, because the results come out a bit more neatly.

$$MGF = (1 - \theta t)^{-n}$$

Take the first derivative with respect to t, using the chain rule.

$$-n(1 - \theta t)^{-n-1}(-\theta)$$

Evaluate where $t = 0$, to find the **expected value.**

$$E(X) = n\theta$$

Take the second derivative; the derivative of the derivative above, again using the chain rule.

$$n\theta\bigl(-(n + 1)\bigr)(-\theta)(1 - \theta t)^{-n-2}$$
$$= n(n + 1)\theta^2(1 - \theta t)^{-n-2}$$

Evaluate where $t = 0$, to find the **variance.**

$$E(X^2) = n(n + 1)\theta^2$$

Repeat again following the same pattern, for the **skewness.**

$$E(X^3) = n(n + 1)(n + 2)\theta^3$$

Once more, for the **kurtosis.**

$$E(X^4) = n(n + 1)(n + 2)(n + 3)\theta^4$$

As in this instance, knowing the MGF (usually) makes it quite straightforward to find the moments of a distribution.

Key points:

- The MGF of a binomial is $(pe^t + q)^n$.

- The **expected value**, or the mean, is the **first moment** of a distribution. It is sometimes symbolised by m_1'.

- The mean of a binomial distribution is np.

- The **second moment**, or **variance**, gives a measure of the average of the square of the distance of each data point from the mean. That makes the formula

$$m_2' = \sum \frac{\left(x_i - E(X)\right)^2}{n}$$

- It can also be expressed as

$$m_2' = E\left[\left(x - E(X)\right)^2\right]$$

$$= E(x^2) - \left(E(x)\right)^2$$

- The variance of a binomial distribution is npq.

- It cannot be assumed that variance will have the property of linearity. When x is multiplied by a constant, for example, the constant needs to be squared when taken outside the variance calculation.

- $variance(X + V) = variance(X) + variance(V)$ if and only if the variables are independent.

- Variance is always positive: $Var(-Y) = Var(Y)$

- $Var(X - Y) = Var(X) + Var(Y)$

- The third central moment of a distribution is the skewness. In a uniform distribution the skewness should equal zero. If a unimodal distribution is right skewed it has a positive skew, whereas a left skew is a negative skew.

- $Skewness = \dfrac{m_3'}{(m_2')^{\frac{3}{2}}}$

- The fourth central moment of a distribution is the kurtosis. This is a measure of how thick the tails are of a distribution. $\gamma = \dfrac{m_4'}{(m_2')^2} - 3$

- Distributions with thicker tales than a normal distribution are leptokurtic. Distributions with similar tales to the normal are mesokurtic, and distributions with thinner tales are platykurtic.

The Markov and Chebyshev Inequalities

Read this if you aren't familiar with the Markov and Chebyshev inequalities; or you don't know how to work out the probability that a number within a probability distribution will fall outside a certain range, if you know nothing much about the distribution other than the mean or the variance.

Moving on from moments of distributions to Chebyshev's next achievement, it is a bit of a mystery why the first of two inequalities that he initially published was named after his student, Markov. (Recall from page 23 and following that an inequality is like an equation, except that instead of two things being equal, one of the things is greater or less than the other.) It is possible that it was Markov who first gave Chebyshev the idea, and so Chebyshev decided to give him the credit.

Regardless, the question he or they were looking at was: if you have a continuous probability distribution and all you know about it is the expected value, what are the chances that some extreme value falls a long way from the expected value? What is the chance of an extreme event? This is a particularly important question in statistics, where we are looking at outcomes and situations that are likely, rather than situations that are definite. What are the chances that something unexpected is happening? Figure 253 illustrates this situation using a graph we used earlier, for seeing another car on a country road with an exponential distribution.

The Markov inequality says that if you have some specific positive random number (call it x), and another positive number (call it a) then the probability that x will be greater than or equal to a will be less than or equal to the expected value of X divided by a. Mathematically we can express this is:

$$If \; x \geq 0 \; and \; a > 0, then \; Pr(x \geq a) \leq \frac{E(X)}{a}$$

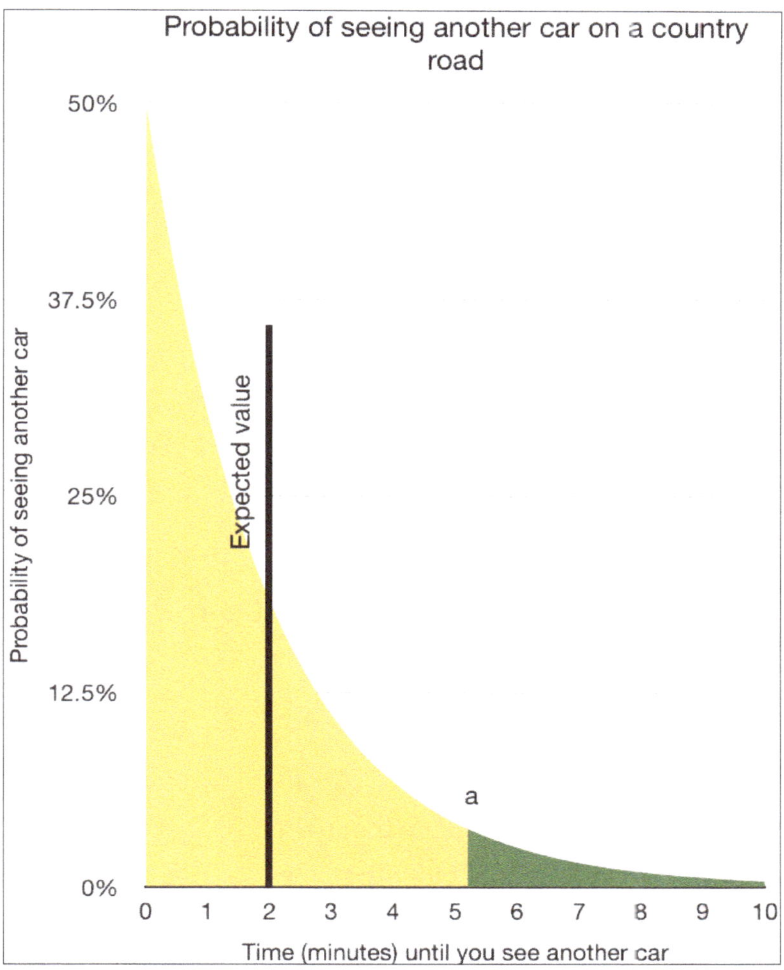

Figure 253

So if a is small, there is a high probability that x is greater than a, whereas if a is large there will be a relatively low probability that x will be larger than it, which makes sense intuitively. To prove it, we can go back to our definition of expected value. In the continuous case:

$$E(X) = \int_{-\infty}^{\infty} xp(X)\,dx$$

In the discrete case it would be a sum, not an integral, but the same reasoning applies.

We have defined x as being positive, and so its expected value must be positive. That means we only need to consider the integral between 0 and infinity.

$$E(X) = \int_{0}^{\infty} xp(X)\,dx$$

We can split this integral into two: the value where x is between 0 and a, and the value where it is between a and infinity.

$$\int_0^\infty xp(X)\,dx = \int_0^a xp(X) + \int_a^\infty xp(X)$$

This whole integral on the left must therefore be larger than just the part on the right.

$$\int_0^\infty xp(X)\,dx > \int_a^\infty xp(X)\,dx$$

x and a are positive numbers, and x is equal to or greater than a. If that is the case, this statement must be true.

$$\int_a^\infty xp(X)\,dx \geq \int_a^\infty ap(X)\,dx$$

Take a (which is a constant) outside the integration.

$$\int_a^\infty xp(X)\,dx \geq a\int_a^\infty p(X)\,dx$$

$$E(X) \geq a\int_a^\infty p(X)\,dx$$

Divide both sides by a, and swap the inequality around. $\int_a^\infty p(X)\,dx$ is the probability that x is greater than a, so we have proved the Markov inequality.

$$\int_a^\infty p(X)\,dx \leq \frac{E(X)}{a}$$

$$Pr(x_i \geq a) < \frac{E(X)}{a}$$

This means we have a way of working out the probability that there will be an extreme event within a distribution, based on the first moment of the distribution (the mean). Markov's inequality uses the mean, and, like the mean, is based on the distance from zero.

The Chebyshev inequality follows from this. Where the Markov inequality uses the first moment of the distribution, the Chebyshev inequality uses the second central moment. Consistently with this, it looks at how far a specific data point is from the mean of the distribution, rather than from zero.

It says that the probability that some value is further from the mean than any particular distance you choose to consider, can be calculated by the second moment of the distribution divided by the square of the distance.

This is the mathematical expression of the probability that the absolute value of the distance of x from the mean is greater than some constant c.

$$P(|x - \mu| \geq c)$$ for some positive constant c.

Square the expression—the probability stays the same.

$$P(|x - \mu| \geq c) = P((x - \mu)^2 \geq c^2)$$

In other words, whatever the chances that absolute value of the difference between x and μ will be greater than a certain constant, there will be the same chance that the probability of the square of that difference will be greater than the square of that constant. There is no need for the absolute value markers now (if x and μ are real numbers), because the square will always be positive. We now have two non-negative numbers: $(x - \mu)^2$, and c^2.

Substitute $(x - \mu)^2$ for x in the formula for the Markov inequality, and c^2 for a. The expected value of some particular value of x will be the same as the expected value of x in general. That means that $E(x - \mu)^2$ will be the variance in general.

$$P((x - \mu)^2 \geq c^2) \leq \frac{E(x - \mu)^2}{c^2}$$

This gives us the Chebyshev inequality.

$$P(|x - \mu| \geq c) \leq \frac{E(x - \mu)^2}{c^2}$$

Remarkably, these two inequalities provide a way of working out the probability that a number within a probability distribution will fall outside a certain range, without knowing anything much about the distribution other than (in the case of the Markov) the mean, or (in the case of the Chebyshev), what was soon to be called the variance.

Key points:

- The Markov inequality says that:

$$Pr(x \geq a) < \frac{E(X)}{a}$$

  where x and a are positive.

- The Chebyshev inequality says that:

$$P(|x - \mu| \geq c) \leq \frac{E(x - \mu)^2}{c^2}$$

Confluence: Pearson and Friends

Figure 254
Karl Pearson
1857–1936

All of this wonderful work just might have remained an obscure branch of mathematics (like anabelian geometry and other branches that most of us have never heard of) if it wasn't for English academic Karl Pearson (1857–1936), possibly one the last of the great polymaths*, who not only developed Pearson's correlation, standard deviation, chi square distribution and tests, z scores and a large number of other statistical measures, but who also wrote a passion play, a romantic novel and socialist hymns. Vladimir Lenon (a materialist) described him as a "conscientious and honest opponent of materialism"[64]. He was a feminist, who employed women to do scientific work

*A **polymath** is someone whose expertise spans a number of subjects. Up until the knowledge explosion from around 1900 (which was partly facilitated by statistics) most of the great leaps forward were made by polymaths. Notable examples (many of whom we have met in this book) include: Leonardo da Vinci, Rene Descartes, Blaise Pascal, Isaac Newton, Leonhard Euler, Benjamin Franklin, Thomas Jefferson, Pierre-Simon Laplace, and Thomas Edison. There isn't space to go into their achievements, but they are all worth looking up for the breadth of their contributions.

(which was highly unusual at the time) and gave us the word "sibling", because he didn't think there was a reason that names depicting family relationships should be gender-based[65].

Pearson was *professionally* engaged in structural engineering, the law, astronomy, philosophy, physical anthropology, history, teaching and academia, as well as running academic institutions. The space of his published works on a bookshelf can be measured in metres, which is extraordinary for someone whose writings were mostly scientific and mathematical. The volume of his writings even exceeds that of the great Leonhard Euler[66] (whose writings could also be measured in metres). When asked later in life how he achieved all that he did, he reportedly answered that, while Americans would never understand it, he never attended meetings or answered the telephone.

Pearson wasn't always a fan of statistics. He started writing about statistics in 1893. Before that he had already published 9 books and had 91 other publications. (The eventual total of publications was 648). In 1889 his response to Galton's research was cautious; and it is a caution that those who use statistics today should keep in mind.

> *"Personally I ought to say that there is, in my own opinion, considerable danger in applying the methods of exact science to problems in descriptive science, whether they be problems of heredity or of political economy; the grace and logical accuracy of the mathematical processes are apt to so fascinate the descriptive scientist that he seeks for sociological hypotheses which fit his mathematical reasoning and this without first ascertaining whether the basis of his hypotheses is as broad as that human life to which the theory is to be applied."*[67]

Somewhere between 1889 and 1893, possibly through the work of Francis Edgeworth, Pearson became equated with Chebyshev's concept of moments of distributions, had some brilliant ideas about it, and became a powerful advocate for the use of statistics.

Pearson had a network of highly-talented friends. He wrote a biography of Francis Galton, and he (sadly) followed Galton in chairing the Eugenics Society. (He also wrote that child labour laws should never have been introduced, because he thought that preventing children from working placed an unfair burden on their parents, inadvertently demonstrating—if demonstration were needed—that intellectual brilliance need not equate to moral insight.)

We saw earlier than another friend, Raphael Weldon, sought his help in analysing the results of his data from marine biology. Some of his friends were also his students, and followed after him extending his work.

Figure 255
Francis Edgeworth
1845–1926

Francis Edgeworth (1845–1926) was another friend of Pearson's, and another great polymath, as well as a distant cousin of Francis Galton. He was unquestionably a genius, who was president of the Royal Society from 1912 to 1914. Edgeworth was born and educated by tutors in Ireland, studied the classics at Trinity College Dublin, studied ancient and modern languages at Balliol College Oxford, and only started teaching himself mathematics and economics after he graduated, while at the same time studying to become a barrister. (He qualified, but never practiced law). Edgeworth is now best known for his contributions to economics. Unfortunately, much of his writing is inaccessible to many readers, partly because he would use a number of different languages in his books, partly because his ideas were very advanced, and partly because he just wasn't very good at explaining himself. It does appear, however, as though—through his skill in languages and through the originality of his mind—he was a link in bringing together the disparate streams of statistics into a cohesive whole. He read the work of Chebyshev in Russian, and incorporated Chebyshev's thinking into trying to get a better formula for the normal distribution. It may even have been he, not Pearson, who developed the modern formulation, but if so, his communication was so difficult for others to follow that even now it is hard to be sure.

Unlike Edgeworth, one of Pearson's interests was in making difficult concepts easier to understand. From 1893 onwards he gave free public lectures, first at Gresham College, then at University College in London. By the time he started he was enamoured with statistics and strove to introduce the ideas to a much wider audience.

His lectures became increasingly popular, among *"clerks and others engaged during the day in the City"*. He once scattered 10,000 pennies over the lecture room floor and asked his students to pick them up and arrange them according to whether they landed on a head or a tail, to demonstrate the effect of probability with large samples. During his series of free lectures the number of students increased *"five to ten-fold"*. As he put it:

> *"As I was not able to lead my audience through the mazy and not over-sure paths of mathematics I adopted the experimental method of an appeal to statistics and the deduction by easy arithmetic."*

We could therefore surmise that Pearson was an exemplar of the Law of Unintended Consequences. He—probably and hopefully—never envisaged the Nazi death camps as an outworking of his belief in eugenics[68], and he also could not have foreseen that by making it easy for people to use statistics without an understanding of the underlying mathematical complexity, he may have paved the way for a great deal of research to be published that purports to have good statistical support, and that actually has nothing of the kind.

Standard deviation and the *z* score

> **Read this if** you thought that variance and standard deviation were basically there to tell how widely data are spread; you thought that standard deviation was an annoying measure that probably shouldn't exist; you don't know what a *z* score is; or you do know, but can't imagine 6 separate benefits in using it, even without the normal distribution.

Standard deviation

Pearson used a short cut for Chebyshev's second central moment of a distribution—the mean of a sum of squared distances—with the lower-case form of the Greek letter sigma, squared (σ^2). While this later became known as the **variance**, Pearson always referred to it as the second moment. To avoid confusion, as we did earlier when we were looking at the second moment, we will just refer to it as the variance.

The formula for the variance where there are separate data points is:

$$\sigma^2 = \frac{(x_1 - \mu)^2}{n} + \frac{(x_2 - \mu)^2}{n} + \cdots + \frac{(x_n - \mu)^2}{n}$$

where μ is the mean and n is the number of data points. Or simply:

$$Var = \sigma^2$$

$$= \sum_{i=1}^{n} \frac{(x_i - \mu)^2}{n}$$

$$= \frac{1}{n} \sum_{i=1}^{n} (x_i - \mu)^2$$

Or as Chebyshev expressed it even more simply (but perhaps less informatively):

$$E(x - \mu)^2$$

It's the mean or expected value of squared distances from the mean of individual data points in a whole data set.

Once Pearson wrote the variance as σ^2, it was an obvious step to stop squaring it and just have σ. This created the standard deviation, which is the square root of the variance. Its formula is:

$$std = \sigma$$

$$= \sqrt{\sum_{i=1}^{n} \frac{(x_i - \mu)^2}{n}}$$

That may not seem like any momentous achievement. The standard deviation is much-misunderstood and maligned[69]. Unhappily, standard deviation has often been introduced to students as primarily a way to describe the spread of data. It doesn't actually give a particularly intuitive sense of the spread of data at all.

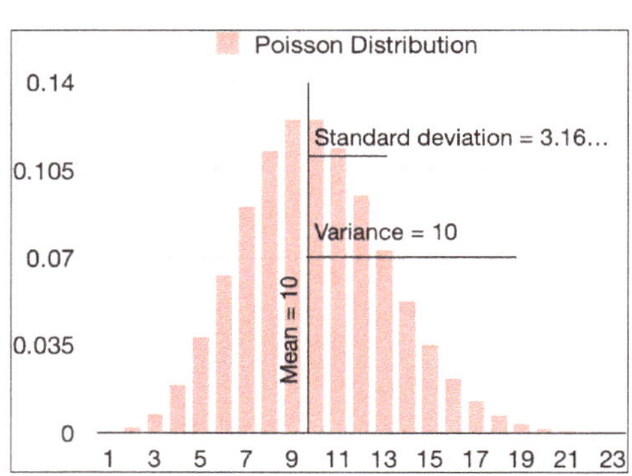

Figure 256

Figure 256 shows a graph of a Poisson distribution with a λ (lambda) of 10. In this instance the variance of 10 (for a Poisson distribution the variance equals the expected value, λ) does give a reasonable approximation of the spread of the data, whereas the standard deviation ($\sqrt{10}$) does not.

On the other hand, Figure 257 shows an exponential distribution with roughly the same spread of data. It has a mean of 5. In this case the standard deviation of 5 isn't a terrible estimate of the spread of the data, whereas the variance of 25 is quite literally off the scale. But unless you knew that these distributions were Poisson and exponential respectively (and you knew what that meant) knowing the standard deviation and the variance would not give you a sense of how the data was spread. (If you want to communicate the spread of a data set without a graph, it's better to use the minimum, the maximum and quantiles such as the median and quartiles.)

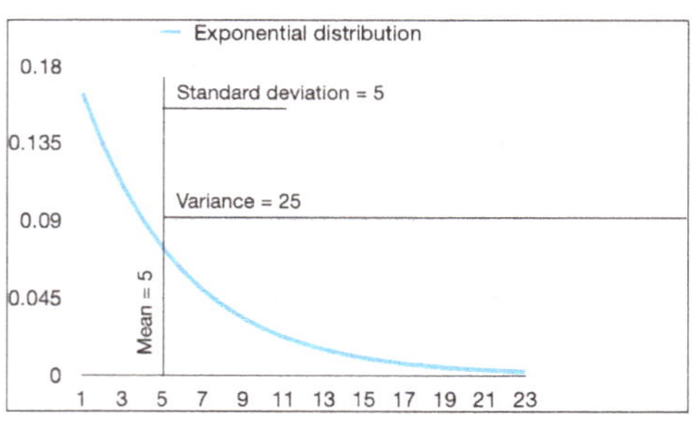

Figure 257

Furthermore, a huge amount of work had already been done on probability distributions before Pearson came along, and, until Chebyshev, not many people had thought much about the value of taking the mean of the squares of the differences from the mean (although people had done a lot of work on finding the minimum possible value of the sum of squared differences from the mean). It wasn't seen as a particularly important aspect of understanding a probability distribution, and Chebyshev seems to have discovered it more or less by accident, through playing around with transforms.

Pearson didn't invent the standard deviation as a descriptive statistic. What he did, brilliantly, was use a standard deviation as a way of comparing one distribution to another and using that to develop a number of formulae that made statistical analysis much more widely applicable. He was very aware that statistics could be useful for much more than it had been to that point, and he set about to make it happen using his incredible brain and his enormous energy.

The *Z* score

Pearson called

$$\frac{x - \mu}{\sigma} = z$$

the **z score**: it is the difference between an x value and the mean, divided by the standard deviation. It is also an extraordinary piece of elegant simplicity. A z score tells you how many standard deviations you are from the mean, in any distribution. While it is most useful with normal distributions (for reasons we will come to), converting distributions to a z variable isn't just useful for normal distributions. It has at least six properties that can make it useful regardless of the distribution.

1. It takes units of measurement out of the picture. Suppose we were looking at people's income, and also their ages in months. Pearson was English, so we will use pounds as our unit of currency.

$$\frac{x \ pounds - \mu \ pounds}{\sigma \ pounds} = \frac{x(\cancel{pounds}) - \mu(\cancel{pounds})}{\sigma(\cancel{pounds})}$$

$$= \frac{x - \mu}{\sigma} = z_{income}$$

$$\frac{x \ months - \mu \ months}{\sigma \ months} = \frac{x(\cancel{months}) - \mu(\cancel{months})}{\sigma(\cancel{months})}$$

$$= \frac{x - \mu}{\sigma} = z_{age}$$

It is now possible to compare income with ages using a similar scale, showing where each score fits within its own distribution; we don't have to try to deal with ages and incomes with different units. If we had divided by the variance rather than the standard deviation, we would have to square the units on the bottom, making it messy again. Dividing by the standard deviation makes it easy.

2. The mean is always zero.

$$\frac{x - \mu}{\sigma} = z$$

If you are looking at the value of the mean; that is, the point where

$$x = \mu$$

then z must be equal to zero.

3. It standardises the variability between different distributions. If, for example, you want to compare students who sit for the same examination but are marked differently by different examiners, and one examiner gives a wider range of marks than the other, converting scores to z scores puts the variation of scores onto a consistent footing.

4. Related to point 3, one standard deviation above the mean will always be $1z$. That is, in fact, why the square root of the variance is called the **standard deviation**. One squared is, of course, one, which means that the square of the standard deviation—the variance—will also always equal one. To understand this, remember that the standard deviation won't change if the mean changes, provided that the whole distribution moves as much as the mean. Changing the mean in this way just slides it left and right along the horizontal axis. So if we set the mean to zero, and assume that all the other parts of the distribution move with it, then we go to the point where our x is one standard deviation from the mean, we can write this as:

$$x = \sigma$$

$$z = \frac{\sigma - 0}{\sigma}$$

$$= 1$$

5. For a discrete, finite distribution, the sum of squared z scores will always equal the total number of data points. That isn't immediately obvious; here's the proof:

Let n equal the total number of data points. Use the definition of z values, square each and sum the results.

$$\sum_{i=1}^{n} z_i^2 = \sum_{i=1}^{n} \left(\frac{x_i - \mu}{\sigma}\right)^2$$

The square of a fraction is equal to a fraction of the squares.

$$= \sum_{i=1}^{n} \frac{(x_i - \mu)^2}{\sigma^2}$$

Using the linearity property of series, take the reciprocal of the variance (which is a constant) outside the summation.

$$= \frac{1}{\sigma^2} \sum_{i=1}^{n} (x_i - \mu)^2$$

Replace σ^2 with the definition of the variance.

$$= \frac{1}{\frac{1}{n}\sum_{i=1}^{n}(x_i - \mu)^2} \sum_{i=1}^{n} (x_i - \mu)^2$$

Multiply the numerator and denominator by n. The rest cancels out.

$$= \frac{n}{\sum_{i=1}^{n}(x_i - \mu)^2} \sum_{i=1}^{n} (x_i - \mu)^2$$

$$\boxed{\sum_{i=1}^{n} z_i^2 = n}$$

6. These properties make it possible to convert a raw score to a **test statistic**. That is, scores can be standardized so that test procedures that have been developed against standardized data can be used. We will see quite a bit of this as we proceed.

We will see an example of how this thinking is used soon when we look at **correlation**.

Key points:

- Standard deviation wasn't developed as a way to communicate the spread of data.

- It is the square root of the variance, and is used to standardise different distributions using the z score.

$$Var = \sigma^2$$

$$= \sum_{i=1}^{n} \frac{(x_i - \mu)^2}{n}$$

$$= \frac{1}{n} \sum_{i=1}^{n} (x_i - \mu)^2$$

$$std = \sigma$$

$$= \sqrt{\sum_{i=1}^{n} \frac{(x_i - \mu)^2}{n}}$$

$$z = \frac{x - \mu}{\sigma}$$

- There are six benefits to the z score:

    1. It takes units of measurement out of the picture.

    2. The mean is always zero.

    3. It standardises the variability between different distributions.

    4. One standard deviation above the mean will always be $1z$.

    5. For a discrete, finite distribution, the sum of squared z scores will always equal the total number of data points.

    6. These properties make it easier to convert raw data to a **test statistic**.

Correlation and covariance

> **Read this if** you don't understand the difference between correlation and covariance, you have no idea where the formulae for either came from, or you assumed that using r^2 as a measure must be a good idea because it gets applied so widely.

We've already started to look at the relationships between distributions (page 397 and following). Now we are thinking about the *strength* and *direction* of the relationship between two distributions.

Correlation coefficient

When we looked at using the least squares method (page 496), we left open the question of how close the residuals are to the line of best-fit (the regression line). In other words, if there is correlation between paired sets of data, how good is it?

When people were using the least squares approach to find the equation for the most likely celestial path based on a set of error-prone measurements, the question of how close the measurements were to the line was secondary; once you had more readings you could refine your calculation that way. But when Francis Galton published his hypothesis on regression to the mean, and challenged *"a competent person who cares to investigate it"* to develop the mathematics around correlation, Pearson decided to be that competent person. He used the formula for the regression coefficient as his starting point, together with his new invention of the z score. He took the same conceptual model that Gauss (or perhaps someone else) had used for a least squares calculation (Figure 258), but rather than thinking of the points as measurements, he thought of them as representing paired variables: for example the number of widgets that a worker could produce on one axis, and the number of hours of training that same person had received on the other. (He didn't use those examples, but that is the idea).

The first step therefore was to use the z scores to standardise the measurement units, so that he would have an appropriate basis for comparing the x and the y components.

Recall that the formula for the slope of the regression line (the m in the equation of the red line $y = mx + b$ in Figure 258) is:

$$m = \frac{\Sigma(x_i - \bar{x})(y_i - \bar{y})}{\Sigma(x_i - \bar{x})^2}$$

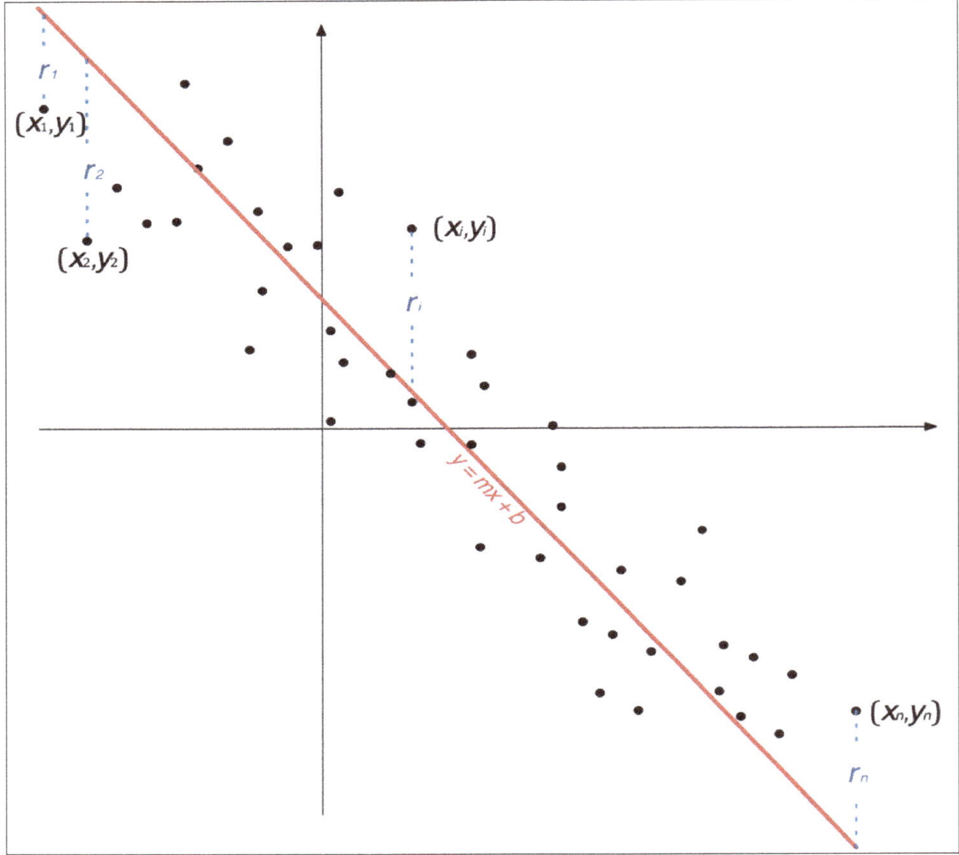

Figure 258

As a reminder, $\bar{x}$ and $\bar{y}$ are the means of the x and y values and are constants. x_i and y_i represent each instance of the x and y values. Now suppose you divide the numerator and denominator by the variance of the x values. (This isn't exactly what Pearson did, but stay with it, because it's instructive. We'll get to what he actually did soon.) Remembering that the standard deviation is the square root of the variance, and that it will be a constant that can be distributed within the summations. We'll use σ_x^2 to represent the variance of the x values (a constant), and z_{xi} to represent the z score of each instance of the x values. This gives us:

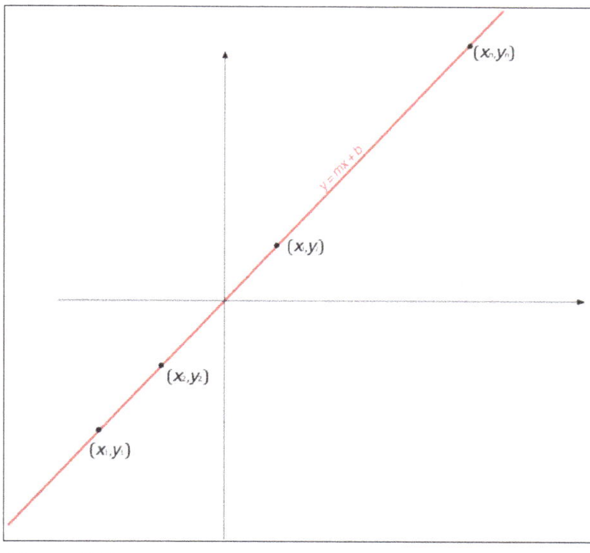

Figure 259

$$m = \frac{\frac{1}{\sigma_x{}^2}\sum(x_i - \bar{x})(y_i - \bar{y})}{\frac{1}{\sigma_x{}^2}\sum(x_i - \bar{x})^2}$$

$$= \frac{\sum\left(\frac{x_i - \bar{x}}{\sigma_x}\right)\left(\frac{y_i - \bar{y}}{\sigma_x}\right)}{\sum\left(\frac{x_i - \bar{x}}{\sigma_x}\right)^2}$$

$$= \frac{\sum z_{xi}\left(\frac{y_i - \bar{y}}{\sigma_x}\right)}{\sum(z_{xi})^2}$$

Now imagine that there was perfect agreement between the standardised x scores and the standardised y scores (Figure 259). That would mean:

$$\frac{x_i - \bar{x}}{\sigma_x} = \frac{y_i - \bar{y}}{\sigma_x}$$

which would mean that

$$m = \frac{\sum\left(\frac{x_i - \bar{x}}{\sigma_x}\right)\left(\frac{x_i - \bar{x}}{\sigma_x}\right)}{\sum\left(\frac{x_i - \bar{x}}{\sigma_x}\right)^2}$$

$$= 1$$

So this will draw a line with a slope of 45 degrees which is a slope of 1: for every distance you move to the right, you move the same distance vertically.

Alternatively, imagine that they were exactly **negatively correlated** or **inversely proportional.** For each unit forward of x there would be a drop of one unit in y. Mathematically we could express this as

$$\frac{x_i - \bar{x}}{\sigma_x} = -\frac{y_i - \bar{y}}{\sigma_x}$$

$$m = \frac{\sum -\left(\frac{x_i - \bar{x}}{\sigma_x}\right)\left(\frac{x_i - \bar{x}}{\sigma_x}\right)}{\sum\left(\frac{x_i - \bar{x}}{\sigma_x}\right)^2}$$

$$= -1$$

Now the line will have a slope of 45°, but will slope backwards. (Figure 260).

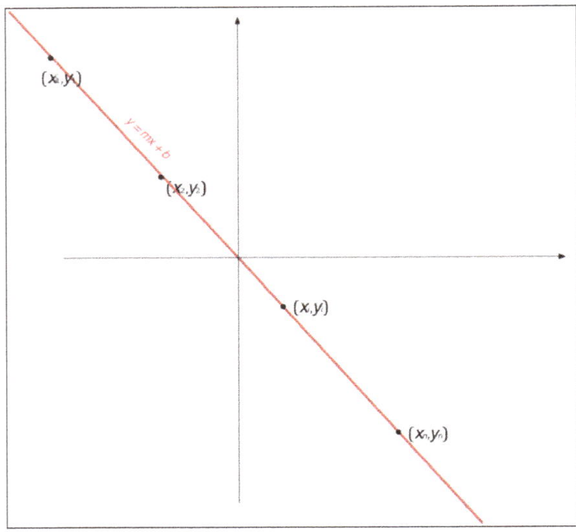

Figure 260

Now assume that the line had a positive slope, but the y values, on average, were a little further away from the mean than the x values; they were a bit more spread out. That would increase the value of m and make the regression line steeper. If the y values were closer to their mean than the x values were to theirs, that would decrease the value of m.

Which is all fine if you are trying to understand how the spread of the values changes the slope of the regression line, but what Pearson wanted to know, and what is usually a more important question is: how good is the line as a description of the relationship between the x and y values? That is, now that we have a line, how close are the points to it? He achieved this by a couple of small modifications to the equation we have been looking at.

We saw previously that the sum of squared z values is equal to the number of data points. That means we can exchange the denominator for n, which gives us:

$$m = \frac{\sum z_{xi}\left(\frac{y_i - \bar{y}}{\sigma_x}\right)}{n}$$

If he kept the n in the numerator (if the variables are paired then the number of x values will equal the number of y values) but divided the $y_i - \bar{y}$ by the standard deviation of the y values (which makes more sense) he would have:

$$\frac{\sum z_{xi}\left(\frac{y_i - \bar{y}}{\sigma_z}\right)}{n} = \frac{\sum z_{xi} z_{zi}}{n}$$

This no longer equals m, of course. He designated this **correlation coefficient** with the Greek letter **rho (ρ)**, which looks like and English p, but isn't. You will also see it written using the English letter r. Quite commonly ρ will represent the correlation of the whole population (the parameter) while r will be used for the correlation of a sample (the statistic). So the formula, in full is:

$$\text{Correlation coefficient} = \rho \ (rho)$$

$$= \frac{\sum_{i=1}^{n} z_{xi} z_{yi}}{n}$$

where the z values are the distances of the x and y values from their respective means divided by their respective standard deviations, and n is the number of subjects (i.e. the number of *pairs* of data points).

This simple formula gave a standardised measure of how far the x and y values were from one another. It also tells you whether the relationship is positive or negative.

It can be expressed in several different (much less simple-looking) ways, and turns out to be a remarkably convenient formula. Here's one example, but there are others you might see:

This is based on the formula above, and the definition of z values:

$$\rho = \frac{\Sigma \left(\frac{x_i - \bar{x}}{\sigma_x} \right) \left(\frac{y_i - \bar{y}}{\sigma_y} \right)}{n}$$

The reciprocals of the standard deviations are constants and can be taken outside of the summation. Then multiply the numerator and denominator by their product.

$$\rho = \frac{\Sigma(x_i - \bar{x})(y_i - \bar{y})}{n \sigma_x \sigma_y}$$

We saw above that perfect agreement gives a slope of the regression line of one: for each distance to the right there is a corresponding vertical rise, creating a slope of 45 degrees. The same thing happens when using the correlation coefficient to determine how close the points are to the regression line. If there is perfect correlation between the x and y values, that gives us: $(x_i - \bar{x}) = (y_i - \bar{y})$ and also $\sigma_x = \sigma_y$.

On that basis,

$$\rho = \frac{\Sigma(x_i - \bar{x})^2}{n} \frac{1}{\sigma_x{}^2}$$

which is another way of saying it is the variance multiplied by the reciprocal of the variance, which is equal to 1. So a really helpful thing about **Pearson's correlation** is that perfect correlation equals 1.

Furthermore, if there is a completely perfect negative, or **inverse correlation,** that is the same as saying $-(x_i - \bar{x}) = (y_i - \bar{y})$. Remember that standard deviation involves squaring then taking the positive square root, which means that standard deviations are always positive. That makes the formula

$$\rho = \frac{-\Sigma(x_i - \bar{x})^2}{n} \cdot \frac{1}{\sigma_x{}^2}$$

$$= -1$$

So a perfect negative correlation is equal to minus one. Other values fall in between. In other words:

$$-1 \leq \rho \leq 1$$

Here's a proof that values without perfect correlation fall between minus one and one:

The proof is about the product of z's, but we start by looking at the average of their squared sums. Don't worry about where this equation comes from: it just exists for the proof. The reason will be clear soon. The square can't be negative for any real numbers, and n, as the count of the number of data points, must also be positive. That means the sum must be positive.

$$\frac{1}{n}\sum_{i=1}^{n}\left(z_{xi} + z_{yi}\right)^2 \geq 0$$

$$for\ all\ real\ z_{xi}\ and\ z_{yi}$$

Expand out the binomial expression.

$$\frac{1}{n}\sum_{i=1}^{n}(z_{xi}^2 + 2z_{xi}z_{yi} + z_{yi}^2) \geq 0$$

Use the linearity property to express this as separate sums, and to distribute $\frac{1}{n}$ to each of them.

$$\frac{1}{n}\sum_{i=1}^{n}z_{xi}^2 + \frac{2}{n}\sum_{i=1}^{n}z_{xi}z_{yi} + \frac{1}{n}\sum_{i=1}^{n}z_{yi}^2 \geq 0$$

We saw (page 567) that $\sum_{i=1}^{n}z_{xi}^2 = n$. We can therefore say that $\frac{1}{n}\sum_{i=1}^{n}z_{xi}^2 = 1$ (dividing both sides by n). This also has to mean that $\frac{1}{n}\sum_{i=1}^{n}z_{yi}^2 = 1$

$$1 + \frac{2}{n}\sum_{i=1}^{n}z_{xi}z_{yi} + 1 \geq 0$$

Subtract 2 from each side, and divide each side by 2.

$$\frac{1}{n}\sum_{i=1}^{n}z_{xi}z_{yi} \geq -1$$

$\frac{1}{n}\sum_{i=1}^{n}z_{xi}z_{yi}$ is the definition of ρ, so ρ can't be less than –1.

$$\rho \geq -1$$

Back to (almost) our starting point, change the plus sign to a minus sign. Because the total expression is squared, the inequality still holds up.

$$\frac{1}{n}\sum_{i=1}^{n}\left(z_{xi} - z_{yi}\right)^2 \geq 0$$

for all real z_{xi} and z_{yi}

Expand out the binomial expression.

$$\frac{1}{n}\sum_{i=1}^{n}(z_{xi}^2 - 2z_{xi}z_{yi} + z_{yi}^2) \geq 0$$

As per the arguments above.

$$1 - \frac{2}{n}\sum_{i=1}^{n}z_{xi}z_{yi} + 1 \geq 0$$

Subtract 2 from both sides and divide by 2. Call this line inequality A.

$$-\frac{1}{n}\sum_{i=1}^{n}z_{xi}z_{yi} \geq -1$$

Multiply both sides by –1. The inequality is reversed; now the left side is less than or equal to the right. Note that the equality will be reversed when you multiply by minus one, whether or not ρ is negative. If you are unsure about this, remember that, as we saw above, ρ has to be greater than -1. Let's imagine, in inequality A, that $\rho = \frac{1}{2}$. That means that inequality A is saying $-\frac{1}{2} \geq -1$, which is true. Now if we multiply both sides by -1, and reverse the inequality, we have $\frac{1}{2} \leq 1$, which is also true. Now imagine that $\rho = -\frac{1}{2}$. That means, according to inequality A, that $-\frac{1}{2} \geq -1$, which is true. Multiplying both sides by –1 means that $-\frac{1}{2} \leq 1$, which is also true.

$$\rho \leq 1$$

And so the only values that ρ can have lie between perfectly negatively and perfectly positively correlated.

$$\boxed{-1 \leq \rho \leq 1}$$

Unfortunately, once you know the direction of the slope of the line, the amount of information given by ρ is somewhat limited. You know that a correlation of 0.75 is closer than 0.25, but you don't know how much closer. You cannot assume, for example, that 0.75 is 3 times closer than 0.25. That is the issue that what we are about to look at is supposed to address.

The cautionary tale of r^2: the "coefficient of determination"

You will probably come across something called the **coefficient of determination**, or r^2 if you spend much time looking into statistics or reading professional journals. In the medical world, for example, Figure 261 shows the trend in its usage, based on the number of times it appears in a search on the medical database PubMed, by year. (To keep it in perspective it is still less than 3 in every 10,000 articles on PubMed). Popular statistics and spreadsheet packages don't offer a huge amount of statistical options (unless you load up extra features) but if you set up a scatter diagram, use the program to generate a trend line, and select the option to format the trend line, you will probably find an option to display the r^2 value.

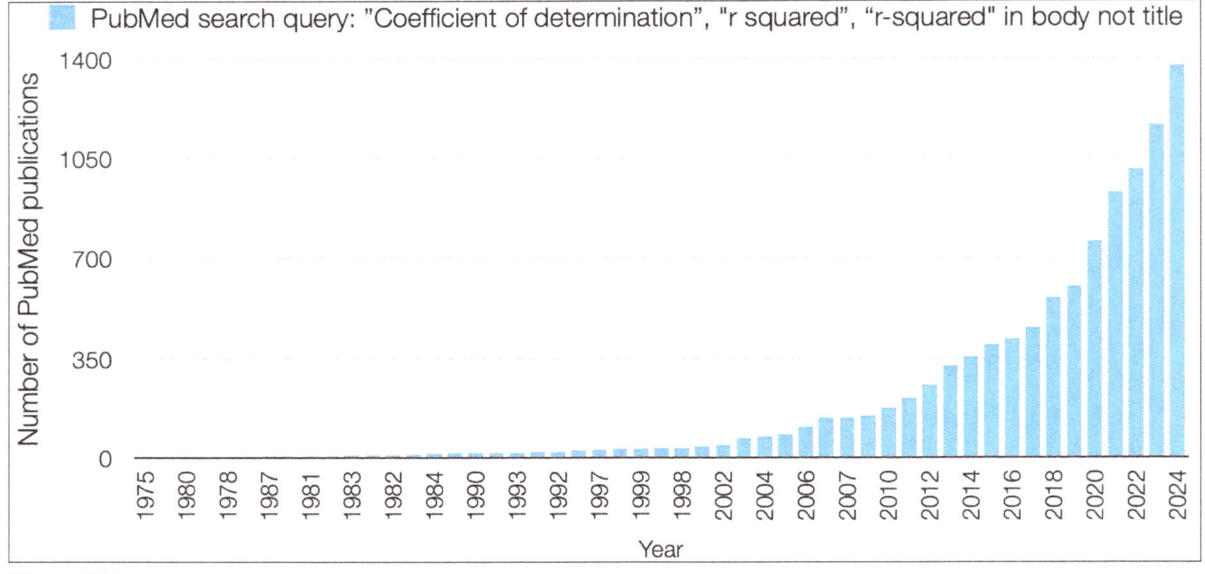

Figure 261

The set up for this statistic is that there is a regression line through a set of data points (Figure 262). There is a sum of squared distances to the regression line, which we'll designate as SS_{rl} (red lines) and a different sum of squared distances to the mean of the y values, which we'll designate as $SS_{mean\,y}$ (green lines). (If $SS_{mean\,y}$ is divided by the number of data points, you have the variance of the y values.)

If you divide SS_{rl} by the $SS_{mean\,y}$ that is supposed to tell you what percentage of the total variation is *not* described by the regression line. If you then subtract that from 1, it is supposed to tell you the percentage of total variation that *is* described by the regression line, which is called the coefficient of determination:

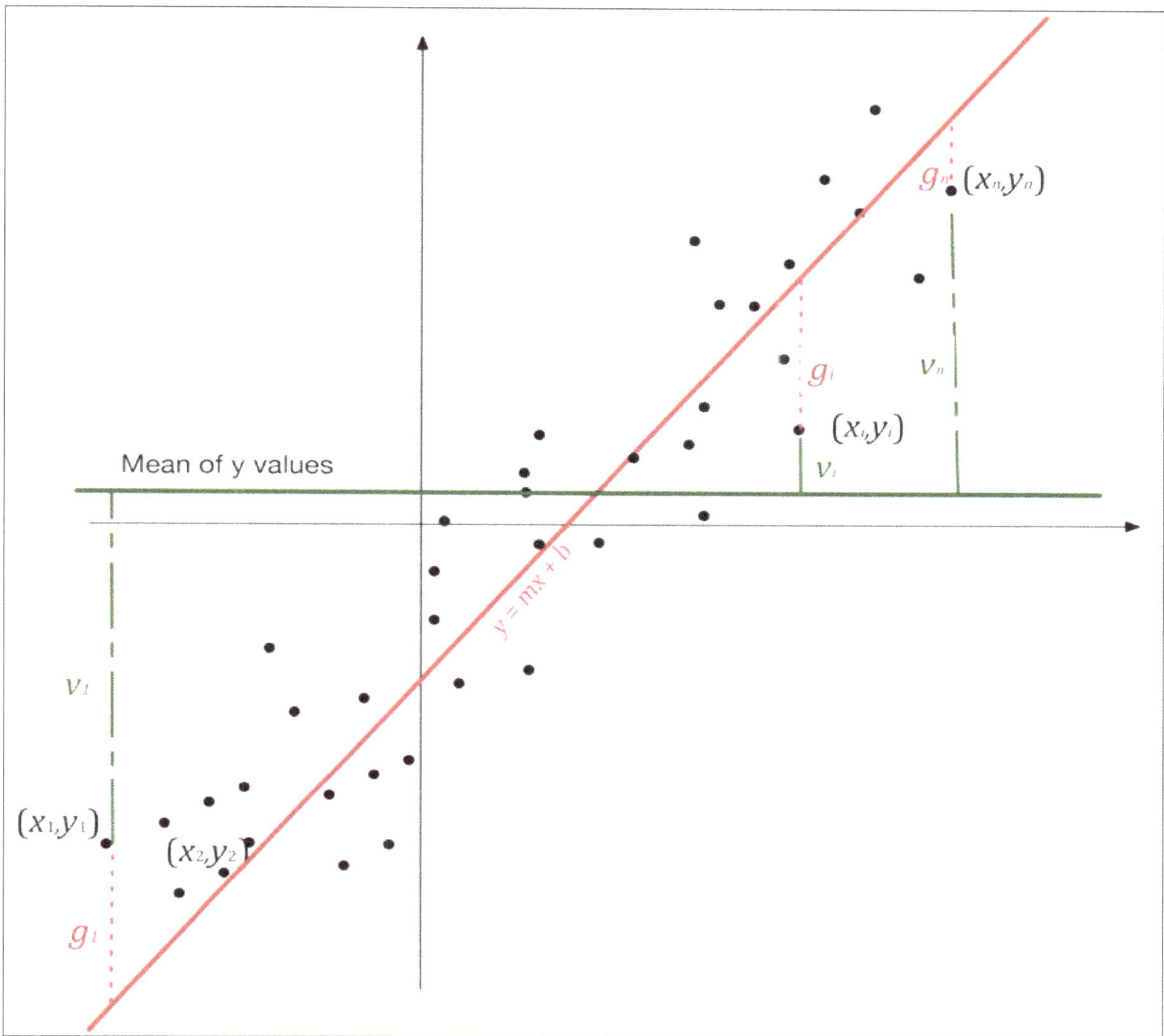

Figure 262

$$Coefficient\ of\ determination = 1 - \frac{SS_{rl}}{SS_{mean\ y}}$$

Conveniently, that is supposed to be the result you get when you square Pearson's ρ (rho) correlation.

This is often used as an indication of the extent to which the correlation with the x value accounts for the variance in the y values, which is a nice idea.

There are, however, three major problems with the coefficient of determination.

The first is that the mathematics doesn't work out. You probably won't be able to find a derivation proving that this ratio of squared differences is the same as the square of the correlation, and if you can, there will probably be a problem with the set up. It would be extremely neat if the mathematics worked out, but it doesn't. They can be made to come close, but they don't quite make it.

This results in the second problem. As Wikipedia (at least at the time of writing) expresses it:

"There are several definitions of r^2 that are only sometimes equivalent."[70]

Which is another way of saying: "if you look at it, you don't necessarily know what you are looking at". This issue is so extreme that some definitions suggest that answers can range from $-\infty$ to 1, while it is also supposed to be obtained by squaring Pearson's ρ correlation which would mean, of course, that it shouldn't be able to be less than zero[71].

There may, in fact, be one or more versions of this formula that do make sense, but if you don't know what the basis of calculation was, and you can't tell which version is being used, you won't have a way to know whether what you are looking at is trustworthy.

The third problem is that the logic doesn't work out. Even if it behaved as it is purported to: giving the ratio between the sum-of-squared-distances from the regression line and the sum-of-squared-distances from the mean, it still wouldn't be all that helpful, because a ratio of the sum-of-squared-distances shouldn't be confused with the ratio of the sum of distances. For example:

$$\frac{2+3}{1+9} = \frac{5}{10}$$

$$= \frac{1}{2}$$

Whereas if we square each of these numbers before adding them, then look at the ratio, we get:

$$\frac{2^2 + 3^2}{1^2 + 9^2} = \frac{4+9}{1+81}$$

$$= \frac{13}{82}$$

One half is obviously not remotely close to thirteen over eighty-two.

There are, as we have seen, some good reasons for working with the sum of squares: but using them in the sort of ratio that r^2 is supposed to give isn't one of them.

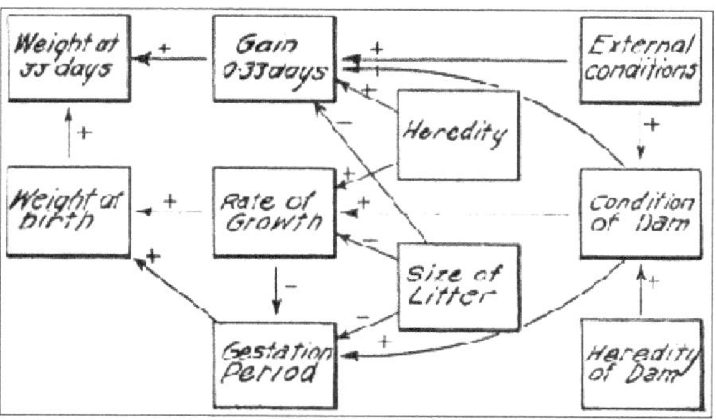

Figure 263

At this stage you may be wondering where this came from, and who originally thought it was a good idea. The answer to this is probably nobody knows. It has been attributed to Sewell Wright, a geneticist working for the US Department of Agriculture in the 1920s. But the paper others have referenced for it is actually dealing with the interesting issue of using correlation to make assumptions about the extent of causation where there are multiple, interlinked causes[72] (Figure 263). In this paper Wright did, in fact, mention r^2, but in the context of multiplying r along the branches of the tree, similar to the approach we saw earlier with tree charts and the multiplication rule. He also used a number of ratios of different aspects of variances. At no stage did he equate the two. But somewhere along the line, it appears as though someone was exposed to his paper, misread it, and used it to successfully promulgate a statistic that should never have been used in that way. It has become popular.

In a situation where the majority of people learning statistics, and even many of those teaching it and writing text books about it, don't really understand the mathematical rationale behind what they are doing; this sort of situation is the likely result. Not only that: it has definitely happened.

Other cautions with correlation

1) Pearson's correlation can't be interpreted as a percentage. A correlation of 0.2 isn't a 20% correlation, and it also isn't half of a correlation of 0.4. (This is the problem that the r^2 statistic was supposed to resolve and, of course, it fails miserably.)

2) Correlation, as you probably know, shouldn't be confused with causation. The fact that two variables are closely correlated doesn't tell you whether one is the cause of another, or whether there is something else causing both of them. More than one paper has been written, for example, suggesting that specific environmental factors must cause specific health conditions because they are geographically associated, without taking into account the concentrations of population in those regions. (The more people, of course, the more health conditions, but also, very often, the more environmental pollution.) That doesn't mean that

statistics can't help us to intelligently estimate causation; it sometimes can, but you generally need more techniques and more data before reaching any sensible opinions about it.

3) Pearson's correlation is a measure of the straight-line relationship between sets of paired variables. It isn't designed to look at non-linear relationships. If you think your data has a relationship that isn't linear, there are a couple of approaches. You can use the least squares method we looked at earlier to find the curve of best fit, and work out from first principles how close your residuals are, on average, to the curve of best fit. The difficulty here is that you may not have a lot to compare it to, so you may not understand the significance of the number you are looking at. The alternative approach, which is more-commonly employed, is to apply a formula to your data so that what you have is a straight line, to which you can then apply Pearson's correlation. If you think your data shows an exponential relationship for example, you can take the log of the y values, and you will have a straight line you can work with. (You will have to keep in mind that you are talking about correlations of the logarithms, not the raw data.)

4) You may have heard that Pearson's correlation isn't appropriate for small sets of data that are not normally distributed. That isn't correct; there is nothing in our analysis that suggests this is the case. It is, however, possible to use t tests (which we will come to later) to tell you how statistically significant a correlation is, based on the number of data points in your sample. Sometimes a computer will include the t test in its analysis, and it may not be immediately obvious that it has used an additional approach. t tests *do* assume that the underlying data is normally distributed, at least if the data set is small, but Pearson's correlation itself doesn't.

5) Before attempting to use correlation to extrapolate data, it is worth doing an appropriate test (probably a t test if you think your data is normally distributed) to find the statistical significance of your correlation, and you should think carefully about other factors that could come into play, particularly if your extrapolation moves you further along the line than your data points.

Generalising correlation, variance and covariance

To this point when looking at correlation we have been quite specific, considering paired sets of data; mostly relating to individual subjects having more than one random variable assigned to them, and wanting to compare these variables. We can, however, generalise some of this thinking to compare data distributions that aren't paired. A retailer may, for example, want understand the relationship between the number of children living within an area and the number of toys sold. The data isn't paired—the retailer won't know which children bought which toys—but understanding the relationship may still be important, in terms of deciding where to locate stores.

Earlier (page 398 and following) we looked at two distributions at the same time, considering the concepts of joint and marginal distributions. We saw that a joint distribution can only be calculated by multiplying marginal distributions when both marginal distributions are independent. When we think about covariance and correlation, however, we are dealing with situations (much more common in real life) where there is at least some relationship between two variables, and we want to know how much. If X and Y are independent, then they are uncorrelated. (At the same time, just because they are uncorrelated doesn't mean they are independent. There might, for example, be a common driver of both of them, that has a different effect on each.)

One of the ways we expressed the formula for Pearson s correlation was:

$$\rho = \frac{\Sigma(x_i - \mu_x)(y_i - \mu_y)}{n\sigma_x\sigma_y}$$

The part in green is called the **covariance**.

For probability distributions, it's the expected value of the product of the differences between the random variables and their respective means. In other words, it is similar to the variance, except that there are two variables involved instead of one. If X is equal to Y, then the formula becomes

$$\frac{\Sigma(x_i - \mu_x)^2}{n}$$

which is the formula for the variance.

The more general definition of the correlation is the covariance divided by the product of the standard deviations:

$$correlation(X,Y) = \frac{covariance(X,Y)}{\sigma_X\sigma_Y}$$

The standard deviations are constants, so can be distributed within the expected value (by linearity). This is, of course, another way to write Pearson's correlation, and expressed in this way it doesn't need to be restricted to paired variables. You can think of correlation as a way to standardise the covariance using the standard deviations.

Because of its similarity to the variance (second moment) the correlation is sometimes called the **product-moment coefficient**.

Earlier (page 546) we saw that variance can be expressed as

$$E(X^2) - \big(E(X)\big)^2$$

which is less intuitively obvious in terms of what it is about, but can be easier to work with. An analogous expression can be used for the covariance, when we are thinking about probability distributions.

Use the expected value definition of covariance. It is mathematically the same as other definitions above but should, as it is talking about expected value, be limited to probability distributions.

$$COV(X, Y) = E\{[X - E(X)][Y - E(Y)]\}$$

Expand out the terms.

$$= E\{XY - YE(X) - XE(Y) + E(X)E(Y)\}$$

Use linearity to show this as separate expected values

$$= E(XY) - E\big(YE(X)\big) - E\big(XE(Y)\big) + E\big(E(X)E(Y)\big)$$

The expected values are constants, and can be taken outside of the expected values (by linearity).

$$= E(XY) - E(X)E(Y) - E(Y)E(X) + E(X)E(Y)\, E(1)$$

Simplify. $E(1) = 1$

$$\boxed{cov(X, Y) = E(XY) - E(X)E(Y)}$$

So the covariance of two variables is equal to the expected value of their product, less the product of their expected values.

Unlike with variance, there isn't normally a squared term within covariance, which means that linearity can be applied more often. If one of the terms is multiplied by a constant, the constant can be taken outside the covariance, without squaring it.

$$cov(AX, Y) = A cov(X, Y)$$

where A is a constant.

Based on the formula $cov(X, Y) = E(XY) - E(X)E(Y)$, it is easy to demonstrate that

$$cov(X, Y + Z) = cov(X, Y) + cov(X, Z)$$

We saw when we were looking at the second moment of a distribution (pages 544 and following) that, where two variables are *independent*, the variance of a sum is equal to the sum of the variances. We're now in a position to see what happens to the variance of the sum of two *dependent* random variables.

To find the sum of the variance of two random variables:

Use this formula for variance:

$$var(X + Y) = E[(X + Y)^2] - [E(X + Y)]^2$$

Expand out the binomial expression on the left (in red)

$$= E[X^2 + 2XY + Y^2] - [E(X + Y)]^2$$

Use linearity to show this expression as a sum of expected values. We saw earlier that the expected value of the sum of random variables is the sum of their expected values.

$$= E[X^2] + 2E[XY] + E[Y^2] - [E(X) + E(Y)]^2$$

Expand out the binomial expression on the right (blue)

$$= E[X^2] + 2E[XY] + E[Y^2] \\ - \left[(E(X))^2 + 2E(X)E(Y) + (E(Y))^2 \right]$$

Take the blue terms outside the brackets (multiplying each by –1)

$$= E[X^2] + 2E[XY] + E[Y^2] - (E(X))^2 - 2E(X)E(Y) \\ - (E(Y))^2$$

Group the terms differently

$$= \left[E(X^2) - \left(E(X)\right)^2\right] + \left[E(Y^2) - \left(E(Y)\right)^2\right]$$
$$+ \left[2E(XY) - 2E(X)E(Y)\right]$$

$$\boxed{var(X + Y) = var(X) + var(Y) + 2\big(cov(X,Y)\big)}$$

So the variance of a sum of random variables is the sum of the variances, plus twice the covariance. If the two variables are independent, the covariance will be zero, which is why we were able to say earlier that for independent variables the sum of the variances is the same as the variance of the sums.

To generalise this further, it isn't necessary to be restricted to two variables. If you are considering three or more, you can manipulate them algebraically using the approaches above, by just grouping all but one together at one time, then repeating.

Key points:

- *Pearson's correlation* $= \rho$ *(rho)*

$$= \frac{\sum_{i=1}^{n} z_{xi} z_{yi}}{n}$$

$$= \frac{\Sigma (x_i - \bar{x})(y_i - \bar{y})}{n \sigma_x \sigma_y}$$

- $-1 \leq \rho \leq 1$

- The coefficient of determination, or r^2 is flawed because the mathematics doesn't make sense, there are different views about what it actually involves, and it logically doesn't do what it is supposed to. It probably shouldn't be used.

- Pearson's correlation can't be interpreted as a percentage.

- It isn't designed to look at non-linear relationships.

- The data does not have to be normally distributed, unless you are using t test to determine how statistically significant the correlation is.

- $correlation(X, Y) = \frac{covariance(X,Y)}{\sigma_X \sigma_Y}$

- $cov(X, Y) = E(XY) - E(X)E(Y)$

- $cov(AX, Y) = A cov(X, Y)$

- $cov(X, Y + Z) = cov(X, Y) + cov(X, Z)$

- $var(X + Y) = var(X) + var(Y) + 2\big(cov(X, Y)\big)$

Completing the formula for the normal distribution

> **Read this if** you want to learn more about the ideas behind the normal distribution curve, you would like to know why the inflexion is always one standard deviation from the mean, or you want to know where *z* score tables come from.

One of Pearson's (or possibly Edwards's) early achievements was to complete the formula for the normal distribution, incorporating the standard deviation into it. He started with the formula for the second moment of a distribution (the variance), and used this to find the value of A in Laplace's error formula.

As a reminder, this is where Laplace had ended up with his formula for a frequency-distribution curve of errors.

$$p(X) = \frac{\sqrt{A}}{\sqrt{2\pi}} e^{\left(\frac{-A(x-\mu)^2}{2}\right)}$$

Use Chebyshev's formula for the 2nd central moment of a continuous variable:

$$\sigma^2 = \int_{-\infty}^{\infty} (x-\mu)^2 \frac{\sqrt{A}}{\sqrt{2\pi}} e^{\left(\frac{-A}{2}(x-\mu)^2\right)} dx$$

$$\sigma^2 = \int_{-\infty}^{\infty} (x-\mu)^2 p(X) \, dx$$

Take the constant outside the integration.

$$\sigma^2 = \frac{\sqrt{A}}{\sqrt{2\pi}} \int_{-\infty}^{\infty} (x-\mu)^2 e^{\left(\frac{-A}{2}(x-\mu)^2\right)} dx$$

Substitute w for $(x-\mu)$ to simplify the formula (a bit). We can replace

$$= \frac{\sqrt{A}}{\sqrt{2\pi}} \int_{-\infty}^{\infty} w^2 e^{\frac{-Aw^2}{2}} dw$$

dx with dw, because if $w = x - \mu$

then $\dfrac{dw}{dx} = 1$, and so $dw = dx$.

Use integration by parts; one w is used in one part, the other in the other part.

$$= \frac{\sqrt{A}}{\sqrt{2\pi}} \int_{-\infty}^{\infty} w w e^{\frac{-Aw^2}{2}}\, dw$$

$$\int u \frac{dv}{dw}\, \delta w = uv - \int v \frac{du}{dw}\, \delta w$$	This is the formula for integration by parts, replacing x with w.
$$u = w \qquad \frac{dv}{dw} = w e^{\frac{-Aw^2}{2}}$$ $$\frac{du}{dw} = 1 \qquad v = \int w e^{\frac{-Aw^2}{2}}\, \delta w$$	These are the elements to put into the integration-by-parts formula.
$$\text{which means } v = \frac{-1}{A} e^{-\frac{Aw^2}{2}}$$	The formula for v came from integrating $\frac{dv}{dw}$. You can check it by finding the derivative of v using the chain rule.

Substitute back into the formula for integration by parts.

$$\sigma^2 = \frac{\sqrt{A}}{\sqrt{2\pi}} w \left. \frac{-1}{A} e^{-\frac{Aw^2}{2}} \right|_{-\infty}^{\infty} + \frac{\sqrt{A}}{\sqrt{2\pi}} \int_{-\infty}^{\infty} 1 \frac{1}{A} e^{\frac{-Aw^2}{2}}\, dw$$

Now $\left. \frac{-1}{A} e^{-\frac{Aw^2}{2}} \right|_{-\infty}^{\infty} = 0$, because e raised to a negative power is the same as the reciprocal of e raised to the power and as w approaches infinity, the expression approaches zero. The w is squared, so negative infinity becomes positive infinity squared and the same thing applies. The whole first term is multiplied by zero, and so it disappears.

$$\sigma^2 = \cancel{\frac{\sqrt{A}}{\sqrt{2\pi}} w \left. \frac{-1}{A} e^{-\frac{Aw^2}{2}} \right|_{-\infty}^{\infty}} + \frac{\sqrt{A}}{\sqrt{2\pi}} \int_{-\infty}^{\infty} \frac{1}{A} e^{\frac{-Aw^2}{2}}\, dw$$

Take $\frac{1}{A}$ outside of the integration, and put $\frac{\sqrt{A}}{\sqrt{2\pi}}$ inside it.

$$\sigma^2 = \frac{1}{A} \int_{-\infty}^{\infty} \frac{\sqrt{A}}{\sqrt{2\pi}} e^{\frac{-Aw^2}{2}}\, dw$$

Substitute back $(x - \mu) = w$

$$\sigma^2 = \frac{1}{A} \int_{-\infty}^{\infty} \frac{\sqrt{A}}{\sqrt{2\pi}} e^{\frac{-A(x-\mu)^2}{2}}\, dx$$

Now everything included within the integration is the formula for Laplace's error curve, and we already know that this integration = 1, because it is a cumulative probability density.

After all of that, we have an answer which is almost ridiculously simple: *A* is the reciprocal of the variance.

$$A = \frac{1}{\sigma^2}$$

$$\sigma^2 = \frac{1}{A}$$

In statistics, the reciprocal of the variance is called the **precision**. So in the error curve, the constant *A* is equal to the precision.

Once again, this is the formula we started with.

$$y = \frac{\sqrt{A}}{\sqrt{2\pi}} e^{\left(\frac{-A(x-\mu)^2}{2}\right)}$$

Replace *A* with $\frac{1}{\sigma^2}$

$$= \frac{\sqrt{\frac{1}{\sigma^2}}}{\sqrt{2\pi}} e^{\left(\frac{-\frac{1}{\sigma^2}(x-\mu)^2}{2}\right)}$$

Within the exponent, multiply the numerator and denominator by σ^2 and take this inside the brackets, so it can can be shown as σ but then square it when the numerator is squared. This makes it more obvious that it is a *z* score.

$$= \frac{\sqrt{\frac{1}{\sigma^2}}}{\sqrt{2\pi}} e^{\left(\frac{-1}{2}\left(\frac{x-\mu}{\sigma}\right)^2\right)}$$

Show $\sqrt{\frac{1}{\sigma^2}}$ as $\frac{1}{\sigma}$

$$= \frac{\frac{1}{\sigma}}{\sqrt{2\pi}} e^{\left(\frac{-1}{2}\left(\frac{x-\mu}{\sigma}\right)^2\right)}$$

Multiply the numerator and denominator by σ.

$$y = \frac{1}{\sigma\sqrt{2\pi}} e^{\frac{-1}{2}\left(\frac{x-\mu}{\sigma}\right)^2}$$

Pearson (or possibly someone else) arrived at a complete form of the formula for the normal/Gaussian/Bell-shaped distribution curve.

This was huge. Having an idealised probability-density function that would describe the behaviour of virtually any data that was randomly scattered while being clustered around a central point was a big step forward in the ability to not just analyse the data we have, but to make mathematically educated guesses about the data we *don't* have. In yet another statistical paradox, as with the Poisson distribution, the more random the data is, the better the mathematics works.

The *z* score and the normal distribution

The curve falls downward from its peak like an inverted parabola and then swings out again like a hyperbola. The rate of change of the curve—its derivative—is the same as the slope of a line at a tangent to it. It is zero at the mean and approaches zero the further the data points get from the mean (Figure 264).

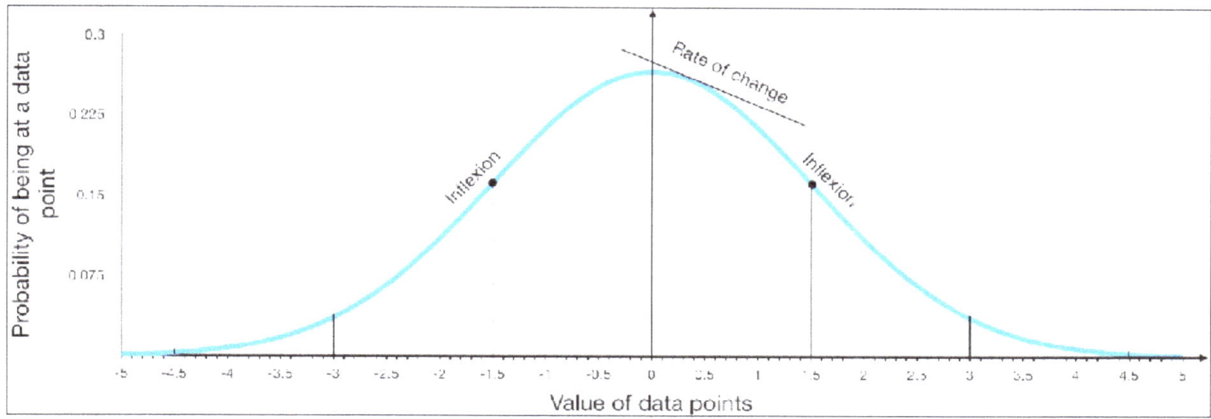

Figure 264

The point where it changes direction—where it is steepest—is called its **inflexion point**. The two inflexion points are where the tangent, or rate of change completely stops moving in one direction and starts to swing back in the other direction. In other words, they are the point where the rate of change *of* the rate of change—the derivative *of* the derivative, or the second derivative—is at zero.

We already know the first derivative of the normal distribution, because that was our starting point:

$$f'(x) = -Axy$$

Use the formula for y

$\left(\dfrac{1}{\sigma\sqrt{2\pi}}e^{\frac{-1}{2}\left(\frac{x-\mu}{\sigma}\right)^2}\right)$ and for A $\left(\dfrac{1}{\sigma^2}\right)$, and the fact that along the way we changed x to $x - \mu$.

$$f'(x) = -\frac{1}{\sigma^2\sigma\sqrt{2\pi}}(x-\mu)e^{\frac{-1}{2}\left(\frac{x-\mu}{\sigma}\right)^2}$$

Use the multiplication rule and the chain rule to find the 2nd derivative. We need the

$$f''(x) = -\frac{1}{\sigma^3\sqrt{2\pi}}\left[1e^{-\frac{1}{2}\left(\frac{x-\mu}{\sigma}\right)^2} - \frac{1}{\sigma^2}(x-\mu)^2 e^{-\frac{1}{2}\left(\frac{x-\mu}{\sigma}\right)^2}\right]$$

$$= 0$$

point where

$$f''(x) = 0$$

Multiply each side by $-\dfrac{\sigma^3\sqrt{2\pi}}{1}$

$$e^{\frac{-1}{2}\left(\frac{x-\mu}{\sigma}\right)^2} - \frac{1}{\sigma^2}(x-\mu)^2 e^{-\frac{1}{2}\left(\frac{x-\mu}{\sigma}\right)^2} = 0$$

Divide each side by $e^{-\frac{1}{2}\left(\frac{x-\mu}{\sigma}\right)^2}$

$$1 - \frac{1}{\sigma^2}(x-\mu)^2 = 0$$

Add $\dfrac{(x-\mu)^2}{\sigma^2}$ to both sides.

$$\frac{(x-\mu)^2}{\sigma^2} = 1$$

Take the square root of both sides.
Note the appearance of the z value!

$$\frac{x-\mu}{\sigma} = \pm 1$$

Multiply both sides by σ.

$$x - \mu = \pm\sigma$$

Add μ to both sides.

$$\boxed{x = \mu \pm \sigma}$$

In other words, extraordinarily, the z score makes another appearance, and we find the inflexion point in a normal distribution, regardless of the values of the mean and the standard deviation, at the points where x is exactly one standard deviation on either side of the mean: that is, $\pm 1z$.

The equation for the normal distribution curve can be written using the z score:

$$\frac{1}{\sigma\sqrt{2\pi}}e^{\frac{-1}{2}\left(\frac{x-\mu}{\sigma}\right)^2} = \frac{1}{\sigma\sqrt{2\pi}}e^{\frac{-z^2}{2}}$$

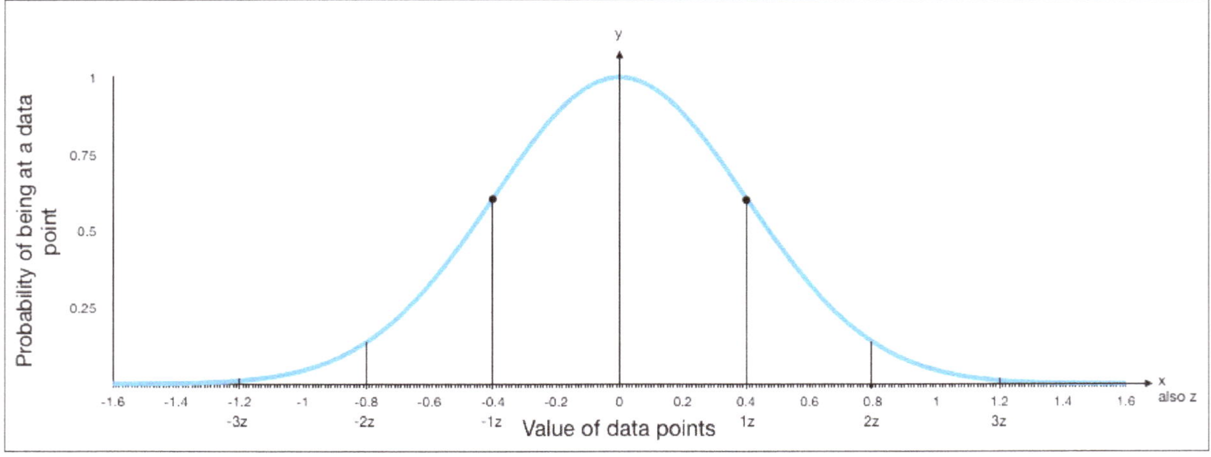

Figure 265

Figure 265 shows the curve with a mean of 0 standard deviation of 0.4, while Figure 266 shows a normal distribution curve for a data set with a

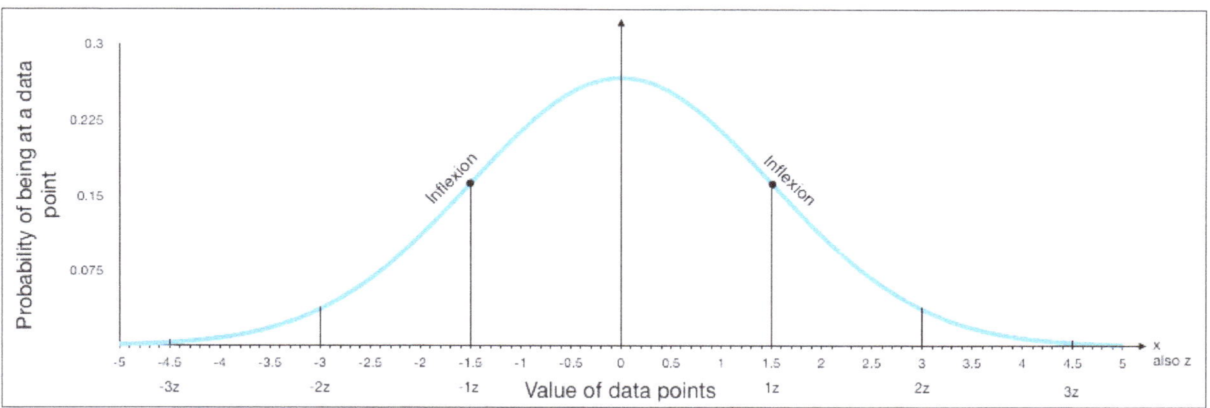

Figure 266

mean of 0 and a standard deviation of 1.5. We can think of the z score as an alternative way of scaling the x axis. You can see that inflexion is at completely different places in terms of the x value, but the same place in terms of the z value. Whatever the actual value of the mean and the standard deviation, the z score will provide a consistent way of identifying what proportion of the curve (and therefore what the probability is) of anything that lies between two values (or of anything outside of those values). It is a way of standardising the measures of proportions of the curve, regardless of the actual values: hence the name "standard deviation" ($1z = 1$ standard deviation from the mean).

The area under the normal distribution curve

The normal distribution can be used for anything from the heights of people in a population to errors made by surveyors to examination scores. All you need, to understand a great deal of information about a data set, is to know that it is normally distributed, and to have its mean and its standard deviation. The big advantage over other probability distributions that had been developed previously was the degree to which it applies to real-world phenomena because, as we saw earlier, it is the natural way that results fall out for phenomena that are randomly clustered and scattered at the same time.

In fact, for experimental purposes we don't even need to know that the data is normally distributed. We can use our understanding of normal distribution to *test* whether our data is consistent with something that is randomly clustered and scattered. If we expect that it should be, and our experiment gives us a different outcome, then we know that something more than random chance is driving the result.

As with other probability density functions, we should be able to integrate the formula to find the cumulative distribution function (CDF). *Except for one annoying thing.*

There is no formula that acts as an anti-derivative to the normal distribution curve, and so there is no straightforward way to integrate it. (Note that we didn't integrate it in the mathematics we used to derive it; we just used integration as a tool to find the value of the constants and significantly manipulated the formula to make that work.) Pearson turned his attention to finding a different way to achieve the same result. To do this, once again he used the z score.

What he was looking for was a way to take a standard z score for any normal distribution, regardless of the actual values, and calculate the probability of having a result less than that score.

First, perform a substitution of an integration, so that the integration will be with respect to z rather than with respect to x.

$$z = \frac{x - \mu}{\sigma}$$

Use the power rule of differentiation.

$$\frac{\delta z}{\delta x} = \frac{1}{\sigma}$$

Then multiply both sides by dx and both sides by σ. Any small changed to x will be scaled by the factor of a standard deviation, to make it a small change to z.

$$\delta x = \sigma \delta z$$

So in the substitution, the standard deviations cancel one another out.

$$\int \frac{1}{\sigma\sqrt{2\pi}} e^{\frac{-1}{2}\left(\frac{x-\mu}{\sigma}\right)^2} \delta x = \int \frac{1}{\sigma\sqrt{2\pi}} e^{\frac{-z^2}{2}} \sigma \delta z$$

$$\boxed{CDF = \frac{1}{\sqrt{2\pi}} \int e^{\frac{-z^2}{2}}}$$

By convention, the symbol Φ, which is the Greek letter for capital phi (not to be confused with lower-case phi, which symbolises the golden ratio) is used for the CDF of the standard normal distribution.

$$\Phi(z) = \Pr(Z \geq z)$$

$$= \frac{1}{\sqrt{2\pi}} \int e^{\frac{-z^2}{2}}$$

We could, of course, have arrived at this result by writing out the original formula in terms of

$$z = \frac{x - \mu}{\sigma}$$

and recalling that when we are using z values the standard deviation is equal to 1. This formula for the normal distribution, written in terms of the z score, is called the **standard normal distribution**. It is a simpler form of the formula and is useful whenever we are talking about a normal distribution in general, rather than a specific one. It is equivalent to the formula for a normal distribution with a mean of 0 and a standard deviation of 1. We write this as $Z \sim N(0,1)$. That is, the random variable Z follows a normal distribution, with parameters $mean = 0$ and $variance = 1$.

Which is fine, except there still isn't a function which, when differentiated, equals this equation. So Pearson found another way.

e^x can be expressed as an infinite convergent Taylor series (page 266 and following):

$$e^x = \sum_{n=0}^{\infty} \frac{x^n}{n!}$$

In this series x can be any number (real or otherwise), including $\frac{-z^2}{2}$, and so we can use this fraction in the equation.

$$e^{\frac{-z^2}{2}} = \sum_{n=0}^{\infty} \frac{\left(\frac{-z^2}{2}\right)^n}{n!}$$

Substitute the Taylor-series version back into the equation for the area under the curve.

$$\frac{1}{\sqrt{2\pi}} \int e^{\frac{-z^2}{2}} \, \delta z = \frac{1}{\sqrt{2\pi}} \int \sum_{n=0}^{\infty} \frac{\left(\frac{-z^2}{2}\right)^n}{n!} \, \delta z$$

Based on linearity, the integral of a sum can be calculated as the sum of the integrals; we can integrate each term, so the integration sign can come inside the series notation.

$$= \frac{1}{\sqrt{2\pi}} \sum_{n=0}^{\infty} \int \frac{\left(\frac{-z^2}{2}\right)^n}{n!} \, \delta z$$

Expand out the terms within the integral.

$$= \frac{1}{\sqrt{2\pi}} \sum_{n=0}^{\infty} \int \frac{-1^n z^{2n}}{2^n n!} \, \delta z$$

Take the constant outside the integration, but not outside the sum. The n can be treated as a constant in terms of the integration, because the integration is with respect to z. However

$$= \frac{1}{\sqrt{2\pi}} \sum_{n=0}^{\infty} \frac{-1^n}{2^n n!} \int z^{2n} \, \delta z$$

the n is a variable with respect to the series: it increases by 1 for each term of the series.

Because $2n$ is a constant in terms of the integration with respect to z, integrating z^{2n} simply involves adding one to the index and dividing by the new index.

$$\Phi(z) = \frac{1}{\sqrt{2\pi}} \sum_{n=0}^{\infty} \frac{-1^{\,n} z^{2n+1}}{2^{\,n} n! \, (2n+1)}$$

This gives us a way to compute the area between the mean and the z score. For each value of z you were looking at, you would need to calculate each of the terms in the sum, starting with $n = 0$ then adding 1 to n each time, until the required level of precision was reached. You would then add them together and divide the result by $\sqrt{2\pi}$. That would give you the answer for a single z score. Trying to do that by hand for any given value of z would be incredibly laborious. These days we would obviously just plug the x value, the mean and the standard deviation into a computer (or just plug all our data in and tell it which value of x we were looking at), and let it do the heavy lifting. In Pearson's time and for most of the 1900s people used tables, which Pearson and friends painstakingly calculated and then published. (This is the way that trigonometry and logarithms were calculated and used). These tables were originally sold under copyright, and the people devising them didn't advertise too loudly about how they calculated them; that was how they made the money to recompense them for the huge amount of work involved. When the copyright ran out they were often printed inside the back or front covers of statistics textbooks.

As per Figure 267, based on the calculated z scores, regardless of the actual numbers involved, if data are in a normal distribution there is approximately a 34.13% chance of any given data point lying in the region between the mean and one standard deviation above it. This means there is a 68.26% chance that it will lie within 1 standard deviation either side of the mean. Because the probabilities will all add to 100%, that means there will be a 31.74% chance that the data point will be more than 1 standard deviation from the mean. There is roughly a 95.45% chance that it will lie within two standard deviations of the mean (4.55% chance it will be outside of this), and a 99.74% chance it will be within 3 standard deviations (0.26% that it will be outside that).

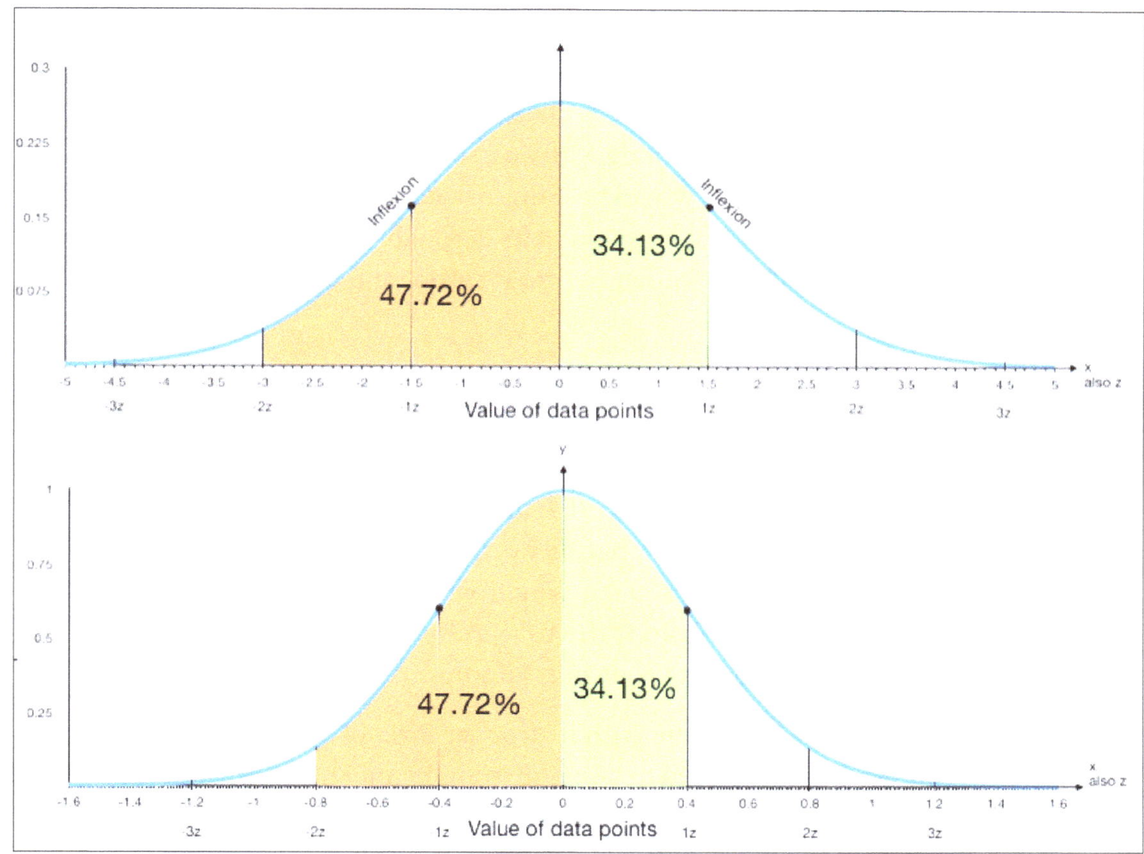

Figure 267

Once you know the probability of getting a result that is between 0 and whatever *z* score you are looking for, and the fact that probabilities will always add to 100% (in other words 1), then finding probabilities outside of that or between two *z* scores is simple arithmetic. If, for example, you know that the life expectancy of a particular piece of critical equipment—say a crucial light-bulb—tends to follow a normal distribution, and you know the mean life-expectancy and the standard deviation for a particular brand, you will be able to put a realistic number value on the chances it will fail, and therefore of the level of risk you are willing to undertake in order to delay some expenditure. Serum cholesterol levels are normally distributed. If you are prescribed a statin drug to reduce your cholesterol, and you know your current level and the mean and standard deviation of the sub-group you belong to, you will be able to tell where your reading places you in relation to others in your sub group.

Statistical significance and P values

As a general rule, the minimum value for **statistical significance** is taken to be 0.05. In other words, there is a 5% chance that the result was obtained by random chance. If we are working with an assumption that data are normally distributed (or that they would be if there wasn't something else going on), this means that a result of less than 0.05 is roughly two standard deviations from the mean. This is known as the **P value**, and expressed in academic papers as $P < 0.05$.

This, of course, isn't a great level of significance. If there is a 5% chance of something occurring randomly, then we would expect it to occur randomly, over time, every 20th time. Researchers sometimes inadvertently do something that statisticians call **P hunting**. If you, for example, ask 20 questions in a survey, you would expect that, purely by random chance, you might get a P value for one question of < 0.05, which will show statistical significance and enable you to publish. If the researcher published that and no other data, that would be an example of P hunting.

Researchers prefer to get results with a stronger level of significance than 0.05. If a result is three standard deviations from the mean in a normal distribution, there is less than a one in 100 chance that the result was random: $P < .01$. Even then, of course, the result would be expected to occur by chance 1 time in 100. Once again, statistics isn't about proof or about certainty; just about understanding the level of uncertainty.

Key points:

- The constant in the formula that Laplace and others were unable to identify turns out to be the reciprocal of the variance (the precision).

- That makes the formula

$$y = \frac{1}{\sigma\sqrt{2\pi}}\, e^{\frac{-1}{2}\left(\frac{x-\mu}{\sigma}\right)^2}$$

- Based on taking the second derivative, the inflexion of the curve is 1 standard deviation, or $1z$, from the mean.

- The standardised normal distribution, with mean of 0, standard variation and variance of 1, is:

$$y = \frac{1}{\sqrt{2\pi}}\, e^{\frac{-z^2}{2}}$$

- It isn't necessary, for experimental purposes, to know that the data is normally distributed. Because normal distribution is the result of random clustering and scattering, it can be used to test whether data is randomly clustered and scattered.

- There is no formula that acts as an anti-derivative to the normal distribution curve, which means that a Taylor series expansion needs to be used to find the area under it.

- This results in

$$CDF = \frac{1}{\sqrt{2\pi}} \sum_{n=0}^{\infty} \frac{-1^{n} z^{2n+1}}{2^{n} n!\,(2n+1)}$$

  which can be calculated to the desired level of precision for any given z. By convention this is written as $\Phi(z)$.

- For a normal distribution approximately **34.13%** of any given data are between the mean and one standard deviation above it. That means there is a **68.26%** chance that it will lie within **1** standard deviation either side of the mean.

- There will be about **31.74%** chance that the data point will be more than **1** standard deviation from the mean.

- There is roughly about a **95.45%** chance that it will lie within two standard deviations of the mean (**4.55%** chance it will be outside of this), and a **99.74%** chance it will be within **3** standard deviations (**0.26%** that it will be outside that).

The normal, the binomial, and beyond

> **Only read the first parts of this section if** you want to see a fairly difficult proof that the moment generating function (MGF) of the standard normal distribution is the same as the MGF of a standard binomial, as $n \to \infty$. Otherwise you can skip through to the subheading that says "Extending the normal distribution" and read the key points at the end of the section.

We saw earlier that one of the ways Laplace (following de Moivre) arrived at the idea of the normal distribution was through thinking about the binomial, as $n \to \infty$. From Pearson's point of view, once he had developed his tables of z scores for the normal distribution, it was much easier to use them to approximate binomials than to calculate binomials using a pencil and paper, and he developed a method for doing this. We won't cover that here, because now that we use computers, approximating a binomial with a normal is redundant; we can just calculate the binomial. (It isn't quite as straightforward as just applying the binomial z to the normal Z tables, because a correction needs to be applied to line up the normal with the middle of the columns on the binomial.)

There are a number of ways that this relationship can be demonstrated mathematically: here's one of them, comparing moment generating functions (MGFs). It relies on the MGF of the normal distribution, so let's look at that first.

Moment generating function of the normal distribution

Put the formula for the normal distribution into the formula for an MGF.

$$MGF = \int_{-\infty}^{\infty} e^{tx} \frac{1}{\sigma\sqrt{2\pi}} e^{\left[\frac{-1}{2}\left(\frac{x-\mu}{\sigma}\right)^2\right]} dx$$

Take the constant outside the integration and add the exponents of e.

$$= \frac{1}{\sigma\sqrt{2\pi}} \int_{-\infty}^{\infty} e^{\left[tx + \frac{-1}{2}\left(\frac{x-\mu}{\sigma}\right)^2\right]} dx$$

Expand out the term in brackets.

$$= \frac{1}{\sigma\sqrt{2\pi}} \int_{-\infty}^{\infty} e^{\left[tx + \frac{-1(x^2 - 2ux + \mu^2)}{2\sigma^2}\right]} dx$$

Use $2\sigma^2$ as a common denominator for the terms in square brackets.

$$= \frac{1}{\sigma\sqrt{2\pi}} \int_{-\infty}^{\infty} e^{\left[\frac{2\sigma^2 tx}{2\sigma^2} + \frac{-1(x^2 - 2ux + \mu^2)}{2\sigma^2}\right]} dx$$

Multiply the numerator and denominator of the expression in square brackets by –1 and rearrange the terms.

$$= \frac{1}{\sigma\sqrt{2\pi}} \int_{-\infty}^{\infty} e^{\left[\frac{x^2 - 2x(\sigma^2 t + \mu) + \mu^2}{-2\sigma^2}\right]} dx$$

Add, then subtract $(\sigma^2 t + \mu)^2$ to the numerator of the exponent of e.

$$= \frac{1}{\sigma\sqrt{2\pi}} \int_{-\infty}^{\infty} e^{\left[\frac{x^2 - 2x(\sigma^2 t + \mu) + \mu^2 + \left(\sigma^2 t + \mu\right)^2 - \left(\sigma^2 t + \mu\right)^2}{-2\sigma^2}\right]} dx$$

Rearrange the terms.

$$= \frac{1}{\sigma\sqrt{2\pi}} \int_{-\infty}^{\infty} e^{\left[\frac{x^2 - 2x(\sigma^2 t + \mu) + \left(\sigma^2 t + \mu\right)^2 + \mu^2 - \left(\sigma^2 t + \mu\right)^2}{-2\sigma^2}\right]} dx$$

Write $x^2 - 2x(\sigma^2 t + \mu) + (\sigma^2 t + \mu)^2$ as $\{x - (\sigma^2 t + \mu)\}^2$.

$$= \frac{1}{\sigma\sqrt{2\pi}} \int_{-\infty}^{\infty} e^{\left[\frac{\{x - (\sigma^2 t + \mu)\}^2 + \mu^2 - \left(\sigma^2 t + \mu\right)^2}{-2\sigma^2}\right]} dx$$

Separate the exponent of e into two different powers of e, multiplied together.

$$= \frac{1}{\sigma\sqrt{2\pi}} \int_{-\infty}^{\infty} e^{\left[\frac{\{x - (\sigma^2 t + \mu)\}^2}{-2\sigma^2}\right]} e^{\left[\frac{\mu^2 - \left(\sigma^2 t + \mu\right)^2}{-2\sigma^2}\right]} dx$$

Move the constant $\frac{1}{\sigma\sqrt{2\pi}}$ inside the integration, and the constant (with respect to x) $e^{\frac{\mu^2 - (\sigma^2 t + \mu)^2}{-2\sigma^2}}$ outside the integration, and multiply the numerator and denominator of both exponents by –1.

$$= e^{\frac{-\mu^2 + (\sigma^2 t + \mu)^2}{2\sigma^2}} \int_{-\infty}^{\infty} \frac{1}{\sigma\sqrt{2\pi}} e^{\frac{-\{x - (\sigma^2 t + \mu)\}^2}{2\sigma^2}} dx$$

What is left inside the integration is a normal probability distribution with a mean of $(\sigma^2 t +$

$$= e^{\frac{-\mu^2 + (\sigma^2 t + \mu)^2}{2\sigma^2}} \int_{-\infty}^{\infty} \frac{1}{\sigma\sqrt{2\pi}} e^{\frac{-\{x - (\sigma^2 t + \mu)\}^2}{2\sigma^2}} dx$$

$$= e^{\frac{-\mu^2 + (\sigma^2 t + \mu)^2}{2\sigma^2}}$$

μ). It therefore integrates to 1 and disappears.

Expand out the term in brackets in the numerator and simplify.

$$= e^{\frac{-\mu^2+(\sigma^2 t)(\sigma^2 t)+2\sigma^2 t\mu+\mu^2}{2\sigma^2}}$$

Which brings us to the MGF.

$$MGF \ of \ N(\mu,\sigma^2) = e^{\frac{\sigma^2 t^2}{2}+t\mu}$$

Note that $N(\mu,\sigma^2)$ is a shorthand way of writing a normal distribution, with mean μ and variance σ^2. To compare this MGF with that of the binomial, we need the MGF of the standard normal, which means it has mean 0 and variance 1, so we can substitute these numbers for the mean and variance into our formula, to simplify it further.

As we've seen, the formula for a **standard normal distribution**, where $\mu = 0$ and $\sigma^2=1$ is:

$$y = \frac{1}{\sqrt{2\pi}} e^{-\frac{x^2}{2}}$$

That simplifies the formula for the MGF.

$$MGF \ of \ N(0,1) = e^{\frac{t^2}{2}}$$

Moment generating function of the standardised binomial

To compare the MGF of the binomial with the MGF of the normal we need to standardise the binomial. We saw when we looked at the moments of probability distributions that the expected value of a binomial distribution is np, and that the variance is npq. This means that the z score for a standardised binomial (subtracting the mean then dividing by the standard deviation) is:

$$Z = \frac{x - np}{\sqrt{npq}}$$

We saw earlier (page 543) that the MGF of the binomial is $(q + pe^t)^n$, where $q = 1 - p$.

Now let's look at the MGF of the standardised binomial distribution, as $n \rightarrow \infty$.

$$MGF_{standard \ binomial} = E(e^{tz})$$

$$= E\left(e^{t\left(\frac{x-np}{\sqrt{npq}}\right)}\right)$$

Distribute the t.

$$= E\left(e^{\left(\frac{tx}{\sqrt{npq}} - \frac{tnp}{\sqrt{npq}}\right)}\right)$$

Use the properties of exponents to express this as two powers of e multiplied together.

$$= E\left(e^{\left(\frac{tx}{\sqrt{npq}}\right)} e^{\left(\frac{-tnp}{\sqrt{npq}}\right)}\right)$$

$e^{\left(\frac{-tnp}{\sqrt{npq}}\right)}$ is a constant with respect to x and can be taken outside the expected value calculation.

$$= e^{\left(\frac{-tnp}{\sqrt{npq}}\right)} E\left(e^{\left(\frac{tx}{\sqrt{npq}}\right)}\right)$$

Now $E\left(e^{\left(\frac{tx}{\sqrt{npq}}\right)}\right)$, when x is distributed binomially, is the moment generating function of the binomial, except that t is replaced by $\frac{t}{\sqrt{npq}}$. That makes the formula: $\left(q + pe^{\frac{t}{\sqrt{npq}}}\right)^n$

Substitute this MGF back into our formular for the MGF of the standard binomial.

$$\text{MGF}_{standard\ binomial} = e^{\left(\frac{-tnp}{\sqrt{npq}}\right)}\left(q + pe^{\frac{t}{\sqrt{npq}}}\right)^n$$

Take the power of n, which both parts of this expression are raised to, outside of the expression.

$$= \left(e^{\left(\frac{-tp}{\sqrt{npq}}\right)}\left(q + pe^{\frac{t}{\sqrt{npq}}}\right)\right)^n$$

Distribute $e^{\left(\frac{-tp}{\sqrt{npq}}\right)}$ to the terms in brackets.

$$= \left(qe^{\left(\frac{-tp}{\sqrt{npq}}\right)} + pe^{\left(\frac{t}{\sqrt{npq}}\right)} e^{\left(\frac{-tp}{\sqrt{npq}}\right)}\right)^n$$

Use properties of exponents to combine these powers of e from the term on the right.

$$= \left(qe^{\left(\frac{-tp}{\sqrt{npq}}\right)} + pe^{\left(\frac{t}{\sqrt{npq}}\right)\left(\frac{-tp}{\sqrt{npq}}\right)}\right)^n$$

Use the common denominator of $\sqrt{npq}$ to further combine the exponent on the right.

$$= \left(qe^{\left(\frac{-tp}{\sqrt{npq}}\right)} + pe^{\left(\frac{t-tp}{\sqrt{npq}}\right)}\right)^n$$

Write $t - tp$ as $t(1-p)$, and note that $1 - p = q$

$$= \left(qe^{\left(\frac{-tp}{\sqrt{npq}}\right)} + pe^{\left(\frac{t(1-p)}{\sqrt{npq}}\right)}\right)^n$$

Recall (page 268) that the Taylor series expansion of $e^x = 1 + x + \frac{x^2}{2!} + ...$

$$= \left(qe^{\left(\frac{-tp}{\sqrt{npq}}\right)} + pe^{\left(\frac{tq}{\sqrt{npq}}\right)} \right)^n$$

$$= \left(q\left[1 - \frac{tp}{\sqrt{npq}} + \frac{t^2p^2}{2!\,npq} - \cdots + \cdots\right] + p\left[1 + \frac{tq}{\sqrt{npq}} + \frac{pt^2q^2}{2!\,npq} + \cdots\right] \right)^n$$

$$= \left(\left[q - \frac{tpq}{\sqrt{npq}} + \frac{qt^2p^2}{2!\,npq} - \cdots + \cdots\right] + \left[p + \frac{tpq}{\sqrt{npq}} + \frac{pt^2q^2}{2!\,npq} + \cdots\right] \right)^n$$

$$= \left((p+q) - \frac{\cancel{tpq}}{\cancel{\sqrt{npq}}} + \frac{\cancel{tpq}}{\cancel{\sqrt{npq}}} + \frac{qt^2p^2 + pt^2q^2}{2!\,npq} + \cdots + \cdots \right)^n$$

Factor out t^2pq within the numerator of the term with t^2 in it.

$$= \left((p+q) + \frac{t^2pq(p+q)}{2npq} + \cdots + \cdots \right)^n$$

Remember that $p + q = 1$.

$$MGF_{standard\ binomial} = \left(1 + \frac{t^2}{2n} + \cdots + \cdots \right)^n$$

Divide the numerator and denominator of the term with $t2$ by 2. Take the limit as $n \to \infty$. All of the terms in the series to the right of the ones shown will have higher powers of n in their denominators, and everything else in those terms is a constant less than 1, so that as $n \to \infty$ all of these terms will $\to 0$

$$\lim_{n\to\infty} \left(1 + \frac{\frac{t^2}{2}}{n} + \cdots + \cdots \right)^n$$

Recall that $\lim\limits_{n\to\infty} \left(1 + \frac{x}{n}\right)^n = e^x$ (page 270)

$$= \lim_{n\to\infty} \left(1 + \frac{\frac{t^2}{2}}{n} \right)^n$$

This, as we saw above, is the MGF of the standard normal distribution.

$$\boxed{\quad = e^{\frac{t^2}{2}} \quad}$$

So the MGF of the standardised binomial, when $n \to \infty$, is the same as the MGF of the standardised normal distribution, which means that the distributions must be identical.

Extending the normal distribution

There are four extensions of the idea of the normal distribution that make it such a vital part of statistical analysis (in addition to the fact that it describes virtually any situation where data are randomly scattered yet clustered):

1) The χ^2 (chi squared, pronounced with a hard 'k' sound) distribution has been called Pearson's most important contribution to statistics (which is saying a lot), and helps us to determine whether one measure is within the margin of error of another.

2) The normal distribution is crucial to our understanding of taking smaller samples from larger data sets and understanding how representative the sample is likely to be (regardless of whether the entire data set is normally distributed).

3) An understanding of samples and the way they behave, leads to thinking about how we can minimize errors when designing research.

4) "Student's" t distribution, which is based partially on the normal distribution, is even more widely used in statistical analysis than the normal distribution, because it is more useful for small samples and equally useful for larger ones.

Let's explore these four extensions in more detail.

Key points:

- $N(\mu, \sigma^2)$ is short hand for a normal distribution, with mean μ and variance σ^2.

- MGF of $N(\mu, \sigma^2) = e^{\frac{\sigma^2 t^2}{2} + t\mu}$

- $N(0,1)$ is a standard normal distribution, with mean 0 and variance 1. Its probability density function (PDF) is

$$y = \frac{1}{\sqrt{2\pi}} e^{-\frac{x^2}{2}}$$

and its MGF is

$$e^{\frac{t^2}{2}}$$

- The z score for a standardised binomial (subtracting the mean then dividing by the standard deviation) is

$$Z = \frac{x - np}{\sqrt{npq}}$$

- The MGF of a standardised binomial, as $n \to \infty$ is the same as the MGF of a standard normal distribution $\left(e^{\frac{t^2}{2}}\right)$ proving that the normal distribution is the binomial, as $n \to \infty$.

Chi square
(first extension of the normal)

> **Read this if** you want to understand what lies behind chi square tests, or you thought they were limited to normally distributed data.

So far we've seen that Karl Pearson had managed to:

- develop the idea of standard deviation,

- use it to come up with a way to standardise random variables,

- invent a neat way to look at the correlation of paired variables,

- complete the formula for the normal distribution curve,

- prove that the inflexion point on a normal distribution curve was exactly one standard deviation away from the mean,

- calculate the volume under the normal distribution,

- demonstrate (although he wasn't the first) that the normal distribution is what you get when the n of a binomial approaches infinity, and

- invent histograms and popularise statistics generally.

He did a lot more than that, but none of those are considered to be his greatest achievement. That happened in 1900, when he developed a way to compare observed with expected results and determine whether the former was within the margin of error of the latter. (A similar formula had actually been developed nearly 30 years earlier, but that discovery didn't go anywhere and Pearson was probably unaware of it.[73])

Pearson called this the **chi square distribution test**, based on the **chi square distribution** (or **chi-squared distribution**; the two terms are used interchangeably), and symbolised it with χ^2. (The Greek letter chi, χ, is pronounced with a hard "k" and a long "i" sound, as in the first syllable of "kayak".) This has become one of the most widely-used approaches in all of statistics in its own right. It also underpins the *t* **distribution** and the *F* **distribution**, which in turn underpins **ANOVA** (all of which we will get to shortly), making it absolutely fundamental to modern statistics.

Imagine that you are asking people about their voting intentions. They could indicate that they planned to vote for 1 of 4 political parties, or they could be undecided. That means that voting intention is a categorical variable with a multinomial distribution, not a numerical variable, but there are still numbers involved: the number of people in each category.

You could compare what people tell you with your assumptions about what you were going to see, or you could compare them with what happened at the previous election, you could list them by age grouping, compare them with what happened in another electorate, or any other comparison you wanted.

The first logical step will be to create a **comparison table** so that you can compare what you expected with what you observed in different categories.

You could standardise your results by doing something like:

$$\frac{Observed\ count - expected\ count}{expected\ count}$$

Party	Expected voting intention, from historical trends	Observed voting intention from questioning	Standardised difference: Observed minus expected, divided by expected
Far right	10	10	0%
Centre right	100	90	−10%
Centre left	100	110	+10%
Far left	10	10	0%
Total	**220**	**220**	

Figure 268

(Astute readers might notice the similarity of this calculation to the calculation of relative risks, and chi square tests can be used to determine the statistical validity of relative-risk and odds-ratio calculations; see pages 406 and following.)

That would give you the percentage by which what you observed differed from what you expected. If 90 people planned to vote for a centre-right party, but you expected (based on your best guess, past experience or behaviour elsewhere) that 100 people would vote that way, this formula would tell you that your result was 10% lower than expected. If 110 people told you they were going to vote for that party, the formula would tell you that the result was 10% higher. All of which is easy enough and doesn't particularly need the sort of statistics that we have been looking at. That *isn't* what Pearson did, though: his approach was cleverer.

As we saw earlier, statistics isn't usually about the information we already have. Statistics is about applying mathematics to uncertainly, so that we can have a better idea about the mathematical degree of uncertainty. He was interested in questions that *aren't* immediately obvious from the data we have, like: is a 10% difference in this situation just a random fluctuation, or should we take it seriously? Is it telling us anything about a longer-term trend? Is it likely to reflect what others are thinking? Is it, in fact, within the **margin of error**, or outside of it?

When we think about questions like this, the immediate point that stands out is: how do we answer questions like that, with so little data to work with? We just have a handful of numbers. What can we do with that?

What Pearson realised is that we don't just have the numbers in front of us: we can work with the distribution of possible numbers that would be random errors around what we expected. Errors, as we have seen, tend to follow a normal distribution, and their mean is likely to be close to the true answer. (We'll examine this issue further in the next section, when we look at taking samples from populations.) So the question is: if there is a normal distribution of possible answers about the expected number, where does what we have observed fall on that distribution? What does that tell us about the probability that we are looking at a random fluctuation?

You may hear some strange (and erroneous) ideas about the χ^2 distribution and tests, like the idea that the underlying data has to have a normal distribution in order for the distribution to be used. As with our example, it often gets used for multinomial distributions. It can, in fact, be used for all sorts of data, including Poisson distributions and even continuous ones. Pearson wrote:

> *"The object of this paper is to investigate a criterion of the probability on* ***any*** *theory of an* ***observed system of errors***, *and to apply it to the*

determination of goodness of fit in the case of frequency **curves**"[74]. [Emphasis added]*

What is important isn't that the underlying data be normally distributed, but that potential errors or random fluctuations should be normally distributed; and, as we have seen, they usually are. They may not be: for example, asking a group of narcissists to estimate their own abilities is likely to lead to errors with a systematic bias in them. In general, though, it is safe to assume that random errors and fluctuations will be normally distributed, regardless of the underlying distribution. So while χ^2 is mostly used for multinomial distributions, it doesn't have to be.

With multinomials there are usually a lot of different possible combinations, and so the chance of getting exactly one particular outcome is usually quite low. (When we looked at multinomial probabilities, on pages 356 and following, we only considered four possible categories, but the probability of getting precisely the number of people in each that we specified was approximately 0.15%, or about 1 in 667. If we had had more categories, the probabilities would have been even lower. For this reason alone it is helpful to have a way to know whether an observed value is close-enough to a theoretical one; whether it is within the margin of error.

Pearson was interested in the total difference of what was observed compared to what was expected. So he calculated each variation, and standardised them by dividing by the expected value, as we did above on page 607. (If it is a bit confusing calling this an "expected" value, you can think about it as "anticipated" but in fact, we can use "expected value" in the way we have been using it in many cases; we will anticipate seeing an "average".) In keeping with the sum-of-squares approach that works so well for other issues he looked at, Pearson then squared the values in the numerator and added them together so that he had a (now familiar to us) sum of squares. (He didn't make the sort of mistake we saw earlier with the coefficient of determination; he knew what he doing, as he did when he developed correlation.) That eliminated negative values and also made it easier to manipulate his terms with calculus.

Using the values from our table of voting intentions on page 607, that makes the χ^2 score for the table:

$$\chi^2 = \frac{(10-10)^2}{10} + \frac{(90-100)^2}{100} + \frac{(110-100)^2}{100} + \frac{(10-10)^2}{10}$$

$$= 0 + 1 + 1 + 0$$

$$= 2$$

We can generalise that as:

$$\chi^2 = \sum \frac{(O_i - E_i)^2}{E_i}$$

where E_i is each expected value, and O_i is each observed value. That gets you to a single number.

That was convenient, except that now that he had squared the values, he couldn't compare them directly to a normal distribution. He had to find a way to square a standardised normal distribution (or, more accurately, to take the formula for a standardised normal distribution, then transform it so that the X was squared).

That might work for a single category, but he was adding up responses from a number of different categories. So, having squared his standard-normally distributed X variable, then he had to add enough of them so that the number of squared, standardised normal distributions he put together was appropriate to the number of categorical variables he was looking at. This is how he went about it. (At this point he did, in fact, make a mistake, which Ronald Fisher picked up about 20 years later, much to Pearson's annoyance. That doesn't affect the workings below; we will see what that error was when we come to look at Fisher.)

The χ^2 (chi-square) distribution

As we saw looking at transformations, it is easiest to do something like this starting with the cumulative distribution function (CDF). There are two problems with this: first, we don't have a formula for the CDF of the normal distribution that doesn't involve an infinite sum. Second, as we also saw earlier, X^2 isn't monotone; X^2 can be created by x and $-x$. Fortunately we can still do the calculations. (Note that in the working that follows, it is important to distinguish X, which is the random variable we are starting with, from the Greek letter χ (chi).

Let X be a random variable with a standard normal distribution (i.e. mean of 0 and variance/standard deviation of 1. The square root of 1 is 1, so the variance and standard deviation are the same). $Let X \sim N(0,1)$

The definition of the CDF is the probability that any *instance* of the random variable will be less than or equal to a specific value of that random variable (and X^2 is a random variable because X is a random variable).

$$CDF(X^2) = P[X^2 \leq x^2]$$

This is the same thing as the line above: the probability that X^2 is less than or equal to x^2 is the same as the probability that X is greater than or equal to $-x$ and less than or equal to x.

$$= Pr[-x \leq X \leq x]$$

$$= Pr[X \leq x] - Pr[X \leq -x]$$

$$= CDF[x] - CDF[-x]$$

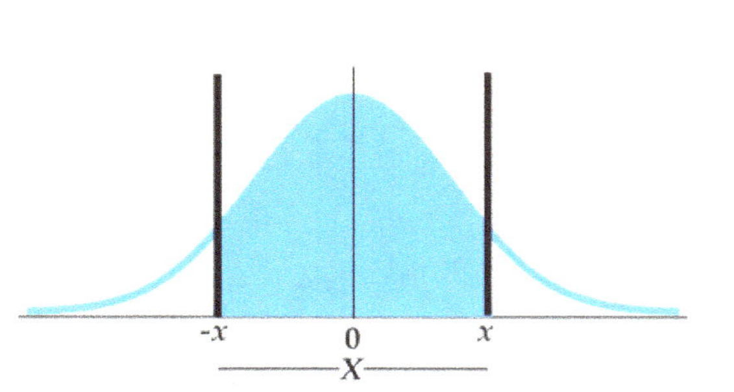

The probability that $-x \leq X \leq x$ (represented by the solid-blue area) is the same as the probability that X lies to the left of x, less the probability that X lies to the left of $-x$. The result below is based on the symmetry of the standard normal distribution. The area to the left of $-x$ (i.e. $CDF[-x]$)is the same as the area to the right of $+x$. The total area under the curve must be 1 because it represents a probability function, so the area to the right of x is $1 - CDF[x]$. That means that the area to the left of $-x$ must also be equal to $1 - CDF[x]$.

$$= CDF[x] - (1 - CDF[x])$$

So now we can relate $CDF(X^2)$ to $CDF[x]$. Now we need the probability density function (PDF), which will be the derivative of the CDF.

$$CDF(X^2) = 2CDF[X] - 1$$

This is the inverse function of $v = x^2$:

Let $V = X^2$

$$v^{\frac{1}{2}} = x$$

$$CDF(V) = 2CDF\left[v^{\frac{1}{2}}\right] - 1$$

Take the derivative of both sides with respect to V. Use the chain rule and the power rule. The outer function is the CDF, and its derivative is the PDF.

$$PDF(V) = 2PDF\left[v^{\frac{1}{2}}\right]\frac{v^{-\frac{1}{2}}}{2} - 0$$

$$= PDF\left[v^{\frac{1}{2}}\right]v^{-\frac{1}{2}}$$

The inner function is $v^{\frac{1}{2}}$ and its derivative is $\frac{1}{2}v^{-\frac{1}{2}}$

$v^{\frac{1}{2}} = X$ which means that the PDF will have the same distribution as the X, which is the standard normal distribution: $\frac{1}{\sqrt{2\pi}}e^{\frac{-x^2}{2}}$

$$PDF(V) = \frac{1}{\sqrt{2\pi}}e^{\frac{-v}{2}}v^{-\frac{1}{2}}$$

Recall from page 305 that $\Gamma\left(\frac{1}{2}\right) = \sqrt{\pi}$

$$= \frac{1}{\Gamma\left(\frac{1}{2}\right)\sqrt{2}}e^{\frac{-v}{2}}v^{-\frac{1}{2}}$$

Write $-\frac{1}{2}$ as $\left(\frac{1}{2} - 1\right)$

$$= \frac{1}{\Gamma\left(\frac{1}{2}\right)\sqrt{2}}e^{\frac{-v}{2}}v^{\left(\frac{1}{2}-1\right)}$$

Write $\frac{1}{\sqrt{2}}$ as $\left(\frac{1}{2}\right)^{\frac{1}{2}}$

$$= \frac{1}{\Gamma\left(\frac{1}{2}\right)}\left(\frac{1}{2}\right)^{\frac{1}{2}}e^{\frac{-v}{2}}v^{\left(\frac{1}{2}-1\right)}$$

Recall from when we looked at the gamma distribution that it actually has two different forms. The PDF of a Gamma(n, λ distribution is either

$$\sim\text{gamma}\left(\frac{1}{2},\frac{1}{2}\right) \text{ or}$$

$$\sim\text{gamma}\left(\frac{1}{2},2\right)$$

$$\frac{1}{\Gamma(n)}\lambda^n x^{n-1}e^{-\lambda x} \text{ or } \frac{1}{\Gamma(\alpha)\beta^\alpha}\frac{x^\alpha e^{-\frac{x}{\beta}}}{x}$$

So what we have is the PDF of a gamma distribution, with $n = \frac{1}{2}$ and $\lambda = \frac{1}{2}$, with respect to, which may be written as either gamma $\left(\frac{1}{2},\frac{1}{2}\right)$ or gamma $\left(\frac{1}{2},2\right)$.

When you think about it that is truly remarkable: another of those awe-inspiring mathematical results. We started with a normal distribution, which is based on what happens when virtually anything is randomly clustered and scattered. We squared the random variable. And suddenly the gamma distribution (which we arrived at by summing exponential distributions) appears, as if from nowhere.

So we have a PDF for X^2 when X is normally distributed, but what Pearson was working with was the *sum* of squares. The easiest way to sum random variables of probability distributions is usually to use the moment generating function (MGF).

When we looked at the MGF of the gamma distribution (page 539 and following) we saw that it can be written as:

$$MGF = (1 - \theta t)^{-n}$$

(if we use the alternative parameterization usually used in statistics).

Putting in the values $\frac{1}{2}$ and 2 into this give us:

$$MGF = (1 - 2t)^{-\frac{1}{2}}$$

As we have seen previously, adding a specified number of MGFs a certain number of times is the same as raising them to that power. We'll use the letter k to represent the number of times we add the MGF. Adding this MGF to itself k times is the same as raising it to the kth power. That gives us:

$$(1 - 2t)^{-\frac{k}{2}}$$

But that is the MGF of a gamma distribution where $\theta = 2$ and $n = \frac{k}{2}$. So when we substitute that back into the equation for the PDF of a gamma distribution, we get:

$$PDF\left(\chi_k^2\right) = \frac{x^{\frac{k}{2}-1} e^{\frac{-x}{2}}}{2^{\frac{k}{2}} \Gamma\left(\frac{k}{2}\right)} \, for \ x \geq 0$$

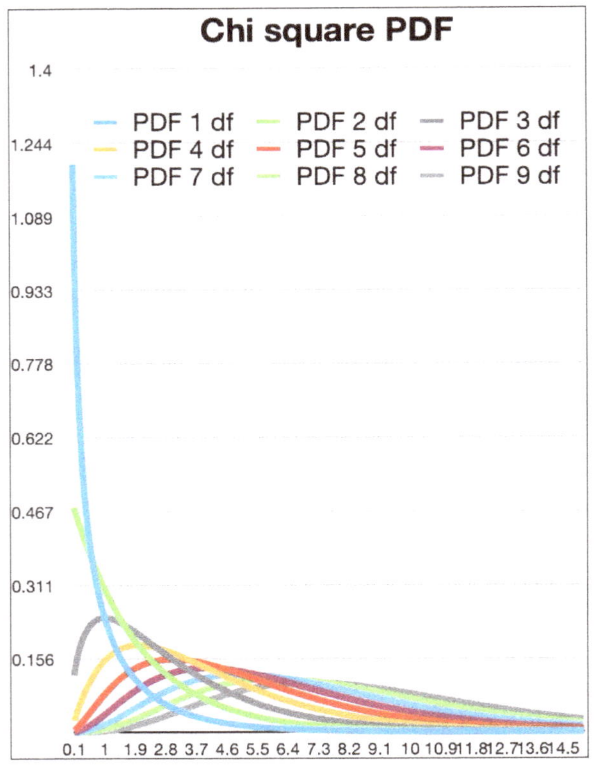

Figure 269

where k is the number of squared, normally-distributed random variables we are looking at. This number isn't quite what Pearson thought it was: he thought it was the number of observations for which he was trying to find the margin of error. It is actually the **degrees of freedom**, and we will see what that means later in this book.

Like the beta and gamma distributions, this actually creates a family of distributions. Different values of k will create different curves (Figure 269).

As you can see from Figure 269, the higher the degrees of freedom, the more closely the shape of the chi square distribution resembles that of a normal distribution.

Take the upper limit of the integration to some constant a. The integral will find the area under the curve up to this point.

$$CDF\left(\chi_k^2\right) = Pr(X \le x)$$

$$= \int_0^a \frac{x^{\frac{k}{2}-1} e^{\frac{-x}{2}}}{2^{\frac{k}{2}} \Gamma\left(\frac{k}{2}\right)} \, dx$$

Take the constants with respect to x outside the integration. What is left is the formula for a lower incomplete gamma function: a gamma function that integrates to some constant less than infinity.

$$= \frac{1}{2^{\frac{k}{2}} \Gamma\left(\frac{k}{2}\right)} \int_0^a x^{\frac{k}{2}-1} e^{\frac{-x}{2}} \, dx$$

Lower incomplete gamma functions are identified using the lower case of the Greek letter gamma, γ (page 304).

$$CDF\left(\chi_k^2\right) = \frac{\gamma\left(\frac{k}{2}, a\right)}{2^{\frac{k}{2}} \Gamma\left(\frac{k}{2}\right)}$$

We know how to calculate the denominator, because k will always be a positive integer (representing the number of normal distributions that are added together). That means we can calculate

$$\Gamma\left(\frac{k}{2}\right)$$

if k is even, by using the relationship between the gamma function and factorials, and if k is odd by using what we learned on page 305 and following. Calculating the value of the numerator, however, isn't quite so straightforward.

We haven't looked at the process for finding this: it used to be done by tables, and you would now do it using a computer. As with the normal distribution, it isn't usually possible to find a value just by working out the integral in the CDF, so a computer will use a Taylor series to approximate it. We covered a way this can be done when we looked at the gamma distribution (page 459).

So to work out whether the difference between an anticipated and observed result is within the margin of error, the process, broadly, is to square the difference of each observed value from the expected value, divide that by the observed value, add these squared, standardised differences together, and use the χ^2_k distribution, with the appropriate degrees of freedom, to work out the probability that that amount of difference could have arisen as a normally-distributed error from the true value.

Example of using a chi square distribution

To see how this can work in practice, imagine you are trying to determine whether a dice is fair.

First, work out your null and alternative hypotheses:

- H_0 : *The dice is fair; the probability of each outcome is the same.*

- H_A : *The dice is biased; the probability of each outcome is not the same.*

To test this, you roll it 60 times and record the results. This is what you get:

Outcome	One	Two	Three	Four	Five	Six
Number of times this was rolled: (Observed (O)	6	9	20	9	8	8
Expected (E)	10	10	10	10	10	10

Note that, even though the outcome is a number, it isn't a numeric variable; it's categorical. (Six isn't worth twice as much as three.) That makes this a multinomial distribution. The probability of each is the same, which means the expected value of each outcome is 10. The results are scattered around this. Does this, or does it not, tell you that the result is fair?

To answer this we need to compute the chi-square statistic:

$$\chi^2 = \sum \frac{(O_i - E_i)^2}{E_i}$$

$$= \frac{(6 - 10)^2}{10} + \frac{(9 - 10)^2}{10} + \frac{(20 - 10)^2}{10} + \frac{(9 - 10)^2}{10} + \frac{(8 - 10)^2}{10} + \frac{(8 - 10)^2}{10}$$

$$= \frac{16 + 1 + 100 + 1 + 4 + 4}{10}$$

$$= 12.6$$

Next we need the number of categories we are interested in. You might think there are 6, but we are actually only interested in 5. This is called the **degrees of freedom**. We'll come to the reason for this when we look at the work of George Fisher in a little while; for now, so that we can stay on track with what we are looking at, just go with it.

When you look up 12.6 with 5 degrees of freedom on the computer, it uses the chi square distribution to work out that the probability of getting that result is 0.02742 ... , or 2.742...% (if the dice is fair).

That result is less than 5%, which means there is more than a 95% chance that the dice is biased with these results. We can reject the null hypothesis on that basis. As we saw earlier, however, you might get that result every 20 times you did the exercise. If the level of certainty you were looking at is that there can't even be 1 chance in 100 that the dice is fair, then you wouldn't have enough data to reject the null hypothesis. So it really depends on just how confident you want to be in any given situation. We will look at this issue soon when we think about sampling.

MGF and moments of the χ^2 (chi-square) distribution

Because the χ^2 distribution is a particular instance of the gamma distribution, we can plug in the relevant values and just write cut the MGF and the moments, based on the proofs we looked at earlier. Using the parameterization that shows the chi square as $gamma\left(\frac{1}{2}, 2\right)$, and where k represents the degrees of freedom, is:

$$MGF = (1 - 2t)^{\frac{-k}{2}}$$

The moments are:

$$E(X) = k$$

$$E(X^2) = k(k + 2)$$

$$E(X^3) = k(k + 2)(k + 4)$$

$$E(X^4) = k(k + 2)(k + 4)(k + 6)$$

Key points:

- χ^2 uses the fact that errors are normally distributed to determine whether research findings are within the margin of error of something else.

- It is a myth that they should be applied to normally distributed data; they aren't usually used that way. What is necessary is that it is reasonable to assume that *errors* in the data will be normally distributed.

- The tests are normally used on categorical/multinomial data, but don't have to be.

- They work by summing standardised squared differences from an anticipated value, and comparing that to a distribution made by summing squared, normally distributed, random variables: $\chi^2 = \sum \frac{(O_i - E_i)^2}{E_i}$

- $\mathrm{PDF}(\chi_k^2) = \dfrac{x^{\frac{k}{2}-1} e^{\frac{-x}{2}}}{2^{\frac{k}{2}} \Gamma\left(\frac{k}{2}\right)} \, for \; x \geq 0$ where k is the number of squared, random variables we are looking at.

- Pearson made a mistake, later corrected by Fisher, in terms of understanding how many variables were actually being looked at.

- $\mathrm{MGF} = (1 - 2t)^{\frac{-k}{2}}$

Taking samples
(second extension of the normal)

> **Read this if** you aren't clear about how confident you can be that a sample used for research is reflective of the underlying population; you don't know the difference between a sampling distribution and a sample; you don't understand the central limit theorem (or you do understand it and assumed it had been mathematically proven); you thought the central limit theorem only applied to samples above a certain size; you don't know what the standard error of the sample mean actually means; you aren't sure why people say that a sample should have a minimum of **30** or **50** subjects or you don't know what the law of large numbers refers to (in either the strong or weak version).

Statistics is about mathematically measuring uncertainty, and an area where this commonly used is when we take a **sample** from a **population**. A population, in the statistical sense, doesn't have to be made up of people (although that is where the word originally came from). It can be any complete set of anything—animate or inanimate—that we are interested in learning about. It often isn't realistic to try to measure much about a whole population, so we need to look at a smaller subset of the population—a sample—and use it to infer some truths about the whole population. This is, not surprisingly, known as **inferential statistics.** It applies, for example, to market research, when we need to talk to a manageable number of consumers because we can't talk to all of them, but it also applies to many scientific experiments, when we are looking at what happens in a specific situation and attempting to infer what that means more broadly.

The fundamental question about sampling is, therefore: "How likely is it that *any* randomly chosen sample is representative of the wider population?"

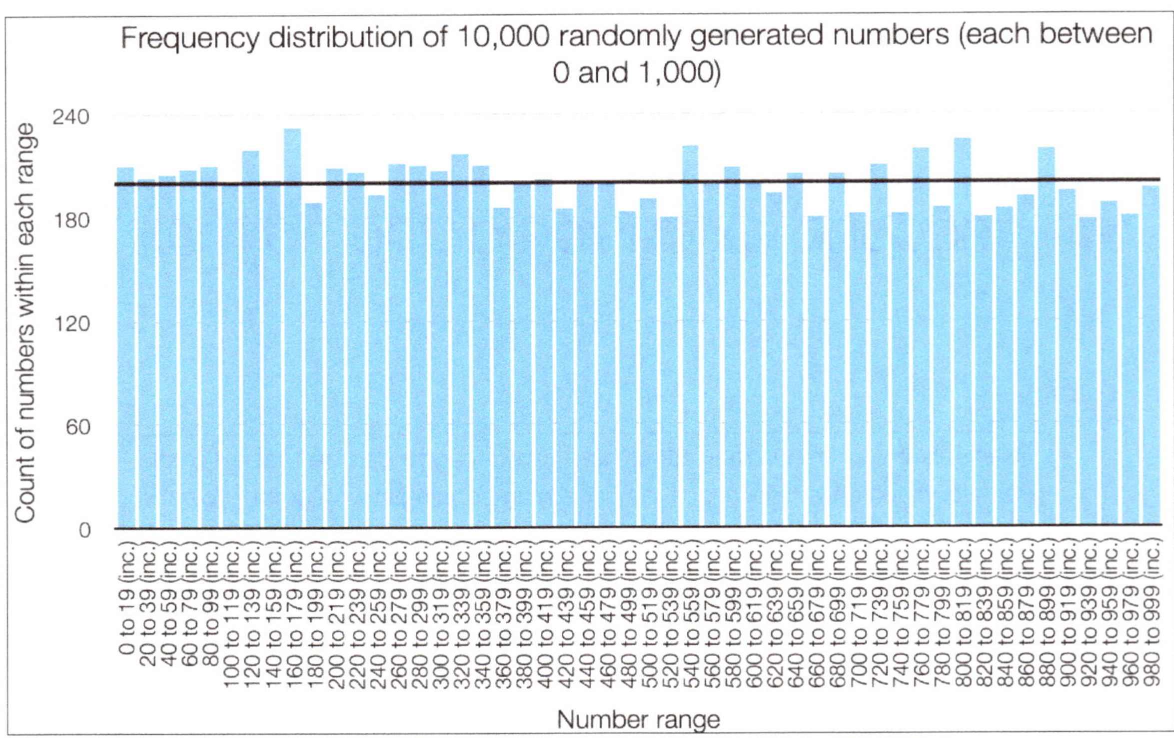

Figure 270

At first this may seem like an impossible thing to determine, but it gets easier when we realise that if we think about all possible samples of a given size from a broader population, those samples will, themselves, have their own probability distribution.

Let's see what that looks like. A programmer asked a computer to pick a random number between 0 and 1,000. This was then repeated 10,000 times. The numbers were sorted into different number ranges, and the results were graphed. That created that the fairly uniformly-distributed frequency graph at Figure 270.

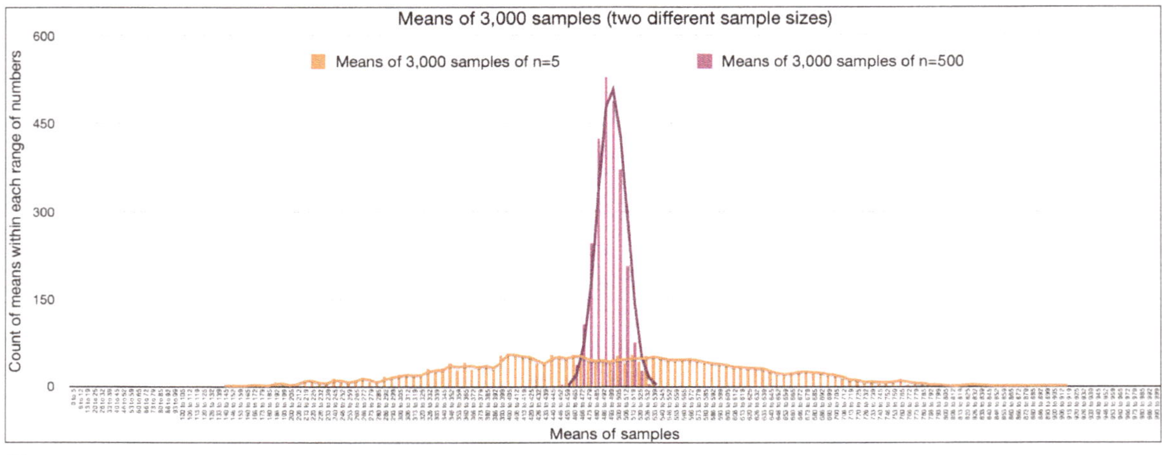

Figure 271

After that, the programmer told the computer to take a random sample of five numbers from this population of 10,000 numbers and calculate the mean of this sample. It stored that number and repeated the process 3,000 times. It then had a list of 3,000 means. The orange part of Figure 271 shows the graph of these means from the 3,000 samples. The computer then repeated the process, but the second time the size of each sample was 500 rather than 5. Again, the computer took the sample, calculated its mean, stored that results and repeated the process 3,000 times. The pink part of Figure 271 shows the result. These graphs are *frequency* distributions: they show the frequency with which the means of the samples occurred within particular ranges of numbers.

There are some things to notice about the graphs of the means of samples in Figure 271, compared to the graph of the underlying population these samples were taken from (Figure 270). First, whereas the population graph was relatively uniform, the graphs of the means of samples are curved. Each is, in fact, close to a normal distribution curve. The second is that the means of these distributions of sample means are extremely close to the mean of the population. (In fact, the mean of the population came out approximately 492.05. The mean of the means of samples of size 5 was approximately 493.62, while the mean of the means of samples of size 500 was approximately 492.01.) The third thing to notice is that the larger the sample size, the more tightly the means of the samples are clustered around the mean of the population.

While the graph at Figure 271 shows a frequency distribution of the means of samples, with different scaling it could actually represent the *probability* distribution of the mean of any sample. In other words, the probability that the mean of *any* sample from this population will fit into one of the number ranges shown at the bottom of the graph. As you can imagine, while there was a good chance that any of the numbers in the original population would be at either end of the graph, once we took means of groups of numbers, it was more likely that the numbers would be clustered towards the centre.

This probability distribution is the **sampling distribution of the sample mean**. It can guide us about the probability that any random sample we take will represent the population it is drawn from.

Now we'll look at how we can work out this information, and how it applies to taking samples.

Mean of a sampling distribution

It's intuitively obvious that the mean of a sampling distribution is the same as the mean of the underlying population, and the proof isn't difficult, but it can be challenging to keep track of just which variables and distributions are being talked about. If you are getting confused at any point, refer back to Figure 270 and Figure 271.

These are the mathematical symbols we will use to represent different concepts. Individuals from within the population will be represented as x_1, x_2, x_3 etc., or more generally as x_i.

Set	Mean
Whole population (Figure 270)	μ
Any potential sample of size n	$\bar{x}$
Sampling distribution of the sample mean (all potential samples of size n) (Figure 271). Note that we are using the expected value notation for this, because we are talking about a continuous probability distribution; but the random variable we are finding the expected value of is, itself, a mean.	$E(\bar{x})$

Find the mean of any potential sample of size n.

$$\bar{x} = \frac{x_1 + x_2 + \cdots + x_n}{n}$$

Then find the expected value, within the sampling distribution *of* means of potential samples, of this expected value: an expected value of an expected value.

$$= E[\bar{x}]$$

$$= E\left[\frac{x_1 + x_2 + \cdots + x_n}{n}\right]$$

Factor out $\frac{1}{n}$

$$= E\left[\frac{1}{n}(x_1 + x_2 + \cdots + x_n)\right]$$

Take the constant $\frac{1}{n}$ outside the expected value calculation. In other words, divide the expected value of the sum of the values by the number of values. (We can do this by linearity.)

$$= \frac{1}{n}E\left[(x_1 + x_2 + \cdots + x_n)\right]$$

The calculation above doesn't just apply to a sum of specific values of x. It applies to all possible samples of size n taken from our population. So

$$E(\bar{x}) = \frac{1}{n}E\left[\sum_{i=1}^{n}x\right]$$

the expected value of the mean of possible samples (left side of the equation) is the same as $\frac{1}{n}$ times the expected value of the sum of **any** n random variables. This is the key part of the proof: it doesn't matter which values of x we choose, which means that the equation applies to the random variable x in general.

By linearity, the expected value of a sum of values is the same as the sum of the expected values of each of the values.

$$= \frac{1}{n}\left[\sum_{i=1}^{n} E[x]\right]$$

Adding $E[x]$ together n times is the same as multiplying $E[x]$ by n.

$$= \frac{nE[x]}{n}$$

The n's cancel out. So the expected value of the means of potential random samples is the same as mean of the underlying random variable, regardless of the sample size.

$$E(\bar{x}) = E[x]$$
$$= \mu$$

This makes perfect sense when you think about it; it is hard to see how the expected value of the expected values of samples of a given size could be anything *but* the mean of the population as a whole.

Central limit theorem

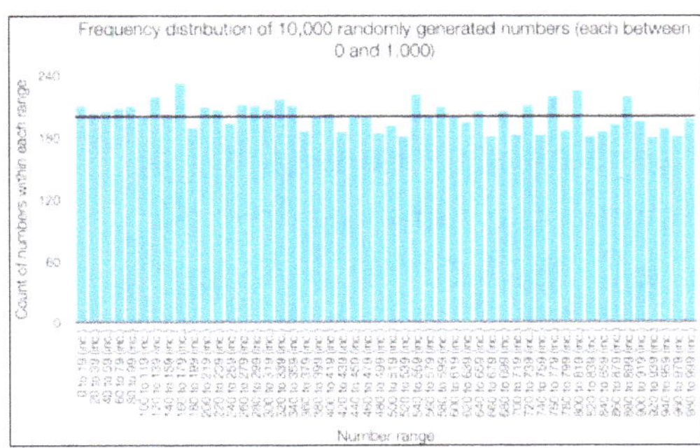
Figure 272

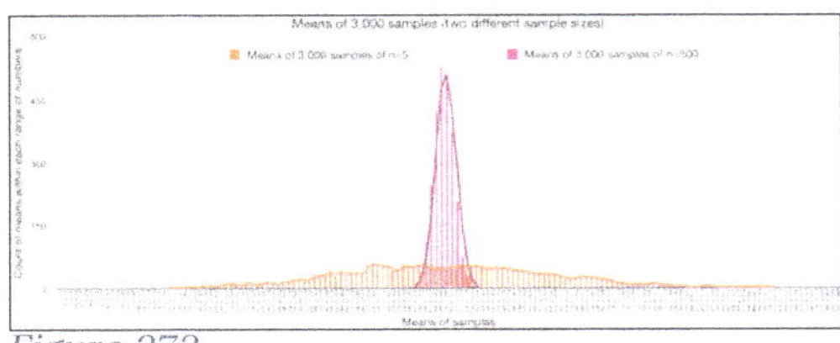
Figure 273

That's useful information, but much more so when we understand the shape of the probability distribution of potential means of samples.

Returning to the diagrams of the underlying data set and the sampling distributions which came from it (Figure 272 and Figure 273), you can see that even though the underlying data is distributed in a way that bears no relation to a normal distribution, the means of potential samples are scattered yet clustered around the middle in a way that roughly approximates the normal distribution. This is *always* the case.

You can think of the difference between the mean of any given sample and the mean of an underlying population as being an error and, as we have seen, errors tend to follow a normal distribution. Means of potential samples are clustered yet randomly scattered around the mean of the population, which means they tend towards a normal distribution. The fact that the distribution of sample means will always be roughly normally distributed is called the **central limit theorem.**

This concept applies regardless of the shape of the underlying distribution. It can be uniform, multimodal, or skewed strongly in either direction. It doesn't matter. If you take a lot of samples, then you find their means and then graph the results, the distribution *of those means* will always tend towards a normal distribution, the mean of which will be very close to the mean of the underlying population.

Despite what you may have heard, however, no one has actually proven mathematically that this is what will happen for samples of all sizes. (You might be the first to do so.) Once we recognise that means of samples will be clustered around the mean of the distribution, but at the same time randomly scattered around it, we can see why it will tend to be normally distributed. A sort-of proof of the central limit theorem has been around since the early 1900s, but it actually proves that *as the sample size approaches infinity*, the sampling distribution will be normally distributed. It doesn't deal with sample sizes that aren't approaching infinity, unfortunately; and sample sizes never remotely approach infinity. Because this mathematical proof doesn't actually rigorously prove the central limit theorem in the way it is commonly used, you can find it in an appendix on page 766, rather than in the body of the text here. It uses moment generating functions (MGFs), and shows that the MGF of a standardised version of the sampling distribution of the sample mean (as $n \to \infty$) is the same as the MGF of a standardised normal distribution, which shows that the curves must be the same.

Putting this together with what we already saw about the normal distribution on pages 591 and following, we can tell that there will be a 68.26% chance that the mean of *any* sample we take will be within 1 standard deviation *of the sampling distribution of the sample mean* of the true mean of the population. There will be a 95.45% chance that it will lie within two standard deviations of the mean (4.55% chance it will be outside of this), and a 99.74% chance it will be within 3 standard deviations (0.26% that it will be outside that).

That information becomes more useful when we know what the standard deviation of the sampling distribution of the sample mean will be.

Common misconception

You may have heard that the central limit theorem is the reason that so many phenomena are distributed according to the normal distribution. Having read this far, you can probably see why it has to be the other way around, for the following five reasons:

1. **Historically**, the central limit theorem was first proved by Russian mathematician Aleksandr Lyapunov (1857–1918)—a student of Pafnuty Chebyshev and friend of Andrey Markov—in 1901, while, as we've seen, the thinking about the normal distribution began much earlier, as an explanation for errors.

2. **Logically**, the means of samples can be thought of as a type of error; but errors, in general, can't be thought of as an example of sampling distributions. It doesn't work to think of them in this way. The central limit theorem applies to specific statistics (usually the means) of samples of a given size taken out of a population. There's no particular reason for assuming that it should apply to all data in the same population. In terms of set theory, sampling distributions are a subset of phenomena that are normally distributed, not the other way around (Figure 274).

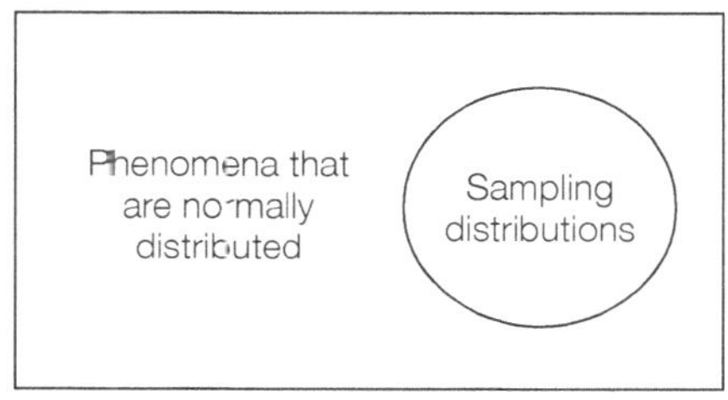

Figure 274

3. **Rationally**, the central limit theorem is applicable to the sampling distribution applied to any other distribution of any shape, while normal distribution only applies to data which are...well... normally distributed. If the central limit theorem accounted for all data that are normally distributed, then normal distribution would have to apply to all populations, just as the central limit theorem does.

4. In terms of **explanation**, saying that many phenomena are normally distributed as a result of the central limit theorem doesn't really explain anything, and doesn't provide us with a sense of when it is likely to be helpful to think about a normal distribution.

5. **Mathematically**, as mentioned above, the central limit theorem has only been proven for sample sizes approaching infinity. There isn't a mathematical basis for applying it to any phenomenon that happens to

be normally distributed. Also mathematically, if you are interested enough to study the proofs of the MGF of the normal distribution (page 600), then the proof of the central limit theorem (page 766) you will see that there is no easy way to work back from the MGF of a sampling distribution, in order to develop the formula for the normal distribution.

Standard deviation ("standard error") of the sample mean

As mentioned earlier, we can see from our graph of sampling distributions (Figure 275) that the larger the sample size, the more closely this sampling distribution will be clustered around the mean of the sampling distribution. We also saw earlier that is the same as the mean of the population. As sample sizes get larger, **the standard deviation of the sample mean** within the **sampling distribution of the mean** (probable distribution of mean of samples; and yes, the terminology can get confusing) gets smaller. In other words, the distribution of probable means-of-samples (**sample means**) becomes more tightly clustered around the mean of the underlying data set. It becomes less probable that the mean of any sample you choose will be

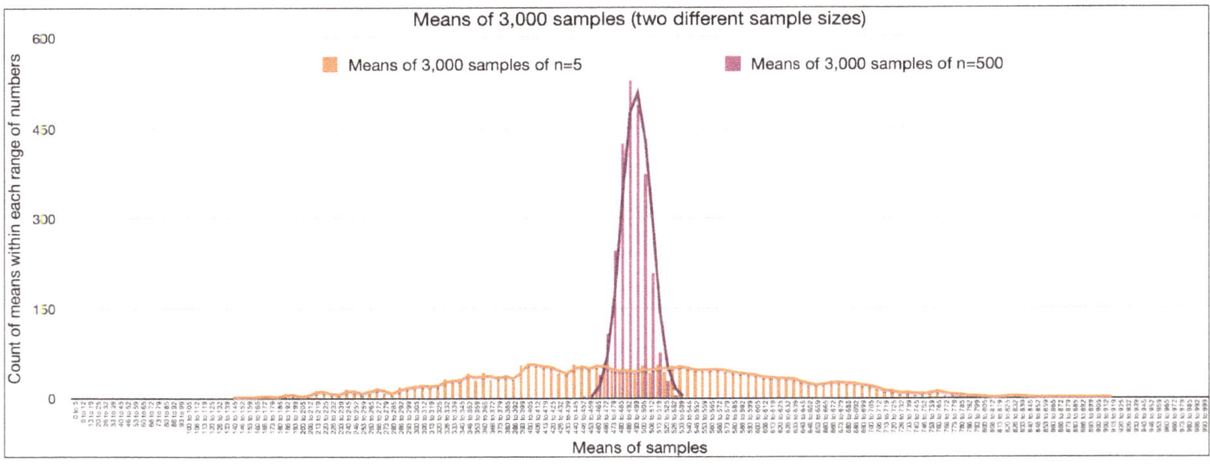

Figure 275

very far from the mean of the underlying population. (We saw in the introduction to the section on probability on page 334 and following that smaller schools were disproportionately represented in lists of better-performing schools, but also in lists of worse-performing schools. All that was actually happening is that the larger the size of the school, the closer its mean was to the mean of all students.)

The standard deviation of the sampling distribution indicates how tight this clustering is, and is the **standard error of the mean**: the extent to which taking the mean of *any* given sample will be close to the mean of the population. This is, of course, picking up on the idea that the difference between the mean of a sample and the mean of the underlying population can be thought of as an error, and errors tend to follow a normal distribution.

There is a way to mathematically represent this relationship. As with the proof above about the mean of a sampling distribution, this proof isn't difficult, except that it can be challenging to keep track of which distribution is referred to by which symbols. These are colour coded in a similar way, to try to make it a bit easier.

Set	Mean	Variance	Standard deviation
Whole population:		σ^2	σ
Any potential sample of size n:	$\bar{x}$		
Sampling distribution of the means of potential samples (all potential samples of size n):		$s_{\bar{x}}^2$	$s_{\bar{x}}$

Note that the following proof assumes that the variable x is i.i.d.: it is independent (knowing something about one x_i tells us nothing about another one) and identically distributed (the same function describes any instance of x).

By definition:

$$s_{\bar{x}}^2 = variance(\bar{x})$$

$$= variance\left(\frac{\sum_{i=1}^{n} x_i}{n}\right)$$

Square the constant $\frac{1}{n}$ when it is taken outside of the variance calculation. (Refer back to page 550 if you are unsure about this.)

$$= \frac{1}{n^2} variance\left(\sum_{i=1}^{n} x_i\right)$$

This is the tricky bit, which might look a little like sleight-of-hand, but really isn't. There is nothing, in this instance, to distinguish any specific x_i from the random variable x. This is where the condition that x is identically distributed comes into play. Any version of x will have the same variance.

$$= \frac{1}{n^2} variance\left(\sum_{1}^{n} x\right)$$

The variance of the sum is the sum of the variances, as we saw on page 549 and following.

$$= \frac{1}{n^2} \Sigma_1^n \, variance(x)$$

The variance of x is the same thing as the variance of the population.

$$= \frac{1}{n^2} \Sigma_1^n \, \sigma^2$$

Adding an identical variance n times is the same as multiplying it by n so $\Sigma_1^n \sigma^2 = n\sigma^2$

$$= \frac{n\sigma^2}{n^2}$$

Divide the numerator and denominator by n. So the variance of the sampling distribution is the same as the variance of the population divided by the sample size.

$$s_{\bar{x}}^2 = \frac{\sigma^2}{n}$$

Take the square root of both sides. The standard deviation of the means of a potential distribution of possible samples of size n is the standard deviation of the population, divided by the square root of the sample size.

$$s_{\bar{x}} = \frac{\sigma}{\sqrt{n}}$$

So we've arrived at an incredibly useful formula: the **standard error of the mean:**

$$s_{\bar{x}} = \frac{\sigma}{\sqrt{n}}$$

$$= SE(\bar{x})$$

where $s_{\bar{x}}$ is the standard deviation of the probability distribution of potential means of samples, σ is the standard deviation of the population, n is the sample size, and $SE(\bar{x})$ is the standard error of the sample mean.

It is not intuitively obvious that the standard deviation of the sampling distribution will be inversely proportional to the *square root* of the sample size; but it will be.

The standard deviation of our original population of 10,000 numbers was approximately 288.96. Thinking about our samples of 5, if we apply this formula we get:

$$\frac{288.96}{\sqrt{5}} \approx 129.23$$

The standard deviation of the sampling distribution of the sample means for samples of this size was approximately 129.05.

Thinking about our samples of 500, if we apply this formula we get:

$$\frac{288.96}{\sqrt{500}} \approx 12.92$$

The standard deviation of the sampling distribution of the sample means for samples of this size was approximately 12.50. So both times, the predicted value was less than 0.5 from the actual value.

Once we have this result, and we accept that the sampling distribution will be a normal distribution, we can apply everything we know about the fixed proportions of a normal distribution to be able to say what the chances are that the mean (or in fact, other statistics of a given sample as well) of a sample will represent the underlying population, based entirely on the size of the sample and the mean and standard deviation of the population.

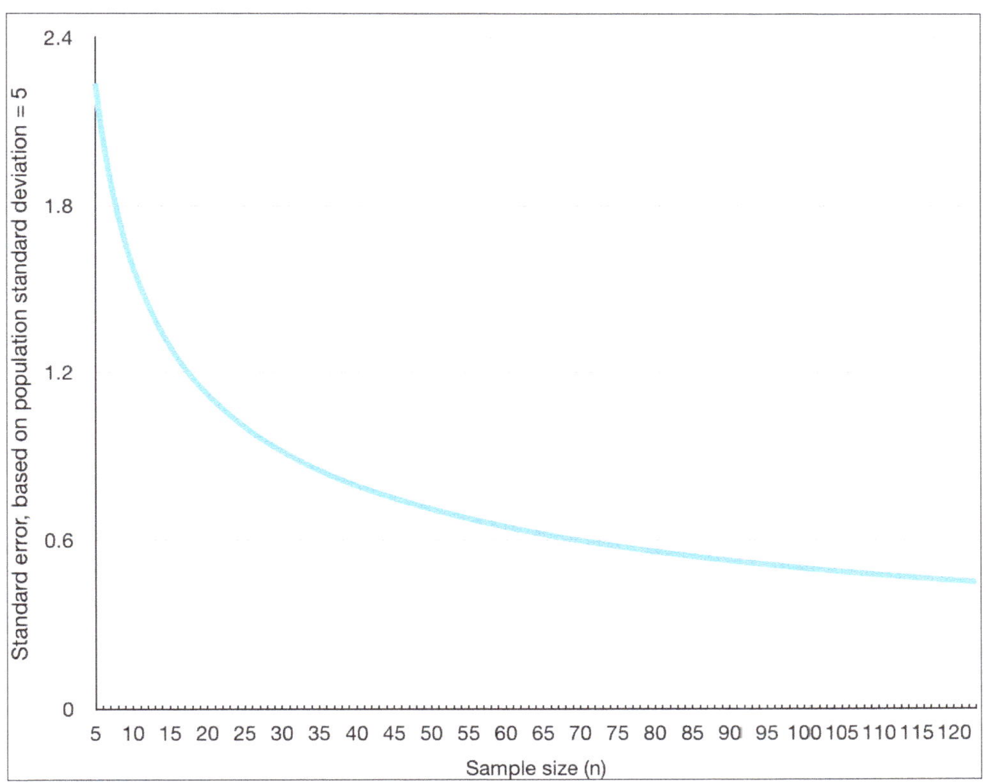

Figure 276

Figure 276 graphs the relationship of sample size to standard error. As you can see, it has a similar shape to an exponential distribution. This has led to a common approach that 30-50 is the minimum size for a sample (depending on who you ask). As samples get larger, there are diminishing returns as far as reducing standard error. (One of the reasons for surveys of $n = 1,000$ or more is to make it easier to analyse sub-populations, while keeping the size of them over 30.) There is a common misconception that the sampling distribution will only follow a normal distribution for samples of 30 or more. As we have seen, this is incorrect. It is the standard deviation that will change with increasing sample size, not the normality of the probability distribution.

While the standard error of the mean is a useful result, and one that really makes statistical sampling possible, there are a couple of important limitations to it. The first is that our result depended on the idea that our samples are independent, but taking multiple samples might not be independent if we do it without replacement. If, for example, we sample 5 people from a population of 100 and then we take another sample but make sure we exclude the initial 5 people, then the next sample will be out of 95 people, not 100. That means our samples won't really be independent: taking one will potentially affect taking the next one. As we saw earlier (page 549), ignoring the condition that our variables must be independent can have a drastic impact on the result of a calculation.

Researchers get around this by either making sure that the population is large enough where taking a sample won't make an appreciative difference to the size of the remaining population, or, if it may, they apply different mathematical corrections.

The second limitation is the big one, though. It presupposes that we know what the standard deviation of the population is and, of course, people normally take samples *precisely because* they don't understand enough about the population. One way to deal with this is by using the standard deviation of the sample to give a guide to the standard deviation of the population. That is, of course, potentially problematic without further analysis. We'll look at how it was addressed later, when we see how "Student" and Ronald Fisher looked at it.

The law of large numbers

We can think about this issue in a related way by using Chebyshev's inequality. As a reminder, it says that:

$$P(|x_i - \mu| \geq c) \leq \frac{E(x - \mu)^2}{c^2}$$

In other words the probability that the distance between any point you choose and the mean will be greater than or equal to any constant you like, will be less than or equal to the variance of the distribution over the square of the constant. Let's see how that relates to what we've been looking at:

We're working with the sampling distribution once again, so our arbitrary point will be one of the potential means of samples, our mean will be the expected value of the sampling distribution, and the variance will be the variance of the sampling distribution.

$$Pr(|\bar{x} - E(\bar{x})| \geq c) \leq \frac{s_{\bar{x}}^2}{c^2}$$

We know from the analysis above that the expected value of the sampling distribution is the same as the mean of the population, and also that the variance of the sampling distribution is equal to the variance of the population, divided by the size of the sample.

$$Pr(|\bar{x} - \mu| \geq c) \leq \frac{\frac{\sigma^2}{n}}{c^2}$$

Multiply the numerator and denominator on the right by n.

$$\boxed{Pr(|\bar{x} - \mu| \geq c) \leq \frac{\sigma^2}{nc^2}}$$

The formula we've arrived at is now looking at the mean of any sample, taken from any population, regardless of how it is distributed; it doesn't have to be remotely normal. The probability that the distance between the mean of any sample we choose to take and the mean of the population we are sampling from will be greater than some arbitrary constant C, will be less than or equal to the variance of the population divided by the sample size multiplied by the square of the constant.

This is a useful result in its own right and is also used to prove the **law of large numbers.** This law possibly gets attention because its name is so interesting, and even more so when you realise that there is a **strong law of large numbers** and a **weak law of large numbers**, which are even more interesting names. (The weak law isn't really any weaker than the strong law except that there may be some rare situations in which the weak law doesn't quite do the job.)

Mathematically the strong law is written as:

$$\boxed{\bar{x} \to \mu \ as \ n \to \infty}$$

This translates as: the mean of samples of size n approaches the mean of the population, as the size of the sample approaches infinity.

The weak law is written as:

$$\boxed{Pr(|\bar{x} - \mu| \geq c) \to 0 \ as \ n \to \infty}$$

This can be read as saying that, for any fixed threshold, the chance that the sample mean deviates from the population mean by more than that amount vanishes as the sample size grows.

The law of large numbers isn't all that helpful, in fact, because when the size of the sample approaches infinity, the sample will be greater than the population, which makes no sense. Taking that a step further, if the population was literally infinite it would be impossible count all the members, and so a sample mean would be the only mean. If the population was large but not infinite, then a sample size that approaches infinity won't exist, because you can't draw a sample larger than the population. But the idea that larger samples will give you more confidence that you are making accurate assumptions about the possibility is useful; particularly now that we can quantify the relationship using the Chebyshev inequality.

The story of sampling so far…

In summary, all of the potential samples of size n taken from a population will have their own probability distribution, which is called the **sampling distribution**. We have largely been looking at the means, variances and standard deviations of these potential samples, although we could have looked at other parameters. This distribution is called the **sampling distribution of the mean** or the **sampling distribution of the sample mean.** This distribution will tend to be normal, even with small sample sizes, and if it were possible (which it obviously isn't) to take a sample of infinite size, its sampling distribution would be exactly normal.

The expected value of a sampling distribution of the mean will be the same as the mean of the population. Unfortunately this is a long way from being able to say that the mean of a sample will be the exact mean of the population. That will rarely be true.	$E(\bar{x}) = \mu$
The variance of the sampling distribution of the mean will be the variance of the population divided by the size of the sample.	$s_{\bar{x}}^2 = \dfrac{\sigma^2}{n}$
The standard deviation of the sampling distribution of the mean will be the standard deviation of the population divided by the square root of the sample size. This is known as the standard error of the mean.	$s_{\bar{x}} = \dfrac{\sigma}{\sqrt{n}}$ $= SE(\bar{x})$

The larger the sample size, the higher the probability that the mean of any given sample will be close to the mean of the population, but this still depends on the variance of the underlying population.	$Pr(|\bar{x} - \mu| \geq c) \to 0$ $as\ n \to \infty$ $Pr(|\bar{x} - \mu| \geq c) \leq \dfrac{\sigma^2}{nc^2}$ Where c is any constant.

If only we knew the mean and the variance of the underlying population, that would be enough information to completely describe the sampling distribution for any n in any population, regardless of its distribution. We could take samples and have a really good sense of how reliable they were as an indication of the population. Unfortunately, without knowing the population mean and variance, none of this information is much better to us than just taking a decent-sized sample and hoping that it is representative. In other words, statistics hasn't helped us much with sampling yet, for all of its clever ideas.

That was a major issue that some of the people we will learn about soon set out to resolve.

> **To cement your understanding:**
>
> Based on what we have covered so far, what are your ideas about how to tell how useful a survey is if you don't know the mean or the variance of the underlying population? Have a think about this problem, before reading on.

Key points:

- A **population**, in the statistical sense is any complete set of anything—animate or inanimate—that we are interested in learning about.

- A **sampling distribution of the sample mean** is a probability distribution, indicating how probable it is that the mean of any randomly-chosen sample of size n will have a particular mean.

- The mean of a sampling distribution of the sample mean is the same as the mean of the underlying population.

- The standard deviation, or **standard error**, of a sampling distribution of the sample mean is given by

$$SE(\bar{x}) = \frac{\sigma}{\sqrt{n}}$$

where σ is the standard deviation of the population, and n is the size of the sample. This applies where samples are independent.

- The **central limit theorem** basically says that sampling distributions are normally distributed, regardless of the underlying distribution of the population. It has only been mathematically proven in the case where the size of a sample approaches infinity (which doesn't happen), but it makes sense conceptually given the nature of normal distribution.

- The mean of a sampling distribution is the same as the mean of a population.

- The law of large numbers states that the mean of any sample will approach the mean of the population as the sample size approaches infinity. According to the Chebyshev inequality, the probability that the mean of a sample will be further than a given amount from the mean of the population is directly proportional to the variance of the population, and inversely proportional to both the size of the sample and the square of the chosen constant.

Errors, power and confidence (third extension of the normal)

Read this if you aren't confident about statistical significance, you don't understand the difference between a type I and a type II error nor how their probabilities are described; you thought that large research studies were likely to be better than smaller ones; you don't know how to perform a power analysis; you don't know what a confidence interval is or you don't understand the difference between a point estimate and an interval estimate.

Types of errors

Statistics doesn't make it possible to be 100% certain of anything. As we've seen, it's all about understanding, and hopefully thereby reducing, the degree of uncertainty. So if we are 95% confident that something is true, we are 5% confident that it isn't. There can be all sorts of errors made in research. The ones we are looking at here are those that are built into the process of using statistics itself (as opposed to errors such as biased study design).

When statistics suggests that something is true but it actually isn't, the error can be classified in two ways. A **type I error**, is one where we think something is important, but it isn't. The probability of a type I is usually indicated by the Greek letter alpha (α), and means we reject the null hypothesis (H_0) and accept the alternate hypothesis (H_A); but we shouldn't. It is a **false positive**.

In other words, statistics might say that a clinical treatment works, but it doesn't. We believe that there is a shift in people's opinions, but there isn't. A **type II error** is one where we don't think something is important, but it is. Its probability is often represented by the Greek letter beta (β), and it means we retain the null hypothesis, but we should reject it. It is a **false negative.** In other words, test results say a treatment doesn't work, but it really does. Results say that people's opinions haven't shifted, but they really have (Figure 277).

	The null hypothesis should be *retained*. What we are looking at *doesn't* achieve what we are testing for.	The null hypothesis should be *rejected*. What we are looking at *does* achieve what we are testing for.
We decide to *retain* the null hypothesis. We *don't* believe there is a statistically significant outcome.	Good decision, with probability $1 - \alpha$ (one minus alpha)	**Type II** error, with probability β (beta). False negative.
We decide to *reject* the null hypothesis. We *do* believe there is a statistically significant outcome.	**Type I** error, with probability α (alpha). False positive.	Good decision, with probability $1 - \beta$ (one minus beta)

Figure 277

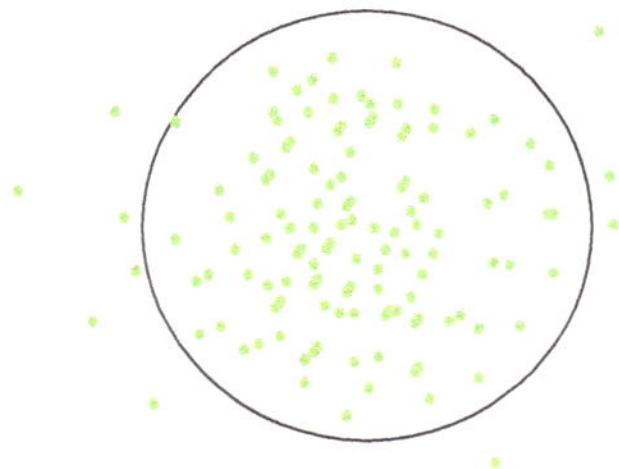

Figure 278

Imagine that a shooter has bought a new scope for her rifle and wants to know whether it is accurate. She already knows how closely her shots are likely to group around a target using an accurate scope (Figure 278). The shots that land outside of the circle are a type I error: they are randomly scattered results which make it appear, on their own, as though her scope isn't accurate, when in fact it is.

She has shot enough in her life to know what that percentage is, and she designates that probability as alpha (α). Probabilities always add to one, which means that her probability of landing within the circle with an accurate scope is $1 - \alpha$. For example, if she normally expects 5% of her shots to land outside the circle, then she will expect 95% of her shots to land inside the circle. Her α in that case is 5%, and her $1 - \alpha = 95\%$

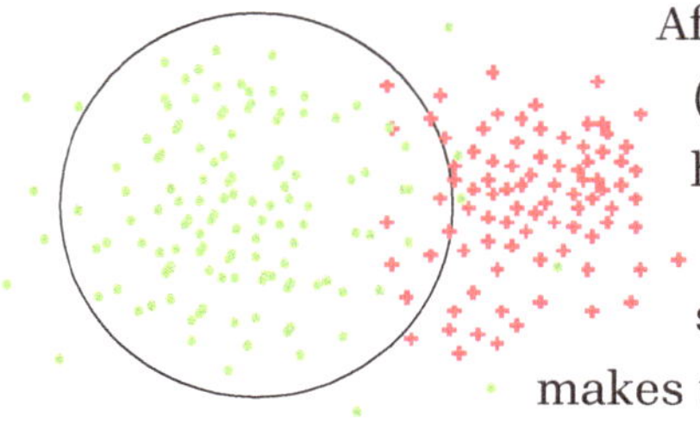

After she does some testing of her new scope (the red crosses in Figure 279), she finds that her new scope isn't accurate, and needs adjusting. But some of the shots with the new scope still land within the original circle. This makes them a type II error, with probability beta (β). That is, they make the scope look accurate when it isn't.

Figure 279

You can see from Figure 279 that if she had drawn a larger circle, she would have *reduced* her chances of getting a type I error; more of the shots would have indicated that the scope was accurate. But it would have *increased* her chances of getting a type II error. More of the shots would have suggested that the scope was accurate even if it wasn't.

Now if we were to ignore the vertical elevation of each shot and just mark their position on a horizontal number line, then graph the number of shots at each point, we would expect to see a normal distribution curve (because errors are normally distributed), as in (Figure 280).

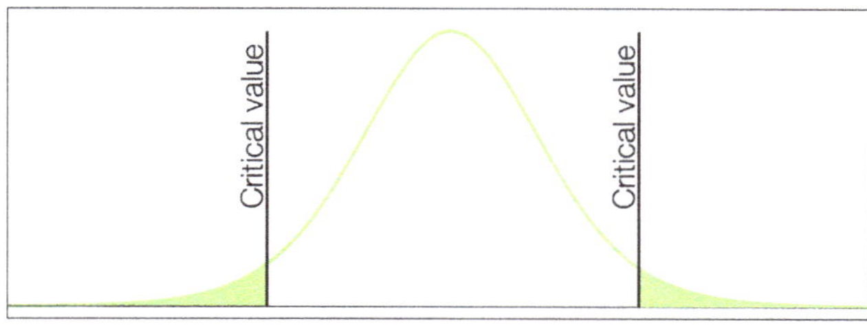

Figure 280

Normal distribution curves extend in an infinite direction on each side. Virtually anything *could* happen. We need to decide a cut-off point. Outside of the cut-off we say that a result is **statistically significant**, while inside of that we say that it isn't. We call this cut off point the **critical value.** (We looked at this issue briefly when we thought about P values, on page 598.) In Figure 280 the shaded green areas are the areas that are statistically significant, and the place between the white and green areas under the curve is the critical value.

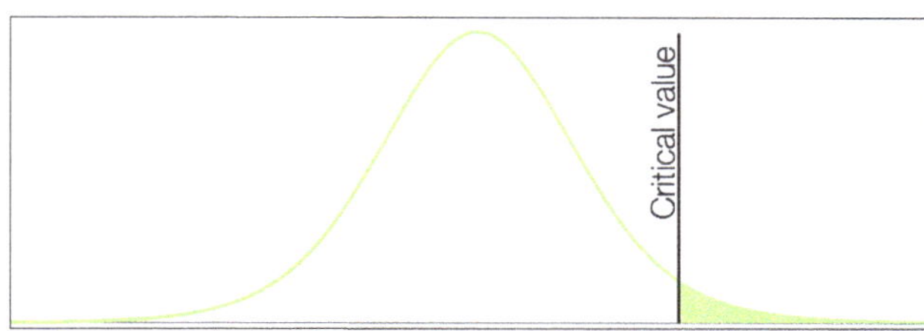

Figure 281

Sometimes researchers are interested in whether there is a deviation from the mean in either direction, or sometimes they are only interested in deviation in one direction (Figure 281). (We will explore the implications of whether there is one or there are two critical values when we come to looking at one-and two-tailed *t* tests.)

If this normal distribution represents the null hypothesis (perhaps based on prior understanding of a population, or on the use of a **control group** in an experiment), and the null hypothesis is true, then anything falling into the green shaded areas—anything suggesting that the result is statistically significant, when it isn't—represents a type I error. It is this area that has a probability of α.

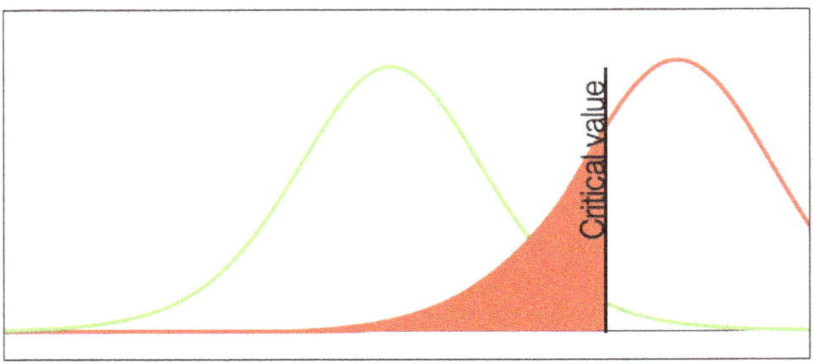

Figure 282

If, on the other hand, the null hypothesis is false and should be rejected, because normal distribution curves stretch an infinite distance to the left and the right, there will always be an overlap of at least some degree between the two distributions, and some of the distribution representing the alternative hypothesis will still fall on the null-hypothesis side of the critical value. This area, shaded red in Figure 282, is the area of type II errors. Even though anything inside of that is part of the probability distribution of the alternative hypothesis, it will be counted as being part of the null hypothesis, because it is on the null-mean side of the critical value. There is a probability of β being within this area, given that the alternative hypothesis is the correct one.

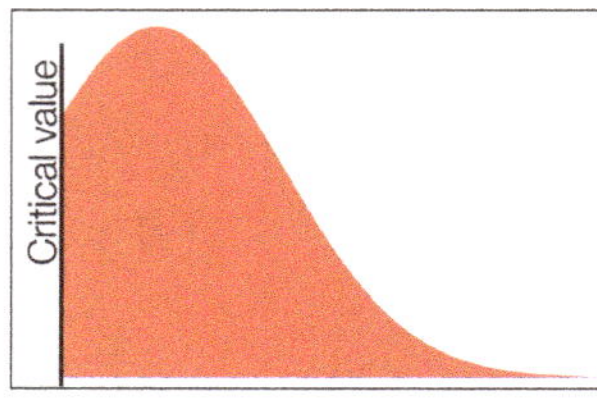

Figure 283

The area of the alternative probability distribution that falls outside of the critical value has a probability of $1 - \beta$. This is the probability of not having a type II error given that the alternative hypothesis is correct, and it is the **power** of the test. (Not to be confused with raising a number to a power, nor a power series; this is a completely different concept.)

There is a trade-off between these two types of error. If you were running an experiment and wanted to be 99.9999% certain that you didn't say something was making a difference if it wasn't—in other words, if you wanted an α of 0.0001%—then you would greatly increase your chances saying that something really wasn't making a difference, when in fact it was.

To cement your understanding:

• You are reading a paper which says that scientists obtained a particular result: they rejected the null hypothesis. You disagree: you think that they didn't have enough evidence to draw the conclusion that they did. In your opinion, did they make a type I error or a type II error?

• The answer to the previous question was, of course, a type I error. On that basis, what Greek letter (alpha or beta) would be used, by convention, to indicate the probability that this is what happened?

• Given that the answer to the last question was alpha, what is the probability that they did not make that particular error? (Hint: top left box on Figure 277.)

> • Being scientifically minded, you are open to the possibility that you are the one who has got it wrong. If you would like to reject the null hypothesis, that is accept the result of the study, and the null hypothesis should in fact be accepted what kind of error are you making? What are the chances that that has happened? (Hint: top right box on Figure 277.)

Type II errors and power analysis

Type II errors are painful for researchers. If your study clearly demonstrates that something doesn't have an effect, that's one thing. It might be harder to get your results published, but at least you've learned something. (In an ideal world it would be just as easy to get negative results published as positive, because negative results are still informative, but unfortunately it doesn't often work that way.)

If, on the other hand, your study is **under-powered**—if you think you could probably have demonstrated statistical significance with a larger study, but you can't do so with the study you have—then you've completely wasted time and money and probably haven't learned much.

> It is poor research practice to have an under-powered study then just keep testing more subjects until you have enough to reach statistical significance. That would make it too easy to manipulate the result; to keep going until you fluked finding the right subjects to give you the answer you wanted. Best-practice research involves starting again with a large-enough sample.

To clarify, when we think about power calculations we are (mostly) thinking about sampling: to what extent will our sample represent an underlying population? That means we are (usually) thinking about a sampling distribution which, thanks to the central limit theorem, will be normally distributed.

At the same time, part of what needs to be taken into account with power analysis is the size of the effect, and the way that is understood depends on the sort of variable(s) and distribution(s) you are interested in. That means the specifics of power calculations can be quite varied, and so rather than covering the mathematics in detail here, we'll think about the principles, which are consistent. The specific ways these are applied depend on the specific situation.

1. **Decide on your significance level (α).** When we first looked at this, we said that, although 5% isn't a great level of statistical significance, it is usually taken as the minimum acceptable level to be able to say you have a positive result, and it is usually a useful place to start calculating power.

2. **Think about the variables and the distribution(s).** These will determine the specifics of the way you calculate power.

3. **Estimate the means and standard deviations for the null and alternative hypotheses.** You may already know what the mean and standard deviation of the null-hypothesis are, or you may be planning to find that information from a control group. You won't know what they are for the alternative hypothesis, and so you will need to base your estimate on your prior understanding of the subject area.

4. **Standardise your scores.** That normally means dividing by the standard deviation. This will tell you how far from the mean you need to be, in standardised terms, in order to reject the null hypothesis. Recall that, for a sampling distribution, the standardised variable will be

$$z = \frac{\bar{x} - \mu_0}{\frac{\sigma}{\sqrt{n}}}$$

Based on the proportions of the standardised normal distribution curve (using the cumulative distribution function (CDF)), a z score of 1.645 (when only testing the effect in one direction, as we are here) will be close to the critical 5% value. For other types of test, other standardised test statistics will be used.

5. **Using whatever cumulative distribution (of the standardised form of the variable) is of interest, calculate the power.** The power is the probability that your results will be on the side of the critical value that indicates the alternative hypothesis is the correct one. You calculate it using the CDF of the *alternative* hypothesis (based on your estimated parameters) but using the *critical value* of the *null* hypothesis. In general the power of a test is lower than the α of the test: in keeping with the use of statistics to be confident that your results aren't a result of random chance. Statisticians generally consider it to be more important to avoid a type I error—avoid overstating the achievement of a test—than to avoid a type II error—failing to achieve statistical significance when the effect is still real. So the lowest acceptable value for the power of a test is usually 0.7 or 70%. The most commonly used figure is 0.8 or 80%. A 0.9 or 90% power, or even higher, can be used with very stringent research, with either a very large sample size or a very strong effect.

6. **Using algebra, isolate the variable of interest.** It is quite likely that you have an idea of the effect size you are expecting to see. It is fairly

standard to opt for 80% as the power of a test. (Again, this is not ideal. If your power is 80%, there is one chance in five that you will have an under-powered test. The test will say that you don't have a positive result, when in fact you do. But 80% is a recognised standard and is generally a good place to start.) Using those numbers, you can solve your standardised score for n; it will tell you the size of the sample you need. Alternatively, you may already know your sample size and the power and want to understand the size of the effect that the test you are using will be sensitive enough to pick up. You might also have an idea of the effect and the sample size and want to use these to calculate the power of the test.

You will also need to run some test scenarios with different outcomes, to make sure that you have a big enough sample to let you know if you have a real effect. If you don't run the test scenarios, the danger will be in putting too much reliance on your mathematics, forgetting that it was only ever a function of your underlying assumptions (which you won't be able to check until you have done the experiment).

There are several broad strategies for improving the power of a study.

1. You can select a larger sample. This will reduce the standard error of the mean, which will effectively narrow the sampling distribution of the alternative hypothesis. We saw previously that the larger the sample, the lower the standard error of the mean. In other words, the greater the chance that the mean of any sample will be close to the mean of the population. This makes intuitive sense, and we saw it in the following three formulae:

$$SE(\bar{x}) = \frac{\sigma}{\sqrt{n}}$$

$$Pr(|\bar{x} - \mu| \geq c) \to 0 \ as \ n \to \infty$$

and

$$Pr(|\bar{x} - \mu| \geq c) \leq \frac{\sigma^2}{nc^2}$$

All three of these tell us that bigger samples will be more helpful in a situation like this, where there isn't a huge difference between two means, or at least not in comparison to the amount of variability. That in turn will create a smaller area of overlap between the probability distributions of your null and alternative hypotheses. This will increase the chance that you will have a good P value and still have a good power of the test.

2. You can have a bigger effect. For example, you may want to change your intervention so that it has a stronger effect (if this is desirable and feasible). This will move the means further apart, also reducing the size of the overlap.

3. You can decrease the amount of probable variation in your alternative hypothesis. If, for example, your alternative hypothesis involves an intervention of some kind, you could increase the rigour of your preparation, to try to achieve more consistent results. Or you could have more strict selection criteria for your subjects, to ensure that more of them will respond consistently.

In terms of selecting the size for a sample the power of the study isn't the only consideration. Another is the expected drop-out rate. If the study is to be conducted over an extended period, and/or if adherence to requirements might be challenging, this may be also need to be considered in terms of sample-size selection. This in turn might be connected to whether you plan to use an **intention-to-treat** or a **per protocol** approach to analysing your results (see text box at Figure 284).

Errors, sampling size and effect size

There is a trade-off between sample size and effect size. As a general rule, the weaker the effect, and the less consistent the effect, the larger the study you need to statistically demonstrate it. When you are reviewing research undertaken by others, depending on what you are investigating, it is usually more important to look at the level of statistical significance and the size of the effect, than the number of subjects/tests carried out. In fact, a large study may be a pointer to a weak or highly variable effect. (You may need to look very closely to realise that in a published article, because the researchers, being human, are unlikely to make a big thing of it if that is the case.)

Thinking back to our earlier example of tossing a coin, imagine someone told you they had a technique that allowed them to increase their chances of getting a head. If you asked them to prove it and they proceeded to throw 15 heads in a row (or, in fact, considerably less), you would be convinced. The chances of that happening are 1 in 32,768. You wouldn't be saying "that study is too small to be credible; I only believe studies with more than 100

Intention-to-treat analysis is an approach where you analyse the outcome of a study against the number of subjects initially involved. In other words, if a study commenced with 100 subjects, 20 dropped out and 2 died, you still measure the final results against 100 people. **Per-protocol** analysis means that you are only assessing your results against the people who completed the entire study protocol. In the situation above you would assess your final results against the 78 people who completed the study, which makes it easier to demonstrate a successful result.

The choice of which to use depends on the study. Intention-to-treat is normally a preferred approach, because when subjects decide not to continue that may be an indication that something won't be adhered to in the real world because it is too difficult, and if subjects die that may also be an indication of the failure of an intervention.

Alternatively, if you are examining the effectiveness of a lengthy complex protocol, and subjects might leave the study due to a range of unrelated reasons, per protocol may be more appropriate. As usual, you need to think about it on a case-by-case basis (before you select your study size) and you need to communicate clearly about how you have conducted it.

Figure 284

subjects". But now imagine that their technique *does* work, but only 6 times out of 10. If they threw the coin 15 times and they achieved 9 heads they could claim that that is the exact result they predicted, but you probably wouldn't be particularly impressed. They would have to throw it over 300 times to claim statistical significance. At that point they could write up a paper about it, but if you read the paper carefully (either in terms of what they were saying, or what they left out) you might realise that even though they can credibly claim their technique works with statistical significance, and they have a large study to prove it, you still probably wouldn't be interested in putting time or money into learning the technique. The outcome just isn't that important.

What statistics in research is largely asking is: "what are the chances that what we are looking at was achieved by random chance?" It is much easier to have wide fluctuations, or a high amount of random chance, in a smaller study than in a larger one. That means it is much harder to demonstrate that you have a statistically significant effect when your sample is small. So if you *do* manage to reach statistical significance with a well-designed, small trial, you have achieved something difficult: probably because you have a big effect, with low variability. Paradoxically, this means that smaller studies are sometimes more valuable than larger ones.

However, a problem which *can* occur with small studies, even when they demonstrate a high degree of statistical significance, is that the effect might be exaggerated. In other words, the result might be a fluke at one end of the probability curve. It might be improbable but still achieved through random chance. If that *does* happen, the effect size is likely to be larger in the sample than it is in the underlying population.

There is an argument, which has some merit, that power calculations are therefore particularly important for small studies. If the power is very low, there may be a high overlap between the probability distributions of the null and alternate hypotheses. Therefore, if a power calculation is undertaken in advance of a small study, it decreases the chance of an accidentally large effect creating statistical significance. This is a sound reason for conducting a power analysis in advance, regardless of the size of the study (which is just a sensible idea, anyway). It is not, however, a compelling reason to dismiss small studies just because they don't have a published power calculation (which many published papers do not). This is because, ultimately, a power calculation is still just a reflection of the size of the effect anticipated by the researchers, relative to the size of the sample and the amount of variation. Researchers in most situations would not put effort into doing a study at all if they thought the effect size would be too small to show in the sample they were using.

What this actually means is, for small studies, care needs to be taken if the size of the effect seems unrealistic; which is ultimately a judgement call. Subsequent research is likely to confirm or throw into doubt whether that is the case.

All of this illustrates an important point that is often missed, when people with little understanding of statistics comment on the quality of scientific evidence for something. It's another area where relying on intuition in terms of probability can lead people astray. Intuitively you may think that a larger study must be a better test than a smaller one.

This is certainly true in some situations (like looking for rare adverse events) but testing is usually expensive, and researchers try to limit the amount they do, while still being able to test for statistical significance. You would hope for and expect a low-level of adverse events. In other words, there will probably be a low effect. But the effects that are there might be extremely serious for a few people; you definitely want to know about them. That means the size of the sample should be large. A study that fails to find adverse effects could well be under-powered.

The reality is that well-designed studies are difficult and expensive to conduct, and researchers don't go generally out of their way to include more subjects unless they have a reason to. An issue related to this is that larger studies are likely to have more funding behind them, and if that funding comes from a vested interest looking to profit as a result of the research (like a company selling a medical solution) then other biases may come into play. There is no shortage of research that indicating that funding can be a source of research bias[75].

If you see a large study—large or small, but especially if it is unusually large—which claims something, there are three questions to ask:

    1) How large was the effect size?

    2) How great was the variability in the results? and

    3) Who was paying for this?

Type I errors and confidence intervals

Much of the early work on power analysis was done by **Jerzy Neyman** (1894-1981) (Figure 285), a Polish[76] statistician who came to London work with Karl Pearson's son **Egon Pearson** (1895–1980).

Egon was a great statistician in his own right (Figure 286). Karl's energy was such that no single person was able to do his job, and when he retired in 1933, his job was split between Ronald Fisher (whom we will meet shortly) and Egon. Egon pursued ways to use existing probability distributions to gain more information about underlying populations, and to make the use of statistics more practical. Egon was a team player, and reached out to a number of people to help him formulate his ideas more fully: most notably Jerzy Neyman.

Figure 285
Jerzy Neyman (1894-1981)

Figure 286
Egon Pearson (1895–1980).

As well as thinking about power analysis to reduce type II errors, Neyman looked at type I errors and found a way to provide a richer understanding from the available data.

Figure 287 represents a sampling distribution of sample means. Its mean will be at the apex of the curve, and the mean of any given sample will occupy a position on the curve. (Remember that the mean of the sampling distribution is equal to the mean of the population, μ, and that the sampling distribution will be normal, even though the distribution of the underlying population might have quite a different shape.) We set the

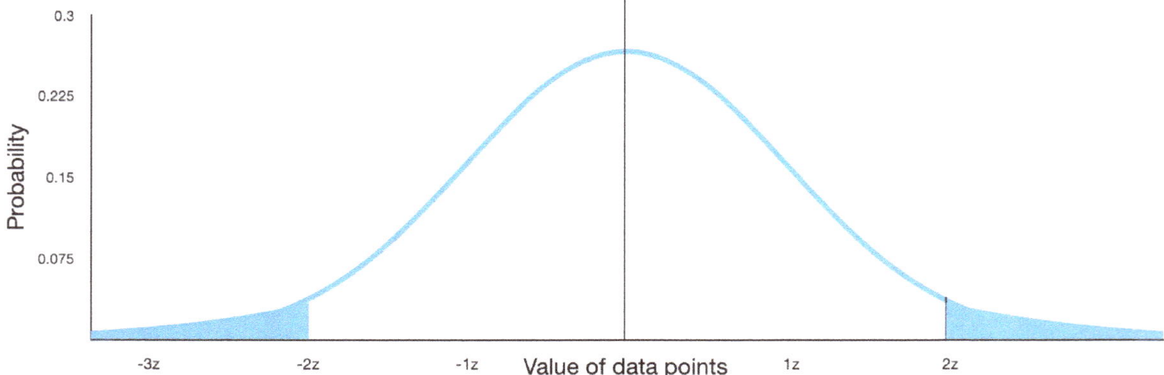

Figure 287

areas on the edges of the curve, at roughly two or more standard deviations from the mean, as being those areas where a result—usually the mean of a sample—is unlikely to have been achieved by random chance. The lowest value of this is typically set at a point where 95% of means of samples of size n will be within the white area, so that results within the blue area are statistically significant; unlikely to have been achieved by random chance. The point selected (arbitrarily but commonly) is actually 1.96 standard deviations on either side of the mean, to get 95% of probable means within the white area. This is a P value of 0.05. Another commonly used P value is 0.01, meaning that there is a 1% chance that the result was achieved by random chance. This actually occurs at 2.576 standard deviations from the mean in a normal distribution. More extreme P values can be used, giving even less probability that the result was achieved by random chance.

Remember when we looked at errors that a type I error—one that involved thinking a result was significant when it wasn't—was allocated a probability of α (alpha). On our curve at Figure 287 the blue areas at either end represent the percentage of the curve that involves this type I error, if there really wasn't a significant effect.

Neyman started with this thinking but looked at it differently. "Suppose", he thought, "instead of asking how far the mean of a given sample probably is from the mean of the population, we ask how far the mean of the population probably is from the mean of the sample?" It is the same question, yet a profoundly different way of thinking about it, that yields a richer result. If we think about it this way, it is the mean of a given sample that will be at the apex of a normal distribution curve (because sampling distributions are normally distributed regardless of the distribution of the population), and the population mean will be somewhere on the curve. The population mean was one of the things that he and Egon Pearson were looking for.

His starting point was to standardise the curve, by thinking of it in terms of z scores. Recall that when Karl Pearson developed the z score for a normal distribution, he came up with this formula:

$$\frac{x - \mu}{\sigma} = z$$

In Neyman's thinking about the sampling distribution curve, the mean of his sample ($\bar{x}$) was treated as conceptually fixed, and the mean of the population, which is really a population parameter, was treated as the variable. The standard deviation for a sampling distribution is, as we have seen (page 625), the population standard deviation divided by the square root of the sample size. (It is also known as the standard error of the mean). So that made his formula:

$$z = \frac{\mu - \bar{x}}{\frac{\sigma}{\sqrt{n}}}$$

where μ is the mean of the population, $\bar{x}$ is the mean of the sample, σ is the standard deviation of the population and n is the size of the sample.

Subtracting $\bar{x}$ from μ and not the other way around meant, of course, that if μ was greater than $\bar{x}$ then z would be positive, and vice versa.

Now suppose, for example, he chose an alpha (α) of 0.05. This meant he could say that 95% of all possible population means would fall within ± 1.96 standard errors of the sample mean.

This expresses mathematically what we said above: there is a 95% probability that the population mean, expressed as a z score, will be between -1.96 and 1.96 .

$$Pr(-1.96 \leq z \leq 1.96) = 0.95$$

Express this using the definition of the z score.

$$Pr\left(-1.96 \leq \frac{\mu - \bar{x}}{\frac{\sigma}{\sqrt{n}}} \leq 1.96\right) = 0.95$$

Multiply the three parts of the inequality by $\frac{\sigma}{\sqrt{n}}$ which will be positive, so doesn't change the inequality.

$$Pr\left(-1.96\frac{\sigma}{\sqrt{n}} \leq \mu - \bar{x} \leq 1.96\frac{\sigma}{\sqrt{n}}\right) = 0.95$$

Add $\bar{x}$ to each part which, again, doesn't change the inequality.

$$Pr\left(\bar{x} - 1.96\frac{\sigma}{\sqrt{n}} \leq \mu \leq \bar{x} + 1.96\frac{\sigma}{\sqrt{n}}\right) = 0.95$$

We write this as the 95% confidence interval of the population mean, and it is written as:

$$95\% \text{ CI of } \mu = \bar{x} \pm 1.96\frac{\sigma}{\sqrt{n}}$$

Likewise

$$99\% \text{ CI of } \mu = \bar{x} \pm 2.576\frac{\sigma}{\sqrt{n}}$$

This means, as you would expect, that the more rigorous you want to be, the harder it is to be precise in your confidence of the result.

So with some simple algebra and a profound change in perspective (inspired by Egon Pearson), Neyman was able to give an indication of a range of values (called an **interval estimate**, as opposed to a P value which is called a **point estimate**) that the population mean would be likely to lie within.

To generalise this, it is normally written as:

$$\text{CI of } \mu = \bar{x} \pm z_{\frac{\alpha}{2}} \frac{\sigma}{\sqrt{n}}$$

The $z_{\frac{\alpha}{2}}$ or $z_{\alpha/2}$ is commonly-accepted shorthand code. In Figure 287 the white area represents the probability of $1 - \alpha$; the probability of *not* having a type I error. $z_{\frac{\alpha}{2}}$ represents the z score corresponding to half of this area because you only move in one direction from the mean at 0 to calculate it; in other words, the 1.96 or 2.576 z values we used above. (So it should probably be written as $z_{\frac{1-\alpha}{2}}$ to increase the clarity, but it's harder to write that way.)

You can see from this formula that, as you would expect, the smaller the population standard deviation, and the larger the sample size, the narrower the confidence interval.

We still have the problem, of course, that this calculation assumes we don't know the mean of the population, and yet we do know its standard deviation. That might happen occasionally, but not very often. We'll soon see how this was dealt with. (It wasn't a problem to Neyman and Egon Pearson, because they actually did their work a lot later than "Student", whom we will meet shortly. We've jumped forward in time because it makes more sense to understand it this way, but soon we'll jump back again.)

Confidence intervals started with thinking about normally-distributed sampling distributions, but over the years developed a wider range of applications. In general it is seen as best practice, in studies, to quote both P values and confidence intervals.

Bayesian critique

Not everyone has been happy with Neyman's achievements. At this point it is important to recognise that statistical ideas had started to drift away from the ideas espoused by Pierre-Simon Laplace, that we would now call **Bayesian statistics**. As we saw earlier, a Bayesian approach today would say that probability is subjective, and that what we are doing in statistics is taking our prior beliefs about a probability, gaining some evidence about it, then using that to create a new (posterior) probability distribution that we (subjectively) believe to be a more appropriate representation. The whole approach involves creating, rather than finding, a distribution, and using likelihood calculations to find the new population parameters, which are essentially treated as being variables.

So a Bayesian approach talks more about **credible intervals** than confidence intervals, and while some confidence interval calculations can also be used as credible intervals, the concept of a credible interval is broader. (In an academic journal you will, of course, see confidence intervals and credible intervals both abbreviated to "CI".)

Power calculations are less relevant with a Bayesian approach. Without P values, the danger of "P hunting" isn't the same, and there is no major problem with just incrementally adding more subjects and using them to continuously refine the model.

Those who strongly support a Bayesian approach often suggest that Jerzy Neyman was the main person who brought about the (now dominant) frequentist approach, largely through his approach to understanding underlying populations. Neyman himself rejected this, believing that purported differences between his approach and those of Bayesians were largely semantic[77]. The debate goes on but, as we saw when we briefly looked at Bayesian statistics, the differences between Bayesian and frequentist approaches are probably changing from fixed positions to decisions on a case-by-case basis, about which approach is more appropriate.

Key points:

- A **type I error**, whose probability is often indicated by the Greek letter alpha (α), would mean that we reject the null hypothesis (H_0) and accept the alternate hypothesis (H_A) when we shouldn't.

- A **type II error**, whose probability is often represented by the Greek letter beta (β), means we retain the null hypothesis, but we should reject it.

- The **power** of a test relates to the probability of not having a type II error; not missing an effect if it is there in the data. It is usually safer for researchers to run scenarios of their data and expected results, to ensure that they have a large-enough sample given the variables involved.

- Paradoxically, small studies with statistically significant results can be more important than large ones, because achieving statistica significance with a small sample size normally requires a pronounced effect and low variability in the results. (There is also the issue that larger studies will have required more funding, and the source of funding has been repeatedly shown to potentially bias research results.)

- However, small studies that still show statistical significance may exaggerate effect sizes.

- P values provide a point estimate, while confidence intervals provide an interval estimate.

- 95% CI of $\mu = \bar{x} \pm 1.96 \frac{\sigma}{\sqrt{n}}$ and 99% CI of $\mu = \bar{x} \pm 2.576 \frac{\sigma}{\sqrt{n}}$

- A confidence interval doesn't work well with a Bayesian approach. Bayesians tend to talk about credible intervals, while still using the same abbreviation (CI). The concept of a credible interval is broader than a confidence interval.

"Student" and the *t* distribution (fourth extension of the normal)

Read this if you are curious about how statisticians started to solve the problem of sampling without knowing the population standard deviation; or you want to know where *t* distributions come from.

At this point we know a lot about sampling, but not, unfortunately, enough to be practically useful in most situations. We've seen that if we take a sample from a population, then the collection of all possible samples of the same size will have its own probability distribution. It will be normally distributed, its mean will be the same as the mean of the population and its standard deviation will be the same as the standard deviation of the underlying population divided by the square root of the size of the sample:

$$SE(\bar{x}) = \frac{\sigma}{\sqrt{n}}$$

We also saw that the probability that the mean of any sample is further from the mean of the underlying population than any given constant c is less than or equal to

$$\left(\frac{\sigma}{\sqrt{n}c}\right)^2$$

That information is enough to tell us the probability that the mean of our sample will be within a certain number of standard deviations (of the population), of the mean (of the population), which is a useful thing to know if we want to know how confident we can be that our sample is telling us anything meaningful.

The usefulness of this is limited, as we have seen, by the fact that we take samples precisely because we don't know as much as we would like to about the population, and all of that information requires us to know the population's standard deviation. This is particularly so if we take a small, single sample. One way to address this is to take a large sample and assume that the standard deviation of the sample will represent the standard deviation of the population, but that isn't a great assumption to work with; particularly if you use a smaller sample.

Figure 288
William Sealy Gosset
1876–1937

This is a problem that a student of Pearson's, working closely with him, set out to solve in 1908. The paper he wrote on this, called "The Probable Error of a Mean" was signed "Student"[78]. We now know that "Student" was William Sealy Gosset (1876–1937), and that he was a master brewer for the Guinness brewing company. He didn't use his own name because the Guinness company wouldn't let him at the time. (They are happy to acknowledge him now.) Other employees had published and given away trade secrets; the Guinness company understandably didn't want that to keep happening. The competitive advantage open to Guinness through the use of statistics was huge (and one many corporations could still learn from), and they didn't want to give their secret away. Gosset's papers don't contain his name, and they don't give any brewing-related examples. There was therefore no danger that competitors would twig to the way that Guinness was using statistical analysis to improve their business.

At that time the annual production of Guinness was approaching an incredible one billion pints. It was important to maintain and improve quality, and testing every batch wasn't feasible. Gosset, who had studied mathematics as well as chemistry at Oxford before joining Guinness, recognised that he needed a mathematical solution to find out how to make testing as efficient as possible, and went to London to study with Pearson, in order to work out how best to go about it.

Gosset is now famous as the developer of the t distribution and t tests, which are among the most widely-used statistical tools. They weren't exactly what he worked on; but his thinking took us most of the way there. What is now used as a t distribution and t

While Guiness used statistics in secret from about 1906, the use of statistics for process improvement was probably first introduced *publicly* by the US Department of Agriculture in the 1920s, about 20 years after Gosset was working on this problem. One of the people who was part of that was J Edwards Deming (1900-1993). Deming became famous as a leading figure behind the turnaround of the quality of Japanese manufacturing in the 1950s and 60s, through his management ideas and through teaching about the use of statistics to improve industrial processes. His ideas, in turn, went on to inform the Total Quality Management (TQM) and Six Sigma approaches.

tests were developed by Ronald Fisher, who we'll meet soon. But Fisher gave full credit to Gosset for his part, so we'll start with him.

On the probable error of a mean, and the variance of a sample

In working out the probable error of a mean (the distance of the mean of a sample from the mean of a population), Gosset's thinking started in a similar way to this. What follows isn't exactly the way that he addressed the problem, because the calculations in his paper are hard for modern readers to follow—they assume a lot of prior understanding which isn't explained in the paper—but this is the gist of where he was going with it. He wanted to understand the relationship of the variance of a sample to the variance of the underlying population. (We have looked at the variance of a sampling distribution in relation to the variance of the population, but we haven't looked at the variance of the data points in any given sample yet.)

We'll use the same symbols we've already seen, for this proof:

Set	Mean	Variance	Standard deviation
Whole population:	μ	σ^2	σ
Any potential sample of size n	$\bar{x}$	s^2	s
Sampling distribution of the means of potential samples (all potential samples of size n):	$E(\bar{x})$	$s_{\bar{x}}^2$	$s_{\bar{x}}$

Apply the formula for variance that we have been using so far to the variance of a given sample of size n.

$$s^2 = \frac{\sum_{i=1}^{n}(x_i - \bar{x})^2}{n}$$

Take the expected value of both sides.

$$E(s^2) = E\left(\frac{\sum_{i=1}^{n}(x_i - \bar{x})^2}{n}\right)$$

Take the constant $\frac{1}{n}$ outside of the expected value calculation (by linearity).

$$= \frac{1}{n}E\left(\sum_{i=1}^{n}(x_i - \bar{x})^2\right)$$

Expand out the binomial.

$$= \frac{1}{n}E\left(\sum_{i=1}^{n}(x_i^2 - 2x_i\bar{x} + \bar{x}^2)\right)$$

Show each term as its own summation (by linearity). In the middle term the -2 and the $\bar{x}$ are both constants and can be taken outside the summation. $\bar{x}^2$ is also a constant, so summing it n times is the same as multiplying it by n

$$= \frac{1}{n}\, E\left(\sum_{i=1}^{n} x_i^2 - 2\bar{x}\sum_{i=1}^{n} x_i + n\bar{x}^2\right)$$

Multiply the middle term by $\frac{n}{n}$

$$= \frac{1}{n}\, E\left(\sum_{i=1}^{n} x_i^2 - 2n\bar{x}\frac{\sum_{i=1}^{n} x_i}{n} + n\bar{x}^2\right)$$

$\frac{\sum_{i=1}^{n} x_i}{n} = \bar{x}$ so in the middle term we are multiplying $\bar{x}$ by $\bar{x}$

$$= \frac{1}{n}\, E\left(\left(\sum_{i=1}^{n} x_i^2\right) - 2n\bar{x}^2 + n\bar{x}^2\right)$$

$-2n\bar{x}^2 + n\bar{x}^2 = -n\bar{x}^2$

$$= \frac{1}{n}\, E\left(\left(\sum_{i=1}^{n} x_i^2\right) - n\bar{x}^2\right)$$

Treat this as the sum of two expected values (by linearity). In the first term take the expected value inside the summation (again by linearity), and in the second term take the constant n outside of the expected value (yet again by linearity).

$$= \frac{1}{n}\left[\left(\sum_{i=1}^{n} E(x_i^2)\right) - nE(\bar{x}^2)\right]$$

Now the variance of x_i is equal to $E(x_i^2) - (E(x_i))^2$ which we saw earlier. We haven't specified whether x_i is from a sample or from the whole population; it is identically distributed, so it doesn't matter. That means that the variance of x_i is the variance of the population: σ^2. In the same way $E(x_i) = \mu$, the mean of the whole population. This tells us that $\sigma^2 = E(x_i^2) - \mu^2$, and therefore (adding μ to both sides, that $E(x_i^2) = \sigma^2 + \mu^2$

Substitute $\sigma^2 + \mu^2$ for $E(x_i^2)$ in our equation.

$$E(s^2) = \frac{1}{n}\left[\left(\sum_{i=1}^{n}(\sigma^2 + \mu^2)\right) - nE(\bar{x}^2)\right]$$

The variance of $\bar{x}$ is equal to $E(\bar{x}^2) - (E(\bar{x}))^2$. We saw earlier that $E(\bar{x}) = \mu$. That is, the variance expected value of the means of potential samples is equal to the expected value of the population as a whole. We also saw earlier that the variance of $\bar{x} = s_{\bar{x}}^2 = \frac{\sigma^2}{n}$. The variance of the means of potential samples of size n is the variance of the population divided by n. That means $\frac{\sigma^2}{n} = E(\bar{x}^2) - \left(E(\bar{x})\right)^2 = E(\bar{x}^2) - \mu^2$. We can therefore say that $E(\bar{x}^2) = \frac{\sigma^2}{n} + \mu^2$ (adding μ^2 to both sides).

Substitute $\frac{\sigma^2}{n} + \mu^2$ for $E(\bar{x}^2)$

$$E(s^2) = \frac{1}{n}\left[\left(\sum_{i=1}^{n}(\sigma^2 + \mu^2)\right) - n\left(\frac{\sigma^2}{n} + \mu^2\right)\right]$$

σ^2 and μ^2 are both constants, and so summing them n times is the same as multiplying them by n.

$$= \frac{1}{n}\left[n\sigma^2 + n\mu^2 - \sigma^2 - n\mu^2\right]$$

$+n\mu^2 - n\mu^2 = 0$

$$= \frac{1}{n}\left[n\sigma^2 - \sigma^2\right]$$

Factor out σ^2

$$\boxed{E(s^2) = \frac{\sigma^2}{n}(n-1)}$$

That is, the expected value of the variance of a sample is the variance of the population, divided by the sample size and multiplied by the sample size minus one.

We'll leave Gosset at this point, because he took this in one direction, but Fisher, in an inspired piece of thinking, took it in a different one, and that's the one we follow today.

Key point from this section:

$$E(s^2) = \frac{\sigma^2}{n}(n-1)$$

Depth:
Fisher's Refinements

Figure 289
Ronald Fisher
1890-1962

Ronald Fisher (1890-1962) warrants his own major heading, partly because his contributions to statistics have been so widespread and important, and partly because the major heading was about the achievements of Karl Pearson and his friends, and including Fisher as a "friend" of Karl Pearson would be a stretch. "Mutual annoyance and grudging respect" might be a better description of their relationship. Fisher was another eugenicist who took over part of Pearson's job at the University College, London, when Pearson retired in 1933. (His title was Francis Galton Professor of National Eugenics.) Happily, the idea of eugenics was losing its popularity by that time (except, tragically, in Germany), and he ended up being the only staff member in the Department of Eugenics. The fact that his relationship with Pearson was probably closer to mutual annoyance than outright enmity is illustrated by the fact that Pearson supported Fisher as his successor, believing he was clearly the best person for the job.

Fisher disagreed with people. A lot. He argued with Karl Pearson over a range of issues. After saying that Pearson's greatest achievement was the chi square distribution (thus also damning him with faint praise, and in another side-swipe, saying that Pearson was following on from someone else when he developed it) he went on to claim there were errors in his thinking about it[79]. He also said that William Sealy Gosset developed his work independently of Pearson, whereas Gosset concluded his paper by saying

> *"I should like to express my thanks to Professor Karl Pearson, without whose constant advice and criticism this paper could not have been written."* [80]

so his development was hardly independent. (In fact, what Gosset came up with was extremely close to something Pearson had already published, as his Type 4 distribution.) Fisher also argued with Gosset over the importance of randomisation in studies (Gosset wasn't a fan, Fisher was). Fisher's office was upstairs from Egon Pearson's who, as we saw earlier, took over the other part of Karl Pearson's role. By all accounts Egon Pearson was likeable, easy to get on with and a great collaborator, but Fisher still managed to fight with him.

Fisher did, though, have three redeeming qualities. First, (although possibly most annoyingly for Karl Pearson) Fisher was usually right when he criticised those who had gone before him. He just thought about things more deeply. (He did make his own little mistakes, for which he has largely escaped criticism[81]. And the world has ended up following Egon Pearson's approaches rather than Fishers, in most areas where they disagreed.) Fisher made small but vital changes to the work of those who had gone before him and developed many of the forms in use today.

Second, while he was somewhat disparaging of Karl Pearson's work, he did give credit to Gosset for *t* tests, even though their final form belonged to Fisher, so it can't be said that he was completely driven by self-aggrandizement.

Third, he went on to develop his own important and original contributions to the field. Like Karl Pearson, he was a genius, and the world would be a poorer place without his contributions.

Degrees of freedom

> **Read this if** you want to see a piece of truly original and elegant thinking; if you have never been able to work out why you are supposed to subtract 1 from the denominator in standard distribution of a sample; or if you find the whole idea of degrees of freedom confusing.

An unorthodox and elegant solution

Fisher returned to the problem that Gosset had addressed but had a more elegant approach. We saw that, in trying to define the variance of a sample in terms of the variance of the population, Gosset calculated that

$$E(s^2) = \frac{\sigma^2}{n}(n-1)$$

where s^2 is the variance of a sample, sigma squared (σ^2) is the variance of the underlying population, and n is the size of the sample.

At this point Fisher thought about things differently. Thinking back to Gosset's calculations, he realised that if, instead of being n in the denominator there was $n-1$, then it would cancel out in the $n-1$ in the numerator, and we would be able to say that the expected value of the variance of a sample was the variance of the population. The only reason it is n instead of $n-1$ is that we defined the variance of a sample as

$$\frac{\sum_{i=1}^{n}(x_i - \bar{x})^2}{n}$$

in the first place. The n was factored out and stayed untouched throughout our calculations. If, he reasoned, we had simply *defined* the variance of a sample as

$$s^2 = \frac{\sum_{i=1}^{n}(x_i - \bar{x})^2}{n-1}$$

instead of over n, then an elegant solution would present itself. Here are the calculations again, done Fisher's way:

Define the variance of a sample in Fisher's way.

$$E(s^2) = E\left(\frac{\sum_{i=1}^{n}(x_i - \bar{x})^2}{n - 1}\right)$$

Take the constant $\frac{1}{n-1}$ outside of the expected value calculation (by linearity).

$$= \frac{1}{n - 1}\, E\left(\sum_{i=1}^{n}(x_i - \bar{x})^2\right)$$

Follow exactly the same steps that Gosset used to get to this point, except that $\frac{1}{n-1}$ is outside the brackets now, rather than $\frac{1}{n}$

$$= \frac{1}{n - 1}\, [n\sigma^2 - \sigma^2]$$

Factor out σ^2. The $n - 1$ in the numerator and denominator then cancel out.

$$E(s^2) = \frac{\sigma^2}{n - 1}\, (n - 1)$$

$$\boxed{E(s^2) = \sigma^2}$$

Defining sample variance in this way, the expected value of the variance of a sample is the variance of the population. So that's what Fisher did. That's what we continue to do. We define the variance and the standard deviation of a *population* in the way we have up to this point. However when we are talking about the variance and standard deviation of a *sample*, and when (this bit's important) it is calculated using the mean *of the sample*, not the mean of the population, then we calculate it with $n - 1$ in the denominator.

$$\boxed{s^2 = \frac{\sum_{i=1}^{n}(x_i - \bar{x})^2}{n - 1}}$$

That is, the variance of a sample is the sum of squared differences of each data point from the mean of that sample, divided by one less than the number of data points.

From there we can take the square root of both sides, to give a new definition of the standard deviation of a sample:

$$\boxed{s = \sqrt{\frac{\sum_{i=1}^{n}(x_i - \bar{x})^2}{n - 1}}}$$

This doesn't mean that the variance of any sample (calculated with $n - 1$) will equal the variance of the population, of course. It means that, on average, it will. In statistical terms this makes it what is called an **unbiased estimator** of the population standard deviation, and it can be used (with care) in probability calculations.

What all that means is that we can now pick a sample, any sample and, re-markably, we can estimate the probabilities that its mean represents the mean of the population. Its standard error will be given by:

$$SE(\bar{x}) = \frac{s}{\sqrt{n}}$$

where $SE(\bar{x})$ is the standard error (the extent to which the sample mean will be close to the mean of the population), s is the standard deviation of the sample (calculated on the $n-1$ basis) and n is the size of the sample.

If we subtract one from the denominator of the variance of a sample, it will make the value of the fraction larger. This makes intuitive sense, because we have less precision in using a sample to guess at the variance and stand-ard deviation of the whole population, and so it is appropriate to calculate using a larger variance, which in turn will give us a larger standard error. The effect of this becomes more pronounced with smaller sample sizes. As n gets larger, subtracting 1 from it makes less relative difference. This also makes sense; more allowance for variance will have to be made for smaller samples.

We'll return to Fisher's approach to Gosset's work in a little while, but now let's follow Fisher down a different path, which started with his change to the definitions of sample variance and standard deviation.

Expanding the concept

In pondering precisely *why* it makes sense to subtract one from the denom-inator of the variance and standard deviation of a sample (apart from the fact that the mathematics makes it more elegant that way) Fisher came up with the idea of **degrees of freedom.**

Imagine a statistics lecturer tells the class that 84% of people passed a test. That's enough information to know that 16% of people didn't. Explaining to the class about the 16% would be redundant information. So if the lecturer explains that he has two pieces of information to give the class, one of those pieces is not actually new information at all. What Fisher recognised (con-ceptually, and with mathematical support) is that if you include redundant information, particularly in terms of the number of distinct pieces of information you think you are dealing with, then you bias the estimations you are making.

If you are calculating the variance of a sample, and you already know the mean of that sample, then all the available information can be calculated from one less than the size of the sample. Suppose we have a list of six numbers: $1, 7, 8, 11, 12, x$. Under these circumstances x could be anything at all. Once we know that this group of numbers has a mean of 7, however, x can only be equal to 3. It doesn't have the freedom to be anything else:

$$\frac{1 + 7 + 8 + 11 + 12 + 3}{6} = 7$$

We saw earlier (page 494) that the sum of differences from the mean will always add to zero. That means that if we know the values of all but one of the items and we know the mean, we can calculate the exact value of the last one. The value of the last one is fixed, and knowing it adds no value to our calculation.

So Fisher said that the standard deviation and the variance of a *population* (normally indicated by sigma (σ) and sigma squared (σ^2)) have n degrees of freedom, but the standard deviation and the variance of a *sample* (normally indicated by s and s^2) have $n - 1$ degrees of freedom.

This line of reasoning in turn led Fisher to think of the importance of understanding the degrees of freedom in the t test he was working on (modifying Gosset's work) and also on the chi square distribution and regression.

Imagine you're comparing the mean test scores of two groups. If Group A has 10 members and Group B has 10 members, the total sample size is 20. However, when calculating degrees of freedom for comparison of means it's not 20: it's $(10 - 1 + 10 - 1) = 18$.

	Group 1	Group 2	Total
Choice A	35%	20%	55%
Choice B	10%	6%	16%
Choice C	25%	4%	29%
Total	70%	30%	100%

Figure 290

This is like saying, "I can change the score of 9 people in each group, and that will determine the score of the 10th person in that group". (In terms of how to go about comparing the means of two groups, we'll look at that when we come to t tests.)

Now suppose you are using a chi-square test to look at 3 choices across 2 groups (Figure 290). Here, the degrees of freedom will be

$$(3 - 1)(2 - 1) = 2$$

(Keep in mind that this assumes you know the totals, for both the categories and the choices. Any 2 from any one group will determine the other choices, if the totals are fixed.)

In terms of **regression** and **correlation,** as you may expect the degrees of freedom are determined by subtracting 2 from the total number of pairs of data; one for each of the data sets, because the last item in each of the two data sets can be determined if you know the total or the average of that set. (If that isn't clear you may want to revise the material from page 571.)

Key points:

- By redefining the variance and standard deviation of a sample (basing them on the degrees of freedom), Fisher was able to take Gosset's work and prove that the **expected value of the variance** of a sample is the **variance of the population.**

- The standard deviation and the variance of a population have n degrees of freedom, but the standard deviation and the variance of a sample have $n - 1$ degrees of freedom.

- $SE(\bar{x}) = \frac{s}{\sqrt{n}}$

- In regression and correlation the degrees of freedom are determined by subtracting 2 from the total number of pairs of data; one for each of the data sets.

Fisher and the *t* distribution

> **Only read the green parts of this section if** you are really feeling brave. The mathematics is, frankly, challenging. Otherwise let your eyes glaze over the green-shaded detailed workings. It's here if you ever need it, but the remainder of the material should give you the idea of where the *t* distribution comes from.

Probability density function of the *t* distribution

We're now ready to explore the distribution Gosset and Fisher were looking for. As a reminder of where we are up to, they wanted to determine if the observed difference between a sample statistic (like a sample mean) and the same statistic under a null hypothesis is statistically significant, given the sample's standard deviation, without knowing the population's standard deviation.

The next step was to assume that the underlying population was normally distributed and to see where that led.

We can start with a normally distributed random variable:

$$X \sim N(\mu, \sigma^2)(\text{i.i.d.})$$

As you know the mean of any given sample of size *n* is:

$$\bar{x} = \frac{\sum_{i=1}^{n} x_i}{n}$$

The standard deviation of any given sample of size *n* is:

$$s = \sqrt{\frac{\sum_{i=1}^{n}(x_i - \bar{x})^2}{n-1}}$$

The standard error of the mean—the standard deviation of the mean within a distribution of means of possible samples—is:

$$SE(\bar{x}) = \frac{\sigma}{\sqrt{n}}$$

where $SE(\bar{x})$ is the standard error of the mean within a sampling distribution, σ is the population standard deviation and n is the sample size.

But Fisher had established that the expected value of a sample standard deviation (defined his way) was the same as the population standard deviation. When dealing with the standard error of the mean we are only dealing with averaged values anyway, which means he could say:

$$SE(\bar{x}) = \frac{s}{\sqrt{n}}$$

where s is the sample standard deviation, calculated with a denominator of $n - 1$.

That was enough to give him the formula for a standardised random variable, which he called T that followed the probability distribution of possible means of samples (sampling distribution of the sample mean), and didn't rely on knowing the standard deviation of the underlying population:

$$T = \frac{\bar{x} - \mu}{SE(\bar{x})}$$

$$= \frac{\bar{x} - \mu}{\frac{s}{\sqrt{n}}}$$

In other words, it's just like any other standardised random variable: we take the difference from the mean and divide by the standard deviation. (As we have seen the population mean is the same as the mean of the sampling distribution of sample means. Which is a confusing sentence but should be meaningful if you've been following a long to this point; although you may need to read the sentence a couple of times to get it.)

Fisher did have a problem, though. He wanted to end up with a distribution that had one random variable, but at this stage he had two to deal with: the mean of the sample, and the standard deviation of the sample. These are independent from one another, so he wasn't going to get anywhere by expressing one in terms of the other. What he *was* able to do, following Gosset, was to think this though in a situation where the *underlying population*, and not just the sampling distribution, was normally distributed.

Multiply the numerator and denominator by the population standard deviation.

$$T = \frac{\bar{x} - \mu}{\frac{s}{\sqrt{n}}} \left(\frac{\sigma}{\sigma} \right)$$

Exchange the positions of the population and sample standard deviations in the denominator (which doesn't change anything, just represents it differently).

$$= \frac{\bar{x} - \mu}{\frac{\sigma}{\sqrt{n}}} \left(\frac{\sigma}{s} \right)$$

Note that

$$\frac{\bar{x} - \mu}{\frac{\sigma}{\sqrt{n}}}$$

has a standard normal distribution where the underlying population has a normal distribution. This isn't the central limit theorem; the n does not need to approach infinity. The proof uses moment generating functions (MGFs) and is a bit involved. This current proof is quite long enough already, so it's in an appendix on page 766 covering this proof if you'd like to see it.

Divide the numerator and denominator of $\frac{\sigma}{s}$ by σ

$$T = \frac{\bar{x} - \mu}{\frac{\sigma}{\sqrt{n}}} \left(\frac{\frac{1}{s}}{\frac{s}{\sigma}} \right)$$

Let $\frac{\bar{x} - \mu}{\frac{\sigma}{\sqrt{n}}} = W$, where W is a random variable with a standard normal distribution.

Square the part of the expression on the right, and then take the positive square root, so that nothing changes.

$$= W \sqrt{\left(\frac{\frac{1}{s}}{\frac{s}{\sigma}} \right)^2}$$

$$= W \sqrt{\frac{1}{\frac{s^2}{\sigma^2}}}$$

Multiply the numerator and denominator under the square root sign by $n - 1$.

$$= W \sqrt{\frac{n - 1}{(n - 1)\frac{s^2}{\sigma^2}}}$$

Let $(n - 1)\frac{s^2}{\sigma^2} = V$. V has as a chi square probability distribution. That also has its own proof, which isn't as long as the one for W, but is long enough to be moved to another appendix (page 773).

$$T = W \sqrt{\frac{n - 1}{V}}$$

Divide the numerator and denominator under the square root sign by $n-1$

$$= W\sqrt{\dfrac{1}{\dfrac{V}{(n-1)}}}$$

$$T = \dfrac{W}{\sqrt{\dfrac{V}{(n-1)}}}$$

where W has a standard normal distribution, V has a chi square distribution and n is the sample size.

That means that we can express T in terms of variables that have distributions that we already understand. (We still have work to do before arriving at the probability density function (PDF), but this is a helpful start).

This suggests, at least conceptually, that what we will be looking at with a T distribution is a standard normal distribution with a chi square distribution factored in. In other words, it's a bit like a normal distribution with a margin of error factored in, and that is close to how it tends to be used in practice.

Multiply the numerator and denominator by $\sqrt{(n-1)}$. Note that this is a constant.

$$T = \sqrt{(n-1)}\,\dfrac{W}{\sqrt{V}}$$

We've simplified this quite a bit, but now we're going to simplify it further. The $n-1$ represents the degrees of freedom, so we'll just use the letter k for this. We saw that V has a chi square distribution. It's worth noting at this point that it has k degrees of freedom. The proof that this particular chi square distribution has that many degrees of freedom is quite complex, and involves moment generating functions. There is nothing in the mathematics of it that we haven't covered, but you probably don't need the proof for our purposes. If you want it, you can find a reference to it in the end note[82]. We'll also let $\sqrt{V} = M$. Finally, we'll let $\dfrac{T}{\sqrt{k}} = S$

$$S = \dfrac{W}{M}$$

For convenience in this proof we'll use the parameterization of the gamma function which has the chi squared distribution of k degrees of freedom as $\chi_k^2 = \Gamma\left(\dfrac{k}{2}, \dfrac{1}{2}\right)$. That means the PDF of V is

$$\Pr(V = v) = \frac{\left(\frac{1}{2}\right)^{\frac{k}{2}}}{\Gamma\left(\frac{k}{2}\right)} v^{\frac{k}{2}-1} e^{-\frac{1}{2}v}$$

$$= f(v), for \; v > 0$$

But what we are after is the PDF of M which is the square root of V. That means that $V = M^2$. Thinking about this in terms of the cumulative distribution function (CDF), we can say that $\Pr(M \leq m) = \Pr(V \leq m^2)$.

That makes the CDF of M:

$$Pr(M \leq m) = \int_0^{m^2} \frac{\left(\frac{1}{2}\right)^{\frac{k}{2}}}{\Gamma\left(\frac{k}{2}\right)} v^{\frac{k}{2}-1} e^{-\frac{1}{2}v} \, dv$$

Now to find the PDF, we need to take the derivative of this. When we see an integral sign it is automatic to think that we need to solve the integral, but in this case we don't, because we want the derivative of the integral. If we were calculating it with an upper value of m it would be straightforward; we would just have to remove the integral sign and evaluate it where $v = m$. (Where $v = 0$ the whole expression equals 0, so we can ignore it.) We are, however, differentiating it where $v = m^2$. When we substitute that into the equation we have a function-within-a-function situation. The inner function is m^2, and the outer function is what is inside the integral. That means we need to use the chain rule to differentiate it, multiplying the derivative of the inner function by the derivative of the outer function. The effect of this is that we need to multiply what is inside the integral by the derivative of m^2, which is $2m$.

We also need to replace the v with m^2. Once again, because we are not calculating the integral but are finding its derivative we don't need to worry about the dv on the end.

$$\Pr(M = m) = 2m \frac{\left(\frac{1}{2}\right)^{\frac{k}{2}}}{\Gamma\left(\frac{k}{2}\right)} (m^2)^{\frac{k}{2}-1} e^{-\frac{1}{2}m^2}$$

Distribute the 2 in the exponent of the second m. $\left(\text{i.e. } (m^2)^{\frac{k}{2}-1} = m^{\frac{2k}{2}-2}\right)$

$$= m^1 \frac{2 \left(\frac{1}{2}\right)^{\frac{k}{2}}}{\Gamma\left(\frac{k}{2}\right)} m^{\frac{2k}{2}-2} e^{-\frac{1}{2}m^2}$$

Gather the exponents of m.

$\left(\text{i. e. } m^1 m^{\frac{2k}{2}-2}\right) = m^{k-1}$

$$\text{PDF}(m) = \frac{2\left(\frac{1}{2}\right)^{\frac{k}{2}}}{\Gamma\left(\frac{k}{2}\right)} m^{k-1} e^{-\frac{1}{2}m^2}$$

To take stock of where we are up to, we are trying to find the PDF of S, which we've simplified down to the expression: $S = \frac{W}{M}$. We now have a PDF for M. The PDF of W is also easy: it has a standard normal distribution, so it's $Pr(W = w) = \frac{1}{\sqrt{2\pi}} e^{-\frac{w^2}{2}}$. The problem is that we still have two variables in the equation, and we need to find a way to get it down to one. These two variables are independent: one is based on the distribution of mean values, and the other on standard deviations. On a graph, changing the mean would push the graph to the right or left along the horizontal axis, without changing its shape. The other describes how widely spread the values are, around the mean; but they act independently of one another. We can find their joint distribution by multiplying them together. That doesn't get us where we want to be; we want to end up dividing them. But it will help us to get there. We need to create a joint distribution, then select from it those values that are relevant to what we are looking for.

This is the formula for our joint PDF, multiplying the two independent PDFs. (If you aren't sure why we are multiplying, this is just to create the joint distribution. Selecting what we actually need to from within the joint distribution when we divide W by M will happen a bit later.)

$$\frac{1}{\sqrt{2\pi}} e^{-\frac{w^2}{2}} \frac{2\left(\frac{1}{2}\right)^{\frac{k}{2}}}{\Gamma\left(\frac{k}{2}\right)} m^{k-1} e^{-\frac{1}{2}m^2}$$

Combine the powers of e and factor out $-\frac{1}{2}$ within the exponent.

$\left(\text{i. e.: } e^{-\frac{w^2}{2}} e^{-\frac{1}{2}m^2} = e^{-\frac{1}{2}(m^2+w^2)}\right)$

$$= \frac{1}{\sqrt{2\pi}} \frac{2\left(\frac{1}{2}\right)^{\frac{k}{2}}}{\Gamma\left(\frac{k}{2}\right)} m^{k-1} e^{-\frac{1}{2}(m^2+w^2)}$$

It will be easier to work with the CDF, so we need to perform a double integration.

$$\text{CDF} = \iint \frac{1}{\sqrt{2\pi}} \frac{2\left(\frac{1}{2}\right)^{\frac{k}{2}}}{\Gamma\left(\frac{k}{2}\right)} m^{k-1} e^{-\frac{1}{2}(m^2+w^2)} \, dw \, dm$$

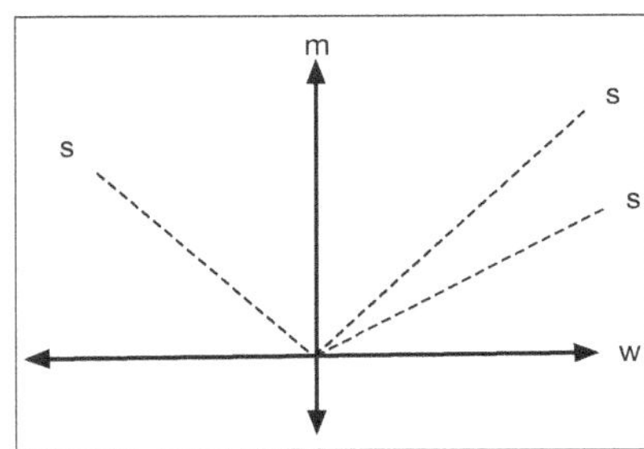

Figure 291

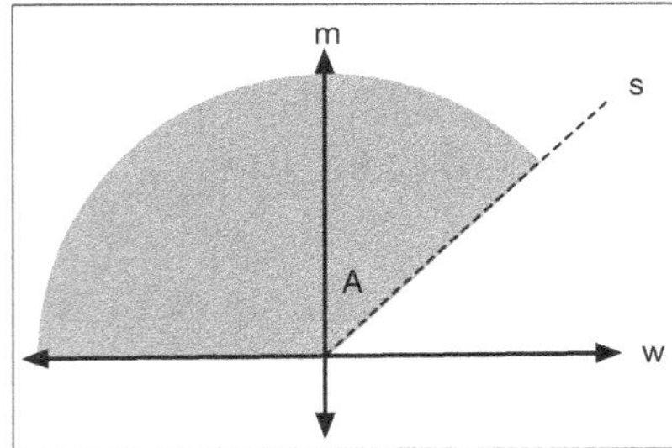

Figure 292

Now let's visualise what we have in a different way. Think of our joint PDF of the variables, m and w, as having its own rectangular plane, with w on the horizontal axis and m on the vertical (Figure 291).

It would be more usual to think of this as a three-dimensional space (given that it is a function made of two variables), but that's hard to draw on a page. We can get away with keeping it to a two-dimensional drawing if we think about each value of s as being represented by a point on the plane, defined by two variables. (We actually did something similar when we looked at regression and correlation: the two random variables we were comparing were identified by points on a plane, with each point representing a pair of variables.)

We know that $s = \dfrac{W}{M}$, which means that $Ms = W$. In other words, this is an equation for a line starting at the origin. Or actually an infinite number of lines, depending on the values of our variables. The M value will always be positive because it is based on the positive square root of a chi square distribution. That means that we aren't interested in anything in the part of the plane below the horizontal axis. The W value will be able to take any value from negative infinity to positive infinity, because it is based on a standard normal distribution. This means that the area of the integration we are interested in is the grey area on Figure 292, with W taking values between negative infinity and the value of the line $Ms = W$.

We are actually integrating over an angle, so it makes sense to switch to polar coordinates. The angle we are interested in, from our diagram, is

$$\frac{\pi}{2} + A$$

This will still work if the angle between the negative horizontal axis and the line s is less than $\frac{\pi}{2}$ or 90 degrees; in that instance A will be negative. To find the value of angle A, when $m = 1$, $w = s$ (because $Ms = W$) (Figure 293). That means that A will be the inverse tan of s (opposite over adjacent). So we are integrating the area from the left-hand horizontal axis, to an angle of $\frac{\pi}{2} + A = \frac{\pi}{2} + \tan^{-1}(s)$.

As is always the case in converting to polar coordinates, the vertical value (in our case m) will be $r \sin \theta$, $m^2 + w^2 = r^2$, and there will be a Jacobian of r multiplying the $drd\theta$.

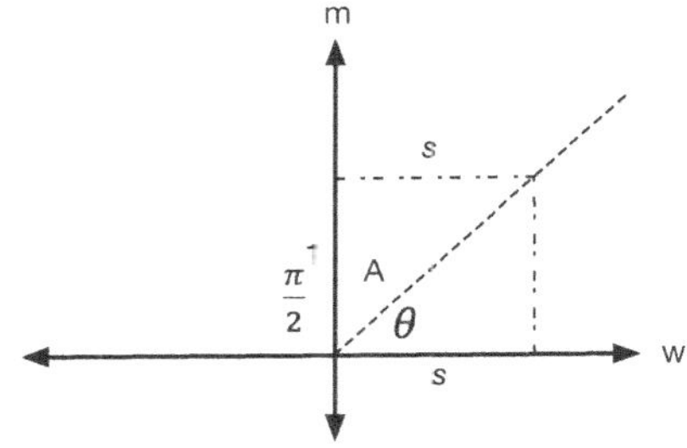

Figure 293

$$CDF = \iint \frac{1}{\sqrt{2\pi}} \frac{2 \left(\frac{1}{2}\right)^{\frac{k}{2}}}{\Gamma\left(\frac{k}{2}\right)} (r \sin \theta)^{k-1} e^{-\frac{1}{2}(r^2)} r\, dr d\theta$$

The limits of integration are, for the outer (θ) integral, from 0 to $\frac{\pi}{2} + \tan^{-1}(s)$, and for the r integral, from 0 to infinity.

$$CDF = \int_0^{\frac{\pi}{2}+\tan^{-1}(s)} \int_0^{\infty} \frac{1}{\sqrt{2\pi}} \frac{2 \left(\frac{1}{2}\right)^{\frac{k}{2}}}{\Gamma\left(\frac{k}{2}\right)} (r \sin \theta)^{k-1} e^{-\frac{1}{2}(r^2)} r\, dr d\theta$$

Take the constants with regard to r outside of the inner integral.

$$= \int_0^{\frac{\pi}{2}+\tan^{-1}(s)} \frac{1}{\sqrt{2\pi}} \frac{2 \left(\frac{1}{2}\right)^{\frac{k}{2}}}{\Gamma\left(\frac{k}{2}\right)} \sin^{k-1}\theta \int_0^{\infty} e^{-\frac{1}{2}(r^2)} r^{k-1} r\, dr d\theta$$

Now we'll do a substitution: Let $u = \frac{1}{2}r^2 \quad \frac{du}{dr} = r$ which means $\delta u = r \delta r$

$$= \int_0^{\frac{\pi}{2}+\tan^{-1}(s)} \frac{1}{\sqrt{2\pi}} \frac{2 \left(\frac{1}{2}\right)^{\frac{k}{2}}}{\Gamma\left(\frac{k}{2}\right)} \sin^{k-1}\theta \int_0^{\infty} r^{k-1} e^{-\frac{1}{2}(r^2)} r\, dr d\theta$$

Therefore also $r = (2u)^{\frac{1}{2}}$. Apply the substitution.

$$= \int_0^{\frac{\pi}{2}+\tan^{-1}(s)} \frac{1}{\sqrt{2\pi}} \frac{2 \left(\frac{1}{2}\right)^{\frac{k}{2}}}{\Gamma\left(\frac{k}{2}\right)} \sin^{k-1}\theta \int_0^{\infty} (2u)^{\frac{k-1}{2}} e^{-u}\, du d\theta$$

Take $2^{\frac{k-1}{2}}$ outside of the inner integral, and

$$= \int_0^{\frac{\pi}{2}+\tan^{-1}(s)} \frac{1}{\sqrt{2\pi}} \frac{2 \left(\frac{1}{2}\right)^{\frac{k}{2}}}{\Gamma\left(\frac{k}{2}\right)} \sin^{k-1}\theta\, 2^{\frac{k-1}{2}} \int_0^{\infty} u^{\frac{k+1}{2}-1} e^{-u}\, du d\theta$$

write $u^{\frac{k-1}{2}}$ as

$u^{\frac{k+1}{2}-1}$ (which is the same thing).

Our inner integral is now a gamma function:

$$= \int_{0}^{\frac{\pi}{2}+\tan^{-1}(s)} \frac{1}{\sqrt{2\pi}} \frac{2\left(\frac{1}{2}\right)^{\frac{k}{2}}}{\Gamma\left(\frac{k}{2}\right)} \sin^{k-1}\theta \; 2^{\frac{k-1}{2}}\Gamma\left(\frac{k+1}{2}\right) d\theta$$

$$\int_{0}^{\infty} u^{\frac{k+1}{2}-1} e^{-u} du$$

$$= \Gamma\left(\frac{k+1}{2}\right)$$

This makes sense, considering the diagram we saw earlier. The distance from the origin on any given line (defined by a given angle) doesn't affect the value of s.

Now we can perform some simplification.

$$= \int_{0}^{\frac{\pi}{2}+\tan^{-1}(s)} \frac{1}{\sqrt{2\pi}} \frac{2\left(\frac{1}{2}\right)^{\frac{k}{2}}}{\Gamma\left(\frac{k}{2}\right)} \sin^{k-1}\theta \; 2^{\frac{k-1}{2}}\Gamma\left(\frac{k+1}{2}\right) d\theta$$

Write $2^{\frac{k-1}{2}}$ as $2^{\frac{k}{2}}\frac{1}{\sqrt{2}}$ (same thing).

$$= \int_{0}^{\frac{\pi}{2}+\tan^{-1}(s)} \frac{1}{\sqrt{2\pi}} \frac{2\left(\frac{1}{2}\right)^{\frac{k}{2}}}{\Gamma\left(\frac{k}{2}\right)} \sin^{k-1}\theta \; 2^{\frac{k}{2}}\frac{1}{\sqrt{2}}\Gamma\left(\frac{k+1}{2}\right) d\theta$$

$\left(\frac{1}{2}\right)^{\frac{k}{2}} 2^{\frac{k}{2}} = 1$ and

$\frac{1}{\sqrt{2}}\frac{1}{\sqrt{2\pi}} = \frac{1}{2}\frac{1}{\sqrt{\pi}}$. The 2 in the denominator cancels with the 2 in the numerator.

$$= \int_{0}^{\frac{\pi}{2}+\tan^{-1}(s)} \frac{1}{2\sqrt{\pi}} \frac{2\left(\frac{1}{2}\right)^{\frac{k}{2}}}{\Gamma\left(\frac{k}{2}\right)} \sin^{k-1}\theta \; 2^{\frac{k}{2}}\frac{1}{\sqrt{2}}\Gamma\left(\frac{k+1}{2}\right) d\theta$$

Take the constants outside of the integration. So that gives us the CDF. Now, to find the PDF, we need the derivative of this integral (which is with respect to θ) with respect to s.

$$CDF = \frac{\Gamma\left(\frac{k+1}{2}\right)}{\sqrt{\pi}\,\Gamma\left(\frac{k}{2}\right)} \int_{0}^{\frac{\pi}{2}+\tan^{-1}(s)} \sin^{k-1}\theta \; d\theta$$

Once again, the constants can stay outside of the derivative, as well as outside the integration.

$$PDF = \frac{\Gamma\left(\frac{k+1}{2}\right)}{\sqrt{\pi}\,\Gamma\left(\frac{k}{2}\right)} \frac{d}{ds}\left(\int_0^{\frac{\pi}{2}+\tan^{-1}(s)} \sin^{k-1}\theta \, d\theta \right)$$

Now we have a similar situation to what we saw earlier. We're differentiating an integral, so we could normally just ignore the integral sign and the job is done. In this case, though, we're taking the derivative of an integral evaluated at a function.

When we replace θ with

$$\frac{\pi}{2} + \tan^{-1}(s)$$

we have a function within a function, so when we take the derivative, the chain rule applies. That means we need to multiply our result by the derivative of the inner function: $\frac{\pi}{2} + \tan^{-1}(s)$. Now $\frac{\pi}{2}$ is a constant, which means its derivative is 0. As we saw on page 253, the derivative of

$$\tan^{-1}(s) = \frac{1}{1+s^2}$$

$$= (1+s^2)^{-1}$$

Now $\sin^{k-1}\theta = (\sin\theta)^{k-1}$. Evaluating this at the upper integral gives

$$\left(\sin\left(\frac{\pi}{2} + \tan^{-1}(s)\right)\right)^{k-1}$$

Recall from when we looked at trigonometric identities that

$$\sin(\theta + \zeta) = \sin\theta\cos\zeta + \cos\theta\sin\zeta$$

That gets us to

$$\sin\frac{\pi}{2}\cos(\tan^{-1}(s)) + \cos\frac{\pi}{2}\sin(\tan^{-1}(s))$$

Now $\cos\frac{\pi}{2} = 0$, so the second term disappears. $\sin\frac{\pi}{2} = 1$ so we are left with $\cos(\tan^{-1}(s))$.

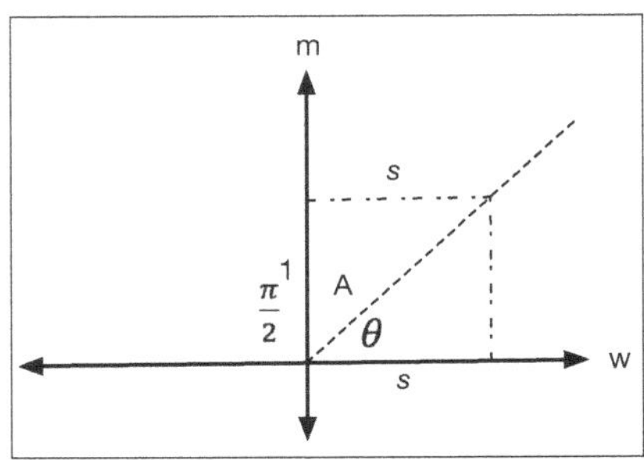

Figure 294

Thinking back to our diagram (Figure 294), $\tan^{-1}(s) = A$. The cosine of A is the adjacent over the hypotenuse, or in this case 1 over the hypotenuse, and by Pythagoras' theorem the hypotenuse is $\sqrt{1 + s^2}$. That means that

$$\cos(\tan^{-1}(s)) = \frac{1}{\sqrt{1 + s^2}}$$

In terms of the outer function we were actually looking at $(\sin \theta)^{k-1}$. That evaluates to

$$\left(\frac{1}{\sqrt{1 + s^2}}\right)^{k-1} = \left((1 + s^2)^{-\frac{1}{2}}\right)^{k-1}$$

So putting all of that into our original formula, we get:

$$PDF = \frac{\Gamma\left(\frac{k+1}{2}\right)}{\sqrt{\pi}\,\Gamma\left(\frac{k}{2}\right)}(1 + s^2)^{-1}(1 + s^2)^{\frac{k-1}{2}}$$

$$= \frac{\Gamma\left(\frac{k+1}{2}\right)}{\sqrt{\pi}\,\Gamma\left(\frac{k}{2}\right)}(1 + s^2)^{-1}\left((1 + s^2)^{-\frac{1}{2}}\right)^{k-1}$$

Adding the indices of $1 + s^2$, we have

$$-1 - \frac{(k-1)}{2} = -\frac{2}{2} - \frac{k}{2} + \frac{1}{2} = -\left(\frac{k+1}{2}\right) \qquad PDF(s) = \frac{\Gamma\left(\frac{k+1}{2}\right)}{\sqrt{\pi}\,\Gamma\left(\frac{k}{2}\right)}(1 + s^2)^{-\left(\frac{k+1}{2}\right)}$$

We originally said that $\frac{T}{\sqrt{k}} = S$. We therefore need to do a transformation of a random variable in a PDF, which means we will replace S with $\frac{T}{\sqrt{k}}$, and multiply by $\frac{1}{\sqrt{k}}$

That gives us the PDF of the T distribution; or family of distributions, because each value of k will be a different distribution.

$$Pr(T = t) = \frac{\Gamma\left(\frac{k+1}{2}\right)}{\sqrt{k\pi}\,\Gamma\left(\frac{k}{2}\right)}\left(1 + \frac{t^2}{k}\right)^{-\left(\frac{k+1}{2}\right)}$$

This isn't actually as bad as it looks, because everything on the left is a constant, and because, whatever the degrees of freedom, k will be an integer. Remember that, where n is a positive integer, $\Gamma(n) = (n - 1)!$, $\Gamma(x + 1) = x\Gamma(x)$, where $x > 0$, and $\Gamma\left(\frac{1}{2}\right) = \sqrt{\pi}$.

Alternative forms of the *t* distribution PDF

There are other ways to express the PDF, which are useful in different contexts. One way uses the beta function. This is popular with people using a Bayesian approach because, as we saw earlier, the beta distribution is particularly useful in Bayesian statistics, and so expressing the *t* distribution using a related function can make the (generally fairly challenging) mathematical analysis easier.

We've seen (page 443) that the beta function (not to be confused with the beta distribution) can be expressed in terms of the gamma function:

$$\beta(x,y) = \frac{\Gamma(x)\Gamma(y)}{\Gamma(x+y)}$$

Multiply both sides by $\Gamma(x+y)$ and divide both sides by $\beta(x,y)$.

$$\Gamma(x+y) = \frac{\Gamma(x)\Gamma(y)}{\beta(x,y)}$$

Set $x = \frac{1}{2}$ and $y = \frac{k}{2}$

$$\Gamma\left(\frac{1+k}{2}\right) = \Gamma\left(\frac{1}{2}\right)\Gamma\left(\frac{k}{2}\right)\left(\beta\left(\frac{1}{2},\frac{k}{2}\right)\right)^{-1}$$

Substitute this into our formula for the *t* distribution.

$$\frac{\Gamma\left(\frac{1}{2}\right)\Gamma\left(\frac{k}{2}\right)\left(\beta\left(\frac{1}{2},\frac{k}{2}\right)\right)^{-1}}{\sqrt{k\pi}\,\Gamma\left(\frac{k}{2}\right)}\left(1+\frac{t^2}{k}\right)^{-\left(\frac{k+1}{2}\right)}$$

Replace $\Gamma\left(\frac{1}{2}\right)$ with $\sqrt{\pi}$, and simplify.

$$= \frac{\sqrt{\pi}\,\Gamma\left(\frac{k}{2}\right)\left(\beta\left(\frac{1}{2},\frac{k}{2}\right)\right)^{-1}}{\sqrt{k\pi}\,\Gamma\left(\frac{k}{2}\right)}\left(1+\frac{t^2}{k}\right)^{-\left(\frac{k+1}{2}\right)}$$

This form of the PDF actually looks simpler than the more usual form.

$$\Pr(T=t) = \frac{1}{\sqrt{k}\,\beta\left(\frac{1}{2},\frac{k}{2}\right)}\left(1+\frac{t^2}{k}\right)^{-\left(\frac{k+1}{2}\right)}$$

Where has *e* disappeared to in the *t* distribution?

We first saw *e* appear as if from nowhere when we looked at the Poisson distribution, and it has been present in the distributions we've looked at since then. Now suddenly it seems to have disappeared again. It is present in the gamma and beta functions, but they form part of the coefficient, not the variable part of the formula.

To clarify, write the constant coefficient as

$$C_1 = \frac{\Gamma\left(\frac{k+1}{2}\right)}{\sqrt{k\pi}\,\Gamma\left(\frac{k}{2}\right)}$$

$$\text{PDF} = C1\left(1 + \frac{t^2}{k}\right)^{-\left(\frac{k+1}{2}\right)}$$

Factor out $-\frac{1}{2}$ from the index, and express it in this way.

$$= C_1\left(\left(1 + \frac{t^2}{k}\right)^{(k+1)}\right)^{-\frac{1}{2}}$$

Now take the limit as $k \to \infty$. In other words an infinite sample size.

$$C_1\left(\lim_{k\to\infty}\left(1 + \frac{t^2}{k}\right)^{(k+1)}\right)^{-\frac{1}{2}}$$

Recall that $\lim_{n\to\infty}\left(1 + \frac{x}{n}\right)^n = e^x$. Also note that as $k \to \infty$, so $k + 1$ becomes indistinguishable from k.

$$C_1\left(e^{t^2}\right)^{-\frac{1}{2}}$$

Our friend e has reappeared again, and as the degrees of freedom approach infinity, the PDF takes on the form of a standardised normal distribution, which has the form $C_2\left(e^{x^2}\right)^{-\frac{1}{2}} = C_2 e^{-\frac{x^2}{2}}$. Because these are both valid PDFs the coefficients can be thought of as normalising constants, that will make the area under the curve equal 1.

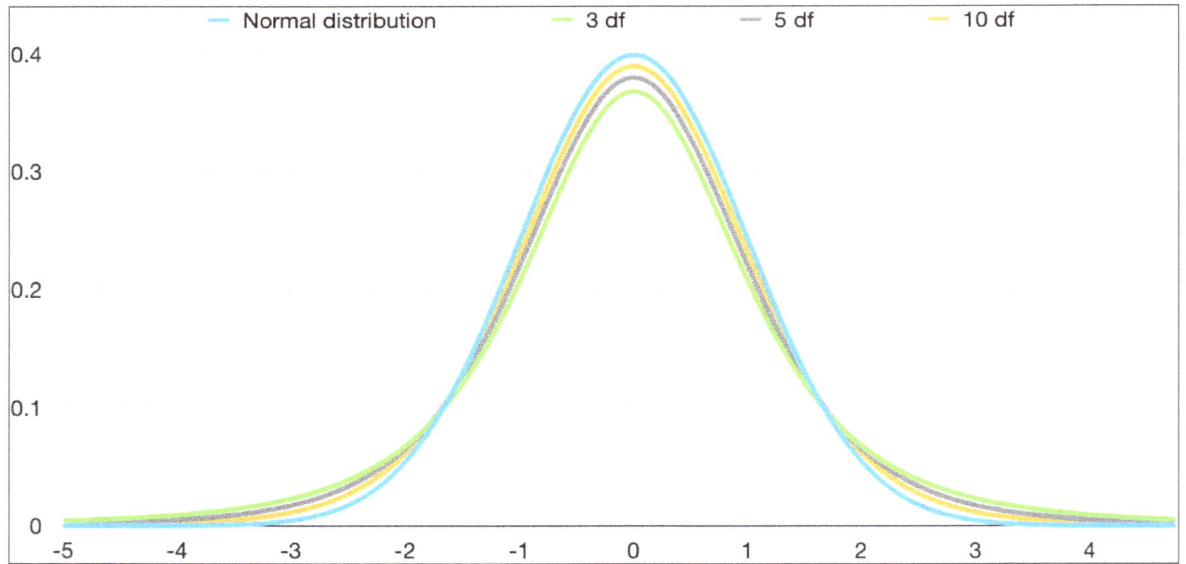

Figure 295

Figure 295 shows the graph of t distributions with different degrees of freedom ('df' on the graph). As you can see, they are very close to the normal distribution. The main difference is that the smaller the degrees of freedom, the greater the excess kurtosis: that is, the thicker the tails. This means that for smaller degrees of freedom—smaller sample sizes—there is less certainty about the result, but a probability distribution can still be used, if the underlying data is normally distributed.

This was exactly what Gosset and Fisher were after; something close to a normal distribution curve, that could be useful with smaller sample sizes. From about 30 degrees of freedom and higher, the distributions are close enough to be indistinguishable for practical purposes.

That means that the t distribution has virtually replaced the normal distribution. The t works better for smaller samples (provided that the underlying distribution is normal), and in virtually the same way for larger samples.

Moment generating function and moments

The easiest way to get to the cumulative distribution function (CDF) is through Fisher's F distribution which we haven't come to yet, but regardless, the values for $\Pr(T \leq t)$ need to be approximated with a Taylor series.

The t distribution does not have a moment generating function (MGF) in the traditional sense. To find it we would normally calculate

$$E[e^{tX}]$$

but unfortunately, with the t distribution, the expected value doesn't exist for $t = 0$, so it doesn't work. Likewise the characteristic function, which does exist, needs to be expressed using a hypergeometric function, which we aren't covering.

The first moment—the mean—is 0, because the distribution is standardised to make it so. The third moment—the skewness—is also 0, because the distribution is symmetrical. The second and fourth moments—the variance and the kurtosis—are dependent on the degrees of freedom. The formula for the variance is:

$$\sigma^2 = \frac{k}{k - 2}$$

Where k is the degrees of freedom (one less than the sample size). You can see from this that the higher the degrees of freedom, the closer the variance is to 1, which is, of course, the variance for a standardised normal distribution.

The formula for the kurtosis is:

$$excess\ kurtosis = \frac{6}{k - 4}$$

The excess kurtosis will always be positive (it isn't applicable when $k \leq 4$), meaning, as mentioned earlier, that the t distribution has thicker tails than the normal distribution. But the higher the degrees of freedom, the closer the excess kurtosis is to 0, as the t distribution gets closer and closer to the normal distribution.

Key points:

- The **t distribution** was designed to get around the problem of having to already know the standard deviation of the population.

- It is based on an assumption that the underlying population has a normal distribution.

- Fisher was able to demonstrate that the standard deviation of a sample divided by the square root of the size of the sample was equal to the standard deviation of the sampling distribution (or standard error) of the mean:

$$SE(\bar{x}) = \frac{s}{\sqrt{n}}$$

- The T statistic is a standardised form of the mean of a sample taken from a normally distributed underlying population:

$$T = \frac{\bar{x} - \mu}{SE(\bar{x})}$$

$$= \frac{\bar{x} - \mu}{\frac{s}{\sqrt{n}}}$$

- The T statistic has two separate and independent variables in it: the mean of the sample and the standard deviation of the sample. To get around this problem in a probability distribution, Fisher was able to use two known distributions as part of his formula: the normal and the chi square. $T = \dfrac{W}{\sqrt{\frac{V}{(n-1)}}}$ where W has a standard normal distribution, and V has a chi square distribution with k degrees of freedom.

- The PDF is $\Pr(T = t) = \dfrac{\Gamma\left(\frac{k+1}{2}\right)}{\sqrt{k\pi}\,\Gamma\left(\frac{k}{2}\right)}\left(1 + \dfrac{t^2}{k}\right)^{-\left(\frac{k+1}{2}\right)}$ where k is the degrees of freedom.

- There are alternative ways to express this PDF, including one that uses the beta function: $\Pr(T = t) = \dfrac{1}{\sqrt{k}\,\beta\left(\frac{1}{2}\frac{k}{2}\right)}\left(1 + \dfrac{t^2}{k}\right)^{-\left(\frac{k+1}{2}\right)}$

- When k approaches infinity the t distribution becomes a standard normal distribution. In practice, with sample sizes of 30 and above, the t distribution is indistinguishable from the normal. For smaller sample sizes it has thicker tails (excess kurtosis).

- For samples below 30 a t distribution is better than a normal distribution, and for samples of more than 30 it is indistinguishable, so its use has largely replaced that of the normal distribution in practical statistics.

Applying the *t* distribution in practice

Read this if you want to know how to make use of *t* tests, in terms of situations where you aren't sure that the underlying data is normally distributed, deciding on (or evaluating) one-tailed or two-tailed *t* tests, working out whether the mean of a single sample is significantly different from a known or hypothesised population mean, calculating P values and confidence intervals, comparing two groups when they are paired together, comparing two groups where they aren't paired together, or if you think using *t* tests for proportions is a good idea (because it is done so commonly).

The *t* distribution assumes that the underlying data is normally distributed. It is the assumption of normality that makes it useful for sample sizes of under 30. If you are using the distribution to make assumptions about sampling, the central limit theorem comes into play. In other words, the sampling distribution of the sampling mean will be at least roughly normally distributed. In practice the distribution is reasonably forgiving of distributions that aren't strictly normal, if they have a bell-shaped curve. The *t* distribution should not, however, be used for situations where there is no reason to believe that the underlying data is likely to be normally distributed. It's always best to think of it, conceptually, as a normal distribution with a margin of error built in to cover smaller sample sizes.

We saw earlier that for a normal distribution, approximately 68% of the data falls within 1 standard deviation of the mean (i.e. between $\mu - \sigma$ and $\mu + \sigma$). Approximately 95 % of the data falls within 2 standard deviations of the mean (i.e., between $\mu - 2\sigma$ and $\mu + 2\sigma$). Approximately 99.7% of the data falls within 3 standard deviations of the mean (i.e., between $\mu - 3\sigma$ and $\mu + 3\sigma$). These percentages are often referred to as the **68-95-99.7** rule or the **empirical rule for normal distributions**. With a *t* distribution roughly the same proportions will be covered by the same number of standard deviations from the mean, although this varies a little depending on the degrees of freedom.

How many tails?

Figure 296 shows what is indicated by a **two-tailed *t* test**, and that is the situation where the empirical rule is roughly appropriate. The border of the blue area, showing the area of a potential type I error, is known as the **critical *t* value**.

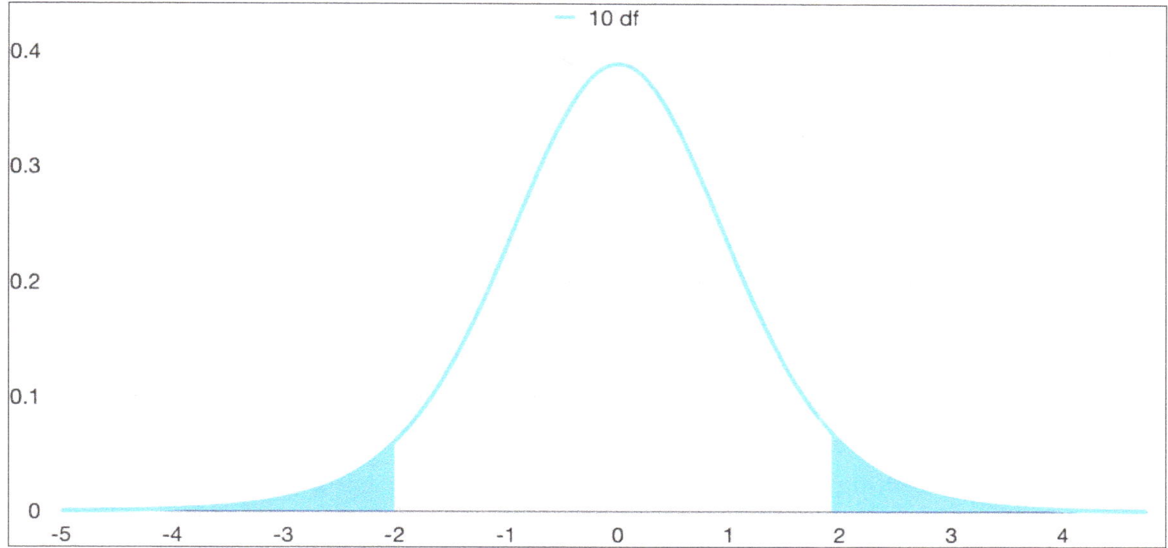

Figure 296

Figure 297 shows what is used for a **one-tailed *t* test**. This applies when you don't care how far your result deviates from the mean in one of the directions; you are only interested in how far from the mean you are in one direction. In a one-tailed test, the probability of being in the blue area will

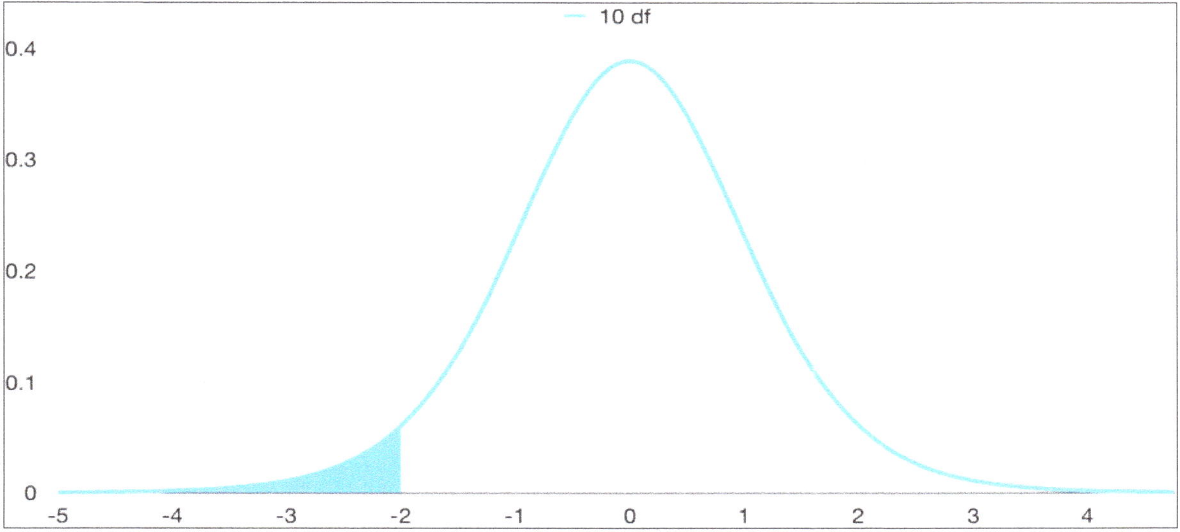

Figure 297

be half of what it is for a two-tailed test (because the distribution is symmetrical). An alternative way to think of this is that if you are doing a one-tailed test and accepting a particular level of significance (say 5%), you will be able to say that your result is significant when it is closer to the mean than you could with a two-tailed test. This makes the two-tailed test a more conservative choice.

Some statisticians believe that researchers should only ever use two-tailed tests, because if a result is more than two standard deviations from the mean, in a direction you didn't expect, that is still a finding worth acknowledging and reporting on[83].

As we saw earlier, saying that something has a P value of 0.05—that there is a 5% chance that it was a random event—may sound impressive, until you realise that this means there is one chance in twenty of saying that a result is significant when it was actually obtained by random chance. This in turn means that for every 20 studies published reporting a result that is significant to the 0.05 level, you would expect one to report a significant result for something purely random. If researchers are using one-tailed tests, it is important to be sure that that isn't making this problem worse.

Single-sample *t* test

If you want to know whether the mean of a single sample is significantly different from a known or hypothesised population mean, you can use a **single-sample *t* test** (not to be confused with whether you are using one or two tails; that is a separate issue).

You may see this written in an academic paper as:

$$H_0: \mu = \mu_0, H_A: \mu \neq \mu_0 \text{ (two tailed)}$$

(Instead of H_A you might also see the alternative hypothesis written as H_1.) That is, the null hypothesis is that the mean of your sample equals the mean of the population you are comparing to (or is within two or three standard deviations of it, meaning that the result could have been achieved by random chance), while the alternative hypothesis is that the mean of your sample is significantly different from the mean of the group you are comparing it to.

Alternatively, you may see this written as either:

$$H_A: \mu > \mu_0 \text{ (one tailed right)}$$

or

$$H_A: \mu < \mu_0 \text{ (one tailed left)}$$

The formula used for a one-sample test is:

$$t = \frac{\bar{x} - \mu_0}{\frac{s}{\sqrt{n}}}$$

That is, you standardise your result using a **t statistic**: take the difference between the mean of your result and the mean of the real or hypothesised group you are comparing it to, and divide this by the sample standard deviation divided by the square root of the sample size.

This is the same as dividing the difference between the means by something similar to the **standard error of the mean**: the extent to which taking the mean of *any* given sample will be close to the mean of the population. We aren't quite dividing by the standard error of the mean because that involves using the population standard deviation. We are using the sample standard deviation. The t statistic is what we originally used as the basis for calculating the probability density function (PDF) of the t distribution. In other words, using a one-sample t test involves finding out if your t score is different from the t score of a different distribution.

Examples where this test may be appropriate include: testing the effectiveness of a drug against a known or widely accepted value, comparing a product feature to accepted industry standards, and comparing student performance to a target. (Always provided, of course, it is reasonable to assume that the underlying data is normally distributed.)

The formula for a **confidence interval** for a single-sample t test is, as you might expect from all that we have seen, the same as the formula for a confidence interval that we saw earlier, except that the z is replaced with a t, and the population standard deviation is replaced by the sample standard deviation.

$$CI = \bar{x} \pm t\frac{s}{\sqrt{n}}$$

where $\bar{x}$ is the sample mean, t is the t score for the desired confidence level and the appropriate degrees of freedom, s is the sample standard deviation and n is the sample size.

Paired *t* test

A paired *t* test is used when you are comparing two different groups, but they are paired together. This often occurs in before-and-after studies, where each subject acts as its own control, and in studies where two different interventions are used with the same subjects (such as a placebo and a treatment, on the same individuals, at different times).

The procedure is very like a single-sample test. You calculate the difference for each pair, and the difference is the random variable you work with. You work out the mean and the standard deviation, then treat the result as a single-sample *t* test.

Two-sample (unpaired) *t* test

It is probably more common to do tests with two samples than with one. This can include:

- comparing a group before and after an intervention;
- comparing the effect of an intervention in one group, with no intervention in another;
- comparing the results of two interventions; or
- just comparing some aspect of two populations.

What we are really doing in this case is subtracting the variables, then working with the distribution that results. At first this appears trickier than the paired *t* test because... well... the data points aren't paired. We have, however, learned enough to this point to make it straightforward. We saw earlier that the mean of the difference between two random variables is the same as the difference between the means (page 527). And we also know that the variance of the difference between two random variables is the sum of their variances, not the difference (page 551). Therefore, the formula for the *t* value for our samples, when we are looking for the difference between them, is:

$$t = \frac{\bar{x}_1 - \bar{x}_2}{\sqrt{\dfrac{s_1^2}{n_1} + \dfrac{s_2^2}{n_2}}}$$

where $\bar{x}_1$ and $\bar{x}_2$ are the means of the two samples, s_1^2 and s_2^2 are the two sample variances, and n_1 and n_2 are the sample sizes. In other words we have standardised our score by taking the difference between the means and dividing by the standard deviation (the square root of the variance).

t tests for proportions

There is a widely-used *t* test for a proportion. Say, for example, that 54% of voters indicate in a poll that they will vote for a particular candidate, and you want to know, based on the number of responses, what the reliability of that statistic is. The problem here is that the *t* test was designed for normally or approximately normally distributed data, whereas a decision about whether or not to vote for a particular candidate is actually a **binomial distribution**. If it is a choice between several candidates it is a **multinomial distribution**. We saw earlier that a chi-square test is ideal for a multinomial distribution like this.

In the days before computers, the reason for using a *t* test for proportions was convenience. It was quite challenging to calculate binomial probabilities for large samples, with all of the factorials and powers involved in each term. It was, of course, even more challenging to manually calculate the areas under the curve for standard normal and for *t* distributions, but there were tables that could be looked up for this, making it reasonably straightforward. Karl Pearson had developed a formula for approximating a binomial distribution with a normal one. His approach allows the *t* test to be used for proportions.

Today, of course, using a computer there is no more difficulty generating a binomial distribution than there is generating any other kind of distribution. For testing the statistical significance of something with a binomial or multinomial distribution it makes more sense to avoid the *t* distribution, not because it won't be accurate, but because it obscures what is really happening, making it easier to make mistakes.

Using *t* tests with regression and correlation

Earlier we thought about regression and correlation in terms of a fixed number of values but, of course, it is more common to apply them when we are taking a sample that involves paired variables, and we want to know not just the slope of the line (regression) and how close the points are to it (correlation), but also whether the result we have is statistically significant. Fisher found that *t* tests were also useful in this situation. The principles are the same as for the applications we have seen so far.

Where x and y had a linear relationship, we said that the relationship between any matched pair of observed values, x_i and y_i was given by the equation

$$y_i = mx_i + b + g_i$$

where m is the **slope of the regression line**, b is the is the value of y when $x = 0$ (the **y intercept**), and g_i is the **residual** (or gap). We also saw that the value of m where the sum of the squared residuals was as low as possible was given by

$$m = \frac{\Sigma(x_i - \bar{x})(y_i - \bar{y})}{\Sigma(x_i - \bar{x})^2}$$

where $\bar{x}$ and $\bar{y}$ are the means of the x and y values respectively.

To calculate the confidence interval we want something in the form of

$$CI = m \pm tSE(m)$$

where t is the critical t value, as per the work we did earlier. For example, for a confidence interval of 95%, $t = 1.96$, and for a confidence interval of 99%, $t = 2.576$.

To see how this works, we will go back to the terminology we were originally using when we looked at least squares regression. One of the things we said then was that

$$g_i = y_i - \hat{y}$$

where $\hat{y}$ is the value of y on the **regression line** (the line of best fit), y_i is the value of each observed value of y_i, and g_i is each **residual**, or the gap between the observed and value of y_i and its predicted value based on the regression line.

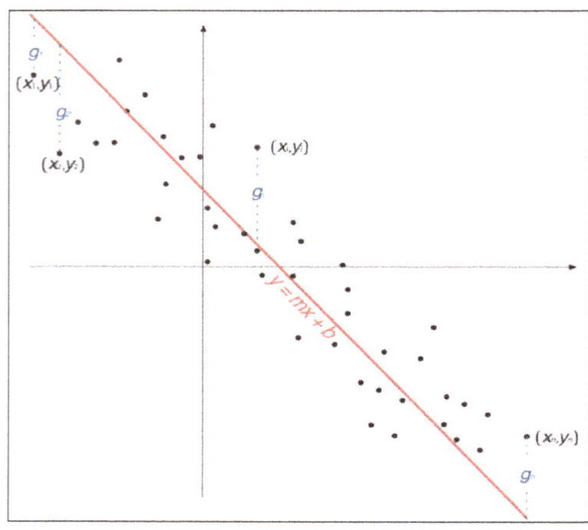

Figure 298

For our confidence interval, we want to modify the formula for m. We keep the denominator: $\Sigma(x_i - \bar{x})^2$. But for the numerator, instead of the covariance of X and Y we want the standard deviation of the residuals, or the g values.

To start with we need the variance of g_i. We usually calculate a variance from the mean of values but, as we saw when we first looked at the moments of a distribution, the 2nd central moment can be calculated from any value we choose; it is just that the mean usually works best. In this instance, though, the value we are interested in calculating the variance from is the value of y on the regression line. If we were dealing with the whole population, the variance would be given by

$$\sigma_g^2 = \frac{\Sigma(y_i - \hat{y})^2}{n}$$

As we are dealing with a sample, however, we need to divide by the degrees of freedom rather than by the number of data points. That makes the sample variance of the residuals

$$s_g^2 = \frac{\Sigma(y_i - \hat{y})^2}{n - 2}$$

(We saw earlier that, when looking at the correlation of two variables, there are $n - 2$ degrees of freedom.) That is our numerator. So to find the standard error of m we put that over the sum of squares of the x values and take the square root. We can multiply the numerator and denominator by $n - 2$, so that we get

$$Standard\ error = \sqrt{\frac{\Sigma(y_i - \hat{y})^2}{(n - 2)\Sigma(x_i - \bar{x})^2}}$$

which makes the confidence interval

$$CI = m \pm t\sqrt{\frac{\Sigma(y_i - \hat{y})^2}{(n - 2)\Sigma(x_i - \bar{x})^2}}$$

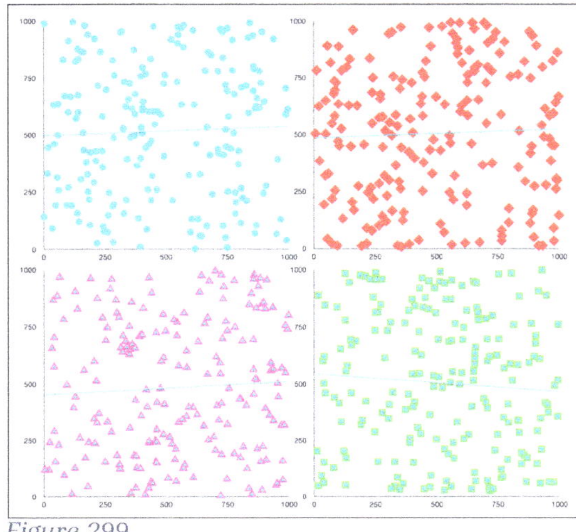

Figure 299

In simple linear regression the null hypothesis is that the slope of the line is 0; it is flat. Figure 299 shows four examples of where 200 computer-generated random x values of between 0 and 1,000 paired with 200 computer-generated random y values, also between 0 and 1,000. In other words, there should be no correlation between them; they are independent.

As you can see, the regression line is almost flat in every case. Fisher was at pains to point out that it may be important in this situation to distinguish the independent variable (the x) value from the y value. The way the mathematics of this is set up, they can't always be used interchangeably. This isn't the same as saying that one of the variables is caused by the other (although it may be). The example he gave was that of children's ages and their heights. Age is the independent variable, and height is the dependent one. If there are some errors made in measurement of height then it probably won't change the values very much, because it is reasonable to assume that errors made in one direction will be balanced out by errors made in the other direction. On the other hand, if you make an error in recording an age, it is likely to create a more significant problem[84]. (The heights of children was, as we saw earlier, the issue that first caused Francis Galton to look at, and to name, regression.)

So in other words, if the two variables we are looking at are uncorrelated, the x values will be what they are, and the y values will be scattered around them in a random way. That will give the line a slope of roughly zero; it will be horizontal.

So, if we are performing a test of significance, the t statistic will be given by

$$t = \frac{m - 0}{standard\ error}$$

$$= \frac{m}{\sqrt{\dfrac{\Sigma(y_i - \hat{y})^2}{(n-2)\Sigma(x_i - \bar{x})^2}}}$$

Correlation and t tests

When it came to statistical testing for correlation, Fisher found, and overcame, a problem: the underlying assumption that the observed values will be normal, while it holds true in relation to the slope of the line, does not hold true in relation to the correlation. In order to be able to use a t test he had to transform the r values into z values, which are normally distributed. To do this he needed to use something now known as the **Fisher z transformation**. This uses the probability density function (PDF) of the **bivariate normal distribution**, which we aren't covering in this book.

Key points:

- The *t* test can be somewhat forgiving if the underlying population is not normally distributed, but only to a point. It shouldn't be used where it isn't reasonable to assume there is a normal distribution.

- The **empirical rule** is most appropriate for two-tailed *t* tests. One-tailed *t* tests apply where you are only interested in assessing probabilities in one direction.

- If you want to know whether the mean of a single sample is significantly different from a known or hypothesised population mean, you can use a **single-sample *t* test**.

- For one sample *t* tests, standardise data using a *t* statistic:

$$t = \frac{\bar{x} - \mu_0}{\frac{s}{\sqrt{n}}}$$

- The formula for confidence interval for a single-sample *t* test is

$$CI = \bar{x} \pm t\frac{s}{\sqrt{n}}$$

  where *t* is the *t* score for the desired confidence level and the appropriate degrees of freedom.

- Paired *t* tests are used for comparing two different groups, when they are paired together. The difference between each pair is used as the random variable.

- The test statistic for two-sample unpaired *t* tests is

$$t = \frac{\bar{x}_1 - \bar{x}_2}{\sqrt{\frac{s_1^2}{n_1} + \frac{s_2^2}{n_2}}}$$

- Binomial or chi square tests are usually a better option than *t* tests for a proportion, because what is being assessed is clearer.

- When performing a test of significance for regression, the *t* statistic will be given by

$$t = \frac{m}{\sqrt{\frac{\Sigma(y_i - \hat{y})^2}{(n-2)\Sigma(x_i - \bar{x})^2}}}$$

ANOVA and the F distribution

> **Read this if** you want to know why type one errors are almost inevitable if you use *t* tests when comparing multiple variables one pair at a time, you want to know how ANOVA addresses the problem, you want to know what the F distribution has to do with it and where it comes from, or you want an overview of ideas that extend the use of ANOVA.

So far we have looked at using *t* tests and chi square tests when we are considering one or two variables. When Fisher started thinking about how a range of different variables related to one another, he realised there was a potential problem. If, for example, you use a *t* test to compare 5 different random variables, you will need to perform 10 tests to compare them all in pairs, and the number of tests goes up faster than the number of variables. If there are 7 variables, you would need to test 21 pairs. The amount of work in doing the tests isn't the problem—it's actually trivial when you use a computer.

The problem is that if you have a P value of 0.05, meaning there is one chance in 20 that that you will have a type I error—i.e. it will look as though you have a statistically significant result when you don't—then, by the addition rule, these probabilities get added together for each test you do.

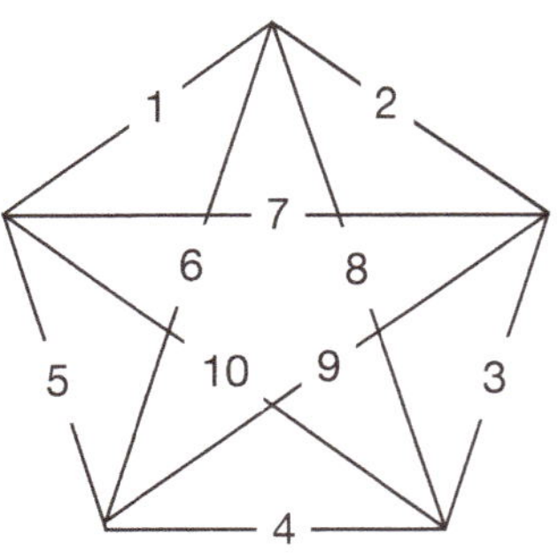

Figure 300

> The **10 tests** needed for pair-wise comparison of **5** variables are show in Figure 300.
>
> The number of pair-wise tests for any number of variables is given by our old friend n choose r, when n is the number of variables, and $r = 2$, (because we are finding the number of possible pairs between these variables). And that means, believe it or not, we have another connection with Pascal's triangle: you can read the number of variables from the second column of the triangle, and the number of pair-wise tests needed to compare them all from the third column (Figure 301).

1						
1	1					
1	2	1				
1	3	3	1			
1	4	6	4	1		
1	5	10	10	5	1	
1	6	15	20	15	6	1
1	7	21	35	35	21	7
1	8	28	56	70	56	28
1	9	36	84	126	126	84
1	10	45	120	210	252	210
1	11	55	165	330	462	462
1	12	66	220	495	792	924
1	13	78	286	715	1287	1716
1	14	91	364	1001	2002	3003
1	15	105	455	1365	3003	5005

Figure 301

If you test 7 variables and so test 21 pairs, it is highly probable that you will get a type I error. Your statistical test will probably indicate that there is a positive result, when there isn't. Even if you use a P value of 0.01, meaning a 1 in 100 chance of getting a statistically significant result, if you have 15 variables you will need to do 105 tests to compare them all. Type I errors become virtually inevitable, and so Fisher needed a better solution.

What he came up with is called the **analysis of variance**, or **ANOVA.** We'll start with the most basic form.

One way ANOVA

An underlying assumption of ANOVA is that the observations and the groups are independent from one another. (Other related approaches assume that there is a degree of dependence.) Even though the ANOVA is called analysis of *variance*, and it uses the variances, it is actually a way of comparing means.

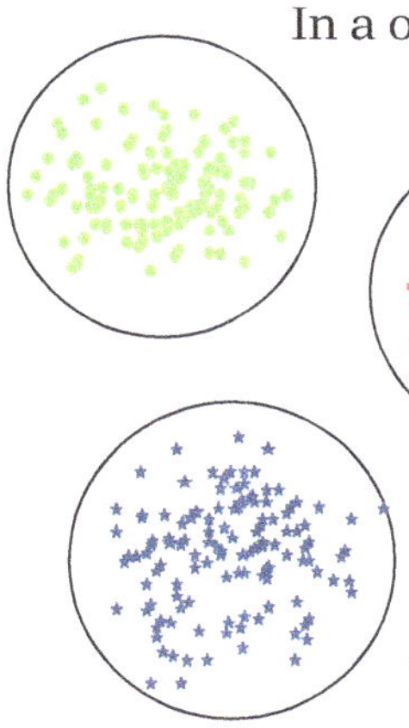

Figure 302

In a one-way analysis of variance all you are asking is: when comparing samples from three or more populations, is there a significant difference between at least one of the means and the others? Once you know that there isn't, the null hypothesis is retained and you don't need to do any more. If there is a significant difference, you can then use other testing to find out what it is, and which pairs it relates to. The problem of an artificially-inflated probability of finding a statistically-significant result is no longer such an issue. The approach Fisher took was to compare the variability *within* the groups to the variability *between* the groups.

Thinking back to our shooter (page 635), imagine she had three or more scopes, and was testing the accuracy of each of them. Figure 302 shows a situation with relatively low variability within each group, and relatively high variability between them. That is, the shots using each scope are clustered together, but there is no overlap between them. It is easy to tell which scope fired which shots. So it is easy for her to be able to say that there is, overall, a significant difference between the scopes. On the other hand Figure 303 shows a situation where there is high variability within each group, and relatively low variability between them. It is much more difficult to be confident that the different scopes are set differently.

To calculate the ratio of these variances, Fisher used the now-familiar concept of the sum of squares, and looked at the ratio:

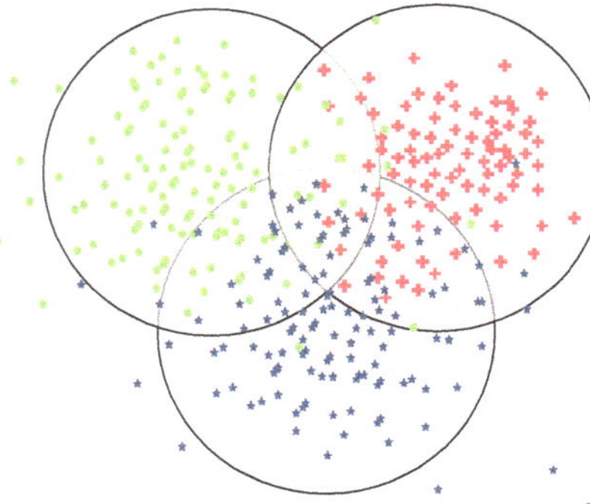

$$\frac{variance\ between\ groups}{variance\ within\ groups}$$

The higher the number, the more likely it is that you can reject the null hypothesis, where the null hypothesis was that there was no difference between the means of the groups.

It's easiest to understand how ANOVA works in practice with an example. Suppose we have three groups that receive different interventions (e.g. different drugs to achieve the same therapeutic outcome), and we want to know whether there was any statistically significant difference between the different interventions. This is the data set:

Figure 303

		Group 1	Group 2	Group 3
Individual results		1	2	3
		5	5	5
		9	8	13
		15		
A. Mean of each group		7.5	5	7
B. Grand mean (mean of all scores)		6.6		
C. Difference between mean of group and grand mean ($A - B$)		0.9	-1.6	0.4
Squared difference between mean of group and grand mean (C^2)		0.81	2.56	0.16
Number of observations in group (n)		4	3	3
Product of number of observations by squared difference from grand mean (nC^2)		3.24	7.68	0.48

Figure 304

As a reminder, the way Fisher defined the variance of a sample was:

$$Variance\ (sample) = \frac{\sum_{i=1}^{n}(x_i - \bar{x})^2}{n - 1}$$

It is the sum of the squared difference of each observation from the mean, divided by the degrees of freedom. Using this approach, the mean sum of squares between the groups is calculated by the sum of the squared difference of the mean of the group of each observation from the grand mean, then divided by the total degrees of freedom of all observations. So for group 1, with a mean of 7.5, the difference from the grand mean (which is 6.6) is 0.9. Square that and you have 0.81. There are four observations in the first group, so that makes the calculation for that group $4 \cdot 0.81 = 3.24$. For group 2 we have 7.68, and for group 3 we have 0.48, giving a total of 11.4.

Now there are three groups we are comparing, which means there are two degrees of freedom. (If you know the total and the scores of any 2 of the groups, you will be able to calculate the third).

That makes the variance between groups (often called the **mean sum of squares between)**

$$\frac{11.4}{2} = 5.7$$

To calculate the mean sum of squares within groups, we are again working with each of the observations, but this time taking the difference of each observation from the mean of its own group (rather than the total of all the groups), squaring that, then adding all of these across all the groups before dividing by the degrees of freedom. The first score in the first group (Figure 304) was 1, the mean of that group was 7.5, so the difference is 6.5, which, when squared, is 42.25.

> **To cement your understanding:** calculate a few more of these results for yourself.

When we add that sum of squares, we get 181. The degrees of freedom for each group is calculated by the number of observations in the group, less one. That is:

$$(4 - 1) + (3 - 1) + (3 - 1) = 7$$

(Note that this is the same calculation as adding up all of the observations and subtracting the number of groups; they both involve doing the same thing.)

That gives the total of the variance within groups (often called the mean within group sum of squares) as

$$\frac{181}{7} \approx 25.857 \ldots$$

So the ratio Fisher was looking for is, in this case,

$$F = \frac{5.7}{25.857}$$

$$\approx 0.22$$

The degrees of freedom for between the groups was $k - 1$ where k is the number of groups, while the degrees of freedom with each group was $n - k$ where n is the total number of observations and k is the number of groups. That means that the total number of degrees of freedom was

$$n - k + \; k - 1 = n - 1$$

So that gave him a basis for calculating the ratio of between group variation to that of within group variation. What he then needed was a way to determine what would happen if all of this variation was down to random chance. That way he could tell what was and was not statistically significant.

"So", he said to himself, "if all of that variation is down to random chance, then all of the variation should follow a normal distribution. If it does, then the expression

$$\frac{mean\ sum\ of\ squares\ between\ groups}{mean\ sum\ of\ squares\ within\ groups}$$

will actually be a ratio between two chi square distributions, because that's what a chi square distribution is: a sum of squared normal distributions". That was the basis for his new distribution which he called the F distribution (and yes, it does stand for "Fisher").

The F distribution

The mathematics of working out the ratio of two chi square distributions isn't trivial, but the overall approach is the same as the one he used to calculate the t distribution, except this is slightly easier. That is, he worked with the cumulative distribution functions (CDFs), created a joint distribution by multiplying the two chi squared distributions together (which he could do because of the assumption of independence we started with), then selected out that part of the joint distribution that related specifically to the ratio he was looking at, converted it to polar coordinates, managed to get rid of one of the variables that way so that he only had one random variable to work with, took the derivative to find the probability density function (PDF), and wound up with the PDF for the F distribution, to work out whether the results of ANOVA calculations were statistically significant. Here's the working for all of that.

Define the F statistic (and don't confuse it with function notation).

$$F = \frac{sum\ of\ squares\ between\ groups/df}{sum\ of\ squares\ within\ groups/df}$$

Let *sum of squares between groups* $= Y$

Let *sum of squares within groups* $= X$

Let j and k be the degrees of freedom of Y and X respectively. Recall that the chi square distribution is a particular form of the gamma distribution and note the form of parameterization of the gamma distribution that we are using.

$$Y \sim \chi_k^2 = \Gamma\left(\frac{k}{2}, \frac{1}{2}\right)$$

$$X \sim \chi_j^2 = \Gamma\left(\frac{j}{2}, \frac{1}{2}\right)$$

Show F in terms of the variables we have created and multiply the numerator and denominator by the degrees of freedom.

$$F = \frac{\frac{Y}{k}}{\frac{X}{j}}$$

Shift the degrees of freedom to the left side of the equation by multiplying both sides by $\frac{k}{j}$. This is just to get them out of the way, for now, so that we can concentration on the relationship between the two chi-square distributed variables.

$$F\frac{k}{j} = \frac{Y}{X}$$

As we did when we looked at the t distribution, we now want to create a joint distribution by multiplying X and Y. Once we have that done and can represent it on a Cartesian plane we will be able to deal with the ratio between them. So we start by multiplying the two gamma distribution PDFs.

$$PDF(X) = \frac{x^{\frac{j}{2}-1}e^{\frac{-x}{2}}}{2^{\frac{j}{2}}\Gamma\left(\frac{j}{2}\right)}\ for\ x \geq 0$$

$$PDF(Y) = \frac{y^{\frac{k}{2}-1}e^{\frac{-y}{2}}}{2^{\frac{k}{2}}\Gamma\left(\frac{k}{2}\right)}\ for\ y \geq 0$$

$$PDF(X, Y) = \frac{x^{\frac{j}{2}-1}e^{\frac{-x}{2}}}{2^{\frac{j}{2}}\Gamma\left(\frac{j}{2}\right)}\frac{y^{\frac{k}{2}-1}e^{\frac{-y}{2}}}{2^{\frac{k}{2}}\Gamma\left(\frac{k}{2}\right)}$$

Rearrange. Combine the powers exponents of 2 and the exponents of e.

$$= \frac{x^{\frac{j}{2}-1}y^{\frac{k}{2}-1}e^{\frac{-(x+y)}{2}}}{2^{\frac{j+k}{2}}\Gamma\left(\frac{j}{2}\right)\Gamma\left(\frac{k}{2}\right)}$$

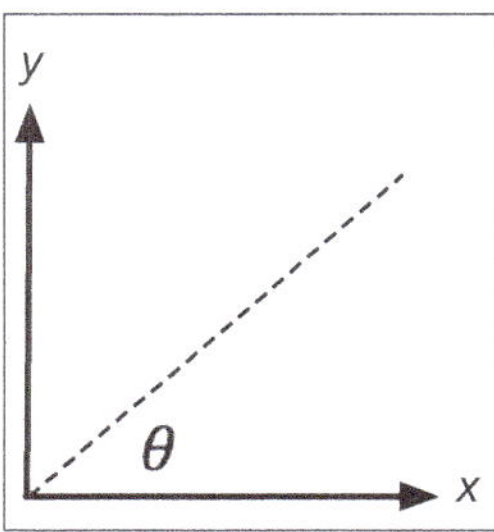

Figure 305

Now we have a joint PDF that we can draw on a plane. Chi square distributions are undefined for negative values, so neither x nor y can be negative. That means that every value of the joint PDF will be in the upper right quadrant of the plane (Figure 305). Thinking about which values of this joint PDF are relevant to our equation, we can treat $F\frac{k}{j}$ as a constant; let's call it C.

So $C = \frac{y}{x}$ $\quad y = Cx$. That means the points that are defined by our equation—that is, the value C—all fit into a line in the top right quadrant, starting at the origin. The perfect scenario to convert to polar coordinates! Thinking about the CDF we are going to be interested in the area under the line; that is, indicated by the angle θ.

We need a double integral for the joint CDF, and now we will convert it to polar coordinates.

$$\iint \frac{x^{\frac{j}{2}-1} y^{\frac{k}{2}-1} e^{\frac{-(x+y)}{2}}}{2^{\frac{j+k}{2}} \Gamma(\frac{j}{2}) \Gamma(\frac{k}{2})} \, dx \, dy$$

$$\frac{1}{2^{\frac{j+k}{2}} \Gamma(\frac{j}{2}) \Gamma(\frac{k}{2})} \iint (r\cos\theta)^{\frac{j}{2}-1} (r\sin\theta)^{\frac{k}{2}-1} e^{\frac{-r(\cos\theta+\sin\theta)}{2}} r \, dr \, d\theta$$

The constants can be moved outside the integration. As always when we switch to polar coordinates $x = r\cos\theta$ and $y = r\sin\theta$. As always when we convert to polar coordinates, we add the extra r (Jacobian), so we have $dx \, dy = r \, dr \, d\theta$.

Now in terms of the limits of integration, r will be between 0 and ∞. When we move one unit along on the x axis, the value of y at that point will be C. That forms a triangle, so we can say that $\tan\theta = \frac{C}{1} = C$. That in turn means that $\theta = \tan^{-1} C$ (tan inverse of C). That means the limits for the outer integral will be between 0 and $\tan^{-1} C$

$$\frac{1}{2^{\frac{j+k}{2}} \Gamma(\frac{j}{2}) \Gamma(\frac{k}{2})} \int_0^{\tan^{-1} C} \int_0^{\infty} (r\cos\theta)^{\frac{j}{2}-1} (r\sin\theta)^{\frac{k}{2}-1} e^{\frac{-r(\cos\theta+\sin\theta)}{2}} r \, dr \, d\theta$$

For the next step separate the powers of r in the integral from the powers of trig functions in the outer integral, and combine the powers of r.

$$\frac{1}{2^{\frac{j+k}{2}} \Gamma(\frac{j}{2}) \Gamma(\frac{k}{2})} \int_0^{\tan^{-1} C} (\cos\theta)^{\frac{j}{2}-1} (\sin\theta)^{\frac{k}{2}-1} \int_0^{\infty} r^{\frac{j+k}{2}-2} e^{\frac{-r(\cos\theta+\sin\theta)}{2}} r \, dr \, d\theta$$

The Jacobian r multiplies by $r^{\frac{j+k}{2}-2}$ to make $r^{\frac{j+k}{2}-1}$

$$\frac{1}{2^{\frac{j+k}{2}} \Gamma(\frac{j}{2}) \Gamma(\frac{k}{2})} \int_0^{\tan^{-1} C} (\cos\theta)^{\frac{j}{2}-1} (\sin\theta)^{\frac{k}{2}-1} \int_0^{\infty} r^{\frac{j+k}{2}-1} e^{\frac{-r(\cos\theta+\sin\theta)}{2}} \, d$$

Now we will use a substitution for the inner integral. Let $u = \frac{r(\cos\theta+\sin\theta)}{2}$

That means $r = \frac{2u}{(\cos\theta+\sin\theta)}$ and $\frac{du}{dr} = \frac{(\cos\theta+\sin\theta)}{2}$ $\quad dr = \frac{2}{(\cos\theta+\sin\theta)} \, du$

$$\frac{1}{\Gamma(\frac{j}{2})\Gamma(\frac{k}{2})} \int_0^{\tan^{-1}c} (\cos\theta)^{\frac{j}{2}-1}(\sin\theta)^{\frac{k}{2}-1} \int_0^{\infty} \frac{2^{\frac{j+k}{2}-1}2}{2^{\frac{j+k}{2}}}\left(\frac{u}{(\cos\theta+\sin\theta)}\right)^{\frac{j+k}{2}-1} e^{-u}\frac{1}{(\cos\theta+\sin\theta)}\,du\,d\theta$$

Make the substitution, and bring the constant $\dfrac{1}{2^{\frac{j+k}{2}}}$ from outside both integrals, back into the inner integral. Bring the 2s and their powers together, and they all cancel out.

$$\frac{1}{\Gamma(\frac{j}{2})\Gamma(\frac{k}{2})} \int_0^{\tan^{-1}c} (\cos\theta)^{\frac{j}{2}-1}(\sin\theta)^{\frac{k}{2}-1} \int_0^{\infty} \left(\frac{u}{(\cos\theta+\sin\theta)}\right)^{\frac{j+k}{2}-1} e^{-u}\frac{1}{(\cos\theta+\sin\theta)}\,du\,d\theta$$

Simplify further.

$$\frac{1}{\Gamma(\frac{j}{2})\Gamma(\frac{k}{2})} \int_0^{\tan^{-1}c} (\cos\theta)^{\frac{j}{2}-1}(\sin\theta)^{\frac{k}{2}-1} \int_0^{\infty} \frac{u^{\frac{j+k}{2}-1}}{(\cos\theta+\sin\theta)^{\frac{j+k}{2}}} e^{-u}\,du\,d\theta$$

Take $\dfrac{1}{(\cos\theta+\sin\theta)^{\frac{j+k}{2}}}$ outside of the inner integral

$$\frac{1}{\Gamma(\frac{j}{2})\Gamma(\frac{k}{2})} \int_0^{\tan^{-1}c} \frac{(\cos\theta)^{\frac{j}{2}-1}(\sin\theta)^{\frac{k}{2}-1}}{(\cos\theta+\sin\theta)^{\frac{j+k}{2}}} \int_0^{\infty} u^{\frac{j+k}{2}-1}e^{-u}\,du\,d\theta$$

$\int_0^{\infty} u^{\frac{j+k}{2}-1}e^{-u}\,du$ is the gamma function: $\Gamma\left(\frac{j+k}{2}\right)$.

$$= \frac{\Gamma\left(\frac{j+k}{2}\right)}{\Gamma(\frac{j}{2})\Gamma(\frac{k}{2})} \int_0^{\tan^{-1}c} \frac{(\cos\theta)^{\frac{j}{2}-1}(\sin\theta)^{\frac{k}{2}-1}}{(\cos\theta+\sin\theta)^{\frac{j+k}{2}}}\,d\theta$$

We've now been able to eliminate one of our variables. When we think about our joint distribution drawn on a plane, the r value is irrelevant; any point on the line, at any distance from the origin, will still satisfy the equation. The inner integral we were looking at has now become a constant and can be taken outside of the integral.

We've seen that the beta function can be defined in terms of the gamma function: $\beta(x,y) = \frac{\Gamma(x)\Gamma(y)}{\Gamma(x+y)}$

$$= \frac{1}{\beta\left(\frac{j}{2},\frac{k}{2}\right)} \int_0^{\tan^{-1}c} \frac{(\cos\theta)^{\frac{j}{2}-1}(\sin\theta)^{\frac{k}{2}-1}}{(\cos\theta+\sin\theta)^{\frac{j+k}{2}}}\,d\theta$$

That means that the constant outside the integral can be written as the reciprocal of a beta function.

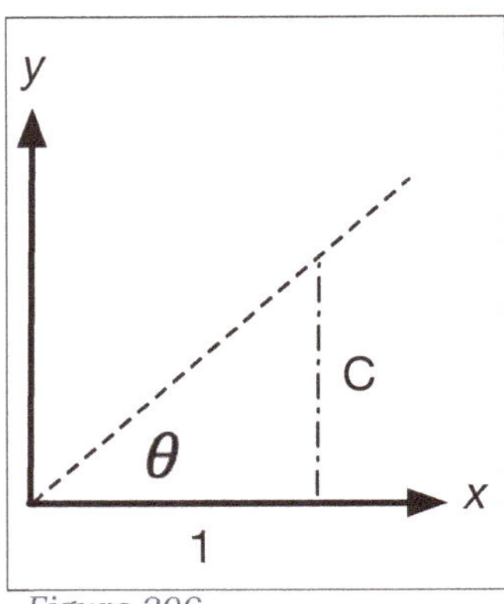

Figure 306

We still have what looks like an ugly integral to deal with, but remember we don't actually have to evaluate the integral, because we are after the PDF. So it would normally be a case of just ignoring the integral sign and evaluating at the limit of integration. However our limit of integration is a function, so we still have some work to do. If we substitute in $\tan^{-1} C$ for θ then we are finding the derivative of a function within a function. The outer function is what we see in front of us, and the inner function is $\tan^{-1} C$. That means, to take the derivative we need to use the chain rule. The derivative of $\tan^{-1} C$ is

$$\frac{1}{1 + C^2}$$

Using the chain rule we need to multiply by this. Now we are evaluating this at the point where $\theta = \tan^{-1} C$ (Figure 306). In other words, the point where the y value is C. We saw earlier that if we move 1 unit along the x axis, the y value will be given by C. That means that the hypotenuse of the triangle is $\sqrt{1 + C^2}$. So $\cos \theta = \frac{1}{\sqrt{1+C^2}}$ (adjacent over hypotenuse), and $\sin \theta = \frac{C}{\sqrt{1+C^2}}$ (opposite over hypotenuse). Putting that all together gives us:

In the denominator,
$$\frac{1}{\sqrt{1+C^2}} + \frac{C}{\sqrt{1+C^2}} = \frac{1+C}{\sqrt{1+C^2}}$$

$$\frac{1}{\beta\left(\frac{j}{2},\frac{k}{2}\right)} \frac{\left(\frac{1}{\sqrt{1+C^2}}\right)^{\frac{j}{2}-1} \left(\frac{C}{\sqrt{1+C^2}}\right)^{\frac{k}{2}-1}}{\left(\frac{1+C}{\sqrt{1+C^2}}\right)^{\frac{j+k}{2}}} \frac{1}{1+C^2}$$

Separate C from $\frac{1}{\sqrt{1+C^2}}$ from the term on the right of the numerator.

$$= \frac{1}{\beta\left(\frac{j}{2},\frac{k}{2}\right)} \frac{\left(\frac{1}{\sqrt{1+C^2}}\right)^{\frac{j}{2}-1} C^{\frac{k}{2}-1}\left(\frac{1}{\sqrt{1+C^2}}\right)^{\frac{k}{2}-1}}{\left(\frac{1+C}{\sqrt{1+C^2}}\right)^{\frac{j+k}{2}}} \frac{1}{1+C^2}$$

Combine the powers of $\frac{1}{\sqrt{1+C^2}}$ in the numerator.

$$= \frac{1}{\beta\left(\frac{j}{2},\frac{k}{2}\right)} \frac{\left(\frac{1}{\sqrt{1+C^2}}\right)^{\frac{j+k}{2}-2} C^{\frac{k}{2}-1}}{\left(\frac{1+C}{\sqrt{1+C^2}}\right)^{\frac{j+k}{2}}} \frac{1}{1+C^2}$$

In the denominator,

$$\left(\frac{1+C}{\sqrt{1+C^2}}\right)^{\frac{j+k}{2}} = (1+C)^{\frac{j+k}{2}}\left(\frac{1}{\sqrt{1+C^2}}\right)^{\frac{j+k}{2}}$$

$$= \frac{1}{\beta\left(\frac{j}{2},\frac{k}{2}\right)}\frac{\left(\frac{1}{\sqrt{1+C^2}}\right)^{\frac{j+k}{2}-2}C^{\frac{k}{2}-1}}{(1+C)^{\frac{j+k}{2}}\left(\frac{1}{\sqrt{1+C^2}}\right)^{\frac{j+k}{2}}}\frac{1}{1+C^2}$$

Separate out $\left(\frac{1}{\sqrt{1+C^2}}\right)^{\frac{j+k}{2}-2}$ in the numerator into what we have.

$$= \frac{C^{\frac{k}{2}-1}}{\beta\left(\frac{j}{2},\frac{k}{2}\right)}\frac{\left(\frac{1}{\sqrt{1+C^2}}\right)^{\frac{j+k}{2}}\left(\frac{1}{\sqrt{1+C^2}}\right)^{-2}}{(1+C)^{\frac{j+k}{2}}\left(\frac{1}{\sqrt{1+C^2}}\right)^{\frac{j+k}{2}}}\frac{1}{1+C^2}$$

Note that $\left(\frac{1}{\sqrt{1+C^2}}\right)^{-2} = 1+C^2$. Cancel out.

$$= \frac{C^{\frac{k}{2}-1}}{\beta\left(\frac{j}{2},\frac{k}{2}\right)}\frac{\left(\frac{1}{\sqrt{1+C^2}}\right)^{\frac{j+k}{2}}(1+C^2)}{(1+C)^{\frac{j+k}{2}}\left(\frac{1}{\sqrt{1+C^2}}\right)^{\frac{j+k}{2}}}\frac{1}{1+C^2}$$

$$= \frac{C^{\frac{k}{2}-1}}{\beta\left(\frac{j}{2},\frac{k}{2}\right)(1+C)^{\frac{j+k}{2}}}$$

Note that $C^{\frac{k}{2}-1} = \frac{1}{C}\,C^{\frac{k}{2}}$

$$= \frac{C^{\frac{k}{2}}}{C\beta\left(\frac{j}{2},\frac{k}{2}\right)(1+C)^{\frac{j+k}{2}}}$$

Note that $\frac{C^{\frac{k}{2}}}{(1+C)^{\frac{j+k}{2}}} = \sqrt{\frac{C^k}{(1+C)^{j+k}}}$ (factor $\frac{1}{2}$ which is,

$$= \frac{1}{C\beta\left(\frac{j}{2},\frac{k}{2}\right)}\sqrt{\frac{C^k}{(1+C)^{j+k}}}$$

of course, the same as the square root, out of each of the exponents).

So we are almost there, but remember that what we have been working with is our F variable multiplied by $\frac{k}{j}$, which are our degrees of freedom.

$$F\frac{k}{j} = \frac{1}{C\beta\left(\frac{j}{2},\frac{k}{2}\right)}\sqrt{\frac{C^k}{(1+C)^{j+k}}}$$

On the right we have a PDF, so we need to transform the PDF to find the PDF of the distribution of F. Our definition was

$$C = F\frac{k}{j}$$

Therefore

$$F = C\frac{j}{k}$$

And

$$\frac{d\mathrm{F}}{dC} = \frac{j}{k}$$

Our Jacobian for the transformation is therefore the absolute value of the reciprocal of this:

$$\left|\frac{k}{j}\right|$$

In this instance we don't need to worry about the absolute value signs; both k and j are natural numbers (the degrees of freedom) which means they must be positive. Rather than use a lower case F as the variable (which, given the way f is used in function notation, could get particularly confusing, we will make the variable in the equation x. This also means the final result will be consistent with common usage. It is important to remember, though, that this variable shouldn't be confused with the use we made of x at the beginning of this proof. We are now using it to represent something quite different. So the substitution we are making is $C = x\frac{k}{j}$

Do some cancelling.

$$\mathrm{F}\frac{k}{j} = \frac{k}{j}\frac{1}{x\frac{k}{j}\beta\left(\frac{j}{2},\frac{k}{2}\right)}\sqrt{\frac{\left(x\frac{k}{j}\right)^k}{\left(1+\left(x\frac{k}{j}\right)\right)^{j+k}}}$$

Multiply the numerator and the denominator under the square root by j^{j+k}

$$= \frac{1}{x\beta\left(\frac{j}{2},\frac{k}{2}\right)}\sqrt{\frac{\left(x\frac{k}{j}\right)^k j^j}{\left(1+\left(x\frac{k}{j}\right)\right)^{j+k} j^{j+k}}}$$

In the numerator, take the j^k inside the brackets, and cancel. In the denominator, take the j^{j+k} inside the brackets.

$$= \frac{1}{x\beta\left(\frac{j}{2},\frac{k}{2}\right)}\sqrt{\frac{\left(x\frac{k}{j}j\right)^k j^j}{\left(\left[1+\left(x\frac{k}{j}\right)\right]j\right)^{j+k}}}$$

In the denominator under the square root, distribute the j and cancel.

$$\boxed{= \frac{1}{x\beta\left(\frac{j}{2},\frac{k}{2}\right)}\sqrt{\frac{(kx)^k j^j}{(j + kx)^{j+k}}}}$$

So that's the one-way ANOVA, the F distribution and the F statistic. There are a number of extensions on this basic idea.

Extending the idea of the ANOVA

Name of test	Description	Use of the F statistic	Example of use
Two-way ANOVA	Evaluates the effect of two independent variables on a dependent variable. It can also assess the interaction effect between the two independent variables.	Has two F statistics: one for each independent variable and one for the interaction between the two variables. Each F statistic is calculated similarly to one-way ANOVA and compared against the F distribution.	Examining the effects of two different teaching methods (Method A, Method B) and two levels of class size (Small, Large) on students' test scores.
Analysis of Covariance (ANCOVA)	Similar to ANOVA, but it includes one or more covariate variables that you believe may affect the dependent variable.	ANCOVA adjusts the dependent variable for differences in covariates before assessing the F statistic. The F statistic for ANCOVA tests the group effects after accounting for covariates and follows the F distribution.	Assessing the effectiveness of a new drug or reducing blood pressure. The study also measures participants' age and body mass index (BMI), as these factors are known to affect blood pressure.
Repeated Measures ANOVA (RM ANOVA)	Similar to one-way ANOVA, but the groups are related. It's used when the same subjects are used for each treatment.	This also involves an F statistic, which compares the variance due to the within-subjects variable (the repeated measures) to the error variance. This F ratio is checked against the F distribution.	Measuring the cognitive performance of participants at three different time points: before, during, and after a specific training program.

Name of test	Description	Use of the F statistic	Example of use
Mixed-model ANOVA (MM ANOVA)	Tests for differences between two or more independent groups while subjecting participants to repeated measures.	The mixed-model ANOVA generates F statistics for between-group differences and within-subjects effects. Each F statistic is compared against the F distribution to determine significance.	A longitudinal study on the effect of a dietary intervention on cholesterol levels involves both between-subject factors (e.g., gender, intervention group vs. control group) and within-subject factors (cholesterol level measured at multiple time points).
Multilevel Model (Hierarchical ANOVA):	Used for data that has a hierarchical structure, like students within classes within schools.	This type of analysis computes an F statistic for fixed effects, which follows the F distribution. It's a more complex model, but at its core, the F test is used to determine the significance of the effects being studied.	Investigating teaching methods across different schools, where schools belong to different districts.

Name of test	Description	Use of the F statistic	Example of use
MANOVA (multivariate-analysis of variance)	Similar to two-way ANOVA, but handles multiple dependent variables.	Does not rely on the F statistic, which, as we have seen, is based on an assumption of independence between the groups. Instead it uses multi-variate test statistics such as Wilks' Lambda, Pillai's Trace, Hotelling-Lawley Trace, or Roy's Greatest Root, which are based on transformed F distributions or approximations.	Evaluating the effect of therapy on several related mental health outcomes, such as anxiety, depression, and stress levels.

Key points:

- When doing pair-wise tests on multiple variables, the number of tests climbs faster than the number of variables. With multiple tests, the chances of a type I error are added with each test, meaning the chances of type I errors climb quickly with more variables.

- All you are asking with ANOVA is: when comparing samples from three or more populations, is there a significant difference between at least one of the means and the others? If there is, you can safely use other tests to analyse it.

- ANOVA uses $\dfrac{mean\ sum\ of\ squares\ between\ groups}{mean\ sum\ of\ squares\ within\ groups}$

- Mean sum of squares between groups is $\dfrac{\Sigma_{Groups}\,n_i(x_i-\bar{x})^2}{degrees\ of\ freedom}$ while mean sum of squares within groups is $\dfrac{\Sigma_{Groups}(n_i-1)s_i^2}{degrees\ of\ freedon}$

- The F distribution compares to chi square distributions, to work out the statistical significance (ultimately based on the normal distribution, via the chi square) of the ratio.

- There are a number of related tests to cope with related but different scenarios.

Distributions at a glance

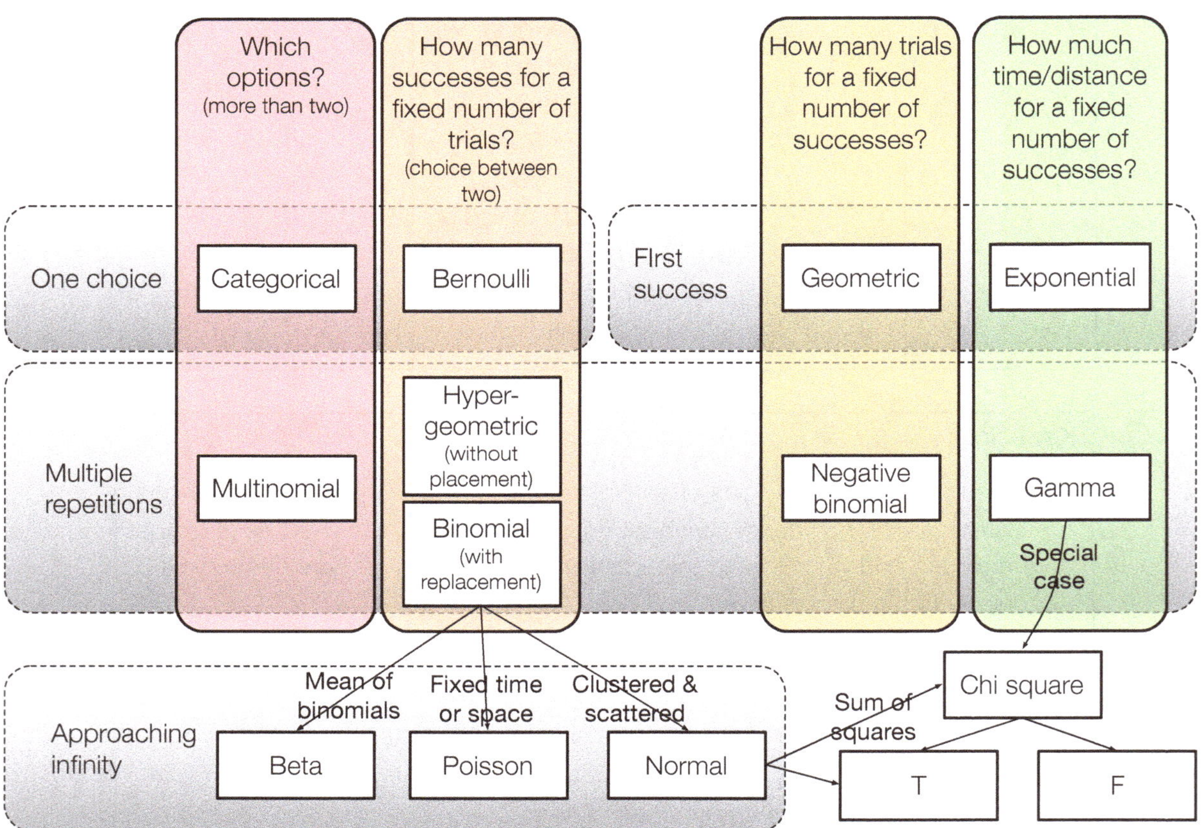

Figure 307

Figure 307 shows the distributions we have explored since we began by looking at the binomial distribution: how they relate to it, and to one another.

You may encounter other distributions, such as:

Distribution	Description	Relationships	Examples of use
Uniform	Every value has an equal probability		Applicable to any situation where you have no idea what the probability is. Sometimes used in Bayesian statistics where there s an uninformative prior (although for mathematical reasons researchers are more likely to use a form of the beta distribution to represent this.)
Cauchy	Created by a ratio of two normal distributions. A special case of the *t* distribution with a degree of freedom.		May be used in physics to describe resonance behaviour.
Bivariate binomial	Like a normal distribution with two variables		Can be used in quality control to model the number of defects in two related production processes.
Multivariate normal	Described using linear algebra (which this book doesn't cover).	Like a normal distribution with more than two variables	Used in finance to model asset returns that are correlated.
Inverse chi square	Distribution of the inverse of a variable distributed according to the chi square distribution.		In Bayesian statistics, it's used as a prior distribution for variance parameters.
Log-normal	If the logarithm of a variable has a normal distribution, then the variable itself is log-normally distributed.		Used to model stock prices in the Black-Scholes model.
Weibull	A flexible distribution used to model time-to-failure data with its shape parameter affecting the distribution form.	Related to the Exponential distribution as a generalization; it becomes exponential when its shape parameter equals 1	Widely used in reliability engineering and failure analysis.

Distribution	Description	Relationships	Examples of use
Rayleigh	A distribution of vectors' magnitudes when their components are independent, normally distributed with equal variance.	A special case of the Weibull distribution when the shape parameter is 2 and related to the chi square distribution with 2 degrees of freedom.	Used to model the lifetime of objects.
Pareto	Describes a distribution that represents the phenomenon of 80/20 rule, where a small percentage holds most of the quantity.	No direct relationship but similar to Beta in that it is often used in Bayesian statistics as a prior distribution.	Used in economics to describe wealth distribution.

Now it's time to step away from the normal distribution and others that were developed from it. First we'll look at something Fisher did with the hypergeometric distribution, then we'll look at what happens when there isn't a straightforward way to work with, or find parameters from, the distribution of an underlying population.

Fisher's exact test

> **Read this if** you want to know how to deal with categorical variables when there are too few samples for a chi square to be used.

Reading through Fisher's work, it's hard to escape the sense that, while statistics is based on the mathematics of uncertainty, Fisher was never completely comfortable with this idea. One of the great strengths he brought to the study of statistics was the precision of his thinking. Indeed, when he developed the idea of degrees of freedom, Karl Pearson's initial response wasn't to disagree; rather, he didn't think that subtracting 1 or 2 from a denominator would make much difference. (It can, of course, with smaller sample sizes). It may have been this relaxed attitude to precise thinking that most annoyed Fisher.

Regardless, he was probably happy to develop a test he could call his **exact test**. The actual reason for calling it "exact" is that, rather than giving some sort of range of probabilities, it gives the exact probability of getting any particular outcome.

This test is used as an alternative to the chi square test when the chi square can't be used. This mostly occurs when the numbers are too low for the chi square to be helpful. The general approach is that when the probable value in any square of a contingency table is less than 5, the chi statistic is inappropriate, and Fisher's exact test is used. It is almost always used with 2 by 2 contingency tables, but it doesn't have to be. That means that, in general, it works with categorical variables, but it doesn't make use of the multinomial distribution; it actually makes use of the hypergeometric.

	Party A	Party B
Female	12	5
Male	8	25

Imagine you took a small sample of voters, and wanted to know whether there was a statistically significant difference in the voting intentions between males and females on a two-party preferred basis.

The null hypothesis is that the proportions will be the same. There is an underlying assumption that the data will be independent. (If, for example, you were asking different members of married couples for their responses, the results might not be independent, and so the test wouldn't be appropriate.)

Recall that the hypergeometric distribution has the same effect as the binomial, except that it is without replacement; once you have made a choice, the number you are choosing from is reduced.

We saw that the calculation for the probability density function (PDF) is

$$\frac{(No.\,of\,combinations\,that\,will\,lead\,to\,success)(No.\,of\,other\,combinations)}{Total\,no.\,of\,samples\,that\,could\,possibly\,be\,chosen}$$

The parameters we start with are: n, the number in the sample, N, the number in the source from which the sample is taken, and R, the number of successful options. The number of options that represent failures is therefore $N - R$. The number of successes in a sample is represented by r.

We also saw that the probability mass function (PMF) is:

$$Pr(X = r) = \frac{\binom{R}{r}\binom{N-R}{n-r}}{\binom{N}{n}}$$

The relationship between the exact test and the hypergeometric distribution works like this:

	Party A	Party B	
Female	$r = 12$ (number of "successes" in sample)	5	$n = 17$ (size of sample)
Male	8	25	$N - n = 33$ (number not sampled)
	$R = 20$ (number of "successes")	$N - R = 30$ (number of "failures")	$N = 50$ (size of group from which the sample was taken)

Figure 308

To perform the test, look at each cell on its own, and call that result a "success", on the assumption that the whole row represents a sub-sample of the entire group. You can then use the parameters from Figure 308 as parameters in the hypergeometric distribution. That way you can work out the exact probability of achieving the result you did in each square. In this instance the probability of achieving 12 females who support party A is:

$$Pr(X = r) = \frac{\binom{R}{r}\binom{N-R}{n-r}}{\binom{N}{n}}$$

$$= \frac{\binom{20}{12}\binom{30}{5}}{\binom{50}{17}}$$

$$= \frac{(125{,}970)(142{,}506)}{9{,}847{,}379{,}391{,}150}$$

$$\approx 0.18\%$$

To cement your understanding: calculate a few more of these results for yourself. (You can use a spreadsheet's HYPGEOM.DIST function, a spreadsheet's COMBIN function to build each n choose r, the statistics software package of your choice or, if you are really courageous, you can work it out manually using factorials

---

Key points:

- Fisher's exact test is primarily used with two-by-two contingency tables, when there isn't enough data for a chi square test to be applicable.

- It is based on the hypergeometric distribution.

Non-Parametric Tests

What follows are sometimes described as **non-parametric tests** or **distribution-free methods**. These descriptions are a bit problematic. In most cases we do, in fact, use a probability distribution to determine a probability with these tests. The reason for the name is they aren't generally based on finding either the shape or any particular parameters of the underlying population. They tend to look at whether two distributions are roughly the same in terms of the median and/or the spread of data.

Of the commonly used so-called non-parametric tests, the majority use ordinal variables: the random variable is ranked. You may collect any set of data, but the process of creating a random variable from the measures is a process of ranking them from 1 to whatever number of them there are.

Note that a ranked variable is still a random variable, because you don't know what rank any particular experimental outcome you do will have, until you undertake the experiment or intervention. You are still converting an outcome to a number.

Examples of where this *may* be appropriate (and you really need to look at it on a case-by-case basis to be sure that it makes sense) are:

1) **customer surveys** such as 'very satisfied', 'satisfied', 'neutral', 'dissatisfied' and 'very dissatisfied';

2) **opinion surveys** with Likert scale items responses, including 'strongly agree', 'agree', 'neutral', 'disagree', 'strongly disagree';

3) **grades** like A, B, C, D, E and F ;

4) **ratings** of products or performance on a star system (1 to 4 stars) or the like;

5) **competitive results** such as 'gold', 'silver' and 'bronze';

6) **symptom severity** such as 'mild', 'moderate', 'severe';

7) **economic classifications** like 'low income', 'middle income' and 'high income';

8) **risk levels** such as 'low', 'moderate' or 'high'.

In ranking the numbers, what we end up with is a distribution of natural (counting) numbers, usually starting from 1.

Some of the
non-parametric methods

Read this if you want to know how to deal with ordinal variables, and some other situations where the distributions we have studied aren't appropriate but you still want to use statistics to determine the probability that results are based on random chance.

There are many non-parametric tests for different situations. These are some common examples. We'll explore those marked with an asterisk in detail.

Test	Purpose	Equivalent
*Spearman's Rank Correlation Coefficient	To assess the strength and direction of the association between two ranked variables.	Pearson's correlation coefficient
Kendall's Tau	To measure the correspondence between two rankings and to assess the significance of this correspondence.	Pearson's correlation coefficient
*Sign test and the Wilcoxon Signed-Rank Test	To compare related or matched samples to detect differences when the distribution is not assumed to be normal.	Paired samples t-test
*Mann-Whitney U Test (Wilcoxon Rank-Sum Test)	To compare differences between two independent groups when the dependent variable is not normally distributed.	Independent samples t-test
Kruskal-Wallis H Test	To compare more than two independent groups on a rank-based approach when the ANOVA assumptions aren't met.	One-way ANOVA

Test	Purpose	Equivalent
Friedman Test	To compare more than two related groups using ranks, often used with repeated measures.	Repeated measures ANOVA
Cochran's Q Test	To test for differences in proportions in related samples.	Repeated measures ANOVA
***Cohen's kappa co-efficient**	To test inter-examiner/inter-rater reliability	Chi-square test

There are more tests than this, of course, and no doubt more will continue to be developed. As a general rule these tests don't have the same power as some of the tests we have seen previously, and so to establish statistical significance you need a clearer result or a bigger sample.

Before we look at those marked with an asterisk, we need to think about mathematically handling ordinal (ordered, or ranked) variables.

Mean, variance and standard deviation of ranks

Reading	Rank (natural number)
43	1
43.7	2
52	3
53.47	4
56	5

Figure 309

The distribution of natural numbers does have parameters. Great care needs to be used when using these parameters to describe data that is ranked, but the parameters do exist.

When we first looked at series (page 134) we saw why the sum of n natural numbers starting with 1 is given by:

$$\sum_{x=1}^{n} x = 1 + 2 + 3 + \cdots + n = \frac{n(n+1)}{2}$$

Therefore the mean of this set of numbers is

$$\frac{\sum_{x=1}^{n} x}{n} = \frac{n(n+1)}{2n}$$

$$Mean = \frac{n+1}{2}$$

For the ranked numbers at Figure 309, the mean ranking will therefore be

$$\frac{5+1}{2} = 3$$

Conveniently, the mean will also be the median, and, if the measures that are collected are mapped one-on-one to the natural numbers, as per Figure 309, then the mean of the ranks will be the median of the underlying data.

To cement your understanding:

Write out a sequence of natural (counting) numbers from 1 to some odd number, and repeat for a sequence from 1 to some even number, to confirm for yourself that this formula will give you both the mean and the median.

The natural numbers also have a variance. In general, a variance can be calculated using the formula:

$$E(x^2) - \left(E(x)\right)^2$$

Substituting $\left(\frac{n+1}{2}\right)$ for the mean, we get:

$$\sigma^2 = \left(\frac{\sum_{x=1}^{n} x^2}{n}\right) - \left(\frac{n+1}{2}\right)^2$$

The formula for the sum of the squares of natural numbers is

$$\frac{n(2n+1)(n+1)}{6} = \frac{2n^3 + 3n^2 + n}{6}$$

where n is the highest number being squared. There is an appendix covering the derivation of this formula, on page 736.

That in turn means we can calculate the variance of an ordered set of natural numbers like this:

This is based on what we did immediately above.

$$\sigma^2 = \left(\frac{2n^3 + 3n^2 + n}{6n}\right) - \left(\frac{n+1}{2}\right)^2$$

Expand out the term on the right.

$$= \frac{2n^3 + 3n^2 + n}{6n} - \left(\frac{n^2 + 2n + 1}{4}\right)$$

Multiply the numerator and denominator of the term on the left by 2, and numerator and denominator of the term on the right by $3n$, to create a common denominator.

$$= \frac{4n^3 + 6n^2 + 2n}{12n} - \left(\frac{3n^3 + 6n^2 + 3n}{12n}\right)$$

$$= \frac{4n^3 + 6n^2 + 2n - 3n^3 - 6n^2 - 3n}{12n}$$

$$= \frac{n^3 - n}{12n}$$

$$\sigma^2 = \frac{n^2 - 1}{12}$$

That makes the standard deviation:

$$\sigma = \sqrt{\frac{n^2 - 1}{12}}$$

Simply describing the mean and standard deviation of ranked data has the potential to lead to statistical errors, but with appropriate care and forethought, these calculations can legitimately be used in some of the techniques that follow.

Spearman's rank correlation coefficient

Spearman's rank correlation coefficient is the equivalent of Pearson's correlation, but uses ranked data. Examples of its use include: comparing rankings for the same subjects in different tests, and comparing the rankings of customer satisfaction with sales figures, by rank, for different stores.

While this is often included with "non-parametric tests" when compared to Pearson's correlation, if this is a "non-parametric test" or a "distribution-free method", then so is Pearson's correlation. (You may sometimes hear that Pearson's correlation is dependent on the distribution of the data, but of course it is not. The distribution only becomes relevant when using a t test to determine the statistical significances of the Pearson correlation.)

As with Pearson's correlation there are a number of expressions of the formula. The calculation (whichever form you choose) is based on starting with Pearson's formula, letting the x and y represent ordered, non-repeating (that bit's important) ordered numbers, then substituting in the values for the mean and the standard deviation that we noted above.

The sign test

The sign test is an alternative to either the single-sample or the paired t test when the necessary assumptions for the t test don't hold up. It either uses paired results (such as before-and-after results for the same subjects) or compares each result to a known population medium. You may hear that the data have to be continuous for it to be used, but this isn't correct. What is important is that there is a numerical difference between either the two results (if it is paired) or the result and the population medium (if it is single-sample).

The variable used is the *sign of the difference* between the two measures; a Bernoulli choice. If the sign is positive it can be counted as $+1$, and if the sign is negative it will be counted as -1. Ties are ignored. For this reason it is preferable, if possible, to use continuous rather than discrete data. In other words, there would be less chance of ties if weight were measured on a continuous scale rather than a discrete, whole-kilogram scale, meaning that more of the collected measures would be usable.

You may think that at this stage you would add the total and work with that, but if you know the total number of results without ties, then one of the other totals (either the total negative or the total positive) is redundant, so you only work with one of these totals.

Because the total of the Bernoulli choices is added, then the relevant distribution is the binomial, with a probability of 0.5 and an n of the number of pairs of results that aren't tied. You can work with either the higher score or the lower one, and you calculate it as a percentage of n. You can then use a binomial distribution to work out the chance of getting that value or lower (if you pick the lower value), or that value or higher (if you pick the higher one. You need to calculate these differently, depending on whether you are doing a one-tailed or a two-tailed test.

The advantages of this are that it can be used in situations where the t test can't, and that what you are doing can be made transparent. Compare this with a t test or a chi square test where there is an underlying assumption of normality in sample sizes or in errors, which may not be obvious as the basis for people using or reading about the test. The major disadvantage—and the reason it is rarely used in published papers—is that a lot of information is lost. The test completely ignores how large the differences are, and only considers whether they are positive or negative. The outcome of this is that a test is likely to require many more subjects, in order to show a significant result; in other words, it is a relatively low-powered test. Just how much power is lost depends, of course, on the magnitude of the differences.

Wilcoxon signed rank test

This was developed by Irish-American chemist/statistician Frank Wilcoxon (1892-1965), and adds additional information to the sign test, by sorting by the absolute value of the differences, then assigning a rank to them based on where they come in the sort, starting from 1 for the smallest difference. In other words, the sort is based on the magnitude of the difference, but not on its direction. Where values are tied they are given an average of the rank and the following rank, so that if two values had the same score which would place them in the 4th position, they will each be assigned a rank of 4.5, and the next rank will be 6. The sign is then reapplied to the ranks. These ranks are then calculated (now by a computer, once based on tables generated by enumerating all possible combinations of ranks). For larger sample sizes an approximation based on the normal distribution is usually used.

Mann-Whitney U test/Wilcoxon rank-sum test

This test is known by two different names because it was developed independently by Frank Wilcoxon who introduced a version of the test in 1945, and by American mathematics professor Henry Mann (1905–2000) and statistician Donald Whitney (1915–2001), who published a similar method in 1947. Consequently, it is often referred to as the Wilcoxon rank-sum test or the Mann-Whitney U test. Whereas the Wilcoxon signed rank test works for paired and single-sample data, this one is the equivalent of the (unpaired) two-sample t test for ranked data.

To start with you combine all observations from the two groups, then rank them from smallest to largest, regardless of which group they belong to. If there are tied values, you assign to each tied value the average of the ranks they would have received if they weren't tied.

Then separately sum the ranks from the different samples. The **U statistic** is the number of times observations in one group precede observations in the other group in the ranked list. This is then compared to a critical value. Like the Wilcoxon signed rank test, this is compared to its own distribution which doesn't have a probability mass function (PMF); it is calculated based on all the possible combinations.

Key points:

- The **mean** of a set of **natural numbers** (which can be used to keep track of the rank of a particular score) is

$$\frac{n + 1}{2}$$

  where n is the highest number.

- The **standard deviation** is

$$\sqrt{\frac{n^2 - 1}{12}}$$

- **Spearman's rank correlation coefficient** assesses the strength and direction of the association between two ranked variables. The calculation is based on starting with Pearson's correlation formula, letting the x and y represent ordered, non-repeating, ordered numbers, then substituting in the values for the mean and the standard deviation that we noted above.

- The **sign test** is an alternative to either the single-sample or the paired t test when the necessary assumptions for the t test don't hold up. It compares the number of positive or negative signs between two variables, and uses a binomial distribution to assess the statistical validity.

- **Wilcoxon signed rank test** adds additional information to the sign test, by sorting by the absolute value of the differences, then assigning a rank to them based on where they come in the sort, starting from 1 for the smallest difference.

- **Mann-Whitney U test/Wilcoxon rank-sum test** (same thing) is the equivalent of the unpaired two-sample t test for ranked data. The U statistic is the number of times observations in one group precede observations in the other group in the ranked list. This is then compared to a critical value. Like the Wilcoxon signed rank test, this is compared to its own distribution which doesn't have a PMF; it is calculated based on all the possible combinations.

Cohen's kappa coefficient (another cautionary tale)

> **Read this if** you have come across Cohen's kappa coefficient as a way of assessing inter-examiner reliability, or you are curious to know how another test that doesn't make sense and doesn't work could come to be widely accepted.

Jacob Cohen (1923-1998) was an American statistician who worked in the psychology department of New York University. He was a pioneer in the mathematics of power analysis, and a co-author of one of the best introductory textbooks on statistics[85]. Being involved with behavioural sciences, one of his interests was in finding a way to check the reliability of two or more different people assessing the same phenomenon.

Reliability and validity are different concepts in research. Reliability refers to the *consistency* of a measure, meaning that the results are repeatable under the same conditions, while validity concerns the *accuracy* of the measure, indicating whether the research actually measures what it intends to measure.

To address this issue he developed Cohen's **kappa coefficient** for use in **inter-examiner reliability** studies: studies that try to provide insight into how much reliance should be placed on a particular test or measure, based on whether different people who are using it can agree about their findings. (Kappa is the Greek letter κ, and the statistic is sometimes written as '$\kappa =$'.) The idea behind the kappa coefficient is that if you ask two people independently about their reading of a particular phenomenon/measure, and simply report on the percentage of agreement and disagreement, you don't consider the chances that their agreement was based on random chance.

The kappa coefficient is supposed to deal with that issue, in relation to categorical data. (Agreement between numeric or ordinal variables can be dealt with by Pearson's correlation and Spearman's rank correlation, respectively.) Examples of categories could include a psychological diagnosis or styles of leadership.

		Judge A			
	Category	1	2	3	
	1	.25 (.20)	.13 (.15)	.12 (.15)	**.50**
Judge B	2	.12 (.12)	.02 (.09)	.16 (.09)	**.30**
	3	.03 (.08)	.15 (.06)	.02 (.06)	**.20**
		.40	**.30**	**.30**	**1.00**

Figure 310

You start by setting up a contingency table, similar to the ones we saw when we were looking at joint and marginal distributions, chi square and Fisher's exact test. Figure 310 shows the one he used in his original paper on the subject, from 1960[86].

In this table the numbers outside the brackets represent the percentage of times that each judge allocated a particular category to whatever they were observing. This is called the **observed percentage.** The numbers inside the brackets represent what you would get if the assessments were completely independent: the joint distribution based on the marginal totals (i.e. multiplying the relevant total of each row and each column). (If you are hazy about this, refer back to page 402). The idea is that if the observations are independent, then that is the same as the result you would expect from random chance. Cohen correctly pointed out that the chi square distribution isn't addressing the issue he was interested in and doesn't give helpful answers in this situation.

The cells in the table he was interested in are those in the top-left to bottom-right diagonal (coloured red); the cells where the judges agreed with one another about the categories they placed what they observed into. His formula involved summing the percentages where they agreed (that is .25 + .02 + .02 = .29), summing the agreement expected by random chance (.20 + .09 + .06 = .35), subtracting the latter from the former, and showing this as a proportion of the result not expected by chance: that is, 1 − .35 = .65

That means his formula was:

$$\kappa = \frac{p_0 - p_c}{1 - p_c}$$

where p_0 is the observed proportion of agreement, and p_c is the proportion of agreement based on chance. Doing it this way it is impossible to get agreement higher than 1, complete disagreement should be -1 (although that is more mathematically difficult) and agreement based entirely on random change would be 0. In other words, the measurements correspond well to other measures of correlation. In his paper he then went on to say that the usual measures of calculating z scores and applying a normal distribution could be used.

All of that seems highly plausible, and it is easy to see how Cohen saw it as a helpful way to solve a challenging and important problem. It has become a widely-used tool and the standard statistic for inter-examiner reliability studies[87]. Figure 311 shows the results of a PubMed (medical database) search in early 2026 for the use of the words "kappa coefficient" (excluding the use of the word "kappa" in the title, to exclude those studies specifically reviewing the use of that statistic). As you can see, it yielded 8,080 papers. Note that this search only looked at papers related to the field of health and medicine. Many fields of study outside of medicine make use of people's ability to accurately observe things, and there are therefore more disciplines where researchers need to study inter-examiner reliability.

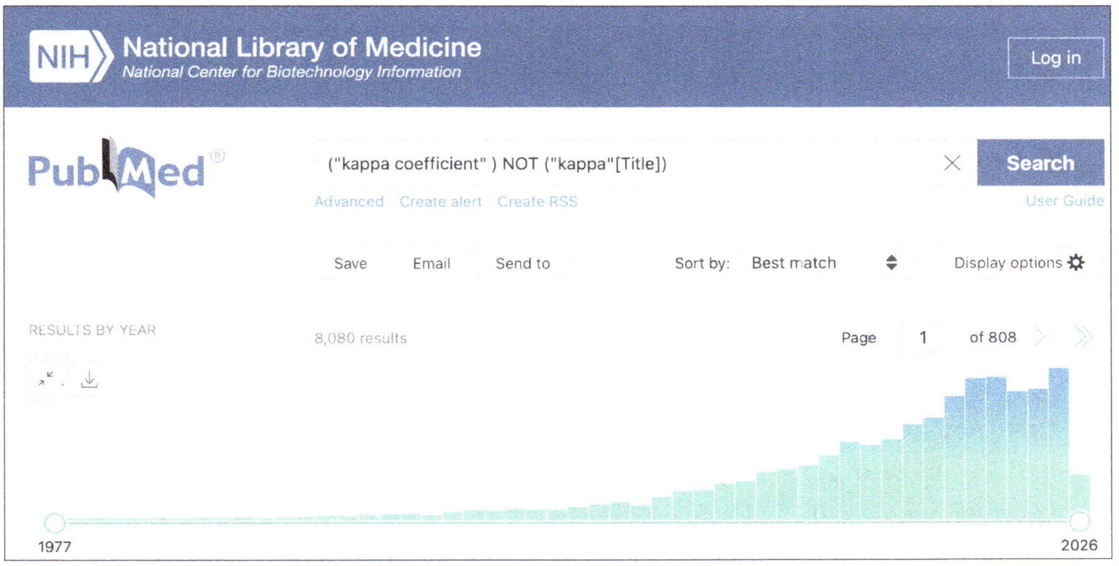

Figure 311

Now that you know more about the mathematics underlying statistics, you may be able to see some fairly large problems with it.

> **To cement your understanding:**
>
> Can you see any flaws in the reasoning behind this? Hint: think about the use of the marginal distributions as the underlying basis for deciding what chance-agreement would look like. Is that really a suitable measure?

As you may have seen, the big problem with the logic behind this measure is that the marginal distributions are not fixed, are not a function of how much agreement there is between the examiners, and changes to them can dramatically distort the statistic. Imagine, for example, that almost all of the subjects being examined for a medical condition actually have it. Or imagine that half of them have it, and half of them don't. If the examiners agree about this when they examine the subjects, these two situations will give completely different values for p_c—what "chance agreement" is supposed to be. You may think that this issue is accounted for by subtracting p_c from both the numerator and denominator, but it is not. If you subtract the same amount from the numerator as you do from the denominator you change the fraction.

$$\frac{2-1}{3-1} = \frac{1}{2}$$

$$\neq \frac{2}{3}$$

If the numerator—the level of observed agreement less the level of chance agreement—remains constant, but the level of chance agreement is different, then the denominator will change when the numerator does not. This means that the same level of agreement can give very different kappa coefficients.

Figure 312 shows a hypothetical range of scenarios in which two examiners report on their level of agreement that a particular phenomenon is positive or negative, over many trials in each scenario[88]. If they agree 80% of the time about a particular finding in each scenario, but with different combinations of what they agree and disagree about, the kappa coefficient will determine that they have anything between, at best, moderate agreement (a score of 0.6154 out of a possible 1.0) to agreement *worse* than random chance (a score of *negative* 0.1111).

It is even more serious than that. This statistical test cannot give a high score if agreement is low, but it can give a low score if agreement is high. In fact, if they agree 98% of the time that something is positive, and disagree just 2% of the time, the reading can still show that their agreement is worse than random chance (−0.0101). This makes it worse than unreliable; it biases test results towards showing that examiners cannot agree, even when they can.

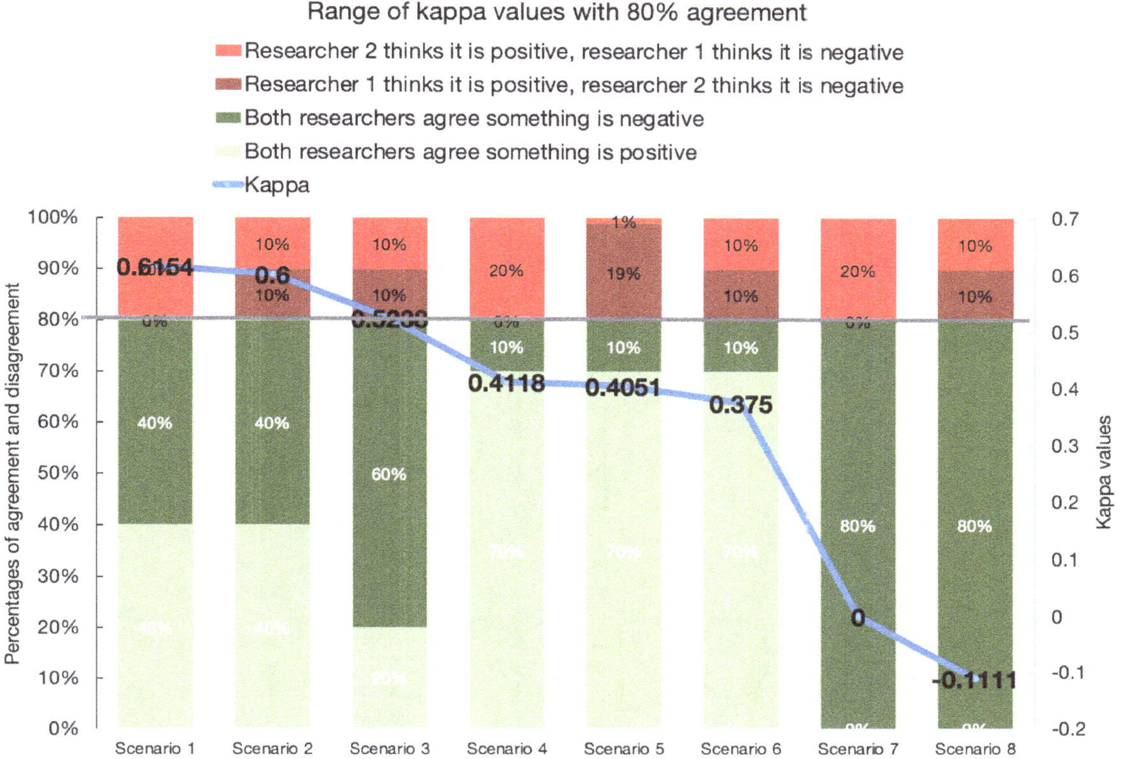

Figure 312

Several writers realised that the statistic was problematic and wrote journal articles saying so, and proposing better solutions, at least back as far as the 1990s[89]. They were largely ignored. It may seem incredible that so many researchers should rely on an unreliable statistic and, ironically, one that is supposed to evaluate reliability. That is, however, what has happened.

If it has happened once, why not more than once? It isn't as though no one has noticed that this is a problem; that has been known, and written about, for decades. Unfortunately the people using it in research must not have understood it; and moreover they presumably haven't been exposed to the statistics papers discouraging its use.

One of the things we have seen repeatedly in our journey through the world of statistics is that plausibility is not the same as trustworthiness. We've seen that some implausible-seeming statistics can be completely right and useful; the kappa coefficient is an example of the opposite. There is no substitute for understanding the underlying mathematical and other ideas, and thinking it through for yourself.

Part 3: Simplicity on the Other Side of Complexity

Putting it together

Whether you are planning a research project, reading research and wondering whether statistics has been used appropriately, or deciding whether statistical analysis can be useful with whatever data you deal with (and with the data you *don't* have, but would like to), the following questions are relevant. It may seem obvious at this stage that these are the important questions, but they are often left as implied, rather than specifically stated, in journal articles. Given the error rates in published statistics, thinking them through for yourself can be valuable.

(If you are attempting to replicate another study in your field of interest it is particularly inappropriate to use the same statistical measures that earlier researchers used, without thinking them through for yourself. The whole point of a replication study is to bring fresh eyes to the subject, to see if the results still stand.)

- What raw data is of interest, and what **type of data** is it? (See pages 360–369.)

- What **conditions and constraints** exist that may affect the probability of what you are looking at? (See pages **402**–404; 414–429.)

- Should this data be **transformed** in some way before being analysed? (See pages 472–481.)

- Once transformed, is there a **probability distribution** that this variable can reasonably be expected follow? If so, why? What do I expect to happen to these variables? Why is it this particular distribution? (See page 704, then refer back as appropriate.)

- What are the important **parameters** of this distribution? Can they be known at this stage? (See pages 492–497; 520–530; 544–554).

- Is there a **single variable,** or a **combination**? If there is a combination, what is the interest in their relationships? (See pages 397–406; 571 and following.)

At this point, you may have the information you need to start working with the data you don't have, and not just the data you do. In a business context, for example, you can start reviewing performance against probable random fluctuations, so that you can ignore random noise but catch real trends quickly. You will be able to use the information you have to assess risk. In any area you are likely to be able to form a view about what will be happening in a wider population of interest.

To cement your understanding:

Read a journal article that uses statistics in an area that interests you and attempt to answer the questions above based on the information in the article and what you can assume from it. Revise anything that you need to. If there simply isn't enough information to answer the questions, be careful about trusting the statistical analysis.

Once you have answered the questions above, if you want to be scientific and to learn more, you will need to do some testing, and further questions will become relevant.

- Based on what we understand now and want to know in the future, is this better approached with a **frequentist or Bayesian approach**? (See pages 429–435.) The questions listed above are relevant for both frequentist and Bayesian thinking.

- Depending on your approach, different questions will become relevant at this stage:

Frequentist approach	Bayesian approach
• What are the null and alternative hypotheses?	• What is the prior distribution?
• What testing would enable me to establish that the alternative hypothesis is *un*likely?	• Having established my prior assumptions about the situation, what data would be the best way to *challenge* my existing beliefs about the subject?
(Addressing these questions honestly is likely to be emotionally challenging for a researcher who is interested in a positive result, but it is this question which statistical analysis is best-equipped to answer. Once you	

Frequentist approach	Bayesian approach
have tried hard to demonstrate that your result is *unlikely*, and failed to do so, you will know, and be able to convince others, that you have found something significant.)	
This leads into experimental design, which is a subject on its own and not one we are covering. But if you start with trying to prove yourself wrong, you give yourself the best chance of coming up with a test design that minimizes bias.	
• Do I understand the mathematical basis of the test? There are many tests available for different situations; and their quality varies. It is important to understand, for yourself, the mathematics and the ideas behind any test you use, so that you avoid using measures that: (1) have been identified as faulty but are still in widespread use, such as the Kappa coefficient (page 719); (2) are widely used but still highly suspect, like r^2 (page 578); or (3) are easy to misinterpret because there are different forms (similar to the gamma distribution (page 459)). If you can't understand them, and you don't want your research efforts to be wasted, it is best not to use them, even if they are popular.	• Do I understand the mathematical basis under which my prior assumptions are going to be incorporated with my data, to form a posterior distribution (pages 429–431)? If the prior, likelihood and posterior distributions work well together mathematically, is that realistic or have they been artificially constrained to work that way (page 470)? If they don't work well mathematically and I'm using a computer to generate a prior distribution, do I at least roughly understand what the computer is doing? Have I tested it with known data to ensure that it is working as expected?

Frequentist approach	Bayesian approach
• Once you know what test you plan to use and what your experimental design will be, it is wise to undertake power testing to determine the sample size that will be appropriate (page 639). It's wise to check this with some test data to ensure that your assumptions will enable you to obtain a clear result.	• Power testing is less important in Bayesian than in frequentist statistics, because there are no constraints to increasing the sample size if you don't have enough information to be clear about the posterior. It is, however, still wise to run test data to check whether the size of your sample seems appropriate.

After that you can analyse your data, knowing that in doing so you really are decreasing the amount of uncertainty involved, and allowing your understanding to progress.

The Beginning

Here are a few more statistics to finish with. Different research studies have found that: less than 40% of physics students, less than 28% of psychology students, and less than 18% of economics students, do all of the course reading they were supposed to[90].

So congratulations! You've made it through some material which was, at times, quite challenging, and you've stayed with it until the end. There is more to learn—in life there is always more to learn—but your statistical journey from now on should be richer and deeper.

We've only touched lightly on situations where there is more than one variable involved. If you want to explore the underlying concepts further, this book plus a dive into the world of linear algebra will equip you to understand more advanced mathematical statistics textbooks, that deal with situations where a large number of variables need to be factored in[91]. If you want to know more about the practical application of statistics in your discipline, an advanced textbook about the use of statistics in your field, and a textbook on the use of the software package of your choice should both be beneficial (and you will probably be able to read them quite quickly). If you want to know more about Bayesian statistics, there is a great deal more to it, but you should now have a realistic chance of understanding it if you read more about it. You now have some solid foundations to take your explorations further.

Hopefully, as well as understanding the specifics we have covered, our journey has illustrated some larger themes.

- Widespread use is not the same as usefulness, and approaches that are trusted are not always trustworthy.

- Sometimes ideas which seem implausible turn out to be true and useful.

- Even the greats can make mistakes. Accepting what you are told without understanding for yourself can be a way to perpetuate errors.

- Randomness can be predictable.

- Uncertainty can be measured.

- Truth is an exploration more than a destination. Science is a journey, not a basis for dogma (which is the opposite of science).

- Knowing *how* to use something is not the same as knowing *why* or *when* you should use it.

- It's often tempting to simplify and smooth over complexity, but sometimes understanding and appreciating variability can yield interesting new perspectives.

- In research, statistics is largely about finding out whether your effect was just a random fluctuation. It is harder to prove that it *wasn't* in a small study than a large one (and particularly if you include a power analysis), which paradoxically means that small studies are sometimes better than large ones. (Look for the size of the effect.)

- All knowledge is conditional. Everything we are certain about is based on our current perspective and assumptions. As more information comes to hand, perspectives can alter radically.

- Sometimes an easy solution is just out of reach, eluding people for years then seeming obvious when it occurs.

- Some of the best developments in history have come from polymaths. Brains work best when more of them is used; over-specialisation can be unhelpful.

- The world is, quite literally, awesome. It can fill us with awe and wonder, and a sense that there is mystery beyond our understanding. Statistics can help us appreciate this. It teases and intrigues us with hints that a deeper, fuller understanding of reality is just beyond our conceptual reach. *"The seeming Paradoxes where with it abounds"* that inspired Abraham de Moivre to write *The Doctrine of Chances* 300 years ago are now more apparent than they were then. Statistics hints at the possibility that there just might be an underlying set of principles at work. Irrational numbers like e (based on compound interest); π (the ratio of the circumference to the diameter of a circle); and $\sqrt{2}$ (an irrational number that intrigued the ancients); keep reappearing in the most implausible situations. Formulae with completely different meanings suddenly coalesce.

You are ready to journey on, to uncover more of the mysteries of statistics, and more of the mysteries that it reveals to us about the world. As is often the case, Albert Einstein said it best:

> *"The most beautiful experience we can have is the mysterious. It is the fundamental emotion that stands at the cradle of true art and true science."*

> *"The important thing is not to stop questioning. Curiosity has its own reason for existence. One cannot help but be in awe when contemplating the mysteries of eternity, of life, of the marvellous structure of reality. It is enough if one tries merely to comprehend a little of this mystery each day."*

Appendices

The appendices contain mathematics that isn't strictly necessary to understand statistics, but that is either of supplementary interest, valuable as a basis for other explorations into higher mathematics, or covers some proofs that would have been distracting if they had been included in the body of the book.

Appendix 1: Using a quadratic equation to find the value of phi

Imagine we have an interval (segment of a line) which has a fixed length of one unit (interval A in Figure 313), and a second interval which is slightly larger (interval B in Figure 313), and a third interval which represents the distance between them (interval C). Interval A is fixed, while interval B has a variable length.

Mathematically we can express this as:

$$A + C = B$$

You can see that the ratio of the length of interval B to the length of interval A will be slightly greater than one, whereas the ratio of the length of interval A to the length of interval C will be much more than one.

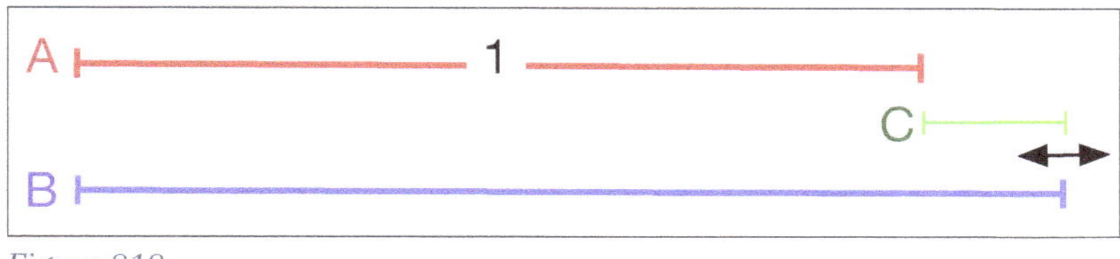

Figure 313

That is:

$$\frac{B}{A} > 1$$

$$\frac{A}{C} > 1$$

Now imagine that we keep interval A at one unit, and stretch interval B longer (and therefore interval C longer), as in Figure 314.

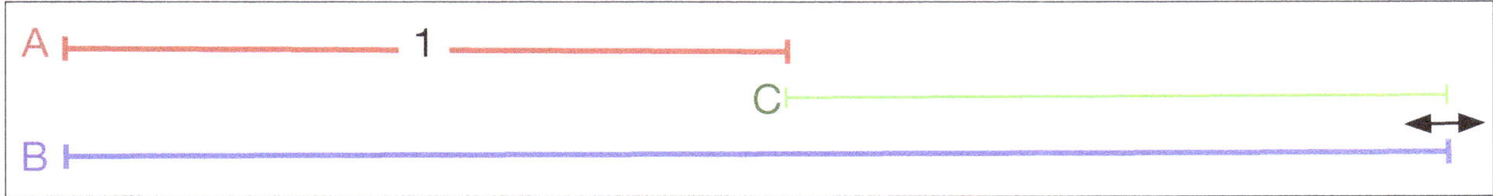

Figure 314

Now the ratios will have changed. The ratio of the length of B to A will be close to 2, whereas the ratio A to C will be slightly more than 1.

$$\frac{B}{A} \approx 2$$

$$\frac{A}{C} \approx 1$$

There is *only one* length of interval B, for which the ratio of B to A will be the same as the ratio of A to C (Figure 315).

$$\frac{B}{A} = \frac{A}{C}$$

This was how the ancient Greeks defined the golden ratio. (They didn't use letters, but this was the ratio they were interested in.) It is roughly a ratio of 1.6:1 and we will use algebra to tell us exactly what it is.

We'll call that length φ (the Greek letter phi). The length of interval C will therefore be $\varphi - 1$.

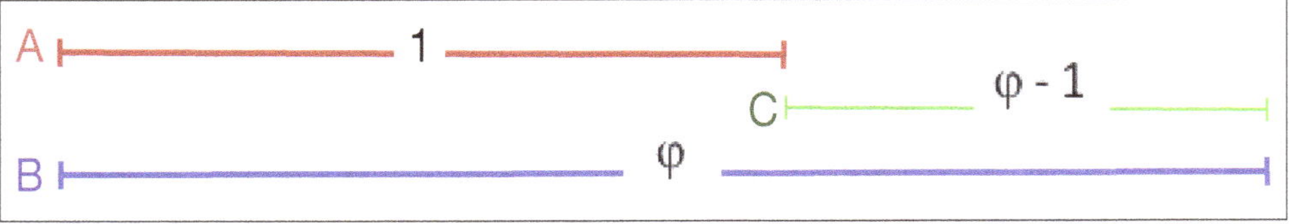

Figure 315

That is the characteristic that makes φ (the golden ratio) so architecturally interesting. It may also be what makes it aesthetically pleasing.

We can use the quadratic formula to find the length of φ. We can mathematically express the ratio of the length of interval B to the length of interval A as $\frac{\varphi}{1}$ and the ratio of the length of interval A to interval C as $\frac{1}{(\varphi-1)}$. We started by saying that these ratios are equal, so that means we can say:

This is a definition of the golden ratio.	$\dfrac{B}{A} = \dfrac{A}{C}$
Plug in the values from the diagram.	$\dfrac{\varphi}{1} = \dfrac{1}{(\varphi - 1)}$
Multiply both sides by $(\varphi - 1)$	$\varphi(\varphi - 1) = \dfrac{1\cancel{(\varphi - 1)}}{\cancel{(\varphi - 1)}}$
Expand out the expression on the left.	$\varphi^2 - \varphi = 1$
Subtract 1 from both sides. This means that we can use these values of: 1 for a, –1 for b and –1 for c, and put them into the quadratic formula.	$\varphi^2 - \varphi - 1 = 0$

Recall that for an equation in the form:

$$ax^2 + bx + c = 0$$

we can find x using the quadratic formula:

$$x = \frac{-b \pm \sqrt{b^2 - 4ac}}{2a}$$

$$\varphi = \frac{-(-1) \pm \sqrt{1^2 - 4(-1)}}{2}$$

$$= \frac{1 \pm \sqrt{1 + 4}}{2}$$

We are looking at the length of an interval, so in this instance we can ignore the negative value; it has to be positive.

$$\varphi = \frac{1 + \sqrt{5}}{2}$$

$$= 1.618033988749895\ldots$$

To cement your understanding:

- Add φ to both sides of $\varphi^2 - \varphi = 1$. Can you see why if you add 1 to φ you get φ^2?

- Divide both sides of $\varphi^2 - \varphi = 1$ by φ. Can you see why when you subtract 1 from φ you get its reciprocal?

Appendix 2: Finding the sum of squares of natural numbers

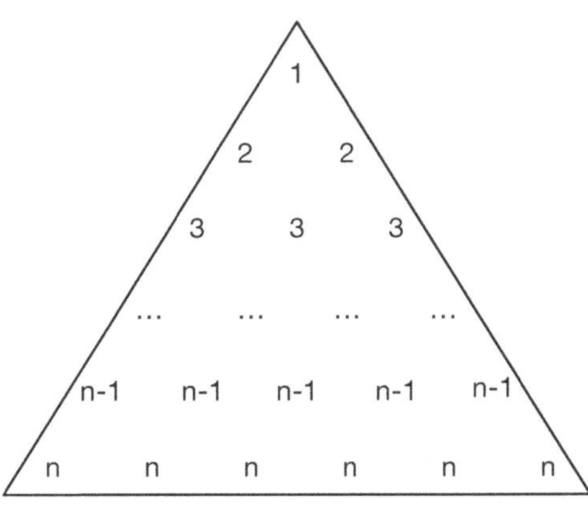

Figure 316

Imagine a triangle of numbers like Figure 316; not, for once, Pascal's triangle but a different one. In this case the number of numbers in each row will be the same as the number that is written there; for example, the third row has the number three, written three times. Two things to notice about this. The first is that the sum of each horizontal row will be the square of the number that appears in that row. For example, the sum of the numbers in the second row will be $2^2 = 4$ and the sum of the numbers in the third row will be $3^2 = 9$. That means the total of all the numbers in the triangle will be the sum of the squares:

$$1^2 + 2^2 + 3^2 + \cdots + n^2 = \sum_{i=1}^{n} i^2$$

What we are looking for is a way to calculate this value, regardless of the size of n.

The second thing to notice is that if you *count* the number of numbers in the triangle (as opposed to adding them), that will be the sum of counting numbers from 1 to n, which, as we saw earlier (page 134) is

$$\frac{n(n + 1)}{2}$$

So if we stopped the triangle at the second row, it would have $1 + 2 = 3$ numbers in it. If we stopped it at the third row, it would have $1 + 2 + 3 = 6$ numbers in it.

Now make two more copies of the triangle, and rotate each of them through a third of a circle, like Figure 317. Finally add a fourth triangle, which contains the sum of the numbers from each position in the first three triangles: see Figure 318. In other words, in the fourth triangle, the number in the middle position of the third row will be the sum of the numbers in the middle position of the third row from the first three triangles.

You can see that the number in each position in the fourth triangle will always be $2n + 1$.

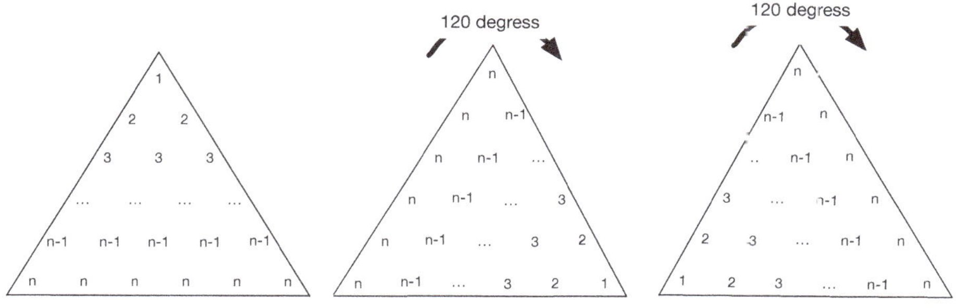

Figure 317

So putting all of that together, we can say that the total of the numbers in the last triangle will be 3 times the total of numbers in each triangle, that is three times the sum of squared sequential natural numbers. It will also be equal to $2n + 1$ multiplied by the count of numbers in the triangle $\left(\frac{n(n+1)}{2}\right)$. Writing that mathematically we get:

$$3 \sum_{i}^{n} i^2 = \frac{n(n + 1)}{2}(2n + 1)$$

Dividing both sides by 3 and rearranging slightly, we get:

$$\sum_{i=1}^{n} i^2 = \frac{n(n + 1)(2n + 1)}{6}$$

$$= \frac{2n^3 + 3n^2 + n}{6}$$

$$= \frac{n^3}{3} + \frac{n^2}{2} + \frac{n}{6}$$

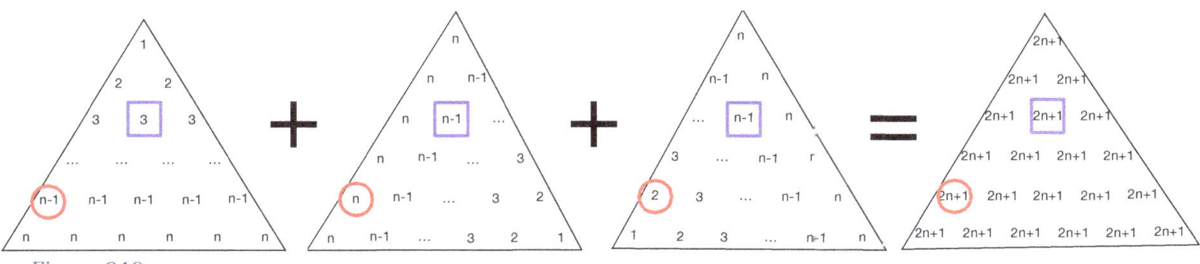

Figure 318

This applies regardless of the size of n. There are other, more mathematically rigorous proofs of this, but this is probably the simplest way to see it.

Appendix 3: Proof of Pascal's identity

For those who would like a more rigorous proof of why the numbers in Pascal's triangle can be calculated using n choose r (Figure 319) he proved this using something called **Pascal's identity**. He didn't write it using the notation shown here, because the factorial notation didn't exist then, but this was the gist of what he was doing.

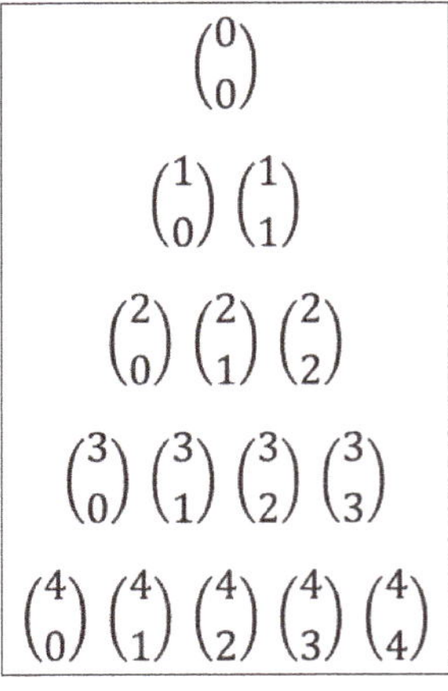

Figure 319

Pascal's identity says that:

$$\binom{n}{r+1} = \binom{n-1}{r} + \binom{n-1}{r+1}$$

That really looks complicated, but it isn't as bad as you might think. It's just expressing mathematically what is shown diagrammatically in Pascal's triangle, where r is the number of places along in the nth row of the triangle. (You can check that by picking any number you like.) It is saying that if you have *any one* number in the triangle that can be calculated by

$$\binom{n}{r} = \frac{n!}{r!(n-r)!}$$

then *every* number can be calculated in this way, because whatever number you chose, the next number you come to $(r + 1)$ in the row where you are (n) is made up by adding the two numbers immediately above it.

We can prove it like this. (Yes, it looks challenging, but it gets easier.) The identity, translated to factorial notation, says:

$$\frac{n!}{(r+1)!\,(n-(r+1))!} = \frac{(n-1)!}{r!\,(n-1-r)!} + \frac{(n-1)!}{(r+1)!\,(n-1-(r+1))!}$$

Start by working with the terms on the right-hand side of the equation above. Simplify the expression of the denominator of the second term.

$$\frac{(n-1)!}{r!\,(n-r-1)!} + \frac{(n-1)!}{(r+1)!\,(n-r-2)!}$$

Find a common denominator. Multiply the numerator and denominator of the first term by $(r + 1)$, and the numerator and

$$\frac{(r+1)(n-1)!}{(r+1)!\,(n-r-1)!} + \frac{(n-r-1)(n-1)!}{(r+1)!\,(n-r-1)!}$$

denominator of the second term by $(n - r - 1)$.

Factor out the $(n - 1)!$ in the numerator, and simplify the numerator.

$$\frac{(n - 1)!\,(\cancel{r} + \cancel{1} + n - \cancel{r} - \cancel{1})}{(r + 1)!\,(n - r - 1)!}$$

$$\frac{n(n - 1)!}{(r + 1)!\,(n - r - 1)!}$$

This is now equal to the term on the left of the equation, so we have proved Pascal's identity.

$$\frac{n!}{(r + 1)!\,(n - (r + 1))!}$$

To complete the proof, pick any number at all in Pascal's triangle, show that it can be represented as

$$\frac{n!}{r!\,(n - r)!}$$

where n is the row of the triangle and r is the number of places from the left side, and, as we have just proven, the next number will be made from summing the two numbers above.

This approach is called a **proof by induction**.

Appendix 4: The golden ratio in a pentagram within a pentagon

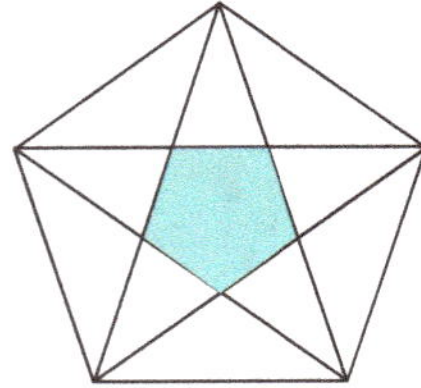

Figure 320

This is a proof that all of the intervals of a pentagram (five-pointed star) within a pentagon (five-sided figure) are either equal, or are in the ratio of $\varphi : 1$ (phi to one), if you are comparing the larger to the smaller.

The proof involves isosceles triangles. An isosceles triangle is one where two, and only two, of the angles are the same. (In an equilateral triangle the three angles will be the same.) If two of the angles are the same in a triangle, that means two of the sides must be the same length.

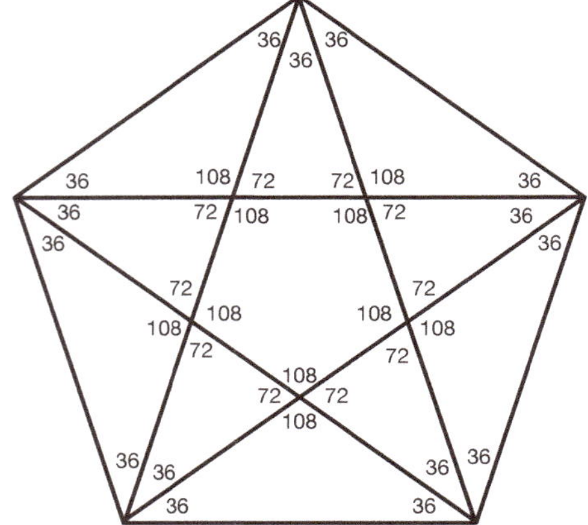

Figure 321

Think about the inner pentagon (Figure 320). We know that the angles of a five-sided figure add to 540 degrees (page 183 and following), and in this figure the angles will all be equal. That means each angle will be 108 degrees. Using that information, together with what we know about the angles of a straight line, the angles of a triangle and isosceles triangles, we can calculate all of the angles in Figure 321.

Looking closely at the angles, you will see that our figure has 20 triangles within it, all of which are isosceles. There are only two shapes of these. One of the shapes has angles of 108 degrees, 36 degrees and 36 degrees. The other shape has angles of 72 degrees, 72 degrees and 36 degrees. That means that all the triangles are either congruent or similar with a number of other triangles. If they are similar, their sides will have the same ratios.

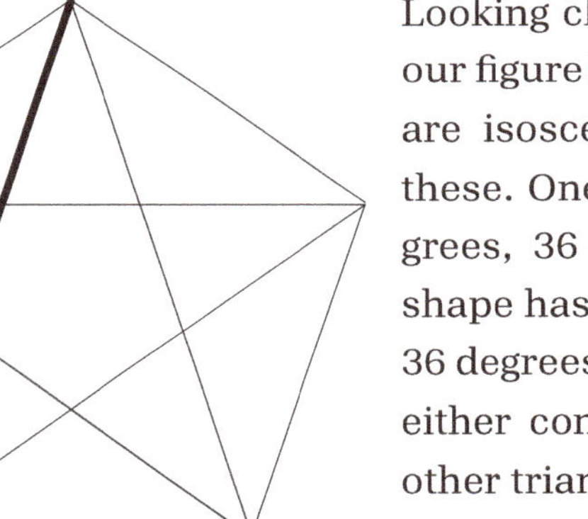

Figure 322

Let's make the length of an edge equal to 1, and the length of a diagonal equal to x (Figure 322).

Looking at Figure 323, the triangle abc is isosceles (in the angle at the bottom right hand side, $36 + 36 = 72$); that means that the interval ab must have the same length as interval ac: that is,

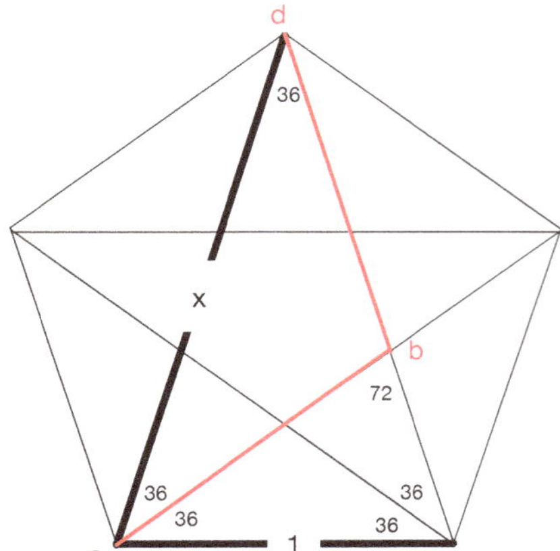

Figure 323

1 unit. Triangle abd is also isosceles, which means that ab = bd, so bd = 1. Triangle adc is isosceles, so ad = dc, therefore dc = x. Since bd = 1, it means bc = $x - 1$ (Figure 324).

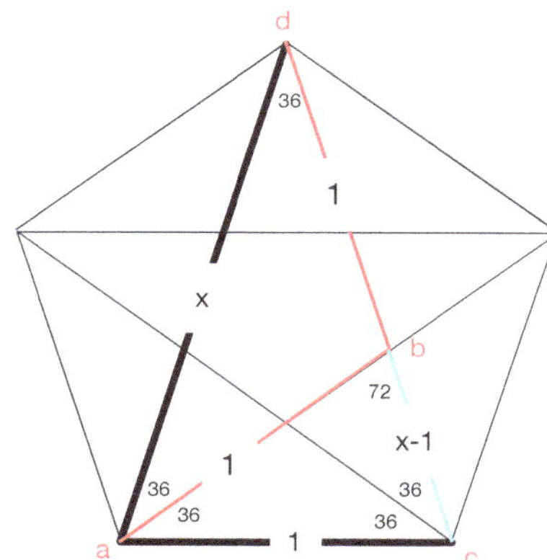

Figure 324

Now triangle adc is similar to triangle abc. That means the ratio of ad:ac (that is, x:1) must be the same as the ratio of ac:bc. That in turn means that

$$\frac{x}{1} = \frac{1}{x - 1}$$

This a definition of the golden ratio, and means $x = \varphi$, because that is the only number for which this is true. Once we know that, and we understand that the ratios of the sides of similar triangles are consistent, we can see that all of the intervals are either equal, or in a ratio of $\varphi : 1$.

Appendix 5: Stricter definitions of conic sections

The parabola

A parabola is the set of points that are equi-distant from some point—called the **focus point**—and some line. The line, called the **directrix** is outside the parabola, while the focus point will be inside it.

One of the interesting things about a parabola is that parallel lines that go into it and reflect off a tangent to the parabola will all intersect at the focus point. This means that parabolic surfaces are used in devices that receive energy and need to focus it (like antenna dishes). They are also used in devices that need to collect emissions from a single source and focus them out in a directed beam. If the bulb of a torch or flashlight could produce light from a single point and its mirror was a perfect parabola, the beam of the torch would be perfectly parallel.

The ellipse

An ellipse is a closed curve with two focus points. It is defined as the set of points that are equidistant from the sum of the distances from the ellipse to each focus point. In other words, if you put two pins in a board, tie a string loosely between them, and use a pencil to stretch the string to its limit, then move the pencil everywhere it is free to go while keeping the string taut, you will draw an ellipse.

The hyperbola

The mathematical definition of a hyperbola is similar to that of an ellipse. It also has two focal points, but, whereas an ellipse is the set of points for which the sum of distances to the focal points is constant, with a hyperbola what is constant is the absolute value of the *difference* between the distances to each of the focal points.

Appendix 6: Some more trigonometric identities

These identities are in an appendix because the working out is more complex than the earlier ones we saw. We'll use Figure 325 to demonstrate them. We are looking for a formula for the sine of the sum of two angles: in this case θ and ζ (the Greek letters theta and zeta).

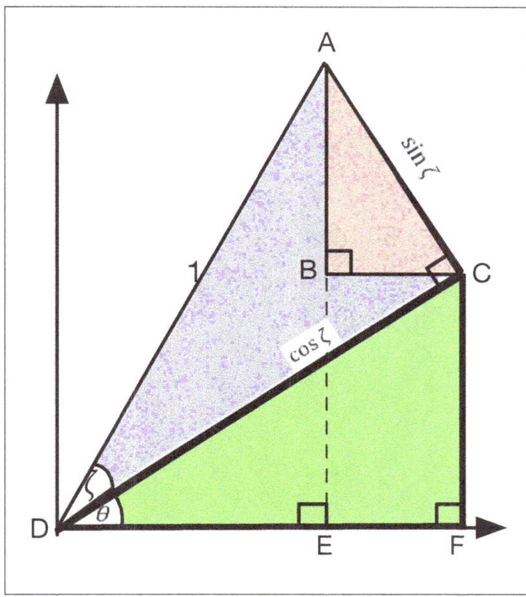

Figure 325

To get this, you'll have to patiently look at the figures to follow through on the thinking.

Triangle ADC (purple and orange) has a right angle at C and a hypotenuse of 1. That makes $\sin \zeta = \frac{AC}{1} = AC$ and $\cos \zeta = \frac{CD}{1} = CD$.

Interval DC of length $\cos \zeta$ is also the hypotenuse of triangle DFC (green). The key to this is that triangle DFC is similar (it has the same angles, but its sides are different lengths) to triangle ABC (orange). We can write this as DFC ~ ABC. That is, it has the same shape.

To see why this is so (Figure 326), based on the fact that there are $\frac{\pi}{2}$ radians in a right angle and the angles of a triangle add up to π radians, the angle at C made by DCF must be equal to $\frac{\pi}{2} - \theta$ radians or $90° - \theta$ (same thing). Therefore the angle at C formed by DCB must be equal to θ. That means that angle at C formed by BCA must be $\frac{\pi}{2} - \theta$ which means that the angle at A formed by BAC therefore also has to equal θ. So with the same angles, the orange and green triangles have to be similar, which means the ratios of their sides are the same.

Based on that, we can work out a formula for $\sin(\theta + \zeta)$.

$$\sin(\theta + \zeta) = \frac{AE}{1} = AE$$

Now AE = AB + BE, and BE = CF. That means we need to find the length of AB and the length of CF, and add them together. In the green triangle:

$$\sin \theta = \frac{CF}{\cos \zeta}$$

$$CF = \sin\theta\cos\zeta$$

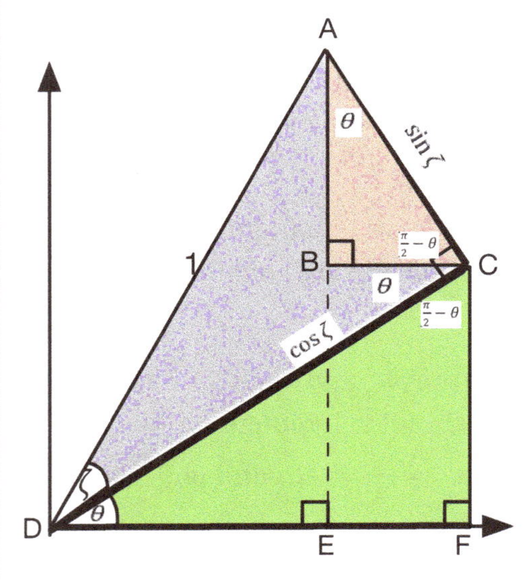

Figure 326

Meanwhile in the orange triangle:

$$\cos\theta = \frac{AB}{\sin\zeta}$$

$$AB = \sin\zeta\cos\theta$$

So putting all that together, we have:

$$\boxed{\sin(\theta + \zeta) = \sin\theta\cos\zeta + \cos\theta\sin\zeta}$$

Which is both very neat, and very useful when it comes to calculus and other areas of higher mathematics. Now to find the cosine of the sum of two angles: $\cos(\theta + \zeta)$

Back to Figure 326:

$$\cos(\theta + \zeta) = \frac{DE}{1} = DE$$

And in turn:

$$DE = DF - EF$$

$$= DF - BC$$

So to find $\cos(\theta + \zeta)$, we need to find the length of DF and the length of BC, and subtract the later from the former.

$$\cos\theta = \frac{DF}{\cos\zeta}$$

$$DF = \cos\theta\cos\zeta$$

Meanwhile up at the orange triangle:

$$\sin\theta = \frac{BC}{\sin\zeta}$$

$$BC = \sin\theta\sin\zeta$$

So putting all that together, we have:

$$\cos(\theta + \zeta) = \cos\theta\cos\zeta - \sin\theta\sin\zeta$$

Appendix 7: More about the complex plane

Complex conjugates

There are a number of ways to move from the complex domain to the real domain. A **complex conjugate** is the negative version of the imaginary part of a complex number. For example, the complex conjugate of $x + iy$ is $x - iy$. When complex conjugates are multiplied together, the result is a real number:

$$(x + iy)(x - iy) = x^2 + \cancel{ixy - ixy} - y^2(-1)$$

$$= x^2 + y^2$$

When they are added, the result is also real $(2x)$.

Just as multiplying the points of a shape by i has the effect of rotating it anticlockwise through $\frac{\pi}{2}$ radians, so taking the complex conjugate of the points has the effect of reflecting the shape across the real axis.

Dividing complex numbers

Adding, subtracting and multiplying complex numbers works the same way that addition, subtraction and multiplication work in real numbers, as in the example above. With division, it is necessary to multiply the numerator and the denominator by the complex conjugate of the denominator, to change it into a real number. This is called **realising** it. (This is the equivalent of finding a common denominator.)

Multiply the numerator and the denominator by the complex conjugate of the denominator.

$$\frac{a + ib}{c + id} = \frac{(a + ib)\,(c - id)}{(c + id)\,(c - id)}$$

Expand out the numerator and the denominator.

$$= \frac{ac + ibc - iad - i^2bd}{c^2 + idc - idc - i^2d^2}$$

Simplify (recalling that $i^2 = -1$), and express the real and imaginary components as separate terms.

$$= \frac{ac + bd}{c^2 + d^2} + \frac{i(bc - ad)}{c^2 + d^2}$$

Square roots of complex numbers

Taking the square root of an *imaginary* number just involves taking i outside the square root, then taking the square root of the positive number.

$$\sqrt{-9} = \sqrt{-1(9)} = (\sqrt{-1})(\sqrt{9}) = i3$$

Taking the square root of a *complex* number is a little more difficult, but can be calculated like this.

The square root of a complex number will itself be a complex number, and so we can express it in these terms. (Note that all numbers can be expressed as complex numbers: it's just that for real numbers the imaginary part is multiplied by 0, while for imaginary numbers the real part is multiplied by 0.)

$$\text{Let } \sqrt{7 + i24} = x + iy$$

Square both sides.

$$7 + i24 = (x + iy)^2$$

Expand out the binomial expression, and keep the real and imaginary parts separate.

$$7 + i24 = (x^2 - y^2) + i2xy$$

The key to this is that the real and imaginary parts of a complex number can be treated separately; they won't mix.

$$x^2 - y^2 = 7 \text{ and}$$
$$i2xy = i24$$

Divide both sides by $i2$. This can be solved as simultaneous equations: find x or y from one equation, substitute into the other to find both values. (If you really want practice in algebra, you can do this for yourself. Or in this example, you might just notice by looking at it that it can be solved by $x = 4$ and $y = 3$).

$$12 = xy$$

And if you still want practice, you can check it by squaring both sides.

$$\sqrt{7 + i24} = 4 + i3$$

Mod-arg form

Multiplying a complex number by i rotates it anticlockwise through $\frac{\pi}{2}$ radians (90°), but more rotational flexibility is available using the complex version of polar coordinates.

Just to be different, on the complex plane the radius is called the **modulus**, and the angle is called the **argument**; hence the name **modulus-argument**, or **mod-arg** form. There are a lot of similarities with polar coordinates.

As with polar coordinates, to convert rectangular coordinates to polar coordinates, the sine of the angle will be the y value over r, while the cosine of the angle will be the x value over r, except that the sine is the imaginary

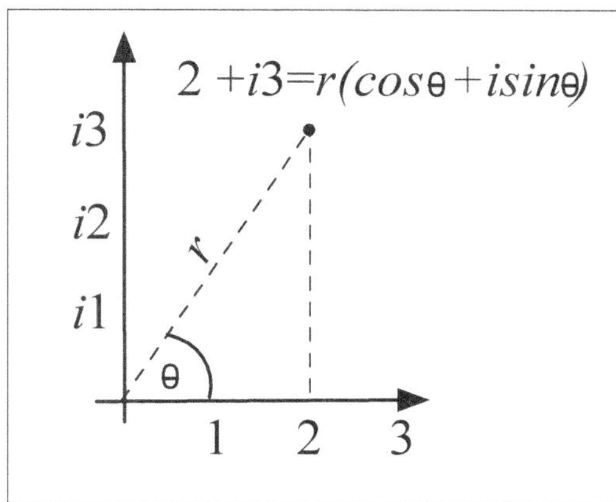

Figure 327

part and so is multiplied by i. This means that the imaginary (vertical) part equals $ri \sin \theta$ and the real (horizontal) part equals $r \cos \theta$ (multiplying both sides of each of these equations by r). While it would be possible to define the position of any point on the plane by $r\theta$, to make it clear that it is the complex plane, it is more common to describe complex numbers as $r(\cos \vartheta + isin \theta)$, as in Figure 327. This is sometimes shortened to **r cis θ**. The distance from the radius is sometimes expressed as an absolute value; because absolute value can be thought of as distance from zero, but until now we have thought of this as only applying to a number line. If a complex number is designated by a single letter, say z, then the modulus can be designated by $|z|$, and so the complex number can be written as

$$z = |z|(\cos \theta + i \sin \theta)$$

$$= a + ib$$

Multiplying complex numbers in mod-arg form works like this:

$$r_1(\cos \theta + i \sin \theta)r_2(\cos \zeta + i \sin \zeta)$$
$$= r_1 r_2 \{\cos \theta \cos \zeta + \cos \theta i \sin \zeta + i \sin \theta \cos \zeta + i \sin \theta i \sin \zeta\}$$
$$= r_1 r_2 \{(\cos \theta \cos \zeta - \sin \theta \sin \zeta) + i (\cos \theta \sin \zeta + \sin \theta \cos \zeta)\}$$

From our trigonometric identities:

$$\cos(\theta + \zeta) = \cos \theta \cos \zeta - \sin \theta \sin \zeta$$

and

$$\sin(\theta + \zeta) = \sin \theta \cos \zeta + \cos \theta \sin \zeta$$

$$r_1(\cos \theta + i \sin \theta)r_2(\cos \zeta + i \sin \zeta) = r_1 r_2 \{\cos(\theta + \zeta) + i \sin (\theta + \zeta)\}$$

So in mod-arg form, to multiply two complex numbers you multiply the moduli and add the arguments, which turns out to be helpful alliteration as well as cool mathematics.

Mod-arg form with Euler's identify

Work on the complex plane was undertaken after Euler's time, but using this formula means that any complex number (and all numbers are technically complex—it's just that either their real or their imaginary part might have a co-efficient of zero—as well as being able to be expressed in the form $a + ib$, can be expressed in the form $|z|e^{ix}$, where $|z|$ is the modulus (distance from zero) and x is the angle. This turns out to be massively useful in practical applications of higher mathematics, particularly electronics and engineering. It provides a way of linking exponential functions with trigonometric functions and cyclical wave forms.

As just one example of the way it can be used, when we looked at multiplying two complex numbers in mod-arg format, we needed to use trigonometric identities to develop the rule that you multiply the moduli and add the arguments. It is easier to see why that works when you express complex numbers in this format:

$$r_1 e^{i\theta} r_2 e^{i\zeta} = r_1 r_2 e^{i(\theta + \zeta)}$$

You add the arguments because that is the way indices work.

Taking the square root of complex numbers in $a + ib$ format involved some algebraic manipulation, and we didn't even attempt to calculate it using mod-arg format, because things might have become messy, whereas taking the square root of a complex number in this format is ridiculously easy:

$$\sqrt{re^{i\theta}} = \left(\sqrt{r}\right)e^{i\frac{\theta}{2}}$$

Likewise:

$$\left(re^{i\theta}\right)^2 = r^2 e^{i2\theta}$$

Working in reverse, we can see that

$$rx^i = re^{i\ln x}$$

$$= r(\cos(\ln x) + i\sin(\ln x))$$

giving us another way to work with complex formulae.

We can also work between exponential and trigonometric formulae in this way:

Euler's formula:	$e^{ix} = \cos x + i\sin x$
Substitute $-x$ for x	$e^{-ix} = \cos(-x) + i\sin(-x)$
Use the identities that $\cos(-x) = \cos x$ and	$e^{-ix} = \cos(x) - i\sin(x)$

$\sin(-x) = -\sin x.$

Add the first and third equations. $+i \sin x$ and $-i \sin(x)$ cancel one another out.

$$e^{ix} + e^{-ix} = 2\cos x$$

Divide both sides by 2. This gives us a complex and exponential definition of cosine.

$$\cos x = \frac{e^{ix} + e^{-ix}}{2}$$

Subtract the third equation from the first.

$$e^{ix} - e^{-iz} = 2i \sin x$$

Divide both sides by 2. This gives us a complex and exponential definition of sine.

$$\sin x = \frac{e^{ix} - e^{-ix}}{2i}$$

Appendix 8: Polynomial long division

With calculus, it is sometimes easier to work with complicated algebraic expressions when they are broken down into their constituent parts. One way to do this *can* be polynomial long division. This is useful when there is a fraction made up of two polynomials, when the numerator is larger than the denominator. We need to work out how many times the denominator will go into the numerator, and what will be left at the end as a remainder.

Take the expression:

$$\frac{6x^3 + 7x^2 + 1}{3x^2 - 8x}$$

Polynomial long division is an iterative process to get closer and closer to the right answer. Start by seeing how many times the term with the highest power of x in the denominator goes into the term with the highest power of x in the numerator. In this case, $3x^2$ goes $2x$ times into $6x^3$. (It's also helpful to write out the expression for the denominator showing all of the powers of x, even where the coefficients are zero. In other words $6x^3 + 7x^2 + 1 = 6x^3 + 7x^2 + 0x^1 + 1x^0$).

$$3x^2 - 8x \overline{\smash{)}6x^3 + 7x^2 + 0x^1 + 1x^0} \quad {}^{2x}$$

Then multiply the denominator by $2x$ to check how close the first estimate was.

$$
\begin{array}{r}
2x \\
3x^2 - 8x \enclose{longdiv}{6x^3 + 7x^2 + 0x^1 + 1x^0} \\
-(6x^3 - 16x^2 + 0x^1 + 0x^0)
\end{array}
$$

Subtract this from the numerator.

$$
\begin{array}{r}
2x \\
3x^2 - 8x \enclose{longdiv}{6x^3 + 7x^2 + 0x^1 + 1x^0} \\
-(6x^3 - 16x^2 + 0x^1 + 0x^0) \\
\hline
23x^2 + 0x^1 + 1x^0
\end{array}
$$

Now repeat the process.

Check how many times the first term of the denominator goes into the first term of the expression we just came up with: $3x^2$ goes 7 times into $23x^2$.

$$
\begin{array}{r}
2x + 7 \\
3x^2 - 8x \,\overline{)\, 6x^3 + 7x^2 + 0x^1 + 1x^0} \\
-(6x^3 - 16x^2 + 0x^1 + 0x^0) \\
\hline
23x^2 + 0x^1 + 1x^0
\end{array}
$$

Add that to the top line.

$$
\begin{array}{r}
2x + 7 \\
3x^2 - 8x \,\overline{)\, 6x^3 + 7x^2 + 0x^1 + 1x^0} \\
-(6x^3 - 16x^2 + 0x^1 + 0x^0) \\
\hline
23x^2 + 0x^1 + 1x^0 \\
-(21x^2 - 56x^1)
\end{array}
$$

Multiply the denominator by 7.

$$
\begin{array}{r}
2x + 7 + \tfrac{2}{3} \\
3x^2 - 8x \,\overline{)\, 6x^3 + 7x^2 + 0x^1 + 1x^0} \\
-(6x^3 - 16x^2 + 0x^1 + 0x^0) \\
\hline
23x^2 + 0x^1 + 1x^0 \\
-(21x^2 - 56x^1) \\
\hline
2x^2 + 56x + 1
\end{array}
$$

$$
\begin{array}{r}
2x + 7 \\
3x^2 - 8x \,\overline{)\, 6x^3 + 7x^2 + 0x^1 + 1x^0} \\
-(6x^3 - 16x^2 + 0x^1 + 0x^0) \\
\hline
23x^2 + 0x^1 + 1x^0 \\
-(21x^2 - 56x^1) \\
\hline
2x^2 + 56x + 1
\end{array}
$$

Subtract that.

The term with the highest power of x from the denominator ($3x^2$) is larger than the term with the highest power of x on the bottom of our working ($2x^2$), but we don't need to stop yet. When we come to partial fraction decomposition (next appendix) we need the highest power of x in the numerator to be lower than the highest term of x in the denominator. So we can repeat the process again.

Divide the term with the highest power of x in the last line of our working by the term with the highest power of x in the denominator.

Multiply the denominator by $\frac{2}{3}$, and subtract.

$$
\begin{array}{r}
2x + 7 + \boxed{\frac{2}{3}} \\[4pt]
3x^2 - 8x \,\overline{\smash{\big)}\, 6x^3 + 7x^2 + 0x^1 + 1x^0} \\
-(6x^3 - 16x^2 + 0x^1 + 0x^0) \\[2pt]
\hline
23x^2 + 0x^1 + 1x^0 \\
-(21x^2 - 56x^1) \\[2pt]
\hline
2x^2 + 56x + 1 \\
-\left(2x^2 - \frac{16}{3}x\right) \\[2pt]
\hline
\frac{184}{3}x + 1
\end{array}
$$

The last line becomes the remainder.

So as a result of our polynomial long division, we can say that

$$
\frac{6x^3 + 7x^2 + 1}{3x^2 - 8x} = 2x + 7\frac{2}{3} + \left(\frac{\frac{184}{3}x + 1}{3x^2 - 8x}\right)
$$

The final term still won't be easy to handle using calculus without further work, but the highest power of x in the numerator is now lower than the highest power of x in the denominator, which makes it a candidate for partial fraction decomposition (next appendix).

We can check our work by multiplying both sides of our equation by $3x^2 - 8x$:

$$
6x^3 + 7x^2 + 1 = (3x^2 - 8x)\left(2x + \frac{23}{3}\right) + (\cancel{3x^2 - 8x})\left(\frac{\frac{184}{3}x + 1}{\cancel{3x^2 - 8x}}\right)
$$

$$
= 6x^3 + 23x^2 - 16x^2 - \cancel{\frac{184}{3}x} + \cancel{\frac{184}{3}x} + 1
$$

$$
= 6x^3 + 7x^2 + 1
$$

Which is the right answer!

To cement your understanding:

Calculate $\frac{4x^3 + 6 + 4}{32 - 3x}$ using the same approach we used above. Or don't, if you prefer not to. There aren't many situations in which you are actually likely to have to do this in statistics; the main point here is just to get the idea that it can be done, and how it works.

Appendix 9: Partial fraction decomposition

To add two fractions, you find a common denominator. There are times though, and particularly in calculus, where it becomes more important to separate common denominators into separate terms. This is called **partial-fraction decomposition**. There are a number of techniques we can use.

The first step is to make sure that the highest power of x in the numerator is lower than the highest power of x in the denominator. (Sometimes you can get away with it being equal, but it can't be higher for this to work). If it isn't use some polynomial long division (page 750).

The second step is to factorise the denominator if this isn't already done. In other words, break it down into a series of expressions that are multiplied together. (See page 91 for ideas on factorising.) If this can't be done, then partial fraction decomposition isn't going to be possible.

Our first example initially looks like a quadratic expression (i.e. the highest power of x is 2), but it can be broken down into linear factors (i.e. the highest power of x in each factor is one). It is relatively simple.

$$\frac{8}{x^2 - 3x - 10} = \frac{8}{(x - 5)(x + 2)}$$

(If you multiply out the denominator on the right, you will see that it equals the denominator on the left.)

To break it down into its component parts, we assume that the numerator can be made into two separate terms. We choose two, because that is now many factors we have on the bottom. We assume it can be broken down into as many different factors as we have in the denominator. This makes it obvious what should be in the denominators of each, but we don't (yet) know what should be in the numerators. We can write this as

$$\frac{8}{x^2 - 3x - 10} = \frac{A}{(x - 5)} + \frac{B}{(x + 2)}$$

Now we need to find A and B.

There are a few ways to do this. The optimum approach depends on the form of the equation we are working with.

Method 1: Insert informative values

For this example we will recreate a common denominator, then **insert values for x** that make it easy to find A and B. Note that we aren't directly trying to break down the equation at this stage. We are only trying to find values for A and B. Once we have them we can return to the equation above to complete our work.

Factorise the expression in the denominator.	$$\frac{8}{x^2 - 3x - 10} = \frac{8}{(x - 5)(x + 2)}$$
Use A and B for the values we are trying to find.	$$\frac{A}{(x - 5)} + \frac{B}{(x + 2)} = \frac{8}{(x - 5)(x + 2)}$$
Find the common denominator for the expression on the left, by multiplying the term with A by the denominator of the term with B and vice versa.	$$\frac{A(x + 2) + B(x - 5)}{(x - 5)(x + 2)} = \frac{8}{(x - 5)(x + 2)}$$

(This might seem a bit weird; you are trying to get rid of a common denominator, so you start by finding a common denominator. It will make more sense soon, though.)

Multiply both sides by $(x - 5)(x + 2)$ to get rid of the denominators.	$$A(x + 2) + B(x - 5) = 8$$
The equation has to be true for all values of x, so we can pick a value that makes it easy to solve the equation for A and B, because they are constants. Letting $x = -2$ makes the term with A equal zero, which allows us to calculate the value of B.	$$\text{Let } x = -2$$ $$A(-2 + 2) + B(-2 - 5) = 8$$ $$A(0) + B(-7) = 8$$ $$B = -\frac{8}{7}$$
Making $x = 5$ allows us to calculate the value of A.	$$\text{Now let } x = 5$$ $$A(5 + 2) + B\cancel{(5 - 5)} = 8$$ $$A = \frac{8}{7}$$
Multiply the numerator and denominator of each term by 7. We've broken the equation down into partial fractions. The final expression may be easier to handle with calculus.	$$\frac{8}{x^2 - 3x - 10} = \frac{\frac{8}{7}}{(x - 5)} + \frac{-\frac{8}{7}}{(x + 2)}$$ $$= \frac{8}{7(x - 5)} - \frac{8}{7(x + 2)}$$

You can check that this gave the correct answer by working backwards, finding the common denominator and simplifying the expressions.

$$= \frac{1}{7}\left(\frac{8}{(x-5)} - \frac{8}{(x+2)}\right)$$

$$= \frac{1}{7}\left(\frac{8(x+2) - 8(x-5)}{(x-5)(x+2)}\right)$$

$$= \frac{1}{7}\left(\frac{8x + 16 - 8x + 40}{(x-5)(x+2)}\right)$$

$$= \frac{1}{7}\left(\frac{56}{(x-5)(x+2)}\right)$$

$$= \frac{3}{x^2 - 3x - 10}$$

Method 2: Repeated terms

The following example has a squared term (or **repeated term**) in the denominator.

$$\frac{8x - 10}{(x+1)^2}$$

The trick here is that the expression $(x+1)^2$ can actually have two possible factors: $(x+1)^2$, and $(x+1)$, so we need to show both of them.

We express this as

$$\frac{8x - 10}{(x+1)^2} = \frac{A}{(x+1)^2} + \frac{B}{(x+1)}$$

They may not ultimately be needed; one of the terms could disappear as we solve the expression.

The common denominator here is $(x+1)^2$. Multiply each term by that.

$$\frac{8x - 10}{(x+1)^2} = \frac{A}{(x+1)^2} + \frac{B}{(x+1)}$$

At this point we could do what we did in the previous example: let $x = -1$ to solve for A, then let $x = 0$ to solve for B. The alternative, though, is to match the powers of x on both sides.

$$8x - 10 = A + B(x+1)$$

Expand out the terms.

$$8x - 10 = A + Bx + B$$

Gather terms with like powers of x, and show all powers of x on each side. This

$$8x^1 - 10x^0 = Bx^1 + (A + B)x^0$$

means that the coefficients of x on each side must be identical.

That gives two different equations: one with B, and one with A (or, if there are more terms, one equation for each term).

$$8x^1 = Bx^1 \text{ which means } B = 8$$

$$-10x^0 = (A + 8)x^0$$

$$-10 = (A + 8)$$

$$A = -18$$

This example worked out easily without simultaneous equations, but simultaneous equations are sometimes required.

$$\frac{8x - 10}{(x + 1)^2} = \frac{-18}{(x + 1)^2} + \frac{8}{(x + 1)}$$

Method 3: Quadratic factors

$$\frac{2x + 1}{x(3x^2 - 1)}$$

In partial fraction decomposition, start by assuming that the numerator of each term has all of the possible powers of x that are lower than the highest power of x in the denominator (whether or not the numerator of the initial fraction does). If that turns out not to be the case, don't worry; the additional terms will disappear as you work it out. That is why the numerator of the final term is $Bx + C$: includes x to the power of 1 and of 0.

$$\frac{2x + 1}{x(3x^2 - 1)} = \frac{A}{x} + \frac{Bx + C}{3x^2 - 1}$$

Multiply both sides by $x(3x^2 - 1)$

$$2x + 1 = \frac{A\cancel{x}(3x^2 - 1)}{\cancel{x}} + \frac{x\cancel{(3x^2 - 1)}(Bx + C)}{\cancel{3x^2 - 1}}$$

$$2x + 1 = A(3x^2 - 1) + x(Bx + C)$$

Expand out the terms.

$$2x + 1 = 3Ax^2 - A + Bx^2 + Cx$$

Gather like powers of x.

$$0x^2 + 2x + 1 = x^2(3A + B) + Cx - A$$

Create separate equations for each of the powers of x.

$$0x^2 = x^2(3A + B)$$

$$0 = 3A + B$$

$$2x = Cx$$

$$2 = C$$

$$1 = -A$$

$$-1 = A$$

Substitute the value for A into the equation with co-efficients for x^2 to find the value of B.

$$0 = -3 + B$$

$$3 = B$$

Substitute these values for A, B and C back into the equation. Once again, if you want the exercise you can work back to confirm that the two expressions are equal.

$$\frac{2x + 1}{x(3x^2 - 1)} = \frac{-1}{x} + \frac{3x + 2}{3x^2 - 1}$$

In this instance we are still left with a final term that will cause some challenges when it comes to calculus. Partial fraction decomposition doesn't always lead to perfect results. What we have is, however, easier to deal with than what we started with.

Appendix 10: The Gaussian integral without polar coordinates

This is the way Laplace tackled the "Gaussian Integral", before Jacobi had developed the idea of the Jacobian.

This is the basic form of the Gaussian integral.

$$\int_{-\infty}^{\infty} e^{-x^2}\,dx$$

It's symmetrical about 0, so the integral from negative infinity to positive infinity is the same as twice the integral from zero to infinity.

$$= 2\int_{0}^{\infty} e^{-x^2}\,dx$$

Square the integral and take the square root.

$$= \sqrt{4\int_{0}^{\infty} e^{-x^2}\,dx \int_{0}^{\infty} e^{-x^2}\,dx}$$

Change the variable in one of the integrals. It's a definite integral, so the variable will disappear when it is calculated, which means that which variable we use doesn't matter.

$$= \sqrt{4\int_{0}^{\infty} e^{-x^2}\,dx \int_{0}^{\infty} e^{-y^2}\,dy}$$

Combine the integrals into a double integral, and add the exponents of e, factoring out -1.

$$= \sqrt{4\iint_{0}^{\infty} e^{-(x^2+y^2)}\,dy\,dx}$$

To this point, the integration is essentially the same as it would be if we were using polar coordinates and a Jacobian. It is from this point on that Laplace did it differently, with a substitution rather than a change of coordinate system.

Let $y = xt$, treating x as a constant and t as a variable. (This is perhaps a bit like a Laplace transpose; introducing a new variable, to take the calculations into another realm in order to be able to make it work.)

$$\frac{dy}{dt} = x$$

$$dy = xdt$$

Make the substitution of $y = xt$ and $dy = xdt$ back into the original equation. The limits of integration of the inner integral don't change, because if $y = 0$ and x is

$$= \sqrt{4\iint_{0}^{\infty} e^{-(x^2+(xt)^2)}xdt\,dx}$$

constant, $t = 0$, and likewise if $y = \infty$ and x is constant, $t = \infty$.

Within the exponent, factor out x^2.

$$= \sqrt{4 \int\!\!\int_0^\infty e^{-x^2(1+t^2)} x\, dt\, dx}$$

Change the order of integration.

$$= \sqrt{4 \int\!\!\int_0^\infty e^{-x^2(1+t^2)} x\, dx\, dt}$$

Now do another substitution. Let $u = -x^2(1+t^2)$ $\frac{du}{dx} = -2x(1+t^2)$

$$dx = \frac{1}{-2x(1+t^2)}\, du$$

Perform the u substitution; the x's cancel out.

$$= \sqrt{4 \int\!\!\int_0^\infty e^u \cancel{x} \frac{1}{-2\cancel{x}(1+t^2)}\, du\, dt}$$

The limits of integration of the inner integral now change. When $x = 0$, $u = 0$, but when $x = \infty, u = -\infty$

$$= \sqrt{4 \int_0^\infty \int_0^{-\infty} e^u \frac{1}{-2(1+t^2)}\, du\, dt}$$

$\frac{1}{-2(1+t^2)}$ is a constant with regard to x but not with regard to t, so it can be taken outside of the inner integral but kept within the outer integral.

$$= \sqrt{4 \int_0^\infty \frac{1}{-2(1+t^2)} \int_0^{-\infty} e^u\, du\, dt}$$

The integral of $e^u du$ is e^u. When $u = -\infty$ this equals 0, and when $u = 0$ this equals 1. We are subtracting 1 from 0, so that gives -1. This negative cancels out with the negative in $\frac{1}{-2(1+t^2)}$.

$$= \sqrt{4 \int_0^\infty \frac{1}{2(1+t^2)}\, dt}$$

Take the $\frac{1}{2}$ outside the integration.

$$= \sqrt{2 \int_0^\infty \frac{1}{(1+t^2)}\, dt}$$

We saw earlier (page 253) that the derivative of $tan^{-1}x$ is $\frac{1}{(1+x^2)}$

$$= \sqrt{2 \tan^{-1} t}\, \Big|_0^\infty$$

$$\tan\theta = \frac{\sin\theta}{\cos\theta} \cdot \lim_{\theta\to\frac{\pi}{2}}\tan\theta = \lim_{\theta\to\frac{\pi}{2}}\frac{\sin\theta}{\cos\theta} = \frac{1}{0} = \infty.$$

$$= \sqrt{2\frac{\pi}{2} - 0}$$

Therefore the value of the arctangent of t as it approaches infinity is $\frac{\pi}{2}$, or $90°$.

The tan of 0 is 0, so the arctangent of 0 is 0.

So without using polar coordinates or a Jacobian, the Gaussian integral works out to be $\sqrt{\pi}$

$$\int_{-\infty}^{\infty} e^{-x^2}\,dx = \sqrt{\pi}$$

Appendix 11: Derivation of the beta function

The derivation of the beta function has some similarities with the Gaussian integral. To start with, we'll multiply two gamma functions:

Replace the dummy variable t with u and v in each of these integrals, because we are multiplying two different functions.

$$\Gamma(x)\Gamma(y) = \int_0^\infty u^{x-1}e^{-u}\,du \int_0^\infty v^{y-1}e^{-v}\,dv$$

As with the Gaussian integral, we can write this as a double integral; in both cases we will undertake the integration with respect to u, then multiply it by the integration with respect to v.

$$= \iint\limits_{0\ 0}^{\infty\ \infty} u^{x-1}v^{y-1}e^{-u}e^{-v}\,du\,dv$$

Now we'll do an even more confusing change of variables, defining two more variables, s and r, such that $s = u + v$ and $r = \frac{u}{s}$. (Stay with it; or jump to the end if you prefer, because it isn't crucial that you understand this.) This means that $u = sr$, therefore that $s = sr + v$ and therefore that $v = s(1 - r)$.

Substitute these values back into the equation.

$$= \iint\limits_{0\ 0}^{\infty\ \infty} (sr)^{x-1}s^{y-1}(1 - r)^{y-1}e^{-sr}e^{-s(1-r)}\,du\,dv$$

Now we want to change the integration, to be with respect to s and r rather than u and v. To do this we need a Jacobian determinate, but it is different to the one we used in the Gaussian integral to change to polar coordinates. To learn how to calculate that will require an understanding of linear algebra (the mathematics of transformation), and we aren't doing that in this book, but just accept it for now. Once we have done that, we find that the Jacobian determinate is s. That means $du\,dv = s\,ds\,dr$.

$$= \iint\limits_{0\ 0}^{1\ \infty} (sr)^{x-1}s^{y-1}(1 - r)^{y-1}e^{-sr}e^{-s(1-r)}s\,ds\,dr$$

Change the limits of integration. The original integral went from 0 to ∞. That means that if $v = s(1 - t)$ then s will go from 0 to ∞. Because v has to be positive, t has to be between 0 and 1, because if it is any larger than 1 then $(1 - t)$ will be negative, which would make v negative.

Likewise t cannot be negative, because $t = \frac{u}{s}$ and we have defined both u and s as being between 0 and ∞.

Simplify, and break apart the two integrals. (Multiplying them will be the same as doing the double integral in this case.)

$$= \int_0^1 r^{x-1}(1-r)^{y-1}dr \int_0^\infty s^{x-1}s^{y-1+1}e^{-s(r-r)}ds$$

$$= \int_0^1 r^{x-1}(1-r)^{y-1}dr \int_0^\infty s^{x+y-1}e^{-s}ds$$

$\int_0^\infty s^{x+y-1}e^{-s}ds = \Gamma(x+y)$, by the definition of the function.

$$\Gamma(x)\Gamma(y) = \int_0^1 r^{x-1}(1-r)^{y-1}dr\,\Gamma(x+y)$$

Divide both sides by $\Gamma(x+y)$

$$\boxed{\frac{\Gamma(x)\Gamma(y)}{\Gamma(x+y)} = \int_0^1 r^{x-1}(1-r)^{y-1}dr}$$

Appendix 12: Using Laplace transforms

Supposing that we have a function that is a derivative of another function: $f'(x)$, but we are having trouble integrating it.

Apply the Laplace transform.

$$\mathcal{L}\{f'(x)\} = \int_c^\infty e^{-sx} f'(x)\, dx$$

Use integration by parts. Let $e^{-sx} = v$, making $\frac{dv}{dx} = -se^{-sx}$ by the chain rule, and $f'(x) = \frac{du}{dx}$.

$$= (f(\infty)e^{-s\infty}) - (f(0)e^0) - \int_0^\infty -se^{-sx} f(x)\, dx$$

$e^{-s\infty} = \frac{1}{e^{s\infty}}$, which approaches 0, and so the first term disappears. $e^0 = 1$

$$= \cancel{(f(\infty)e^{-s\infty})} - (f(0)e^{\cancel{0}}) - \int_0^\infty -se^{-sx} f(x)\, dx$$

Take the $-s$ outside the integration (because it is a constant in an integration with respect to x). But $\int_0^\infty e^{-sx} f(x)\, dx$ is the equation for the Laplace transform.

$$\mathcal{L}\{f'(x)\} = -f(0) + s \int_0^\infty e^{-sx} f(x)\, dx$$

So in a differential equation, the Laplace transform of a differential is equal to the Laplace transform of the original function, multiplied by s and with the value of the function, evaluated at 0, subtracted.

$$\boxed{\mathcal{L}\{f'(x)\} = s\mathcal{L}\{f(x)\} - f(0)}$$

You can keep going with this. For example, if you have the second derivative of a function you can apply the formula above, then repeat it. That will get you to

$$\mathcal{L}\{f''(x)\} = s^2 \mathcal{L}\{f(x)\} - sf(0) - f'(0)$$

You can go further, for any number of derivatives.

For an example of how to use this, take the differential equation

$$f(x) = \frac{f''(x) - f'(x)}{2}$$

when you know that $f(0) = 1$, and $f'(0) = 0$.

Multiply both sides by 2, then subtract $2f(x)$ from both sides.

$$f''(x) - f'(x) - 2f(x) = 0$$

Take the Laplace transform of both sides. The transform of zero evaluates to zero (try it if you aren't sure about this).

$$\mathcal{L}\{f''(x) - f'(x) - 2f(x)\} = \mathcal{L}(0)$$

Because they are based on integration, Laplace transforms are linear in the same way that series and integrals are. The transform of the sum of functions is the same as the sum of the transforms.

$$\mathcal{L}\{f''(x)\} - \mathcal{L}\{f'(x)\} - 2\mathcal{L}\{f(x)\} = 0$$

Substitute the functions we obtained above.

$$s^2\mathcal{L}\{f(x)\} - sf(0) - f'(0) - s\mathcal{L}\{f(x)\} - f(0) - 2\mathcal{L}\{f(x)\} = 0$$

Substitute the values for the conditions we were given ($f(0) = 1$, and $f'(0) = 0$). The original challenging differential equation has now become something we can solve algebraically.

$$s^2\mathcal{L}\{f(x)\} - s - s\mathcal{L}\{f(x)\} + 1 - 2\mathcal{L}\{f(x)\} = 0$$

Factor out $\mathcal{L}\{f(x)\}$

$$\mathcal{L}\{f(x)\}(s^2 - s - 2) - s + 1 = 0$$

Add s to both sides and subtract 1 from both sides.

$$\mathcal{L}\{f(x)\}(s^2 - s - 2) = s - 1$$

Divide both sides by $s^2 - s - 2$.

$$\mathcal{L}\{f(x)\} = \frac{s - 1}{s^2 - s - 2}$$

Use partial fractions to obtain this result.

$$\mathcal{L}\{f(x)\} = \frac{1}{3}\left(\frac{1}{s - 2}\right) + \frac{2}{3}\left(\frac{1}{s + 1}\right)$$

Now we need the inverse Laplace transform.

$$f(x) = \mathcal{L}^{-1}\left\{\frac{1}{3}\left(\frac{1}{s-2}\right) + \frac{2}{3}\left(\frac{1}{s+1}\right)\right\}$$

By linearity. We now have two expressions that are recognisable from tables of Laplace transforms[92]: $\mathcal{L}\{e^{at}\} = \frac{1}{s-a}$

$$f(x) = \frac{1}{3}\mathcal{L}^{-1}\left\{\left(\frac{1}{s-2}\right)\right\} + \frac{2}{3}\mathcal{L}^{-1}\left\{\left(\frac{1}{s+1}\right)\right\}$$

Substitute 2 for a in the first term, and -1 for a in the second.

$$f(x) = \frac{1}{3}e^{2x} + \frac{2}{3}e^{-x}$$

As an exercise you may like to work out this transform ($\mathcal{L}\{e^{at}\} = \frac{1}{s-a}$) for yourself, to check that you understand it. If you want practice with differentiation, you may also like to check that our final answer does in fact solve the original differential equation[93].

Appendix 13: Proof of the Central Limit Theorem

At the time of writing, no one has managed to come up with a proof that says that the possible means of samples of any size are approximately normally distributed. The central limit theorem actually says that the means are normally distributed *as the sample size approaches infinity*. We just find, empirically, that this happens even with quite small sample sizes. There are different versions of the Central Limit theorem, but the one most commonly used works with a random variable that is the mean of a sample of size n.

We'll use the same symbols and colours that we have used elsewhere:

Set	Mean	Variance	Standard deviation
Whole population:	μ	σ^2	σ
Any potential sample of size n	$\bar{x}$	s^2	s
Sampling distribution of the means of potential samples (all potential samples of size n):	$E(\bar{x})$	$s_{\bar{x}}^2$	$s_{\bar{x}}$

We start with a random variable $\bar{x}$ which is the mean of some random sample of size n. We start by standardising it, or creating a Z score of it, by subtracting the mean from it and dividing by the standard deviation. Recall that the standardised form of a random variable has a mean of 0 and a standard deviation and variance of 1.

The expected value of this random variable within its probability distribution, $E(\bar{x})$, is the same as the mean of the underlying population, μ (page 622). The standard deviation of the probability distribution of this random variable is the standard deviation of the population, divided by the square root of the sample size (page 625):

That makes the standardised random variable

$$\frac{(\bar{x} - \mu)}{\frac{\sigma}{\sqrt{n}}}$$

or, multiplying the numerator and denominator by $\sqrt{n}$, we get the statement of the central limit theorem as:

$$\frac{\sqrt{n}(\bar{x}-\mu)}{\sigma} \sim N(0,1) \; as \; n \to \infty$$

where n is the sample size, $\bar{x}$ is the random variable representing the means of possible samples, μ is the mean of the population and σ is the standard deviaton of a population. This is saying that the distribution of this standardised random variable will be a standardised normal distribution as the sample size approaches infinity.

What follows is one proof. (It can also be proven using a Taylor series, and can be proven using the characteristic function, which has the advantage that it always exists. In this instance we have assumed that the moment generating function (MGF) will exist.)

Multiply the numerator and denominator by $\sqrt{n}$

$$\frac{\sqrt{n}(\bar{x}-\mu)}{\sigma} = \frac{n(\bar{x}-\mu)}{\sqrt{n}\sigma}$$

Multiplying by n is the same as adding n lots of the expression.

$$= \frac{\sum_{i=1}^{n}(\bar{x}-\mu)}{\sqrt{n}\sigma}$$

Put $\frac{1}{\sqrt{n}\sigma}$ inside the summation.

$$= \sum_{i=1}^{n}\frac{(\bar{x}-\mu)}{\sqrt{n}\sigma}$$

Find the MGF, using the $E[e^{tX}]$ definition of an MGF.

$$MGF = E\left(e^{t\sum_{i=1}^{n}\frac{(\bar{x}-\mu)}{\sqrt{n}\sigma}}\right)$$

Take the t inside the summation, and show it as the series.

$$= E\left[e^{\left(t\frac{(\bar{x}_1-\mu)}{\sqrt{n}\sigma}\right)+\left(t\frac{(\bar{x}_2-\mu)}{\sqrt{n}\sigma}\right)+\cdots+\left(t\frac{(\bar{x}_n-\mu)}{\sqrt{n}\sigma}\right)}\right]$$

Using the sum of indices.

$$= E\left[e^{\left(t\frac{(\bar{x}_1-\mu)}{\sqrt{n}\sigma}\right)}e^{\left(t\frac{(\bar{x}_1-\mu)}{\sqrt{n}\sigma}\right)}\cdots e^{\left(t\frac{(\bar{x}_1-\mu)}{\sqrt{n}\sigma}\right)}\right]$$

The variables are independent. This means we can use the fact that the expectation of a product is the product of the expectations.

$$= E\left[e^{\left(t\frac{(\bar{x}_1-\mu)}{\sqrt{n}\sigma}\right)}\right]E\left[e^{\left(t\frac{(\bar{x}_1-\mu)}{\sqrt{n}\sigma}\right)}\right]\cdots E\left[e^{\left(t\frac{(\bar{x}_1-\mu)}{\sqrt{n}\sigma}\right)}\right]$$

$\bar{x}_i$ is identically distributed. This means that multiplying different instances n times is the same as raising the general form of $\bar{x}$ to the power of n.

$$MGF = \left\{E\left[\left(t\frac{(\bar{x}-\mu)}{\sqrt{n}\sigma}\right)\right]\right\}^{n}$$

To simplify the notation.

$$Let \; w = \frac{(\bar{x}-\mu)}{\sigma}$$

The limit as $n \to \infty$ is of the form 1^{∞}, which is indeterminate, so we'll need to get this into a form where we can use l'Hopital's rule.

$$\text{MGF} = \left\{ E\left[e^{\left(\frac{tw}{\sqrt{n}}\right)} \right] \right\}^{n}$$

$E\left[e^{\left(\frac{tw}{\sqrt{n}}\right)} \right]$ is the same as the MGF of w, but with respect to $\frac{t}{\sqrt{n}}$ rather than with respect to t. Use this alternative notation of the MGF, to simplify further.

$$\text{MGF} = \left\{ M_w\left(\frac{t}{\sqrt{n}} \right) \right\}^{n}$$

Take the log to base e of both sides.

$$\ln(\text{MGF}) = \ln\left[\left\{ M_w\left(\frac{t}{\sqrt{n}} \right) \right\}^{n} \right]$$

A logarithm of an exponential is the same as multiplying by the exponent.

$$= n \ln\left[M_w\left(\frac{t}{\sqrt{n}} \right) \right]$$

Divide the numerator and denominator by n. At this point we are interested in what happens when $n \to \infty$, but we need to do a bit more before we can easily tell what that will mean.

$$= \frac{\ln\left[M_w\left(\frac{t}{\sqrt{n}} \right) \right]}{\frac{1}{n}}$$

Note: we are using the English letter v here, not the Greek letter μ which we are using as the population mean.

$$\text{Let } v = \frac{1}{\sqrt{n}}$$

$$\ln(\text{MGF}) = \frac{\ln[M_w(tv)]}{v^2}$$

As $n \to \infty$, $v \to 0$ because $v = \frac{1}{\sqrt{n}}$

$$\lim_{n\to\infty} \frac{\ln\left[M_w\left(\frac{t}{\sqrt{n}} \right) \right]}{\frac{1}{n}} = \lim_{v\to 0} \frac{\ln[M_w(tv)]}{v^2}$$

Because we are now interested in the value of the limit, we can use l'Hopital's rule to take the first derivative with respect to v of the numerator and the first derivative of the denominator. For the numerator we need to use the chain rule twice. The derivative of tv is t, the derivative of $M_w(tu)$ is $M_w'(tu)$, and the derivative of the logarithm is the reciprocal.

When $u \to 0$, $M_w'(tv)$ is the first moment evaluated at 0, which is the mean, that we know, from our initial set up, is 0. This is still in the form $\frac{0}{0}$, so we need to simplify then use l'Hopital's rule again,

$$= \lim_{v\to 0} \frac{t M_w'(tv)}{2u M_w(tv)}$$

$\frac{t}{2}$ is constant with regard to the limit, so can be taken outside. As $v \to 0$, $M_w(tv) \to M_w(0) = E[e^0] = E[1] = 1$

$$= \frac{t}{2} \lim_{v \to 0} \frac{M'_w(tv)}{u}$$

Use l'Hopital's rule a second time. $M''_w(tv)$ is the second moment. The variance is the second moment minus the first moment. The First moment is 0, so $M''_w(tv)$ is the variance, which, evaluated at 0, is equal to 1.

$$= \frac{t^2}{2} \lim_{v \to 0} \frac{M''_w(tv)}{1}$$

$$\ln(\text{MGF}) = \frac{t^2}{2} \ as \ n \to \infty$$

Raise both sides to the power of e.

$$\boxed{\text{MGF} = e^{\frac{t^2}{2}}}$$

So as the sample size approaches infinity, the MGF of a standardised distribution of sample means is the same as the MGF of the standardised normal distribution, proving the Central Limit Theorem. But not proving it for sample sizes that don't approach infinity.

Appendix 14: Proof that part of the formula for a t statistic has a normal distribution

For the derivation of the probability density function (PDF) of the t distribution, we need to show that

$$W = \frac{\bar{x} - \mu}{\frac{\sigma}{\sqrt{n}}}$$

has a standard-normal distribution. This looks a lot like what we did in the previous appendix, proving the central limit theorem. The differences are that this time we aren't just proving it as $n \to \infty$; we are proving it for all n. What makes this possible is that we are also assuming that x as being i.i.d., and normally distributed: $X \sim N(\mu, \sigma^2)$. That is, of course, not a requirement for the central limit theorem.

Now we need to look at the moment generating function (MGF) of W.

Use the formula for the MGF, substituting in the formula for W.

$$MGF = E[e^{wt}]$$

$$= E\left[e^{\left(\frac{\bar{x}-\mu}{\frac{\sigma}{\sqrt{n}}}\right)t}\right]$$

In the exponent, multiply the numerator and denominator by $\sqrt{n}$

$$MGF = E\left[e^{\left(\frac{\bar{x}-\mu}{\sigma}\right)t\sqrt{n}}\right]$$

Distribute the $t\sqrt{n}$

$$= E\left[e^{\left(\frac{t\sqrt{n}\bar{x}}{\sigma} - \frac{t\sqrt{n}\mu}{\sigma}\right)}\right]$$

Separate this into the product of two powers of e.

$$= E\left[e^{\bar{x}\frac{t}{\sigma}\sqrt{n}}e^{-\mu\frac{t\sqrt{n}}{\sigma}}\right]$$

Take the power of e which is a constant with regard to x outside the expected value calculation.

$$= e^{-\mu\frac{t\sqrt{n}}{\sigma}}E\left[e^{\bar{x}\frac{t}{\sigma}\sqrt{n}}\right]$$

Use an alternative expression for the mean of x.

$$= e^{-\mu\frac{t\sqrt{n}}{\sigma}}E\left[e^{\frac{(x_1+x_2+x_3+\cdots+x_n)t}{n}\frac{t}{\sigma}\sqrt{n}}\right]$$

Divide the numerator and denominator in the exponent within the expected value, by the square root of n.

$$= e^{-\mu\frac{t\sqrt{n}}{\sigma}}E\left[e^{\frac{(x_1+x_2+x_3+\cdots+x_n)t}{\sqrt{n}}\frac{t}{\sigma}}\right]$$

Distribute $\frac{t}{\sqrt{n}\sigma}$

$$= e^{-\mu\frac{t\sqrt{n}}{\sigma}}E\left[e^{x_1\frac{t}{\sqrt{n}\sigma}+x_2\frac{t}{\sqrt{n}\sigma}+x_3\frac{t}{\sqrt{n}\sigma}+\cdots+x_n\frac{t}{\sqrt{n}\sigma}}\right]$$

Use a substitution. Let

$$\frac{t}{\sqrt{n}\sigma} = q$$

$$= e^{-\mu\frac{t\sqrt{n}}{\sigma}} E[e^{x_1 q + x_2 q + x_3 q + \cdots + x_n q}]$$

Show this as the product of powers of e.

$$= e^{-\mu\frac{t\sqrt{n}}{\sigma}} E[e^{x_1 q} e^{x_2 q} e^{x_3 q} \ldots e^{x_n q}]$$

We defined x as being independent and identically distributed. The independence means that the expected value of the product is the same as the product of the expected values (page 529).

Show this as a product of expected values.

$$= e^{-\mu\frac{t\sqrt{n}}{\sigma}} E[e^{x_1 q}] E[e^{x_2 q}] E[e^{x_3 q}] \ldots E[e^{x_n q}]$$

Because they are identically distributed, different values of x will have the same expected values. That means multiplying them n times is the same as raising a general version of x to the power of n.

$$= e^{-\mu\frac{t\sqrt{n}}{\sigma}} (E[e^{x q}])^n$$

Now the expression $E[e^{x q}]$ is the formula for a moment generating function (MGF) (using q instead of t, which makes no difference; it's just a dummy variable). We've defined x as being normally distributed, and we saw on pages 600 and following that the MGF of a normally distributed random variable (replacing t with q) is: $e^{\frac{\sigma^2 q^2}{2} + q\mu}$

Replace $E[e^{x q}]$ with $e^{\frac{\sigma^2 q^2}{2} + q\mu}$

$$\text{MGF} = e^{-\mu\frac{t\sqrt{n}}{\sigma}} \left(e^{\frac{\sigma^2 q^2}{2} + q\mu}\right)^n$$

Reverse the substitution. Let $q = \dfrac{t}{\sqrt{n}\sigma}$

$$= e^{-\mu\frac{t\sqrt{n}}{\sigma}} \left(e^{\frac{\sigma^2 t^2}{2n\sigma^2} + \frac{\mu t}{\sqrt{n}\sigma}}\right)^n$$

Distribute the n

$$= e^{-\mu\frac{t\sqrt{n}}{\sigma}} \left(e^{\frac{n\sigma^2 t^2}{2n\sigma^2} + \frac{n\mu t}{\sqrt{n}\sigma}}\right)$$

Cancel out:

$$= e^{-\mu\frac{t\sqrt{n}}{\sigma}} \left(e^{\frac{t^2}{2} + \frac{\sqrt{n}\mu t}{\sigma}}\right)$$

Show the negative power of e as the denominator (it means the same thing).

$$= \frac{\left(e^{\frac{t^2}{2} + \frac{\sqrt{n}\mu t}{\sigma}}\right)}{e^{\mu\frac{t\sqrt{n}}{\sigma}}}$$

Write the numerator as the product of two powers of e, and cancel out.

$$= \frac{e^{\frac{t^2}{2}} e^{\frac{\sqrt{n}\mu t}{\sigma}}}{e^{\mu\frac{t\sqrt{n}}{\sigma}}}$$

$$= e^{\frac{t^2}{2}}$$

As we saw on pages 600 and following, this is the MGF of a standard normal distribution. That means that

$$W = \frac{\bar{x} - \mu}{\frac{\sigma}{\sqrt{n}}}$$

has a standard normal distribution ($W \sim N(0,1)$ where x is normally distributed.

Appendix 15: Proof that part of the denominator of a *t* statistic has a chi square distribution

Getting back to Fisher's continuation of Gosset's work, having reached the point of recognising that the expected value of the sample variance is the population variance (for any sort of distribution), he decided to learn a bit more about the relationship between these numbers specifically in relation to a normally distributed population. (Keep in mind that Gosset had looked for a distribution of sample means when the underlying distribution was normal.)

Start with a random variable that follows a normal distribution, with a mean of μ and a variance of σ^2

$$X \sim N(\mu, \sigma^2)$$

This is the standardised form of a normally distributed random variable, as we saw earlier.

$$Z = \frac{(X - \mu)}{\sigma}$$

$$Z \sim N(0,1)$$

Multiply both sides by σ, then add μ to both sides, to express X as a function of Z.

$$X = Z\sigma + \mu$$

This is the formula for the sample variance.

$$s^2 = \frac{\sum_{i=1}^{n}(x_i - \bar{x})^2}{n - 1}$$

Express the sample variance in terms of z_i using the formula above: $x_i = z_i\sigma + \mu$.

$$s^2 = \frac{\sum_{i=1}^{n}(z_i\sigma + \mu - \bar{x})^2}{n - 1}$$

Do the same with the mean, replacing $\bar{x}$ with $\bar{z}$. We can do this because of our definition of the sample mean: $\bar{x} = \frac{\sum_{i=1}^{n} x_i}{n} =$

$$s^2 = \frac{\sum_{i=1}^{n}\left(z_i\sigma + \mu - (\bar{z}\sigma + \mu)\right)^2}{n - 1}$$

$\frac{\sum_{i=1}^{n} z_i\sigma + \mu}{n} = \frac{\sigma\sum_{i=1}^{n} z_i}{n} + \frac{\sum_{i=1}^{n} \mu}{n} = \bar{z}\sigma + \frac{n\mu}{n}$

The $\mu's$ cancel one another out. σ is a constant, so can be taken outside of the summation, but it is within a set of brackets that square it, so when it is taken out it needs to be squared.

$$= \frac{\sigma^2 \sum_{i=1}^{n}\left(z_i + \mu - (\bar{z} + \mu)\right)^2}{n - 1}$$

Multiply both sides by $(n - 1)$ and divide both sides by σ^2.

$$(n - 1)\frac{s^2}{\sigma^2} = \sum_{i=1}^{n}(z_i - \bar{z})^2$$

Recall that $\bar{z} = 0$; the mean of standardised distributions is zero, by definition. Also note that adding up multiple versions of any particular values of z_i, without defining what they are, is the same as adding up the random variable Z.

$$(n-1)\frac{s^2}{\sigma^2} = \sum_{i=1}^{n} Z^2$$

That means that the ratio of $s^2 : \sigma^2$ is proportional to a distribution that is chi squared.

$$(n-1)\frac{s^2}{\sigma^2} = \sum_{i=1}^{n} Z^2 \sim \chi^2$$

Acknowledgements

The following people helped with this book in different ways: some assisted with ideas and with editing, some helped at key points where I was confused and unable to continue, some helped by explaining what isn't commonly understood by many people who work professionally with statistics, some lent texts and some did a combination of these. All gave me their time, and all were encouraging. My wife Tracy's practical support and ongoing encouragement were essential to this project. The following list is alphabetical and doesn't necessarily reflect the relative extent of different contributions: Joyaa Antares, Howard Barnes, Jeanette Barnes, Dr Rebecca Barnes, Roxarne Burns, Penny Christadoulides, Emeritus Professor Geoff Eagleson, Dr Soames Job, Associate Professor Mark Jones, Mamello Lange, Emeritus Professor Juliet Richters, Jeny Shepherd, Lindsay Smartt, Yelena Townsend and Belinda White. None of these people have reviewed the whole text, so any errors or problems that remain are mine. James de Vries gave great input for layout and design.

Ian Lees gave valuable feedback about the book, and also did an extraordinary job helping create the YouTube videos that support it.

This book wouldn't have been possible without the knowledge, communication skills and dedication of many people who freely put information about mathematics and statistics online. There are too many of these to list, but I returned to the following sources repeatedly. Once again, the list is alphabetical rather than any indication of relative importance: Associate Professor Jeremy Balka (JB Statistics), Dr Trefor Bazet, BlackPen RedPen, Professor Joseph Blitzstein (Harvard Statistics), Matthew E Clapham, Professor Iain Collings, Computation Empire, Sal Khan (Khan Academy), Professor Christina Knudson, Ben Lambert (Ox Educ), Elliot Nicholson, Grant Sanderson (3Blue1Brown), Eddy Woo, and Justin Zeltzer (zedstatistics).

Notes

1. De Moivre, A, 1781 **The Doctrine of Chances: or, a method for calculating the probabilities of events in play,** available at https://openlibrary.org/books/OL6239276M/The_doctrine_of_chances

2. The quote "For the simplicity on this side of complexity, I wouldn't give you a fig. But for the simplicity on the other side of complexity, for that I would give you anything I have" is often attributed to Oliver Wendell Holmes, Sr. However, its source is actually a letter written by his son, Oliver Wendell Holmes, Jr., to Sir Frederick Pollock, dated October 14, 1902. In the letter, Holmes, Jr. expressed a similar sentiment: "The only simplicity for which I would give a straw is that which is on the other side of the complex — not that which never has divined it." This quote was included in the **Holmes-Pollock Letters: The Correspondence of Mr. Justice Holmes and Sir Frederick Pollock,** 1874-1932 (2nd edition, 1961), on page 109

3. Both of these quotes are from Hotelling H, 1953 **New Light on the Correlation Coefficient and its Transforms** Journal of the Royal Statistical Society. Series B (Methodological) , Vol. 15, No. 2 Royal Statistical Society Stable URL: https://www.jstor.org/stable/2983768 . This was the paper in which he suggested the approach that is currently used to apply the t distribution to the issue of correlation.

4. This was said by Severus Snape when he was introducing the class to making potions, but it is also a wonderful description of statistics.

5. Thiese M S, Arnold Z. C., Walker S. D. 2015 **The misuse and abuse of statistics in biomedical research.** Biochemia medica 25(1), 5–11 https://doi.org/10.11613/BM.2015.001

6. Nuijten MB, Hartgerink CH, Assen MA, Epskamp S, Wicherts JM, 2016 **The prevalence of statistical reporting errors in psychology** (1985–2013) BRM 48, pages 1205–1226.

7. Gorard, S. 2014. **The widespread abuse of statistics by researchers: what is the problem and what is the ethical way forward?.** Education Section review - British Psychological Society, 38(1), 3-10. (Gorard actually maintained that there were statistical errors in all social science papers he looked at.)

8. Kim JS, Kim DK, Hong SJ. **Assessment of errors and misused statistics in dental research.** Int Dent J. 2011 Jun;61(3):163-7. doi: 10.1111/j.1875-595X.2011.00037.x. PMID: 21692788; PMCID: PMC9374810.

9. For example, R A Fisher was a mathematician who added greatly to our understanding of statistics. In his book **Statistical Methods for**

Research Workers (originally published in 1925) he said "The mathematical complexity of these problems has made it seem undesirable to do more than (i) to indicate the kind of problem in question, (ii) to give numerical illustrations by which the whole process may be checked, (iii) to provide numerical tables by means of which the tests may be made without the evaluation of complicated algebraic expressions." (pp15-16, Indian reprint from 2017, Kalpaz Publications).

10. Rose C, 1985 **Accelerated Learning** Dell Publishing Co page 5

11. Rose C, 1985, op cit

12. Sharma A, 2011 **Regional Cerebral Blood Flow** Encyclopedia of Clinical Neuropsychology 157 https://doi.org/10.1007/978-0-387-79948-3_68

13. Basar E, 2006 **The theory of the whole-brain-work.** International journal of psychophysiology : official journal of the International Organization of Psychophysiology 60(2), 133–138 https://doi.org/10.1016/j.ijpsycho.2005.12.007

14. Bergee M J, Weingarten, K. M. 2021 **Multilevel Models of the Relationship Between Music Achievement and Reading and Math Achievement.** Journal of Research in Music Education 68(4), 398–418 https://doi.org/10.1177/0022429420941432

15. van Elk M, Arciniegas Gomez, M. A., van der Zwaag, W., van Schie, H. T., Sauter, D. 2019 **The neural correlates of the awe experience: Reduced default mode network activity during feelings of awe.** Human brain mapping 40(12), 3561–3574 https://doi.org/10.1002/hbm.24616

16. Chirico A, Gaggioli, A. 2021 **The Potential Role of Awe for Depression: Reassembling the Puzzle.** Frontiers in psychology 12, 617715 https://doi.org/10.3389/fpsyg.2021.617715

17. Thiese M S, Arnold Z. C., Walker S. D. 2015 **The misuse and abuse of statistics in biomedical research.** Biochemia medica 25(1), 5–11 https://doi.org/10.11613/BM.2015.001

18. Nuijten MB, Hartgerink CH, Assen MA, Epskamp S, Wicherts JM, 2016 **The prevalence of statistical reporting errors in psychology** (1985–2013) BRM 48, pages 1205–1226.

19. https://www.goldennumber.net

20. Grant Sanderson presents the wonderful **3Blue1Brown** mathematical visualisations and explanations on YouTube.

21. If you go further looking for powers of 11 this still works, provided you do some "carrying" the way you learned to when you were learning to add and multiply numbers without a calculator at school, so it doesn't work out quite so neatly.

22. Tarek Said has a wonderful video about this at https://www.youtube.com/watch?v=habHK6wLkic

23. For those wishing to explore this further, the areas involved are probably the superior temporal sulcus ventral anterior and posterior (STSva and STSvp), although there may be others as well. These areas are involved in what is known as the "story-math secondary contrast". This is a term from mathematics education, which compares the style of teaching

where there are stories around people and the idea is to provide meaning (which is the approach this book has tried to follow), compared to learning mathematics by repetitive exercises.

24. Box, G, **Science and Statistics**, Journal of the American Statistical Association, Vol. 71, No. 356. (Dec., 1976), pp. 791-799. George Box has been wrongly credited with the quote "all models are wrong, but some are useful"; this is implied, but not stated, in the article cited here. This article was actually about the work of Ronald Fisher.

25. For example, https://tutorial.math.lamar.edu/classes/de/laplace_table.aspx

26. https://www.york.ac.uk/depts/maths/histstat/pascal.pdf

27. This is so argued by Laplace, PS **A Philosophical Essay on Probabilities** https://www.ime.usp.br/~walterfm/cursos/mac5796/Laplace.pdf , page 46

28. Laplace, op cit page 195

29. This analysis first appeared in Wainer, H and Zwerling, HL 2006 **Evidence That Smaller Schools Do Not Improve Student Achievement**, Phi Delta Kappan, ,Vol.88 (4), p.300-303. It was picked up by Kevin deLaplante at https://www.youtube.com/watch?v=sgZQMJQRwRM, which is worth watching for other examples of the same phenomenon.

30. In particular, the factorial sign was invented later, but there are other changes as well.

31. Lewis Carroll was the pen name of Charles Dodgson. As well as being a mathematician and a writer of children's fiction, he was a keen amateur photographer. He has been accused of being a paedophile, but this has been questioned more recently. He enjoyed the company of children, and some of his photographs are nudes of children. Some biographers have suggested that this meant either that he was a paedophile or had tendencies in that direction, and that may well be the case. It is also true, though, that photographs of nude children were viewed in Victorian times as studies of innocence. They were extremely common, and most serious photographers included them in their work. They were, for example, used on Christmas cards from one family to another. For a fuller discussion see Wakely-Mulroney, K, 2021, **The Man Who Loved Children: Lewis Carroll Studies' Evidence Problem** Journal of the History of Sexuality, Vol. 30, No. 3 (SEPTEMBER), pp. 335-362.

32. Gelman A, 2005 **Analysis of Variance – Why it is More Important than Ever** The Annals of Statistics Vol 33, No 1, 1-53

33. Laplace, PS 1814 **A Philosophical Essay on Probabilities** https://www.ime.usp.br/~walterfm/cursos/mac5796/Laplace.pdf

34. Earlier we looked at developments by French mathematicians: Descartes, Pascal, Fermat, Laplace, Legendre and Fourier. We will encounter Laplace again, and we will also meet Poisson.

35. The Huguenots were protestants in a largely-Catholic country. King Henry IV of France was originally a Huguenot before converting to Catholicism to ascend the throne. In 1598 he issued a decree granting protections to the Huguenots. In 1685, King Louis XIV revoked the

protections, which led to renewed persecution of Huguenots, and many fled France to find refuge in countries like England, the Netherlands, Germany, and even as far as South Africa and North America. Paul Revere, who famously rode at midnight to alert the American colonial militia of British invasion was another famous Huguenot.

36. De Moivre A, 1718 The Doctrine of Chances

37. De Moivre A, 1718 op cit

38. ibid

39. It was his older brother, Johann, who came up with l'Hopital's rule. They were rivals.

40. Plumb J H, 1950 **England in the Eighteenth Century**, (Middlesex: Penguin Books Ltd,)

41. Bellhouse DR, **The Reverend Thomas Bayes FRS: a Biography to Celebrate the Tercentenary of his Birth** https://www2.isye.gatech.edu/isyebayes/bank/bayesbiog.pdf

42. Barnard GA, 1958 **Thomas Bayes – A Biographic Note** Biometrika Vol 45, 293-315 https://web.archive.org/web/20110410085940/http://www.stat.ucla.edu/history/essay.pdf

43. https://royalsocietypublishing.org/doi/pdf/10.1098/rstl.1763.0053

44. Stigler SM, 1986 **The History of Statistics: The Measurement of Uncertainty before 1900** Belknap Press

45. Compare, for example, the interpretation found at https://www.youtube.com/watch?v=7GgLSnQ48os with the interpretation in Stigler SM, 1986, op cit.

46. Pascal had previously used an argument from probability as an argument for belief in God, but his argument was still based on gambling. His argument is known as "Pascal's wager".

47. Laplace, op cit, page 189

48. Laplace, op cit, page 1

49. Gigerenzer G, Gaissmaier W, Kurz-Milcke E, Schwartz L M, Woloshin S, 2007 **Helping Doctors and Patients Make Sense of Health Statistics**. Psychological science in the public interest : a journal of the American Psychological Society 8(2), 53–96 https://doi.org/10.1111/j.1539-6053.2008.00033.x

50. http://fitelson.org/probability/good_bayes.pdf

51. Neyman J, 1977 **Frequentst Probabability and Frequentist Statistics** Synthesise 97-131 https://cupdf.com/document/frequentist-probability-and-frequentist-statistics.html

52. You can resolve this problem by putting your data into a software package, and comparing the output for a chi square distribution. That will either match to $\left(\frac{1}{2},\frac{1}{2}\right)$ or $\left(\frac{1}{2},2\right)$. The detail will be clearer when we look at chi square.

53. Stigler SM, 1986, op cit.

54. Stigler SM, 1986, op cit.

55. There are other possibilities. Carl Gustav Jacob Jacobi who we met in part 1 in relation to the Jacobian might have done it. The "Gaussian integral" which is involved only works because of the Jacobian, and not all of Jacobi's works have been published.

56. Laplace, op it, page 35. Laplace stayed with the French usage of "Moivre" rather than "de Moivre", which was the name he adopted when he moved to England.

57. Laplace, op it, page 193

58. Stigler SM, 1986, op cit.

59. Stigler SM, 1986, op cit.

60. Stigler SM, 1986, op cit.

61. https://mathshistory.st-andrews.ac.uk/Biographies/Chebyshev/

62. Butzer P, Jongmans F, 1999 P**. L. Chebyshev (1821–1894): A Guide to his Life and Work Journal of Approximation Theory** January, pages 111-138 https://www.sciencedirect.com/science/article/pii/S0021904598932890

63. For example, Chebyshev doesn't rate a mention in the Wikipedia article on the history of statistics at the time of writing (October 2025): https://en.wikipedia.org/wiki/History_of_statistics . He also isn't mentioned anywhere in Stigler, SM 1986, op cit.

64. Cited in Plackett, RL, 1983, **Karl Pearson and the Chi-squared Test**, International Statistical Review, 51, pages 59-72

65. https://www.gresham.ac.uk/lectures-and-events/karl-pearsons-gresham-lectures-on-geometry-1890- 1894

66. Stigler SM, 2008 **Karl Pearson's Theoretical Errors and the Advances They Inspired** Statist. Sci. 23(2): 261-271 (May 2008) https://projecteuclid.org/journals/statistical-science/volume-23/issue-2/Karl-Pearsons-Theoretical-Errors-and-the-Advances-They-Inspired/10.1214/08-STS256.full

67. Stigler SM, 1986, op cit

68. His overall approach to eugenics was that social welfare should support those whose breeding should be encouraged, rather than, as he saw it, encouraging the worst genetic stock to breed.

69. For example Gorard S, 2015 **An Absolute Deviation Approach to Assessing Correlation** British Journal of Education, Society & Behavioural Science 5(1): 73-81,Article no BJESBS.2015.008 https://dro.dur.ac.uk/13873/1/13873.pdf?DDD29+ded4ss+d700tmt

70. https://en.wikipedia.org/wiki/Coefficient_of_determination

71. Chicco D, Warrens MJ, Jurman G. 2021 **The coefficient of determination R-squared is more informative than SMAPE, MAE, MAPE, MSE and RMSE in regression analysis evaluation** PeerJ Computer Science Jul 5;7:e623 https://peerj.com/articles/cs-623/

72. Wright, S. 1921 **Correlation and Causation**, Journal of Agricultural Research, Vol. XX, No. 7 Jan. 3, Key No. A-55

73. This distribution was actually first dealt with in 1872 in a German book on measuring the surface of the Earth by Friedrich Robert Helmut (1843-1917). (Methode Der Kleinsten Quadrate Anwendungen Auf Die Geodäsie Und Die Theorie Der Messinstrumente: geodesy: **Method of Least Squares applications to geodesy and the theory of measuring instruments.**) His thinking was based on Gauss's work on least squares. It appears that Pearson was unaware of this work, even though he tended to be enthusiastic about all things German.

74. Pearson, K, 1990 **Mathematical contributions to the theory of evolution on the heritance of characters not capable of exact quantitative measurement**, VIII, Ph Trans. Roy. Soc. A195, 79-150.

75. Lundh A, Lexchin J, Mintzes B, Schroll JB, Bero L. (2017). **Industry sponsorship and research outcome**. Cochrane Database of Systematic Reviews. Bekelman JE, Li Y, Gross CP. Scope and impact of financial conflicts of interest in biomedical research. JAMA. 2003. Lesser LI, Ebbeling CB, Goozner M, Wypij D, Ludwig DS. (2007). Relationship between funding source and conclusion among nutrition-related scientific articles. PLoS Medicine.

76. Neyman identified as Polish and came from a line of Polish aristocrats. He was born in Bender (also spelled Tighina). This was part of what was then the Russian empire, and is now normally recognised as part of Moldova, although it is part of Transnistria which is disputed territory close to the border of Ukraine.

77. Neyman J, 1977 op cit.

78. Student, 1908, The Probable Error of a Mean, Biometrika, Vol 6, No1, March, pp1-25.

79. Fisher, R A, 1925, **Statistical Methods for Research Workers**, page 16. The first person to come up with something similar to what Pearson called the chi square distribution was Helmut (op cit)

80. Gosset, op cit

81. As an example of Fisher's errors, he stated that the formula $\frac{\sigma^2}{n}$ gives the variance of means of random samples if the underlying distribution is normal, or at least "does not belong to the exceptional class of distributions for which the distribution of the mean does not tend to normality" (Op cit). As we saw earlier, the formula $\frac{\sigma^2}{n}$ is valid for all distributions, and regardless of the underlying distribution of the population, the distribution of sample means always tends to normality.

82. This is a link to the Computation Empire proof that the chi squared distribution we are using has k degrees of freedom. https://www.youtube.com/watch?v=z00trinIHUA&t=0s

83. See, for example, Welkowitz J, Ewen RB, Cohen J 1976 **Introductory Statistics for the Behavioral Sciences**, Academic Press, pages 129 and following. This is an old text book, but still one of the best introductory statistical texts for people who want to avoid engaging in the mathematics.

84. Fisher, 1925, op cit, pp175ff

85. Welkowitz J, Ewen R, Cohen J, 1976 **Introductory Statistics for the Behavioural Sciences**, Academic Press.

86. Cohen J (1960) **A Coefficient of Agreement for Nominal Scales Educational and Psychological Measurement**, Vol. XX, No. 1, page 37

87. McHugh M L, 2012 **Interrater reliability: the kappa statistic.** Biochemia medica v.22(3) https://www.ncbi.nlm.nih.gov/pmc/articles/PMC3900052/

88. This graph was produced by the author in Microsoft Excel. These findings can be replicated either with statistics software that does Kappa coefficients, or by looking up the kappa formula online, and using it in Excel with the values given in the example.

89. Banerjee M, Capozzoli M, Mcsweeney L, Slnha D 1999 **Beyond kappa: A review of interrater agreement measures** The Canadian Journal of Statistics Vol. 27, No. 1 Pages 3-23

90. These studies were cited in Stratten, G, 2011, **Does Increasing Textbook Portability Increase Reading Rates or Academic Performance?** The Journal of the Virginia Community Colleges

91. Linear algebra can be a bit overwhelming at first, because so many issues can be linked to vectors and matrices. It can help to think about the subject historically, as we did in this book. Many of the concepts (e.g. Gaussian elimination, Jacobians) were actually developed before vectors and matrices. It's best to think of vectors and matrices as convenient, systematised ways of dealing with multiple variables in widely different situations. Once you understand that it is easier to learn each topic, without trying to necessarily relate it to other topics in linear algebra, it gets easier.

92. For example, https://tutorial.math.lamar.edu/classes/de/laplace_table.aspx

93. If you want to explore these transforms more, a good place to start would be those on-line sources mentioned in the acknowledgements of this book. Most of the people mentioned there (in regard to on-line resources) deal with Laplace and/or Fourier transforms, and their relationship with convolution.

Index

A

a priori understanding, 429
absolute probability, 414
absolute value, 26
actuarial, 395, 467
acute angle, 171
addition rule, 128, 235, 236, 248, 280, 282, 326, 345, 346, 348, 401, 424, 690
addition rule for integration, 280
adjacent, 191
alternative hypothesis, 422, 423, 436, 638, 640, 641, 642, 682, 726
alternative parameterization of the gamma distribution, 461, 541
Analysis of Covariance (ANCOVA), 701
analysis of variance. *See* ANOVA
ANOVA, 607, 691, 692, 694, 700, 701, 702, 712, 713; hierarchical, or multilevel, 702; mixed model, 702; one way, 691; repeated measures, 701; two way, 701
anthropology, 514, 562
applied mathematics, 77
approximately equal sign, 29
arc, 183, 184
arc cosine, 196
arc sine, 196
arc tangent, 196
Archimedes, 187
arithmetic triangle, 78
arms, 171
artificial intelligence, 332
astronomy, 332, 394, 466
asymptote, 211
averages, 491, 492, 520

B

barometric readings, 514
base, 53, 126, 127, 128

Bayes, Thomas, 3 335, 410, 411, 412, 413, 419, 425, 441, 450, 779
Bayes's rule, 423, 429
Bayes's theorem, 412, 416, 419, 429
Bayesian probability, 418
Bayesian statistics, 412, 418, 430, 431, 434, 435, 437, 440, 443, 444, 450, 470, 649, 650, 675, 705, 706, 729
before and after an intervention, 684
Bern(p), 380, 468
Bernoulli choice, 453, 506, 716
Bernoulli distribution, 380, 391, 468, 470
Bernoulli trials, 330, 381
Bernoulli trials, 330
Bernoulli, Jacob, 3, 380
Bernoulli, Johann, 3, 255
beta distribution, 432, 443, 445, 446, 447, 448, 450, 459, 466, 470, 537, 540, 549, 614, 675, 705
beta function, 302, 310, 445, 446, 448, 675, 697, 761; incomplete, 449
Beta(a, b, 470
bimodal distribution, 364
Bin(n,p), 372, 469
binomial coefficients, 163, 356
binomial distribution, 372, 373, 374, 378, 379, 380, 382, 384, 385, 386, 391, 402, 429, 430, 434, 435, 443, 444, 452, 470, 494, 506, 508, 523, 537, 544, 546, 548, 600, 602, 685, 704, 716
binomial expansion, 163, 269, 310, 351, 353, 373, 374, 375, 381, 453, 454, 487, 525, 544
binomial formula, 456
binomial probabilities, 360
binomial theorem, 163, 269, 349, 350, 351, 432, 434, 506, 515
biology, 394, 466
bivariate binomial distribution, 705
bivariate distribution, 364
bivariate normal distribution, 688

body mass index, 472, 701

Bombelli, Raphael, 3, 105

Bonacci, Leonardo, 121

braces, 74, 89

brackets, 37, 55, 60, 84, 89

Brahe, Tycho, 3, 493

Brahmagupta, 25

business, 396, 467

C

call centre, 450

Cardano, Gerolamo, 3

Cartesian plane, 199, 200, 203, 204, 205,
213, 225, 228, 297, 328, 498, 695

cartography, 225

case-controlled study, 407

categorical distribution, 390, 391

categorical variables, 360, 366, 370, 607,
616, 707

Cauchy distribution, 705

CDF, 374, 375, 382, 384, 386, 389, 390,
392, 405, 447, 448, 449, 455, 456, 459,
460, 462, 465, 474, 475, 477, 523, 610,
611, 612, 615, 627, 640, 668, 669, 672,
696; joint, 404, 405; normal
distribution, 594; standard normal
distribution, 594

Čebyčev. *See* Chebyshev, Pafnuty

central limit theorem, 624, 625, 639, 666,
680, 766, 770

central moment of a distribution, 545

chain rule, 257, 316, 591

change of base formula, 128

Chebychov. *See* Chebyshev, Pafnuty

Chebysheff. *See* Chebyshev, Pafnuty

Chebyshev inequality, 519, 558, 559, 632

Chebyshev, Pafnuty, 3, 518, 519, 520,
522, 525, 530, 532, 533, 534, 535, 538,
539, 543, 545, 556, 562, 563, 565, 567,
588, 630, 780

Chebyshov. *See* Chebyshev, Pafnuty

chi square distribution, 305, 392, 478,
479, 540, 554, 561, 606, 614, 616, 657,
662, 667, 670, 694, 695, 705, 720, 779,
781

chi square test, 707, 720

chi squared distribution, 605

chord, 174

circle, 174; area of, 187

circumference, 67, 68, 145, 174, 183,
184, 185, 309, 506, 731

clinical trials, 439

closest point of curve to a line, 241

Cochran's Q Test, 713

coefficient, 206

coefficients, 94, 95, 100, 101, 263

Cohen, Jacob, 719, 721

Cohen's kappa coefficient, 713, 719, 722,
723, 727

combinations, 29, 49, 50, 101, 156, 157,
158, 165, 312, 347, 349, 351, 352, 354,
355, 361, 362, 388, 435, 485, 722, 775

common denominator, 46, 47, 49, 65, 97,
251, 601, 603, 714, 738, 745, 753, 754,
755

commutative property of addition, 23,
24, 94, 133

commutative property of convolution,
314

commutative property of
multiplication, 35, 90, 104

comparison table, 607

competitive results, 710

complex conjugates, 226

complex numbers, 77, 106, 226, 316

complex plane, 226, 297

computer science, 394, 466, 780

conditional probability, 342, 414, 421,
424, 430

confidence interval, 436, 649, 650, 683,
686

confounding, 368

congruent, 172, 173, 740

conjugate priors, 434, 470, 471

constants, 85, 90, 94, 140

consumer research, 395, 467

contingency table, 399, 402, 406, 416,
485, 707, 720

continuous distributions, 373

continuous probability distributions,
608

control group, 637, 640, 702

convergent sequences, 267

convergent series, 269

convolution: integrals, 312, 485; theorem
of Laplace transforms, 321

convolutions, 101, 102, 312, 313, 314, 319, 321, 485, 486, 488, 539, 782

correlation, 404, 498, 515, 569, 571, 575, 576, 577, 579, 580, 581, 582, 583, 584, 606, 609, 663, 670, 685, 687, 688, 712, 715, 719, 721, 776, 794; Pearson's, 561

correlation coefficient, 574, 575, 712

correlatisn coefficient, 712

cos, 194, 295, 297

cosecant, 196

cosine, 194, 195, 213, 223, 246

cotangent, 196

Cotes, Roger, 3, 183

counting numbers, 72, 736

covariance, 404, 583, 584, 586, 686

credible intervals, 650

criminology, 394, 466

critical *t* value, 681, 686

crosstab, 399

cross-tabulation. *See* crosstab

cubed root, 59

cumulative distribution function, 374, 382, 401, 454, 457, 536

customer surveys, 710

D

Darwin, Charles, 514

de Moivre, Abraham, 3, 8, 376, 377, 380, 411, 506, 522, 600, 730, 780

deciles, 493

decimal fractions, 63, 73

decimal notation, 65, 82

decimal place, 65

decimal separator, 63

deck of cards, 153, 390

definite integral, 277

degrees of freedom, 614, 615, 616, 617, 661, 662, 663, 667, 674, 676, 677, 678, 680, 683, 687, 693, 694, 695, 699, 700, 706, 707, 781

delta, 207, 208, 229, 232, 233, 238, 249

demographics, 413

denominator, 43, 44, 48, 49, 65, 104, 256

dependent variables, 113

derivative, 233, 261; of arc tangent, 253; of tangent, 253

Descartes, Rene, 3, 127, 199, 225, 328, 409, 778

Dewey, John, 515

diameter, 67, 68, 145, 174, 184, 185, 297, 309, 731

dice, 336, 350, 353, 362, 390

difference of cubes, 91

difference of squares, 91

differential calculus, 223, 233

differential equat:ons, 317, 318

differentiation: acdition rule, 235, 236; chain rule, 249; power rule, 237; product or multiplication rule, 248; quotient rule, 251

dimensions, 169, 170, 188, 285, 373, 508

diophantine equations, 92

discrete distribut:ons, 373

discrete random variable, 405

discrete variable, 366

distribution, 458

distribution-free methods, 710

distributive property of multiplication, 37, 38, 39, 84, 90, 93, 134

divine proportion, 69

division, 50

double integration, 286

double negatives, 84

dummy variable, 301, 302, 486, 533, 761, 771

dummy variables, 301, 486

E

ecology, 394, 466

economic classifications, 711

economics, 410

Edgeworth, Francis, 3, 562, 563

education, 2, 7, 395, 466, 514, 777, 780

effect size, 643

Einstein, Albert, 4, 169, 731

electronics, 106, 748

element of a set, 74

elimination, 109

ellipse, 210, 211, 497, 742

environmental science, 466

epidemiology, 395, 466

equals sign, 28, 29, 68, 83

equation: algebraic, 83

equation, 86

equations: simplifying, 89

equilateral triangle, 173

error-distribution probability, 491

errors in measurement, 332, 413, 505, 508

eugenics, 333, 514, 515, 564, 657, 780

Eugenics Society, 562

Euler, Leonhard, 3, 78, 105, 111, 268, 269, 270, 272, 292, 294, 301, 302, 317, 318, 459, 491, 517, 518, 562, 747, 748

Euler's formula, 297

Eulerian integrals, 302, 310, 497

evolution, 332, 514, 516, 781

excess kurtosis, 678

exclamation mark notation, 150

expanding, 91

expanding algebraic terms, 90

expectation, 377, 378, 522, 529, 533, 547, 767

expected value, 377, 519, 522, 523, 525, 526, 527, 529, 530, 531, 532, 534, 536, 538, 539, 543, 544, 545, 546, 547, 548, 549, 551, 554, 556, 557, 558, 559, 566, 583, 584, 585, 602, 603, 609, 610, 615, 616, 622, 623, 631, 632, 654, 655, 656, 659, 660, 665, 677, 766, 770,771, 773

Expo(λ), 470

exponent, 53, 54, 127

exponential distribution, 457, 459, 461, 462, 466, 470, 471, 473, 474, 520, 525, 534, 535, 536, 537, 539, 540, 549, 556, 566, 629

exponential equations, 128

exponential notation, 55, 64

exponents, 89; multiplying, 54

expressions, 29, 30, 36, 37, 48, 82, 87, 90, 91, 94, 112, 113, 160, 250, 252, 263, 750, 753, 755, 757, 765, 777, 794

F

F distribution, 607, 677, 694, 700, 701, 702; PDF, 694

factorials, 147, 149, 150, 151, 152, 155, 157, 259, 260, 302, 309, 310, 356, 445, 462, 738, 778

factorizing, 91

failure, 350, 373, 378, 443

false negative, 635

false positive, 635

Fermat: last theorem, 326

Fermat, Pierre de, 3, 78, 324, 325, 327, 328, 329, 336, 376, 778; last theorem, 324

Fibonacci, 121, 123

Fibonacci sequence, 122, 123

finance, 395, 467

fingerprints, 394, 514

first moment, 543

Fisher z transformation, 688

Fisher, Ronald, 3, 4, 610, 616, 630, 646, 654, 656, 657, 658, 659, 660, 661, 662, 664, 665, 677, 685, 687, 688, 691, 692, 693, 694, 706, 707, 773, 776, 778, 781, 782, 794

Fisher's exact test, 707, 720

flat plane, 170

forensics, 394, 466

formula, 85, 86, 112

Fourier transforms, 316, 321, 322, 532, 533, 543, 782

Fourier, Jean-Baptiste Joseph, 3, 321, 534

fourth central moment, 553

fractions, 42, 43, 44, 45; adding, 46; division, 50; larger than 1, 48; multiplying, 47, 48, 82; subtracting, 46

frequency distribution, 361, 367, 494, 521, 522, 621

frequentist statistics, 436, 437, 728

Friedman Test, 713

function notation, 112

functions, 112

G

G(p), 469

Galton, Francis, 3, 164, 514, 515, 516, 562, 563, 571, 657, 688

gamma distribution, 459, 460, 461, 462, 463, 466, 470, 471, 479, 481, 537, 539, 540, 541, 549, 554, 612, 613, 614, 615, 617, 695, 727

gamma function, 302, 304, 305, 310, 311, 317, 445, 460, 479, 541, 614, 615, 667, 672, 675, 697; incomplete, 304, 460;

lower incomplete, 304; upper incomplete, 305
Gamma(n, λ), 470
Gauss, Carl Friedrich, 3, 8, 78, 134, 135, 302, 306, 497, 505, 510, 513, 571, 781
Gaussian distribution. *See* normal distribution
Gaussian integral, 306, 307, 309, 512, 758, 760, 761, 780
geometric distribution, 380, 382, 384, 386, 457, 458, 459, 469, 470, 520, 537, 548
Gerolamo Cardano, 77
golden ratio, 69, 70, 99, 124, 183, 594, 734, 740, 741
golden rectangle, 69
Gosset, William Sealy, 3, 653, 654, 656, 657, 658, 659, 660, 661, 662, 664, 677, 773, 781
grades, 710
gradient, 194, 195
greater than or equal to sign, 29
greater than sign, 29
greenhouse effect, 321

H

H_0. See null hypothesis
H_1. See alternative hypothesis
high-pressure systems, 514
Hindu-Arabic numerals, 121
histogram, 362, 606
hyperbola, 211, 237, 591
hypergeometric distribution, 386, 402, 429, 469, 537, 548, 706, 707, 708, 709; PDF, 708
hyperparameters, 429
hypotenuse, 177, 178, 179, 180, 181, 191, 192, 245, 674, 698, 743

I

i.i.d., 368, 372, 381, 627
identity, 91
imaginary numbers, 105, 106
improper fraction, 48
indefinite integral, 277
independent and identically distributed, 368, 372

independent probability, 341
independent variables, 113, 368
index, 53, 125, 127, 292
indices, 89
inequality, 29, 30, 477, 556, 558, 577, 630, 648
inferential statistics, 619
infinite sequences, 269
infinity, 122
inflexion point, 591, 592, 606
integers, 72
integration by parts, 280, 282, 303
integration by substitution, 282
intention-to-treat, 643
inter-examiner reliability, 719, 721
inter-quartile ranges, 493
intersection, 75, 220, 345, 400, 416, 417
interval, 70, 169, 174, 183, 220, 452, 453, 456, 648, 686, 687, 733, 734, 735, 740
interval estimate, 648
inverse chi square distribution, 705
inverse function, 114
inverse functions, 113, 114, 195, 473, 474, 475, 476, 478, 611
inverse operation, 83
irrational numbers, 59, 67, 73, 181, 268, 512
isosceles triangle, 173

J

Jacobi, Carl Gustav Jacob, 3, 290, 306, 309, 758, 780
Jacobian, 289, 290, 306, 307, 309, 476, 478, 512, 671, 696, 700, 758, 760, 761, 780
Jeffries prior, 448
joint distribution, 400, 401, 406, 527, 528, 583, 669, 694, 695, 697, 720

K

Kayyam triangle, 78
Kendall's Tau, 712
Kramp, Christian, 3, 149
Kruskal-Wallis H Test, 712
kurtosis, 553; excess, 553

L

L'Hopital, Guillaume de, 255
l'Hôpital's rule, 255, 256, 257, 258, 303
Laplace transform, 317; inverse, 319, 539
Laplace transforms, 316, 318, 321, 533, 539, 540; **inverse**, 318
Laplace, Pierre Simon, 3, 309, 314, 316, 317, 318, 321, 334, 368, 412, 413, 416, 418, 429, 435, 450, 491, 505, 506, 508, 512, 513, 517, 532, 533, 539, 540, 588, 649, 758, 763, 764, 765, 778, 779, 782
law of cosines, 220
law of large numbers, 631; strong, 631; weak, 631
law of total probability, 426
least squares, 491, 497, 505
least squares method, 498, 571, 582
Legendre, Adrien Marie, 3, 301, 316, 497, 778
Leibniz, Gottfried, 3, 228, 233, 274, 324
leptokurtic, 553
less than or equal to sign, 29
less than sign, 29
likelihood, 430, 431, 432, 433, 434, 435, 444, 649, 727; maximum, 430, 433
Likert scale, 710
limits, 123, 132, 141, 232
limits of integration, 277, 290, 317, 319
line, 70
linear equations, 204
linearity, 134, 280, 318, 456, 463, 500, 502, 522, 527, 528, 529, 547, 549, 551, 569, 576, 584, 585, 595, 622, 623, 654, 655, 660, 765; of series, 133
linguistics, 395, 467
ln, 292
logarithm, 53, 127
logarithm identities, 128
logarithms, 53, 125, 126, 127, 128, 129, 138, 292, 299, 316, 506, 582, 596
logistics, 395, 467
log-normal distribution, 705
lurking variables, 368

M

Maclaurin series, 261

Maclaurin, Colin, 3, 261
Mann, Henry, 3, 717
Mann-Whitney U test, 712, 717
MANOVA, 703
margin of error, 605, 606, 608, 609, 614, 615, 667, 680
marginal distribution, 399, 401, 405, 528, 583
marginal probability, 430
market research, 619
marketing, 395, 467
Markov chain, 471
Markov inequality, 519, 556, 558, 559
Markov, Andrey, 3, 471, 556, 558, 559
maximum, 493
MCMC, 471
mean, 493, 500
median, 492, 493
medicine, 394, 466, 781
mesokurtic, 553
meta-analysis, 429
minimum, 493
mode, 492
moment generating function, 462, 519, 534, 536, 538, 539, 554, 600, 603, 613, 666, 677, 771
moments, 519, 562
monotone, 473, 610
Monte Carlo, 471
Monty Hall problem, 334
multimodal distribution, 364, 624
multinomial coefficients, 356
multinomial distribution, 390, 391, 392, 398, 469, 470, 607, 608, 609, 616, 685, 707
multinomial expansion, 358, 391
multinomial expansions, 358
multinomial probabilities, 356, 358, 609
multiplication rule, 280, 282, 344, 345, 346, 350, 352, 357, 381, 383, 400, 405, 424, 432, 451, 483, 529, 546, 581, 591
multivariate distribution, 364
multivariate normal distribution, 705

N

n binomial r, 157
n choose r, 155, 157, 158, 162, 164, 347, 352, 354, 386, 388, 469, 738

Napier, John, 3, 125, 126, 127, 129
Napier's rods, 129
natural logarithms, 292
natural numbers, 72, 75, 135, 146, 148,
    153, 299, 327, 713
natural selection, 514, 516
navigation, 332
NB(r,p), 469
negative angles, 214
negative binomial distribution, 384, 386,
    459, 469, 470, 537, 548
negative numbers, 25, 30, 32, 39, 46, 72,
    79, 105, 106, 199, 210, 225, 306, 328,
    559
Newton, Isaac, 3, 183, 228, 274, 316,
    317, 324, 376, 517
Neyman, Jerzy, 3, 436, 646, 647, 648,
    649, 650, 779, 781
nominal variable, 360, 366
non-parametric tests, 710, 712, 715
normal distribution, 7, 8, 181, 305, 309,
    471, 479, 505, 510, 512, 513, 515, 516,
    540, 548, 553, 563, 588, 590, 591, 592,
    593, 594, 595, 596, 597, 598, 600, 602,
    604, 605, 606, 608, 610, 611, 612, 613,
    614, 615, 621, 623, 624, 627, 629, 637,
    638, 640, 647, 666, 667, 669, 670, 676,
    677, 678, 680, 688, 694, 705, 706, 717,
    721, 767, 769, 770, 772, 773;
    standard, 610
normal distribution curve, 300
not equal sign, 29
null hypothesis, 436, 482, 486, 616, 635,
    636, 637, 638, 639, 640, 664, 682, 687,
    691, 692, 708
number line, 24, 25, 26, 43, 44, 105, 106,
    199, 225, 328, 409, 637, 747
numerator, 43, 44, 48, 49, 65, 104, 256
numerical variable, 369, 607

O

obtuse angle, 171
odds, 338, 407, 408
odds ratio, 338, 407, 607
operations, 103, 395, 467
opinion surveys, 710
opposite, 191

ordinal variable, 360, 361, 366, 369, 710,
    713, 719
origin, 199

P

P hunting, 598, 650
P value, 598, 642, 647, 648, 682, 690
parabola, 210, 211, 229, 232, 255, 473,
    502, 553, 591, 742
parallel lines, 172
parallelogram, 174; area of, 177
parameter, 363
parameters, 372, 387, 429, 430, 431,
    465, 488, 595, 632, 640, 649, 705, 706,
    708, 709, 710, 713, 725
Pareto distribution, 706
partial derivatives, 285
partial-fraction decomposition, 252
Pascal, Blaise, 3, 78, 183, 199, 324, 329,
    336, 376
Pascal's triangle: rectangular form, 79
Pascal's triangle, 78, 79, 120, 121, 122,
    124, 133, 136, 163, 164, 185, 272, 348,
    349, 351, 362, 453, 506, 508, 736, 738,
    739
patterns of numbers, 116, 117
PDF, 373, 401, 405, 443, 445, 447, 448,
    450, 459, 460, 465, 473, 474, 475, 476,
    477, 478, 479, 485, 486, 488, 529, 535,
    536, 537, 611, 612, 613, 667, 668, 669,
    670, 672, 674, 675, 676, 683, 688, 694,
    695, 698, 699, 770
Pearson, Egon, 3, 646, 647, 648, 649, 658
Pearson, Karl, 3, 8, 362, 516, 561, 562,
    564, 565, 606, 609, 610, 646, 647, 657,
    658, 685, 707, 780
Pearson's correlation, 575, 579, 582, 583
pentagon, 70, 183, 740
pentagram, 70, 183, 740
per protocol, 643
percentages, 42
permutations, 155, 156, 158, 165, 349,
    354, 355, 356, 361, 362
permutations with repetition, 355, 357
phi, 68, 69, 124, 594, 733, 734, 740
philosophers, 74, 78, 199, 410
physics, 111, 259, 319, 332, 363, 413,
    450, 516, 517, 543, 545, 548, 705, 729

pi (π), 145

Pingala, 122

platykurtic, 553

playing cards, 146, 149, 151, 155, 158, 336

PMF, 373, 384, 385, 387, 389, 391, 401, 459, 465, 474, 486, 536, 537, 717; joint, 404, 405

point estimate, 648

Pois(λ), 470

Poisson distribution, 450, 455, 456, 457, 459, 466, 470, 471, 482, 486, 488, 506, 520, 528, 537, 539, 549, 566, 590, 608, 675

Poisson, Simeon Dennis, 3, 413, 450, 451, 452, 454, 487, 510, 518, 549, 778

polar coordinates, 223, 307, 309, 316; changing to, 289

political science, 467

polynomial, 37, 94, 101, 237, 252, 259, 260, 261, 263, 264, 750, 752, 753; multiplication, 312

polynomial long division, 252

polynomials, 36, 87, 100, 101, 102, 259, 264, 750

population, 363, 619, 621

posterior distribution, 431, 433, 434, 435, 447, 727

power, 53, 54, 65, 127, 128, 266; raising to a fraction, 60; raising to a negative, 56; raising to one, 55, 56; raising to zero, 55, 56; subtracting, 57

power of a statistical test, 642

power rule, 257

power series, 462

powers, 127, 139

precision, 590

Price, Richard, 410, 412

prime numbers, 40, 79

prior, 429

prior distribution, 429, 431, 433, 434, 435, 438, 705, 706, 726, 727

prior elicitation, 429, 435

probability density function, 373, 447, 459, 530

probability distribution, 312, 362, 367, 374, 377, 382, 383, 384, 390, 400, 429, 431, 433, 443, 444, 446, 451, 467, 472, 481, 482, 483, 485, 492, 495, 510, 520, 522, 523, 525, 527, 536, 537, 551, 554, 556, 559, 567, 601, 620, 621, 622, 623, 628, 632, 638, 649, 652, 665, 666, 677, 710, 725, 766; continuous, 363, 525; discrete, 363, 521, 523

probability distributions, 302, 312, 363, 368, 372, 373, 375, 380, 390, 397, 399, 400, 404, 405, 416, 429, 456, 465, 483, 517, 518, 519, 545, 546, 567, 583, 584, 593, 602, 613, 641, 644, 646

probability mass function, 373, 382, 447, 455, 459, 523

probability of probabilities, 443, 448, 449

probability of success, 350, 351, 353, 372, 377, 378, 382, 383, 384, 385, 430, 444, 445, 446, 454, 457

process improvement, 413

product, 35

product operator, 145, 146, 148, 150, 151

product-moment coefficient, 584

progressions, 116, 125, 138; arithmetic, 120, 138; geometric, 120, 138, 380

proportional, 68

proportions, 45

psychology, 20, 395, 425, 466, 514, 719, 729, 776, 777, 794

public health, 395, 466

pure mathematics, 77, 259

p-value, 436, 439

Pythagorus's theorem, 179, 193, 218, 220, 221, 224, 307, 325

Q

quadratic equation, 95; minimum value, 240

quadratic equations, 93, 100

quadratic expressions, 95, 204

quadratic formula, 95, 98, 104, 734

quadrilateral, 174

quality control, 395, 439, 467, 705

quantiles, 493

quartiles, 493

quincunx, 515

quintiles, 493

quotient, 72

R

radians, 183, 184

radiation exposure, 450

radiological oncology, 413

radius, 174, 184, 223

random, 363

random variable, 361, 366, 367, 373,
374, 378, 380, 381, 391, 400, 405, 444,
445, 446, 461, 472, 473, 474, 475, 476,
478, 479, 481, 482, 485, 522, 527, 530,
539, 547, 549, 551, 583, 595, 610, 611,
613, 622, 623, 627, 664, 665, 666, 674,
684, 694, 710, 766, 767, 771, 773,774

random variables, 366, 367

randomness, 331, 367

rate of change, 298

ratings, 710

ratio, 174, 195, 229, 256

rational numbers, 72, 75

ratios, 45, 67, 173

raw data, 360, 361, 384, 582, 725

ray, 170

Rayleigh distribution, 706

real numbers, 106

reciprocals, 49, 50, 51, 57, 64, 99, 123,
193, 196, 239, 253, 254, 294, 310, 456,
457, 461, 476, 511, 535, 569, 575, 589,
590, 697, 700, 735, 768

rectangle, 174; area of, 177

reflex angle, 171

regression, 498, 499, 500, 501, 515, 574,
578, 662, 663, 670, 685, 686, 687, 780

regression line, 498, 499, 500, 501, 503,
515, 571, 572, 574, 575, 578, 580, 686,
687; slope, 686

regression to the mean, 571

relative risk, 406, 407, 607

remainder, 50

repeating decimals, 65

residuals, 499, 503, 571, 582, 686, 687

reverse binomial distribution, 429

reverse power rule, 275, 279

rhombus, 174

right angles, 136, 171, 177, 184, 192,
218, 220, 221, 245, 740, 743

rise-over-run, 194, 208, 229

risk levels, 711

root, 73

root symbol, 89

rotating, 226

S

sample, 363, 605, 619, 621

sample size, 643

sample space, 362

sampling, 616, 619

sampling distribution, 626, 632, 641,
654; mean, 622; of the mean, 632; of
the sample mean, 621, 624, 632

secant, 196

second central moment, 546

second derivative, 262, 591

second moment, 545, 584, 585, 588

segment, 170; area of, 188

sensitivity, 420

separable differential equations, 299

separating the terms, 299

sequences, 18, 116, 117, 118, 119, 120,
121, 122, 123, 125, 131, 136, 145, 148,
714; convergent, 122, 123; divergent,
122, 123; finite, 122; infinite, 122

series, 78, 118, 119, 122, 131, 132, 133,
135, 138, 139, 140, 141, 143, 145, 146,
163, 264, 265, 277, 280, 297, 350, 382,
391, 456, 462, 502, 527, 530, 533, 550,
569, 595, 604, 713, 753, 764, 767;
arithmetic, 292; convergent, 141;
divergent, 141; finite, 132; geometric,
139, 140, 142, 145, 292; infinite, 132,
142; of natural numbers, 135; of odd
numbers, 136

sets, 74

sigma notation, 131, 139, 140, 277, 454

sign test, 715, 717

Sign test, 712

signs or operators, 30

similar, 173

similar triangles, 192

Simpson's paradox, 335

simultaneous equations, 108

sin, 194, 297

sine, 194, 195, 213, 223

sine wave, 214, 244

skew: negative, 551; positive, 551

skewed distribution, 364, 378, 382, 386,
448, 494, 520, 551, 552, 624

skewness, 551

slope of a curve, 300

slope of a line, 194, 206, 211, 229, 231, 232, 591

sociology, 396, 467, 514

Spearman's Rank Correlation, 712

Spearman's rank correlation coefficient, 715

specificity, 420

sports analytics, 468

spreadsheet, 36, 54, 69, 153, 194, 196, 221, 434, 435, 465, 469, 578

square root, 59, 103; **of a negative number**, 105

Staircase of Mount Meru, 78

standard deviation, 7, 8, 180, 513, 561, 566, 567, 568, 572, 574, 575, 588, 592, 593, 594, 595, 596, 597, 602, 606, 610, 624, 625, 626, 628, 629, 630, 632, 640, 647, 648, 649, 652, 653, 660, 661, 662, 664, 665, 680, 683, 684, 685, 686, 715, 766

standard error, 661; of the sample mean, 628

standard error of the mean, **626**, **628**, 630, 632, 641, 647, 664, 665, 683

standard normal distribution, 594

statistic, 363

statistical significance, 363, 582, 598, 639, 640, 643, 644, 645, 685, 713

statistical software, 465

Stirling's approximation, 506

straight line, 169, 171, 182, 183, 184, 203, 204, 205, 206, 207, 211, 228, 229, 232, 476, 498, 503, 582, 740, 753

strictly decreasing, 473, 476

strictly increasing, 473, 476

subjective expectations, 418

subscript, 118, 148

subset, 74

substitution, 108

success, 350, 361, 373, 377, 378, 381, 384, 443

sum of cubes, 91

sum of squares, 580, 609, 613, 687, 692, 693, 736; between groups, 693; mean within group, 693; within groups, 693

superscript, 53, 118, 195, 196, 199

support of a variable., 367

surds, 59

surveying, 106, 225

symmetrical distribution, 306, 348, 364, 382, 479, 494, 520, 677, 682, 758

symptom severity, 711

T

t distribution, 605, 607, 653, 675, 677, 678, 680, 683, 685, 695, 705, 770, 776, 794; CDF, 677; PDF, 674, 675, 676, 694

t statistic, 683, 688, 770

t test, 716; for a proportion, 685; one-tailed, 681; paired, 684, 715; single-sample, 682, 683, 684; two sample, 717; two-sample (unpaired), 684; two-tailed, 681

tan, 194

tangent, 175, 194, 195, 196, 197, 213, 214, 229, 231, 232, 240, 241, 254, 256, 267, 274, 298, 300, 430, 502, 509, 591, 742

Tartaglia's triangle, 78

Taylor series, 194, 259, 261, 263, 264, 270, 280, 295, 297, 318, 456, 462, 534, 538, 595, 604, 615, 677

Taylor, Brook, 3, 259, 261, 280

Tchebycheff. *See* Chebyshev, Pafnuty

Tchebychev. *See* Chebyshev, Pafnuty

telephone sales, 481

temperature, 369

terms, 36, 37, 90

the angle of inclination, 195

third central moment, 551

three dots, 147

three-door problem, 334

torque, 517, 532, 543

transcendental numbers, 73

transformations, 610

transversal, 172

trapezium, 174

tree chart, 340

tree charts, 338, 339, 341, 342, 344, 347, 352, 356, 381, 400, 421, 581

trials, 350, 358, 373, 377, 383, 384, 385, 386, 387, 391, 394, 429, 430, 444, 445, 446, 454, 457, 461, 469, 470, 722

triangle, 173

Tschebyschef. *See* Chebyshev, Pafnuty

Tschebyscheff. *See* Chebyshev, Pafnuty
Tschebyschev. *See* Chebyshev, Pafnuty
type I errors, 635, 636, 637, 638, 640,
646, 647, 649, 681, 690, 691
type II errors, 635, 636, 637, 638, 640,
646

U

U statistic, 717
unbiased estimator, 661
under-powered, 639, 641, 645
uniform distribution, 433, 434, 447, 551,
624, 705
unimodal distribution, 364, 494, 551
union, 75, 345, 401
unit circle, 209, 213, 215, 217, 218
unit step function, 320
univariate distribution, 364
unreal numbers, 105, 106

V

variables, 85, 87, 90, 94, 112, 118, 140
variance, 515, 544, 546, 548, 549, 550,
551, 552, 553, 554, 559, 565, 566, 568,
569, 572, 575, 578, 579, 583, 584, 585,
586, 588, 590, 595, 602, 610, 627, 628,
630, 631, 632, 633, 654, 655, 656, 659,
660, 661, 662, 677, 685, 686, 687, 691,
692, 693, 701, 705, 706,714, 766, 769,
773, 781; of a sample, 662; within
groups, 693
vectors, 391
Venn diagram, 74, 75, 76, 344, 345, 417,
419, 420
Venn, John, 3, 74

Voltaire, 441

W

Wallis, John, 3
weather, 368, 466, 514
website, 452
Weibull distribution, 705
Weldon, Raphael, 515, 563
Wessel, Caspar, 3, 225
Whitney, Donald, 3, 717
whole numbers, 72
Wilcoxon rank sum test, 712
Wilcoxon ranksum test, 717
Wilcoxon signed rank test, 712, 717
Wilcoxon, Frank, 3, 717
Wiles, Andrew, 325
wind patterns, 514
without replacement, 386, 394, 395, 396,
402, 469, 630, 708
workforce planning, 413

X

x bar, 494

Y

Yang Hui's triangle, 78

Z

z score, 561, 565, 567, 568, 569, 571,
572, 573, 590, 592, 593, 594, 596, 597,
600, 647, 648, 649, 721
zymology, 332